TS
933.C4
Am

40/2832

Cellulases and Their Applications

Cellulases and Their Applications

A symposium sponsored by
the Division of Cellulose,
Wood, and Fiber Chemistry
at the 156th Meeting of
the American Chemical
Society, Atlantic City, N. J.,
Sept. 11-12, 1968.

George J. Hajny and Elwyn T. Reese
Symposium Co-Chairmen

ADVANCES IN CHEMISTRY SERIES **95**

AMERICAN CHEMICAL SOCIETY
WASHINGTON, D. C. 1969

Coden: ADCSHA

Copyright © 1969

American Chemical Society

All Rights Reserved

Library of Congress Catalog Card 73-108269

PRINTED IN THE UNITED STATES OF AMERICA

Advances in Chemistry Series
Robert F. Gould, *Editor*

Advisory Board

Frank G. Ciapetta

William von Fischer

Frederick M. Fowkes

Edwin J. Hart

F. Leo Kauffman

Stanley Kirschner

John L. Lundberg

William E. Parham

Edward E. Smissman

AMERICAN CHEMICAL SOCIETY PUBLICATIONS

FOREWORD

ADVANCES IN CHEMISTRY SERIES was founded in 1949 by the American Chemical Society as an outlet for symposia and collections of data in special areas of topical interest that could not be accommodated in the Society's journals. It provides a medium for symposia that would otherwise be fragmented, their papers distributed among several journals or not published at all. Papers are refereed critically according to ACS editorial standards and receive the careful attention and processing characteristic of ACS publications. Papers published in ADVANCES IN CHEMISTRY SERIES are original contributions not published elsewhere in whole or major part and include reports of research as well as reviews since symposia may embrace both types of presentation.

CONTENTS

Preface ix

1. A Perspective on Cellulase Research 1
   R. G. H. Siu

2. Enzymes of the Cellulase Complex 7
   K. W. King and Mahmood I. Vessal

3. Estimation of Exo-β-1 → 4-glucanase in Crude Cellulase Solutions   26
   E. T. Reese

4. The Purification and Properties of the C_1-Component of the Cellulase Complex 34
   Keith Selby

5. Cellulase Complex of *Ruminococcus* and a New Mechanism for Cellulose Degradation 53
   J. M. Leatherwood

6. Extracellular and Cell-bound Cellulase Components of Bacteria .. 60
   H. Suzuki, K. Yamane, and K. Nisizawa

7. New Methods for the Investigations of Cellulases 83
   Karl-Erik Eriksson

8. Enzyme-Substrate Relationships Among β-Glucan Hydrolases 105
   D. R. Barras, A. E. Moore, and B. A. Stone

9. Recent Progress on the Structure and Morphology of Cellulose 139
   Bengt Rånby

10. Structural Features of Cellulosic Materials in Relation to Enzymatic Hydrolysis 152
    Ellis B. Cowling and Wynford Brown

11. Constitutive Cellulolytic Enzymes of *Diplodia zeae* 188
    J. N. Bemiller, Diane O. Tegtmeier, and A. J. Pappelis

12. A Mechanism for Improving the Digestibility of Lignocellulosic Materials with Dilute Alkali and Liquid Ammonia 197
    Harold Tarkow and William C. Feist

13. Digestibility as a Simple Function of a Molecule of Similar Size to a Cellulase Enzyme 219
    J. E. Stone, A. M. Scallan, E. Donefer, and E. Ahlgren

14. Utilization of Cellulose by Ruminants 242
    B. B. Baumgardt

15. Lignocellulose in Ruminant Nutrition 245
    W. J. Pigden and D. P. Heaney

16. Composition, Maturity, and the Nutritive Value for Forages 262
    Peter J. Van Soest

17. Use of Wood and Woody By-products as a Source of Energy in Beef
    Cattle Rations 279
    W. D. Kitts, C. R. Krishnamurti, J. A. Shelford, and G. Huffman

18. The Value of Wood-Derived Products in Ruminant Nutrition 298
    William H. Pfander, S. E. Grebing, George Hajny, and Wendell Tyree

19. Hardwood Sawdust as Feed for Ruminants 315
    W. B. Anthony, John P. Cunningham, Jr., and R. R. Harris

20. Effect of Urea Supplementation on the Nutritive Value of NaOH-
    treated Oat Straw 328
    E. Donefer, I. O. A. Adeleye, and T. A. O. C. Jones

21. Development of a Commercial Enzyme Process: Glucoamylase ... 343
    L. A. Underkofler

22. Applications of Cellulases in Japan 359
    Nobuo Toyama

23. The Production of Cellulases 391
    Mary Mandels and James Weber

24. Enzymatic Saccharification of Cellulose in Semi- and Continuously
    Agitated Systems 415
    Tarun K. Ghose and John A. Kostick

25. Utilization of Bagasse 447
    V. R. Srinivasan and Y. W. Han

Index 461

PREFACE

Cellulose is the major constituent of all vegetation, comprising from one-third to one-half of dry-plant material. As such, it is the world's most plentiful renewable resource. Large quantities of cellulose are present in manufactured products such as textiles, paper, and building materials. Enormous quantities of cellulose are consumed by ruminants and serve as their principal source of energy. Still, most vegetation is unused by man or animals and undergoes natural decay through the intervention of micro-organisms capable of producing cellulases. Paradoxically, both the presence and absence of cellulolytic activity are the crux of cellulose use and disposal. In the case of products such as textiles, paper, and lumber, strenuous efforts are made to prevent the action of the cellulases. For ruminant forage or biological disposal processes, a high level of cellulase activity is desired. When one considers the enormous economic importance of both protection against and promotion of cellulase activity, it is indeed surprising that the last symposium on the subject in the United States was held as far back as 1962.

The present symposium was planned to bring together workers from different disciplines so that problems common to such diverse fields as biochemistry, animal nutrition, textiles, and forest products utilization might be looked at from fresh perspectives.

Among the many problems facing the world today, two of the most important are pollution of the environment and the threat of famine. Disposal of lignocellulosic material by burning adds to air pollution. Disposal by biological means, if properly carried out, avoids this pollution but it can be costly. The per capita consumption of paper and paperboard in the United States in 1966 was 530 pounds, much of which is intended for short-term use. Since this is only a minor part of the cellulosic material to be disposed of each year, the pollution potential of the surplus is tremendous.

The population explosion and the worldwide threat of famine have been well publicized. Much effort is being expended to produce higher yielding varieties of grains and to develop better methods of cultivation. Serious work is directed to changing the fishing industry from one of hunting to one of farming. Species previously not eaten are being processed into edible fish meal. Other research is directed toward producing edible proteins from leaves and grasses, and even the possibility of producing single-cell protein from petroleum hydrocarbons is being investigated.

With the tremendous worldwide activity in seeking foodstuffs from conventional and unconventional sources, it is surprising that cellulose and lignocellulose have received so little attention. Here is a material that is present in all vegetation. Lignocellulosic materials are widely dispersed—they are as plentiful in underdeveloped countries as in the developed countries. Yet the amount of scientific research devoted to finding practical means to produce a digestible nutrient from cellulose-bearing materials is negligible. Now is a propitious time for attacking this problem. Success can not only alleviate hunger, but by increased utilization of residues, it can assist in pollution abatement. Several papers in this symposium are devoted to the theme of producing protein-containing foodstuffs from cellulosic materials through the intervention of cellulases.

No attempt was made at this symposium to include all aspects of cellulose decomposition. Surveys of active organisms and interrelationships of organisms growing on complex substrates are certainly important problems. So are the properties of enzymes and the means of enhancing and preventing their action. All impinge on the aim of this symposium which is to stress the practical application of cellulolytic systems to worldwide problems.

The symposium was organized at the suggestion of J. F. Seaman, chairman of the American Chemical Society Division of Cellulose, Wood, and Fiber Chemistry. Special thanks are due to all participants for making this an interesting and informative meeting. J. H. Dusenbury, program chairman, was of great help in insuring that the meetings ran smoothly. R. G. H. Siu, J. F. Saeman, E. T. Reese, and B. R. Baumgardt presided at the sessions. Financial aid for the symposium was contributed by the Masonite Corporation.

G. J. HAJNY

U. S. Forest Products Laboratory
Madison, Wis.

ELWYN T. REESE

U. S. Army Natick Laboratories
Natick, Mass.
November 1968

A Perspective on Cellulase Research

R. G. H. SIU

4428 Albemarle St., N. W., Washington, D. C. 20016

This symposium on cellulases and their applications represents a most worthwhile contribution. Twenty-five papers are presented from 20 laboratories. These groups represent the best talents in the field. They come from six countries distributed over the northern as well as the southern hemispheres, the eastern as well as the western.

The papers are grouped into four sessions. The first emphasizes the nature and components of the cellulase system, including reports on glucotransferase and exoglucanases, the constitutive behavior of cellulases, enzyme-substrate relationships, genic influences, and sites of action. The second session focuses on the question of enzymic accessibility of cellulose, release of cellulase by microorganisms, and association of lignin and cellulose in wood. The third session probes the digestibility and nutritional potential of wood and straw. The fourth session assesses the production of cellulase and commercial applications. Co-chairmen George Hajny and Elwyn T. Reese have organized an excellent conference.

The Bridge

My personal participation in this symposium was most gratifying. It has been quite a number of years since I last fondled a mass of *Trichoderma viride* mycelia or looked a cellulase preparation in the eye. As I caught up with the literature I was greatly impressed with the major progress that has been made in this field. The achievements have been remarkable, and all the research workers are worthy of commendation. This excellent progress paves the way for even greater future developments.

From my distant view, this symposium appears as a bridge. It is a bridge from the past state of hypotheses and potentiality to the future state of theories and applications. Looking backward, we see the extensive investigations on agronomy and animal husbandry going back over eight decades with such illustrious names as Hoppe-Seyler, Omelianski, and

Waksman. We recognize the great effort of two decades ago on the identification of cellulolytic microorganisms by such eminent taxonomists as White, Weston, and Raper. There is also the report of Selliére about 60 years ago on the first observation of an *in vitro* cellulase activity, but it was not until the last 20 years, and particularly the past 10, when cellulase attracted the concentrated attention of a group of very capable microbiologists and biochemists employing the most sophisticated scientific instruments. A new generation of investigators has emerged—Reese, Toyama, King, Nisizawa, Selby, Cowling, Erickson, Mandels, Gascoigne, and others. As a result of their efforts, the underbrush has been cleared away, and we are now beginning to see future salients in much clearer focus.

The terms cellulolytic enzymes and cellulases now enjoy a workable definition. This has been made possible through the separation of the components of the cellulolytic complex, as well as some clarification of their mechanism of action. New viscometric, isoelectric focusing, gel filtration, and other techniques for the determination of the enzymic components have been devised. Detailed observations on the molecular sites of hydrolytic reactions, the susceptibility of the enzymic reactions to metallic ions, proteolytic enzymes, and other chemicals also provided considerable insight into the nature of cellulase.

This summarizes the activities of the past. The papers in this volume bring us up to date as to the current state of the art. To glimpse the future we must ask: Where do we go from here? What is at the other end of this cellulase research bridge?

The other end of the bridge appears to be multiterminal. It leads to a number of fascinating possibilities. Four of these warrant particular attention. Two are direct extensions of current work; the other two are more speculative in nature.

Mechanism of Cellulase Action

The first and most obvious avenue of progress of course is the continuation of excellent work. More sophisticated biochemical techniques will be developed. This in turn will lead to further clarification of the mechanism of cellulase activity. Particular advances will be made in our knowledge of the exoglucanases, the initial reactions on the surfaces of cellulosic substrates, and the nature of the polysaccharide-protein complexes formed as intermediary transitory stages. Much of this is covered by the first eight chapters in this volume. Hence, further discussion is not warranted.

New Source of Food

The second avenue of progress involves the development of new food processing techniques and new sources of food. There is no problem of greater importance to our rapidly expanding human population, expecting to reach the seven billion mark before the year 2000, than increased food supply. Much research is being conducted to offset the threat of food imbalance. Agriculture is being improved through new varieties and cultivation techniques. Farming the sea is making some headway with exploitation of fish meal. Extraction of proteins from leaves and grasses, followed by precipitation into a form suitable for human consumption is receiving serious consideration. So are oil seed cake from the protein rich residues of coconut, sesame, peanut, soybean, and cottonseed being proposed for wide use as food supplements. Microorganisms as protein sources are being studied. Even purely synthetic foods have been suggested.

It is interesting to note that the general scientific and managerial public has rarely mentioned cellulose in a serious vein as a possible source of foodstuffs. Here is a substance which is the most abundant carbohydrate in the plant kingdom. It is the most plentiful of all naturally occurring compounds and comprises at least a third of all vegetable matter. Here is this tremendous store of carbohydrates to be exploited—if only it can be converted into a digestible form through practical means.

We know that horse, cow, goat, rabbit, termite, cockroach, fungi, bacteria, and protozoa are able to digest cellulose to varying degrees—that cellulase can increase the digestibility of wood for animals—that cellulase can hydrolyze cellulose to produce concentrations of glucose as high as 30%—that cellulase can be prepared in fairly large quantities—and finally that a group of very talented and dedicated scientists are working diligently and fruitfully in the area. In the face of all this, I cannot help but take it as a serious proposition that cellulose will one day constitute a very important source of food for human consumption either directly as a processed form or indirectly as converted into simpler carbohydrates or proteins by microorganisms and animals. The pressure of the growing population of the future and the millions of hungry people of today place considerable significance upon this potentiality.

Incidentally, the actual number of senior capable investigators being supported for their work on cellulase approximates only a score, with another score or so of graduate students or laboratory assistants. Contrast this with the thousands of agronomists, who are doing excellent work no doubt, but nevertheless struggling to expand the arable soils to more than 4% of the earth's surface. The only conclusion I can draw from this comparison is that an order of magnitude of added support to the area of cellulase investigations is called for if we are to let social relevancy be a major criterion of investment.

There are many ancillary uses of cellulase in food processing. I need only mention a few examples involving the extraction of essential oils and flavoring materials, tenderizing fruits, clarification of juices, and improvement of filterability of vanilla extracts. Many other examples are given in this volume.

New Chemical Concepts

The third avenue of advancement represents sheer speculation on my part. I am not making any recommendations to any one to follow up on my comments, but I would predict that when the proper time comes, some one will do it because he will have no alternative. This person will have accumulated a formidable mass of failure data, which will force him to challenge the current methods of picturing the structure of chemical molecules. He will find that his observations involving the cellulase-cellulose reactions cannot be understood in the light of the conventional notions. The literature involving enzymic reactions is based on solutions. The cellulase-cellulose system involves an interface between an enzyme in solution and presumably an insoluble substrate surface. The postulation of a C_1 step in the cellulase chain of reactions sharpens our focus. It seems to me that as we bear down on the C_1 activity at that point of the cellulolytic process, we will find ourselves face to face with some very fundamental relationships of such concepts as hydrogen-bonds and crystalline structure as sites of enzymic action. Should this occur, there is no telling as to the greatness of the resulting contributions to the field of chemistry and physics. It may well turn out to be among the most significant of the century.

Flux

The fourth avenue of research is even more conjectural. Here we see the cellulase-cellulose system as a physico-chemical model of flux. It depicts the transition between the structure and the structureless, between the reserve solids and the mobile solutes. Such transitions exist throughout nature.

If our conventional thinking about the reversibility of enzymic reactions in general holds in the case of the cellulase-cellulose system, it would follow that the system would not only hydrolyze cellulose into glucose but also synthesize cellulose from glucose. This would relate the cellulase-cellulose reactions with those of photosynthesis in a most intimate way. Botanical biochemists have been looking at the starch-glucose transformation as the primary surge tank for the glucose of photosynthesis. It seems to me that the cellulose-glucose transformation should be given equal attention. A host of research questions immediately come to the surface: what compound in the cellulose-glucose series of reactions is the primary reserve for metabolic purposes? Is it one of the oligosaccharides or cellobiose, rather than cellulose? What are the crossover compounds from the cellulose-glucose chain to the other metabolic cycles in plants? Can lignin be considered the plant's regulator in protecting the dissolution of the celulosic structure against further metabolic transformation? Is it possible to find lignin-free celluloses, which readily undergo the cellulose-glucose reactions in both directions? And so on.

All of these are fascinating research speculations. But I would like to single out one of them for a few extra words. It concerns the dimensions of structure. At some point in the glucose-to-cellulose transition, structure appears in the elementary form of crystalline cellulose, fibrils, and the like. Are there enzymes which guide the elaboration of physical structure in contrast to the formation of chemical compounds? Or must we seek a new kind of responsible agents? If we extend the implications of our current ambiguous definition of the C_1 step of cellulase, we must conclude that either enzymes *are* responsible for governing the physical structure of cells and organisms through the whole gamut of "Van-der-Waalase," "hydrogen-bondase," "left-handed-twistase," "crystallinase," and so on, or that the C_1 step represents a new class of non-enzymic physico-chemical entity, which works in conjunction with enzymes in living tissues, or that the current notions of the role and nature of enzymes must be drastically extended.

In any case, the deeper study of the cellulase-cellulose system puts us face to face with a most important facet of a most challenging problem of science and philosophy—*i.e.*, putting form into the formless.

The Wild Flower

I would like to emphasize again the social importance of the current research on cellulases and their applications. It is an interesting observation that this field of cellulase research began to gain support because of attempts to prevent the destruction of textile materials. Yet it is now showing its greatest potential contributions in the direction of augmenting the world's food supply and in opening up new salients in fundamental thinking of physics, chemistry, and philosophy. As I gaze intently into the mysteries of the cellulase-cellulose complex, I cannot help but hear the beautiful words of William Blake's most fitting stanza:

> To see the world in a grain of sand,
> And a heaven in a wild flower,
> Hold infinity in the palm of your hand,
> And eternity in an hour.

RECEIVED October 14, 1968.

2

Enzymes of the Cellulase Complex

K. W. KING

Department of Biochemistry and Nutrition, Virginia Polytechnic Institute, Blacksburg, Va. 24061

MAHMOOD I. VESSAL

Pahlavi University, Shirz, Iran

> *The shortcomings of cellulose itself, and of cellulose derivatives as substrates for studying the mechanism of cellulase action are pointed out. The oligosacchardies—prepared with a radioactive, or chemical label—are useful in determining the bond preference of enzymes and hence whether a component is of the endo-β-1 → 4 glucanase or exo-β-1 → 4 glucanase type.*

The first part of this paper will be concerned with an experimental approach to the cellulase complex which involves (1) the effect of supramolecular structure on activity of cellulolytic systems, (2) the effect of chemical modification of cellulose on the activity of cellulase, and (3) the use of oligosaccharides in determination of the mode of action of cellulolytic components.

Effect of Supramolecular Structure on Activity of Cellulolytic Systems. An earlier review (8) dealt with the general problem of determining the position at which cellulases attack their substrates. A number of possible approaches were considered, and some preliminary data from application of certain of these approaches were presented. In the interim we have been able to extend these mode of action or site-of-attack studies to some new highly purified enzyme preparations. In addition the techniques have been used in studying the means by which *Cellvibrio gilvus* conserves the glucosyl bond energy of cellulosic substrates that are being used as energy sources. This discussion is largely limited to our own data, not because we are unaware of the excellent work of others in this field, but because it is extremely difficult to correlate it with what we have found in the organisms we have used.

Present address: Research Corp., 405 Lexington Ave., New York, N. Y. 10017.

For purposes of discussion we can divide the several alternative approaches into those employing long chain substrates as distinct from those using relatively short oligosaccharides. The difference is really more than a matter of size alone for invariably the long chain substrates are also characterized by heterogeneiety in molecular size. In addition, they involve either molecular and supramolecular architectural features that are of enzymatic significance as in the case of the native celluloses, or steric and electrostatic features that modify enzymatic action as in the case of the water soluble derivatives. These two characteristics of the high DP substrates introduce elements of uncertainty into their use to elucidate the site of action of cellulases which are avoided by use of the lower DP water soluble oligosaccharides. It is, therefore, our inclination to use the more precise, analytically quantitative procedures which the oligosaccharides make possible.

That such features of cellulose as the crystal lattice form are significant determinants of cellulase action has only recently been established (19), although a great deal remains to be learned about the enzymatic importance of cellulose fine structure. It is clearly established, however, that each of the three water-stable crystal forms of cellulose is distinct in the rate at which it is hydrolyzed and in its properties as an inducer of cellulase. For example the *Trichoderma viride* cellulase from culture extracts exhibits a lower activation energy in attacking the crystal lattice form used in culture growth, than in attacking the other lattice forms (Table I).

Table I. Activation Energy (calories/mole) in Degradation of Celluloses of Differing Crystal Lattice by Enzymes Produced on the Two Water-stable Lattices

Lattice Used to Produce Enzyme	Lattice Form of Substrate	
	I	II
I	3270	6940
II	6190	5430

In addition the degradation of lattice I and II substrates follows Schutz kinetics quite closely while the kinetics of degradation of lattice IV cannot be identified (19).

There is also real uncertainty about what reaction is being catalyzed by C_1-components in general. Several lines of evidence suggest that the C_1-reaction on hydrocellulose may not be a hydrolytic event at all. The fragmentation process releasing tiny micelles (11, 16) may involve simply a cleavage of the intermolecular hydrogen bonding system. The reaction rate is essentially unaffected by substitution of D_2O for H_2O in the reaction system (10), an observation that is highly improbable for a hydrolytic reaction. The activation energy of the C_1 reaction has been observed to be as low as 3000 calories/mole (19), a value more in line with hydrogen bond cleavage than with hydrolysis, and the bonds that must be broken in order to disaggregate the micelles of native cellulose are hydrogen bonds not glycosidic bonds. Should the C_1-reaction prove in fact to be non-hydrolytic in nature, the whole question of site of action would be altered radically.

With crystalline substrates another factor contributing to uncertainty regarding the site of action can be seen from study of models. Figure 1 represents a Hirschfelder model of a glucosyl, a cellobiosyl, a celloheptaosyl, and an oligosaccharide fragment of DP = 14. All are in the conformation of the dominant natural crystal lattice I form. The black atoms are carbon; the greys are hydrogen; and the white atoms are oxygen. The glycosyl bonds are easily located by the oxygen pairs lying very close together representing the glucosyl oxygens and the ring oxygens. Close study of the oligosaccharide chains shows that only alternate glucosyl bonds are exposed to attack by enzyme, the other glucosyls being heavily masked and facing the opposite direction, into the body of the micelle.

Attack by enzymes onto the mass of such molecules that a crystal lattice represents would therefore be restricted to exposed glucosyl bonds, freeing fragments of DP = 2, 4, 6, 8, etc. These fragments would still be held in the crystal lattice by hydrogen bonds, two such bonds per glucosyl. Kinetic energy sufficient to break four hydrogen bonds would be needed to free a cellobiosyl into solution. Freeing a tetrasaccharide would involve rupturing eight hydrogen bonds.

Considering in addition the increasing mass as the fragment becomes longer, it is evident that a completely non-discriminating random glucanase acting on a crystalline substrate would yield more than twice as much cellobiose as cellotetraose, and the other oligosaccharides would be freed in even lesser amounts. Such a system could give all the appearances of an endwise acting glucanase producing essentially only cellobiose as a reaction product. For this reason the exclusive production of cellobiose from attack on such substrates cannot legitimately be interpreted as evidence of endwise attack. Inherently, then, the nature of the substrate invalidates an experimental approach to cellulase studies which is perfectly valid with the amylases and many other enzymes attacking water soluble polysaccharides.

The compactness and rigid intra- and intermolecular positioning of native cellulose materials also creates problems in interpreting data on the changes in average chain length as expressed by DP measurements and end group analysis. Rigidly endwise glucanases removing terminal glucosyls like the exo-glucanase of *T. viride* (14) fail to attack crystalline hydrocelluoses to a significant degree. Furthermore, a random acting cellulase presented with crystalline substrate would not necessarily find the surface array of atoms on the crystallite correct for effective enzyme attachment at all. Problems of this type are presumably the reason for the variations in activation energy, thermal stability, and reaction rate of the various cellulase components of *T. viride* (*19*).

Clearing, then, crystalline or semi-crystalline celluloses have serious drawbacks for studies of the site of attack by cellulases.

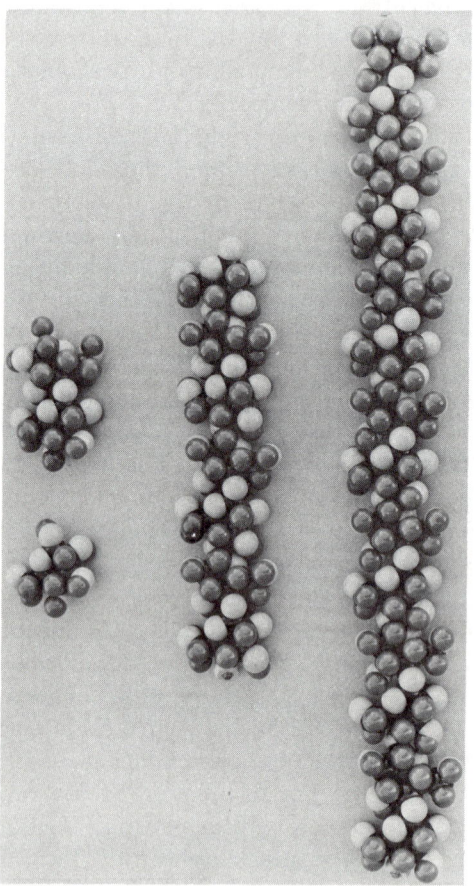

Figure 1. Fisher-Hirschfelder models of glucose, cellulose, and oligosaccharides of 7 and 14 monomers in the boat-chair conformation found in the lattice I structure

Effect of Chemical Modification of Cellulose on the Activity of Cellulase. Soluble cellulose derivatives like carboxymethyl and hydroxyethyl cellulose have an advantage in being randomly coiled, but they are not without limitations. The two basic problems are those created by

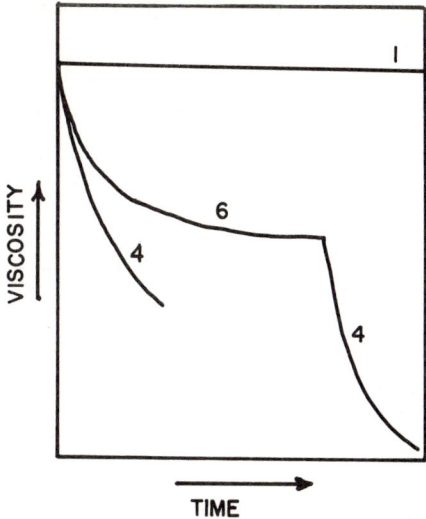

Figure 2. Changes in the intrinsic viscosity of carboxymethylcellulose solutions during hydrolysis by purified β-glucanases of Aspergillus niger (14). Numbers refer to different cellulase components

steric and electrostatic interactions with enzyme. As examples, the highly purified β-glucanases of *Aspergillus niger* show quite well (*11, 14*) the nature of the action on carboxymethylcellulose (Figure 2). The action of Component 6 is of particular interest. After a very high rate of reaction, the velocity slows down to nearly zero far short of completion. Subsequent addition of Component 4 then carries the reaction to essential completion. This behavior appears to indicate that either sterically or electrostatically most of the glucosyl bonds of the carboxymethylcellulose are protected from attack by Component 6, but Component 4 can still make effective contact. The electrostatic blocking of an enzyme component from the same organism is indicated. From the pK_a of the carboxymethyl group one can calculate the percent dissociation of the carboxyl groups at any given pH. If then one measures the rate of hydrolysis of carboxymethylcellulose over a range of pH's covering 0.6 to 90% association of the carboxyls, the data in Figure 3 are obtained. Over a wide range of pH's the reaction velocity is directly proportional to the log of concentration of unionized carboxyls indicating that ionization of the substrate essentially protects it from enzymatic attack. This property is also reflected in the frequent observation that the pH optima for non-ionizing native celluloses differ markedly from the pH optima for attack on carboxymethylcellulose by the same enzyme (Figure 4).

These steric and ionization effects seriously limit the usefulness of substituted celluloses in site-of-action studies. Further uncertainties are introduced by the fact that the substitution reactions used to solubilize cellulose distribute the substituents randomly on the polysaccharide. As a consequence the reaction products become a very large family of variously substituted fragments rather than a single product or a few identifiable oligosaccharides.

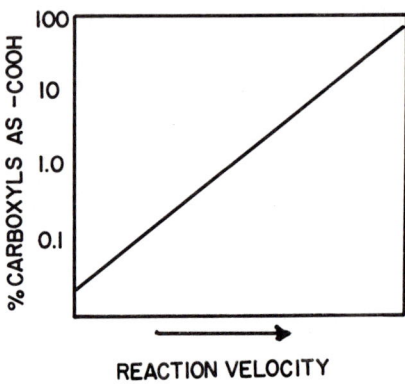

Figure 3. Enzymatic hydrolysis of carboxymethylcellulose as a function of ionization of substrate carboxyls

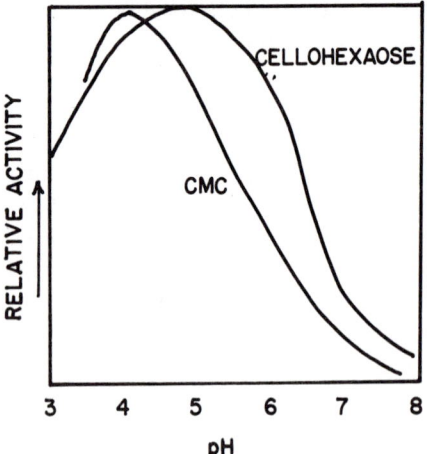

Figure 4. Effect of pH on enzymatic hydrolysis of charged and uncharged cellulosic substrates (12), CMC = Carboxymethylcellulose

Some progress has been possible using carboxymethylcellulose and similar soluble derivatives by relating change in intrinsic viscosity, a parameter related to chain length, to changes in end groups during the course of enzymatic hydrolysis (23). Random cleaving endoglucanases from *Myrothecium verrucaria* and mineral acids catalyze a hydrolysis in which the log of the increment in end groups is proportional to change in average chain length (Figure 5). In contrast the endwise cleaving cellulases of *Cellvibrio gilvus* yielded data showing a linear relationship between the log of the change in chain length to the log of the change in number of end groups (Figure 6). The use of the term "endwise-cleaving cellulases" in this sentence is for enzymes that remove cellobiose (rather than glucose units) from the chain end. This action, unreported for other cellulase systems, is analogous to that of β-amylase, and like the latter it acts by inversion (22). An unfortunate aspect of this approach is that a theoretical basis for these relationships has not been discovered.

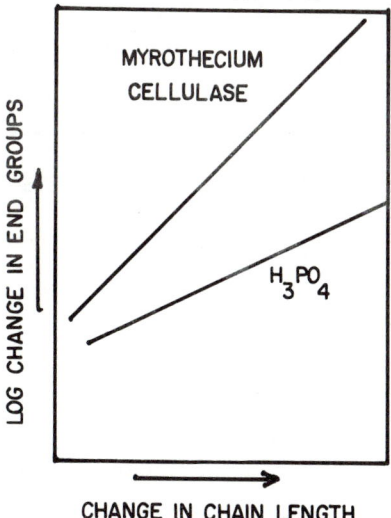

Figure 5. *Relationship of change in number of end groups to change in chain length during random hydrolysis of carboxymethylcellulose by cellulase of* Myrothecium verrucaria *(23)*

The disadvantages, then, of the high molecular weight substrates in study of the mode of action of cellulases are many and derive from inherent properties of the substrate molecules themselves. But in spite of the obvious disadvantages, modified celluloses have been useful in simple assays for determining cellulase activity. Recently, fresh attempts have been made to obtain molecular activity data from viscosity measurements (2, 3). Values of 29 bonds broken per second were obtained. Others have shown that the presence of a substituent on the cellulose interferes with enzyme hydrolysis at adjacent linkages (13). Hydrolysis proceeds much more rapidly where three unsubstituted glucose units occur together, than where there are only two (3).

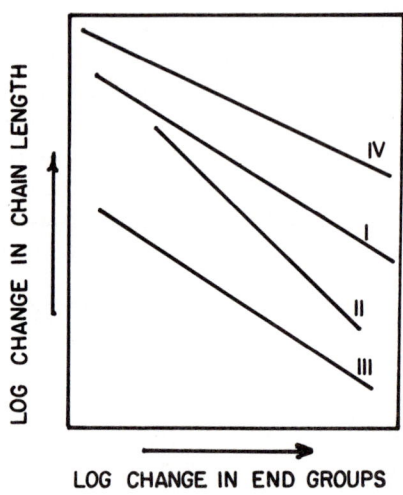

Figure 6. Relationship of change in chain length and change in end groups during hydrolysis of carboxymethylcellulose by exoglucanases of Cellvibrio gilvus (23)

Use of Oligosaccharides in Determination of the Mode of Action of Cellulolytic Components. Broadly the oligosaccharides invite three approaches to determination of the site of action: isotopic labelling, chemical labelling, and use of unlabelled substrates.

Specific labelling of cellobiose can be readily achieved by use of cellobiose phosphorylase (24).

$$\text{glucose-}^{14}\text{C} + \text{glucose-1-PO}_4 \rightleftharpoons \text{cellobiose-}^{14}\text{C} + \text{H}_3\text{PO}_4$$

This reaction goes very well using cellobiose phosphorylase from *Cellvibrio gilvus* yielding cellobiose in which only the reducing glucosyl is radioactive. With the cellodextrin phosphorylase from *Clostridium thermocellum* it should be equally easy to prepare cellotriose, cellotetraose, etc. with the reducing terminal glucosyl labelled. Because of the good yields and the purity of the products, these procedures are far superior to the earlier ones using an *Acetobacter xylinum* enzyme to produce labelled oligosaccharides (9, 22).

Table II. Products of Hydrolysis of End Labelled Cellohexaose-^{14}C in Relation to Bond Cleaved

Bond Cleaved	Products
1	cellopentaose-^{14}C + glucose
2	cellotetraose-^{14}C + cellobiose
3	cellotriose-^{14}C + cellotriose
4	cellobiose-^{14}C + cellotetraose
5	glucose-^{14}C + cellopentaose

With these substrates the approach is direct. Numbering the bonds of cellohexaose 1 through 5 starting from the non-reducing end, the results in Table II are predicted.

The advantage of the labelled over the unlabelled substrates is immediately apparent. Cleavage at bonds 1 and 5 or 2 and 4 yield identical products and without the labelled groups no distinction can be made. The isotopically labelled substrates then permit reliable identification of the site of attack, and where more than one bond is attacked as is usually (4, 9, 22, 23) but not always (4, 11, 14) the case, the relative frequencies of attack at each bond can be quantitatively desired.

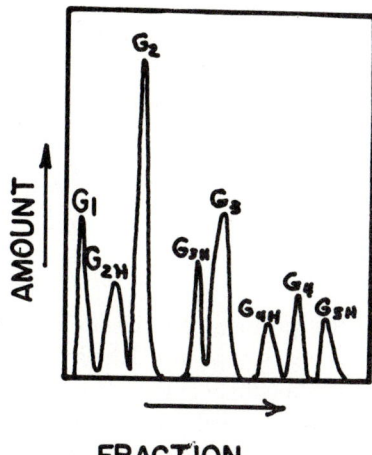

Figure 7. Analysis of normal and reduced members of the cellulose oligosaccharide series by charcoal chromatography. Components from left to right: Glucose (G1); reduced cellobiose (G2H borohydride reduction); cellobiose (G2); reduced cellotriose (G3H); cellotriose (G3); reduced cellotetraose (G4H); cellotetraose (G4), and reduced cellopentaose (G5H)

Chemical labelling of oligosaccharides in our experience has been considerably simpler, less expensive, just as definitive and to date free of steric effects. Reduction of the reducing glucosyl to sorbitol gives a series of non-reducing oligosaccharides in quantitative yield (4, 9, 22) which can be separated chromatographically and analyzed quantitatively very conveniently (Figure 7). In this approach, attack on the various bonds of cellulose is indicated as follows (Table III).

Table III. Relation of Products of Hydrolysis of Reduced Cellohexaose to Bond Cleaved

Bond Attacked	Products
1	glucose + tetraglucosylsorbitol
2	cellobiose + triglucosylsorbitol
3	cellotriose + diglucosylsorbitol
4	cellotetraose + glucosylsorbitol
5	cellopentaose + sorbitol

Table IV. Relative Frequencies of Bond Attack in Hydrolysis of Cellohexaose by Acid and Enzymes

Relative Frequency of Attack

Enzyme	Bond Number				
	1	2	3	4	5
C. gilvus I	0	1	1	0	0
C. gilvus II	0	1	2	0	0
C. gilvus III	0	2	1	0	0
C. gilvus IV	0	3	1	0	0
A. niger I	1	0	0	0	0
HCl Control	1	1	1	1	1

This approach to the problem establishes the site of attack just as reliably as does the use of isotopically labelled substrates and at considerably less expense and inconvenience. The data (Table IV) indicate that the A. niger enzyme is a typical exo-β-1,4 glucanase—i.e., an enzyme which removes single glucose units from the end of the cellulose molecule.

A Consideration of Enzymes Involved in the Breakdown of Cellulose

The **Trichoderma viride Cellulase Complex.** Many cellulolytic organisms secrete cellulase as a multiple component enzyme system. Separation of physically distinguishable components by electrophoretic or chromatographic techniques has been repeatedly demonstrated. It is more difficult to prove that these components are distinct enzymes with different functions in terms of substrate specificity and mode—or mechanism—of action. However, in recent years such distinctions have in fact been shown for a number of cellulase systems. Many of these cell free systems are unable to catalyze hydrolysis of crystalline cellulose to any appreciable degree and are therefore useful only as model systems.

The *Trichoderma viride* complex is a true cellulase in the most rigid sense, being able to convert crystalline, amorphous, and chemically derived celluloses quantitatively to glucose. It has been established that (a) the system is multi-enzymatic, (b) that at least three enzyme components are both physically and enzymatically distinct, and (c) that all three components play essential roles in the overall process of converting cellulose to glucose.

Table V. Amino Acid Composition of the Endo and Exoglucanases of *Trichoderma viride*

Amino Acid	Moles/mole of Protein	
	Endoglucanase[b]	Exoglucanase[a]
Lysine	12	14
Histidine	9	7
Arginine	15	16
Aspartic Acid	74	65
Threonine	44	38
Serine	40	46
Glutamic Acid	46	35
Proline	21	34
Glycine	62	55
Alanine	46	50
Half cystine	0-1	3
Valine	30	25
Methionine	2	5
Isoleucine	30	21
Leucine	36	32
Tyrosine	17	18
Phenylalanine	18	19
Ammonia	93	130

[a] A molecular weight of 76,600, determined by sedimentation velocity-diffusion coefficient measurements, was used in calculation.
[b] A molecular weight of 49,000, estimated from sedimentation velocity data, was used. The presence of cystine is questionable.

The C_1 component is required for attack on crystalline cellulose. It has been considered by some workers to be a hydrolase attacking crystalline cellulose and producing cellobiose as a product (*10, 11*) and by others to be an enzyme of undefined nature converting crystalline cellulose to a form that can be acted on by the glucanases. The glucanase (C_x) components have the capacity to attack amorphous cellulose or cellulose derivatives, and include endo-β-1 $\rightarrow$ 4 glucanase(s) and exo-β-1 $\rightarrow$ 4 glucanase(s). Physically they differ markedly in adsorption characteristics, amino acid composition (Table V), molecular weight, and thermal stability. Enzymologically they differ only slightly in activation energies for hydrolysis of both cellulodextrins and carboxymethylcellulose, but are conspicuously different in mode of action, K_m values for cellulodextrin series (Table VI), optimum substrate chain length, and pH optimum. One, the endoglucanase, is essentially random in its action; the other, the exoglucanase, is endwise. This exo-β-1 $\rightarrow$ 4 glucanase is typical of other exo-glucanases (a) in removing single glucose units from the non-reducing end of the cellulose chain, (b) in its relative affinity for oligosaccharides, and (c) in that it acts by inversion—i.e., it produces α-glucose (Figure 8). The presence of both types of glucanase in filtrates from a single organism was also observed with *A. niger*.

Table VI. Michaelis Constants for Cellulose Polymer Series: Endo and Exoglucanases of *Trichoderma viride*

	Michaelis Constants	
Substrate	Endoglucanase	Exoglucanase
Cellobiose	$190 \times 10^{-4} M$ [a]	$220 \times 10^{-5} M$ [a]
Cellotriose	$31 \times 10^{-4} M$	$18 \times 10^{-5} M$ [a]
Cellotetraose	$28 \times 10^{-4} M$	$6.5 \times 10^{-5} M$ [a]
Cellopentaose	$7.0 \times 10^{-4} M$	$6.0 \times 10^{-5} M$
Cellohexaose	$1.0 \times 10^{-4} M$ [a]	$16.0 \times 10^{-5} M$ [a]

[a] These Lineweaver-Burk plots showed apparent substrate inhibition at concentrations greater than $0.05 M$.

To develop a clear understanding of the role of the three components in degradation, the attack on both crystalline and amorphous substrates with ultimately complete conversion to glucose must be investigated. Attack on the crystalline regions may be attributed to the C_1, the primary product (97%) in our work being cellobiose, at the present stage of purification of that enzyme (11-fold over the crude). Because the endo-enzyme has a greater apparent affinity for substrates of long chain length than the exoglucanase, the endo-glucanase must be assigned the dominant role of initiating the attack on "amorphous" substrate. Here the major products are cellulodextrins and a trace of glucose. The exo-glucanase, then, completes the hydrolysis of the products to glucose. The relation of the roles of the endoglucanase and exoglucanase can be looked at as representing generation of new chain ends by the random acting endo-glucanase, thus increasing the effective concentration of substrate for the endwise-acting glucogenic exoglucanase.

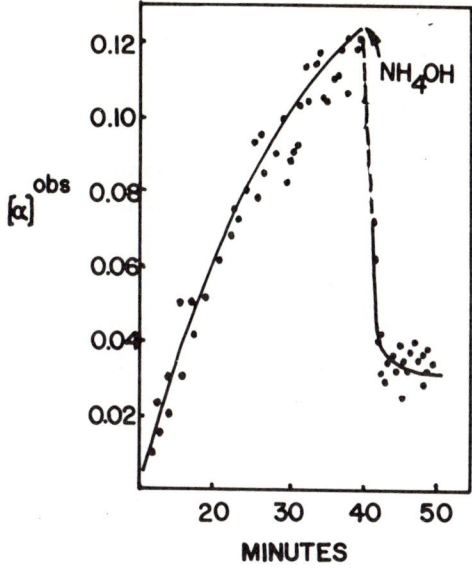

Figure 8. Changes in optical rotation during hydrolysis of reduced cellotetraose by the exoglucanase of T. viride, and after base-catalyzed mutarotation (14)

The system as a whole is therefore not merely multi-enzymic by happenstance, but necessarily so for both maximum rate of attack on native cellulose and for maximum completeness of conversion to glucose.

Metabolism of Cellulose Oligosaccharides by *Cellvibrio gilvus*. In contrast to the cellulase complex of *T. viride* which forms glucose as the principal product of hydrolysis, the hydrolysis of cellulose in *Cellvibrio gilvus* results in the formation of chiefly cellobiose and cellotriose (Table IV). From our studies on the metabolism of these oligosaccharides the following picture has emerged.

C. gilvus grows well on cellulose and on oligosaccharides but poorly on glucose, a phenomenon quite common among polysaccharide-using bacteria and never explained metabolically. Analysis of fermentation products (6, 7) and metabolic load studies indicated that glucose and cellobiose were differently metabolized, and phosphorolysis (7) of cellobiose was demonstrated. Different metabolic routes for the two halves of cellobiose were predicted (7) and later demonstrated (24). The extracellular cellulase system was found to consist of at least 4 C_x components of primarily endwise action (4, 22, 23) and a C_1 component, the major extracellular products being cellobiose and cellotriose. Later it was found that the cells took up all members of the cellulose oligosaccharide family by an active transport mechanism at rates such that the cells were supplied with about 40 million glucosyls/cell/minute (20) and that the metabolic efficiency with which the oligosaccharides were used increased with chain length at a rate suggesting conservation of glucosyl bond energy. The obvious first hypothesis for explaining this phenomenon is that the oligosaccharides are cleaved phosphorolytically as appears to be true for *Clostridium thermocellum* (21). This suggestion, however, is ruled out by the data (Table VII) showing that there is intracellular hydrolysis of cellotriose and phosphorolysis only of cellobiose (25).

Table VII. Fractionation of Phosphorylase and Hydrolase Activities in Cell-free Extracts of *Cellvibrio gilvus*

$(NH_4)_2SO_4$ Precipitate	Hydrolysis (μmoles/hr.) Cellobiose	Cellotriose	Phosphorolysis (μmoles/hr.) Cellobiose	Cellotriose
Crude Enzyme	0.66	1.44	3.44	1.80
0-20 % Satn.	0	0	0	0
20-40 % Satn.	0.16	0	1.36	0
40-50 % Satn.	0	0	1.66	0
50-60 % Satn.	0.22	0.36	0	0
60-65 % Satn.	0	0.20	0	0
65-75 % Satn.	0	0	0	0

Thus, the "apparent" phosphorolysis of cellotriose in the crude cell extracts is in reality the result of hydrolysis of the trisaccharide followed by phosphorolysis of the resulting cellobiose.

How then is the energy of the glucosyl bonds in oligosaccharides to be conserved? There is an active "glucotransferase" located somewhere in the wall of the cell which transfers non-reducing glucosyls from any of the oligosaccharides to any other but apparently not to water so that starting with any one oligosaccharide all of the others are produced without appreciable reduction in the total number of glucosyl bonds through hydrolysis. For example, with cellotriose or cellotetraose as substrates prolonged incubation shows slight if any hydrolysis (Table VIII).

At the same time chemical analyses of this same experimental system shows active transglucosylation (Figure 9). Glucose is the preferential acceptor, the moles of cellobiose far exceeding the value anticipated from purely random transglucosylation.

The transfer reactions follow second order kinetics as expected and when physiological concentrations of oligosaccharides—*i.e.*, those used in growth experiments—are used the transfer product has been characterized as cellobiose. At very low concentrations of oligosaccharide laminaribiose is produced (perhaps indicating non-specific transfer).

Table VIII. Retention of Glucosyl Bonds During Action of Glucotransferase of *C. gilvus* on Cellotriose and Cellotetraose

Incubations, hrs.	Total Glucosyl Bonds	
	Cellotriose (μmoles)	Cellotetraose (μmoles)
0	5.40	4.47
1	5.19	4.43
2	5.10	4.33
3	5.10	4.27
4	4.93	4.37

Table IX. Characteristics of Cellulose Oligosaccharides as Donors to Glucose

Donor	K_m M × 10^{-2}	V_m μg/ml./min.	Relative Rate as Donor
Cellobiose	1.16	167	5
Cellotriose	1.18	38	25
Cellotetraose	1.19	21	28
Cellopentaose	0.36	8	30
Cellohexaose	—	—	36

Table X. Agreement of Experimental Data with Those Predicted by Proposed Sequence of Reactions

Incubations, hrs.	μ moles of Cellobiose	
	Theoretical	Experimental
1	0.40	0.34
2	0.66	0.62
3	0.80	0.81
4	0.95	0.94

Some further characteristics of the oligosaccharides as donors to glucose are summarized (Table IX).

Although the maximum velocity falls rapidly as the chain length increases, the relative rates of donation at physiological concentrations are of the same order to magnitude.

When one assumes the occurrence of the following reactions:

$$G_x + G_x \rightleftharpoons G_{x+1} + G_{x-1}$$
$$G_x + H_2O \rightarrow G_{x-1} + G_1$$
$$G_x + G_1 \rightleftharpoons G_{x-1} + G_2$$
$$G_2 + P_i \rightleftharpoons \text{G-1-P} + G_1$$

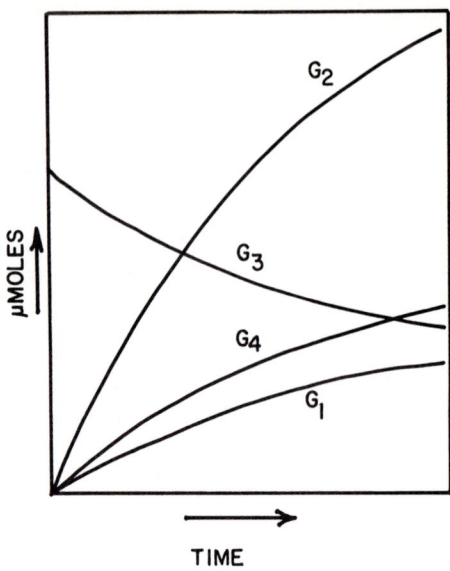

Figure 9. Course of transglucosylation of cellotriose by cell-free extracts of Cellvibrio gilvus *(See legend to Figure 7)*

it is possible to predict the amount of each product to be anticipated during the initial phases of the reaction and to check these experimentally. When such an experiment is performed with cellotriose as substrate the prediction of the cellobiose (or other oligosaccharide) yields is good (Table X).

The combination of the oligosaccharide glucotransferase with the hydrolase and the cellobiose phosphorylase, then accounts fully for the conservation of the glucosyl bond energy predicted by the growth experiments.

Phosphorolysis of Cellulose Oligosaccharides. Although it has not been shown that *Clostridium thermocellum* can freely take cellulose oligosaccharides into the cell, there is reason to believe that this organism, like *C. gilvus*, metabolizes these oligosaccharides intracellularly. An enzyme, cellodextrin phosphorylase, which catalyzes the phosphorolysis of cellulose oligosaccharides, has been found in *Clostridium thermocellum* (21). This enzyme catalyzes the phosphorolysis of cellohexaose, cellopentaose, cellotetraose, and cellotriose, resulting in the formation of cellobiose and glucose 1-P. Cellobiose can be broken down into glucose and glucose 1-P by cellobiose phosphorylase. A somewhat similar situation is found in protozoa where enzymes have been found that catalyze

The Cellulase Complex

As currently understood, the cellulase complex contains the following components (listed in the order in which their action on cellulose occurs):

1. C_1 is an enzyme whose action is unspecified. It is required for the hydrolysis of highly oriented solid cellulose (Cotton, Avicel, etc.) by $\beta\text{-}1 \to 4$ glucanases.

2. **$\beta\text{-}1 \to 4$ Glucanases** ($= C_x$) are the hydrolytic enzymes. In his book, Eriksson (5) uses cellulase as synonymous with $\beta\text{-}1 \to 4$ glucanase. It is preferable to reserve the term cellulase for the $C_1\text{-}C_x$ complex. C_x is an earlier synonym which can be discarded. $\beta\text{-}1 \to 4$ glucanase is usually measured by action on soluble cellulose derivatives, usually carboxymethylcellulose. The "x" in C_x emphasizes the multi-component nature of this fraction.

The $\beta\text{-}1 \to 4$ glucanases are clearly of two types:

(a) exo-$\beta\text{-}1 \to 4$ glucanase, successively removing single glucose units from the non-reducing end of the cellulose chain;

(b) endo-$\beta\text{-}1 \to 4$ glucanases with action of a random nature, the terminal linkages generally being less susceptible to hydrolysis than internal linkages.

3. **β-Glucosidases** vary in their specificities. Those that act primarily on aryl-β-glucosides are not involved in cellulase action. The β-glucosidases that are involved in cellulose breakdown are those highly active on the β-dimers of glucose, including cellobiose. "Cellobiase" is the most descriptive designation for this group (to cellulase workers), but a more accurate name might be "β-glucodimerase," indicating the ability to act on all of the β-dimers of glucose.

β-Glucosidases and exo-$\beta\text{-}1 \to 4$ glucanases have substrates in common, cellobiose to cellohexaose. β-Glucosidases hydrolyze the smaller oligomers most rapidly; exo-glucanases, the larger ones. β-Glucosidases act by retention of configuration; exo-glucanases by inversion. β-Glucosidases are strongly inhibited by gluconolactone, exo-glucanases much less so. Finally, there is a difference in linkage specificity, the exo-enzyme generally being more specific; the β-glucosidase less specific.

Enzyme Commission numbers have been assigned to some of the cellulase components. Endo-$\beta\text{-}1 \to 4$ glucanase is E.C. 3.2.1.4. β-glucosidase is E.C. 3.2.1.21, a designation which does not differentiate between aryl-glucosidase and dimerase (cellobiase). The exo-$\beta\text{-}1 \to 4$ glucanase has sometimes been placed in E.C. 3.2.1.21, but a new number should be assigned to it. The Commission has not yet begun to think about such odd-balls as C_1.

the phosphorolysis of laminaridextrins (*17, 18*). These enzymes catalyze the phosphorolysis of laminaridextrins that are formed by the intracellular hydrolysis of paramylon.

The phosphorolytic reactions catalyzed by these enzymes can be described as: (Glucose)$_n$ + Pi ⟷ (Glucose)$_{n-1}$ + α-glucose 1-P. They resemble the hydrolytic exoglucanases in that they remove one unit at a time from the nonreducing end of the glucan; and at least in the β-series, in that they act by inversion—*e.g.*, to α-glucose-1-P from a β-glucan. At present, knowledge of such phosphorolytic systems are limited to a few microbial species.

These enzymes also resemble the exo-glucanases in their specificity —*i.e.*, they form, or split, a single linkage type—*e.g.*, B-1 → 3, or B-1 → 4. Some, at least, of the phosphorylases show a preference for oligomers over dimers (*20*); others act more readily (or exclusively) on the dimers (like the glucosidases). Most do not act on the long polymers (laminarin, etc.) and in this respect, they differ from exo-glucanases.

Experimentally, the interesting feature of these enzymes is their synthetic action. They make it possible to attach a glucose unit to another glucose, or oligomer, by a specific linkage. The acceptor molecule may possess linkages (and even sugars) that are the same or different from that being added. A variety of products can, thus, be synthesized of predetermined linkage sequence. Such products are useful for determining specificity of hydrolytic enzymes (*1, 7, 21*).

It should be emphasized that these phosphorolytic enzymes are intracellular. They are not found in the extracellular cellulase complex, and they are involved in the cellulolytic systems of relatively few microorganisms.

Literature Cited

(1) Alexander, J. K., *J. Bacteriol.* **81**, 903 (1961).
(2) Almin, K. E., Eriksson, K. E., *Biochim. Biophys. Acta* **139**, 238 (1967).
(3) Almin, K. E., Eriksson, K. E., Jansson, C., *Biochim. Biophys. Acta* **139**, 248 (1967).
(4) Cole, F. E., King, K. W., *Biochim. Biophys. Acta* **81**, 122 (1964).
(5) Eriksson, K. E., ADVAN. CHEM. SER. **95**, 83 (1969).
(6) Hulcher, F. H., King, K. W., *J. Bacteriol.* **76**, 565 (1958).
(7) Hulcher, F. H., King, K. W., *J. Bacteriol.* **76**, 571 (1958).
(8) King, K. W., "Advances in Enzymatic Hydrolysis of Cellulose and Related Materials," E. T. Reese, Ed., p. 159, Pergamon Press, London, England, 1963.
(9) King, K. W., *Virginia Agri. Exp. Sta. Tech. Bull.* (Blacksburg, Virginia) **154**, 1 (1964).
(10) King, K. W., *J. Fermentation Technol.* (Japan) **43**, 79 (1965).
(11) King, K. W., *Biochem. Biophys. Res. Comm.* **24**, 295 (1966).
(12) King, K. W., Smibert, R. M., *Appl. Microbiol.* **11**, 315 (1963).
(13) Klop, W., Kooiman, P., *Biochim. Biophys. Acta* **99**, 102 (1965).

(14) Li, L. H., Flora, R. M., King, K. W., *Arch. of Biochem. Biophys.* **111**, 439 (1965).
(15) Li, L. H., King, K. W., *Appl. Microbiol.* **11**, 320 (1963).
(16) Liu, T. H., King, K. W., *Arch. of Biochem. Biophys.* **120**, 462 (1967).
(17) Marechal, L. R., *Biochim. Biophys. Acta* **146**, 417 (1967).
(18) *Ibid.*, **146**, 431 (1967).
(19) Rautela, G. S., King, K. W., *Arch. of Biochem. Biophys.* **123**, 589 (1968).
(20) Schafer, M. L., King, K. W., *J. Bacteriol.* **89**, 113 (1965).
(21) Sheth, K., Alexander, J. K., *Biochim. Biophys. Acta* **148**, 808 (1967).
(22) Storvick, W. O., Cole, F. E., King, K. W., *Biochem.* **2**, 1106 (1963).
(23) Storvick, W. O., King, K. W., *J. of Biol. Chem.* **235**, 303 (1960).
(24) Swisher, E. J., Storvick, W. O., King, K. W., *J. Bacteriol.* **88**, 817 (1964).
(25) Vessal, M. I., Ph.D. Dissertation, Virginia Polytechnic Institute, Blacksburg, Virginia (1967).

RECEIVED October 14, 1968.

3

Estimation of Exo-β-1 → 4-glucanase in Crude Cellulase Solutions

E. T. REESE

Pioneering Research Laboratory, U. S. Army Natick Laboratories, Natick, Mass. 01760

> *A method has been developed for the estimation of exo-β-1 → 4 glucanase in cellulolytic solutions based on the relative susceptibility of cellotetraose and cellobiose to hydrolysis. By its use we have found that most cellulolytic organisms produce some exo-β-1 → 4 glucanase. Trichoderma viride QM6a is most productive of those tested. (Oddly enough it is also the best producer of C_1, but not of the endo-β-1 → 4 glucanase.) The exo-β-1 → 4 glucanase, like the endo-, appears to be an induced enzyme.*

The exo-polysaccharases, as a group, are relatively new (except for β-amylase). Their investigation goes back less than thirty years to the discovery of the exo-α-1,4-glucanase known also as glucoamylase or amyloglucosidase. Most of these enzymes remove a single sugar unit from the non-reducing end of the polymer. Action is most rapid on the longer chains, such small molecules as the dimer being highly resistant to hydrolysis. Within the past 10 years, additional exo-glucanases have been discovered, the best known of which is the exo-β-1,3-glucanase. The exo-β-1,4-glucanase which is part of the cellulolytic system has received little attention (1, 2). These enzymes are usually found in association with endo-enzymes and glucosidases, and, as yet, little effort has been expended in detecting the presence of the exo-enzyme.

Further work on exo-β-1 → 4-glucanase demands a method for its detection and estimation in crude solutions—*i.e.*, in the presence of both endo-β-1 → 4-glucanase and cellobiase. For simplification and other reasons (*see* below), let us ignore the contribution of the endo-β-1 → 4-glucanases. If our generalizations on exo-glucanases (1, 2) are correct, we may approach the problem in several ways. The one which seems most promising depends upon the differences in hydrolysis rates of dimer

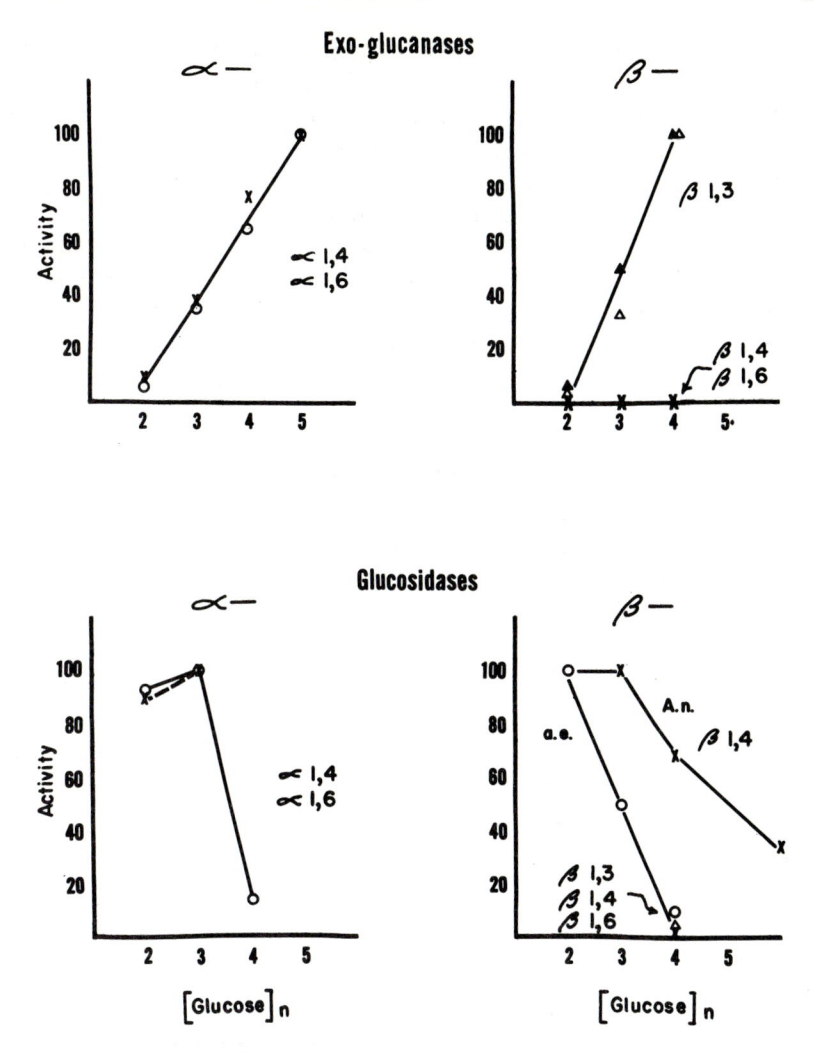

Figure 1. Activity of glucosidases and of exo-glucanases as a function of degree of polymerization (DP) of substrate (Reference 2). The exo-β-glucanases shown here act only on β-1 → 3 linkages

and tetramer by the two enzyme types (Figure 1). Cellotetraose (G_4) is much more rapidly hydrolyzed than cellobiose (G_2) by exo-β-1→4-glucanase (1), and much less rapidly hydrolyzed than cellobiose by cellobiase. Therefore, a high activity ratio (cellotetraose/cellobiose) indicates a predominance of exo-glucanase; a low ratio indicates cellobiase.

Briefly, the procedure is as follows: Substrate (G_4 8 mg./ml., or G_2 4 mg./ml.) in $M/20$ citrate, pH 4.5 (0.5 ml.), is added to an enzyme solution properly diluted (0.5 ml.) and incubated at 40° for 30 minutes.

Three ml. of glucose oxidase solution (Worthington; buffered to pH 7.0) are added, and the incubation continued for 15 minutes. One ml. of 5M HCl is added, and the color measured at 425 mμ. The rate of production of glucose is thereby determined for each substrate.

The values for the ratio G_4/G_2 for pure cellobiase, and for pure exo-glucanase vary depending upon the source of enzyme (2). For our example (Figure 2), we assign for cellobiase a ratio $G_4/G_2 = 0.2$, and for exo-β-1 → 4 glucanase $G_4/G_2 = 100$. These values are based on data with purified enzymes, and approach the low and the high limits which the ratios may reach. The curve (Figure 2) is plotted from the formula:

$$\text{Exo-glucanase, \%} = 100 \times \frac{(Y_T/Y_D) - 0.2}{0.99\,(Y_T/Y_D) + 0.8}$$

Where Y_T = rate of glucose formation from tetramer, and
Y_D = rate of glucose formation from dimer

(This formula was worked out by H. J. Hoge for this specific case, and for the generalized condition.)

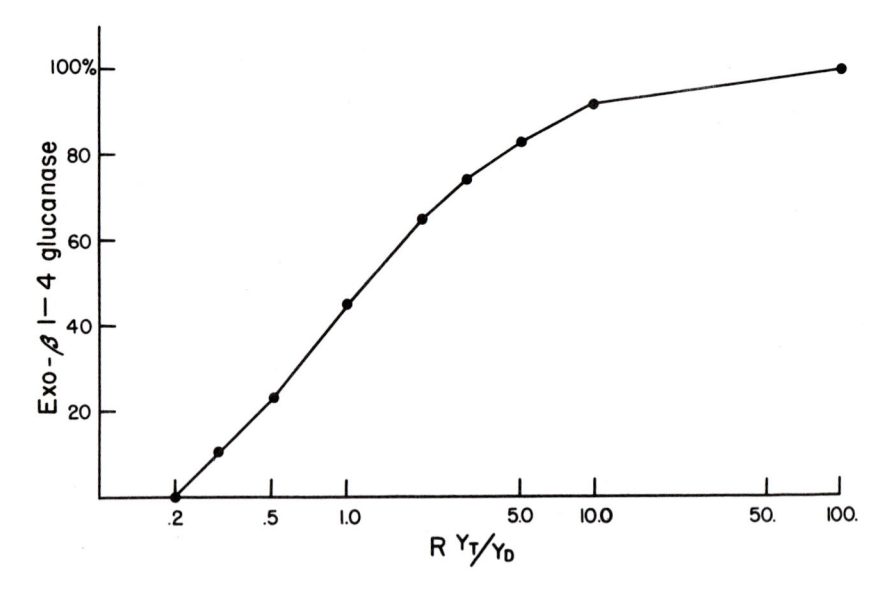

Figure 2. Estimation of exo-β-1 → 4-glucanase in solutions containing cellobiase

Plotted from equation shown in text

The procedure was applied to a variety of available cellulase solutions (Table I). There are two points to consider in estimating exo-β-1 → 4-glucanase. First is the rate at which cellotetraose is hydrolyzed (here indicated in units of enzyme, where one unit yields 0.5 mg. glucose).

Second, is the proportion of this hydrolysis which can be assigned to the exo-enzyme (here represented by the ratio Y_T/Y_D). The product of these two is a measure of exo-glucanase.

The interpretation is not as simple as outlined. There is the complication which we have ignored above—e.g., the contribution of endo-β-1 $\rightarrow$ 4-glucanase to the glucose produced from tetramer. Endo-glucanases act on glucose polymers in a fashion which approaches randomness. End linkages are, however, much less susceptible to attack than internal linkages. This was first observed for the α-amylases (3) and later for the cellulase of *Myrothecium* (4). In some cases, the end linkages are completely resistant, and hydrolysis of the glucan yields only dimer and trimer (salivary α-amylase; endo-α-1 $\rightarrow$ 6-glucanase = dextranase; endo-β-1 $\rightarrow$ 4-glucanase of *Streptomyces*). In other cases, the final hydrolysis products are dimer and monomer, the relative amounts being 5.5:1 for *B. subtilis* α-amylase (3) acting on amylose, and about 7:1 for *Myrothecium* cellulase acting on tetramer (4). There is always the possibility that the latter enzymes may have been contaminated with either exo-glucanase or glucosidase, accounting for the appearance of glucose. This is particularly likely for exo-glucanase since its very existence was unsuspected until quite recently. But until this is proven, we must accept the possibility that some of the glucose produced by the enzyme mixtures acting on tetramer may arise from endo-β-1 $\rightarrow$ 4-glucanase activity. On the other hand, none of the glucose from the dimer hydrolysis is due to endo-enzyme.

Another complication, which we ignore because so few exo-glucanases have been studied, is the possible variability between enzymes from different sources. The basic reaction is common to all, but even now minor differences are recognized, such as the effects of branching of the substrate molecule, and of linkage specificity. There may, also, be "relative" differences in the action on tetramer as compared to action on dimer. These factors will influence the absolute values obtained by the procedure here outlined. (It is quite likely that changes in the method will be required as more information becomes available.)

The only other enzymes which could yield glucose from dimer and tetramer are the phosphorylases (cellobiose phosphorylase, etc.).

$$G - G + P_i \rightleftarrows G + G\text{-}1\text{-}P$$

Fortunately, these enzymes are not found among the extra-cellular cellulases with which we are involved.

"The proof of the pudding is in the eating." If the above procedure can effectively locate good sources of exo-β-1 $\rightarrow$ 4-glucanase, it will have served its purpose.

Table I. Estimation of Exo-β-1 →
Activity in μ/ml.

Organism	Growth[a] Substrate	4 mg./ml. vs. G_2	8 mg./ml. vs. G_4
Trichoderma viride 6a	CB + Lactose	0.24	2.4[+]
Trichoderma viride 6a	Cellulose	0.6	18.0
Basidiomycete sp. 806	CB + Lactose	0.3	0.6
Basidiomycete sp. 806	Cellulose	4.0	8.0
Chrysosporium pruinosum 826	CB + Lactose	0.9	1.0
Chrysosporium pruinosum 826	Cellulose	8.0	13.0
Penicillium pusillum 137g	CB + Lactose	2.6	2.6
Penicillium pusillum 137g	Cellulose	2.4	9.2
Aspergillus foetidus 328	Bran	60.	18.
Aspergillus luchuensis 873	CB + Lactose	15.0	8.
Aspergillus terreus 442	CB + Lactose	5.4	5.4
Aspergillus terreus 72f	CB + Lactose	1.9	2.0
Penicillium brasilianum 6947	CB + Lactose	1.5	1.2
Penicillium funiculosum 474	CB + Lactose	14.0	7.0
Penicillium parvum 1878	CB + Lactose	9.8	5.0
Fusarium moniliforme 1224	Cellulose	2.5	1.5
Lenzites trabea 1009	Cellulose	0.1	2.0
Myrothecium verrucaria 460	Cellulose	0.18	1.0
Pestalotiopsis westerdijkii 381	Cellulose	0.4	0.4
Polyporus cinnabarinus 8846	Cellulose	0.12	0.5
Polyporus versicolor 1013	Cellulose	0.1	0.3
Stachybotrys atra 94d	CB Acetate	0.3	0.5

[a] Growth: CB + Lactose = Cellobiose 0.3% + Lactose 0.3%.
[c] Col. 4 × Col. 6. Activity 40°C., 30 min.; μ = 0.5 mg. glucose.

According to our data (Table I) and our interpretation of them, most cellulolytic preparations contain exo-β-1 → 4-glucanase. *Trichoderma viride* QM6a cultures produce the most exo-enzyme and the best proportion of exo-enzyme to cellobiase. While *Aspergillus foetidus* is just as active as *Trichoderma* on the tetramer, most of this activity is attributable to cellobiase. All good cellulolytic fungi produce much more of the exo-enzyme when grown on cellulose than when grown on cellobiose plus lactose.

The data further suggest that *Chrysosporium pruinosum*, and *Penicillium pusillum* are good sources of exo-β-1 → 4 enzyme, if one does not object to the cellobiase present. *Polyporus cinnabarinus* produces less exo-enzyme, but as in *Trichoderma viride*, it is nearly free of cellobiase. *Streptomyces* sp. B814 is the only organism which appears to lack exo-glucanase in its cellulase complex.

Our concern (above) about interference by endo-β-1 → 4-glucanase was unwarranted. The Y_T/Y_D values are relatively constant for a par-

4-glucanase in Cellulolytic Filtrates

R G_4/G_2	Exo-glucanase % [b]	$\mu/ml.$ [c]	C_x $\mu/ml.$	Lactone Inhib.
10.+	92.	2.2+	2.	
30.	96.	17.3	38.	9. %
2.	65.	0.4	1.	
2.	65.	5.3	32.	68.
1.1	47.	0.5	1.	
1.6	58.	7.5	42.	83.
1.0	45.	1.2	10.	
3.8	80.	7.4	88.	40.
0.3	11.	2.0	22.	81.
0.54	26.	2.1	0.8	95.
1.0	45.	2.4	5.0	
1.1	47.	1.9	3.	
0.75	36.	0.4	2.	
0.5	23.	1.6	0.3	94.
0.5	23.	1.2	3.	
0.6	29.	0.4	2.8	61.
20.0	94.	1.9	5.4	11.
5.	83.	0.8	46.	NT
1.	45.	0.2	50.	72.
4.	80.	0.4	45.	NT
3.	74.	0.2	2.9	19.
1.7	60.	0.3	5.9	NT

[b] From Figure 2.

ticular organism over wide ranges of C_x concentration (Table I). Thus, the R value for *Basidiomycete* 806 is 2, both for a solution containing 1 C_x unit and for a solution containing 32 C_x units per ml. C_x activity is determined with carboxymethyl cellulose as substrate. It is chiefly a measure of endo-β-1 → 4-glucanase activity, since adding substituents— e.g., carboxymethyl—to a substrate inhibits action of exo-glucanase much more than it inhibits action of endo-glucanase. As a result, it should be possible to find much exo-glucanase in a culture filtrate low in endo-glucanase.

An alternate means of estimating exo-glucanase is by means of glucono-lactone. This agent inhibits cellobiase, but is without effect on exo-β-glucanases (2). A ratio of lactone to substrate of 0.1 (by weight) is suitable for making the determination. The results (Figure 3) show that the procedure is useful only for solutions of relatively high exo-glucanase. Cellotriose or cellotetraose may be used as substrates, but cellobiose cannot be used because of its resistance to hydrolysis by the exo-enzyme.

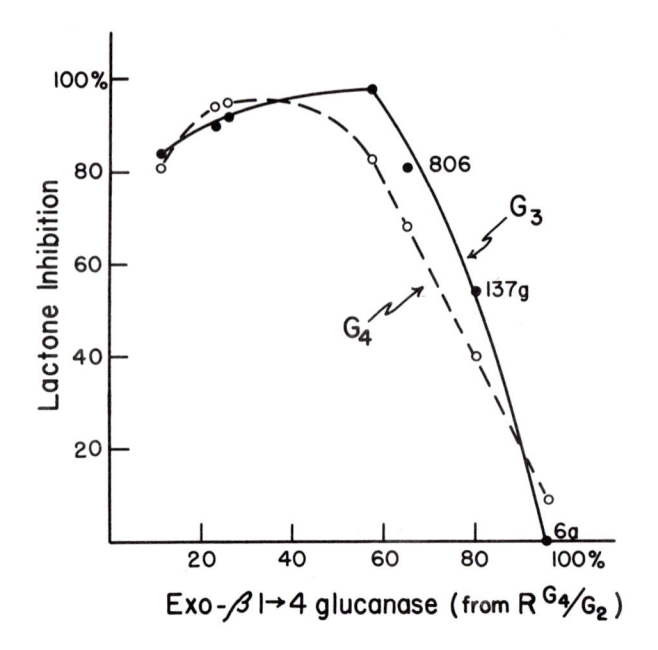

Figure 3. Estimation of exo-β-1 → 4-glucanase from inhibition by gluconolactone. R Lactone/substrate = 0.1 (wt.)
●———● *Cellotriose* (G_3) *as substrate;* ○– – ○ *cellotetraose* (G_4) *as substrate; 6a = Trichoderma viride; 137g = Penicillium pusillum; 806 = Basidiomycete sp.*

Literature Cited

(1) Li, L. H., Flora, R. M., King, K. W., *Arch. Biochem. Biophys.* **111**, 439 (1965).
(2) Reese, E. T., Maguire, Anne, Parrish, F. W., *Can. J. Biochem.* **46**, 25 (1968).
(3) Whelan, W. J., "Handbucher Pflanzenphysiologie," W. Ruhland, Ed., **Vol. VI**, p. 154, Springer-Verlag, Berlin, 1958.
(4) Whitaker, D. R., "Marine Borers," Dixie Lee Ray, Ed., Univ. Wash. Press, Seattle, 1959.

RECEIVED October 14, 1968.

Discussion

Dr. King: "The method appears to be suitable for the detection of exo-β-1 → 4 glucanase. I believe, however, that one needs more data on exo-glucanases from different organisms—*i.e.*, how similar or how different may these be in their relative activities on the tetramer and the dimer ($R\ G_4/G_2$)."

Dr. Reese: "This is quite right. The proposed method is to help us find new exo-glucanases. Then we can determine their $R\ G_4/G_2$ values, and refine the procedure if need be."

4

The Purification and Properties of the C_1-Component of the Cellulase Complex

KEITH SELBY

Shirley Institute, Didsbury, Manchester M20 8RX, England

Certain cellulolytic fungi yield cell-free filtrates capable of extensive degradation of highly ordered forms of cellulosic material. These filtrates have been shown to contain a so-called C_1-component, which, although essential for this type of activity, is virtually without action when freed from the other (C_x) components. C_1-components, with very similar properties, have been isolated from Trichoderma viride *and from* Penicillium funiculosum. *The powerful synergistic action on cotton previously found between the C_1 and C_x components of* T. viride *was also displayed by those from* P. funiculosum; *cross-synergism has also been demonstrated. An attempt has been made to explain the role of C_1 in the solubilization of cotton.*

When Reese, Siu, and Levinson (24) first suggested the existence of a C_1-component in the cellulase system, they did so because certain microorganisms could grow on media containing cellulose as sole carbon source whereas others required more easily assimilable forms of carbon, such as glucose or carboxymethylcellulose. It was suggested, therefore, that although all "cellulase" systems probably contained enzymes called C_x capable of digesting easily accessible forms or derivatives of cellulose, only those that contained a C_1-component could utilize highly ordered forms of cellulose, as found, for example, in cotton hairs and in some other vegetable fibres, against which attack is principally influenced by the supra molecular structure of the substrate. It was further suggested that C_1 acted as a chain-separating enzyme, which at the time Siu, as he said "in semi-jest" (29, 30), referred to as a "hydrogen bondase"; the C_1-component was thus thought of as an enzyme whose presence was required before the depolymerizing enzymes (C_x) could start to break down the cellulose chain. Ten years after this theory was postulated there had been little further progress because no cell-free filtrates had been isolated with sufficient activity towards highly ordered substrates to suggest that they contained a significant amount of the C_1-component. A hopeful sign came from the discovery (26) that filtrates from *Myrothecium verucaria*, suitably manipulated, could produce 30% solubilization of cotton, and this was the stage that had been reached at the time of the last ACS symposium on cellulases held in Washington in 1962 (22). It subsequently became apparent from the high solubilizations of cotton obtained by single treatments with filtrates from *Trichoderma viride* (10) that this would be a better organism to use in the search for C_1-component.

Fortunately, at this time, there were developments in the techniques for the fractionation of macromolecules; at a 1967 ACS Symposium, a review of the existing fractionation procedures for cellulases (4) revealed that, although notable successes had been achieved, better methods for the purification and characterization of the enzyme components were needed. Towards the end of 1961, developments in the production of gel-filtration media and of ion-exchange forms of the same materials made possible the application of these techniques to the separation of the components of the cellulase system.

Over the last five years, the results of many fractionation studies on cellulases have been published; the greatest number of these has been on enzymes from *Trichoderma viride* (and *T. koningii*). The reason for this interest must surely be the demonstration by several groups of workers in 1964 and 1965 (2, 5, 8, 10, 16, 19) that filtrates from this fungus contained the C_1-component. During this period the importance of a discovery reported by Gilligan and Reese 10 years earlier (3), which at the time seems to have attracted little attention, was realized. Fractionation of a filtrate from *T. viride* on calcium phosphate gel had yielded two components between which synergistic action was found. Because this action was tested only on a swollen and partially degraded cellulosic substrate (Walseth cellulose) its potential importance in relation to C_1-action was not appreciated until Mandels and Reese (10) showed that this synergism could best be demonstrated on cotton, with the obvious corollary that C_1 must be involved.

On the basis of fractionation studies made at this time, C_1-type enzyme was envisaged as the only component capable on its own of solubilizing highly ordered forms of cellulose; cotton was used by Mandels and Reese (10) and in some of Flora's work (2) and crystalline hydrocellulose (an acid-hydrolyzed cotton) by Li, Flora, and King (8). Addition of C_x to C_1 caused about threefold increases in solubilization. Contemporary work at the Shirley Institute centered on rigorous fractionation of the cellulase system of *T. viride* into components displaying single enzyme function, which seemed imperative in view of the demonstrable synergism between C_1 and other components of the cellulase system. An account of this work has already been published (28), so that only a brief description of the purification procedure will be given here. A typical fractionation of a filtrate from *T. viride* on Sephadex G-75 is shown in Figure 1. A low molecular-weight carboxymethylcellulase, which apparently did not participate in the solubilization of cotton, was removed at this stage. Although the other components were unresolved, there was

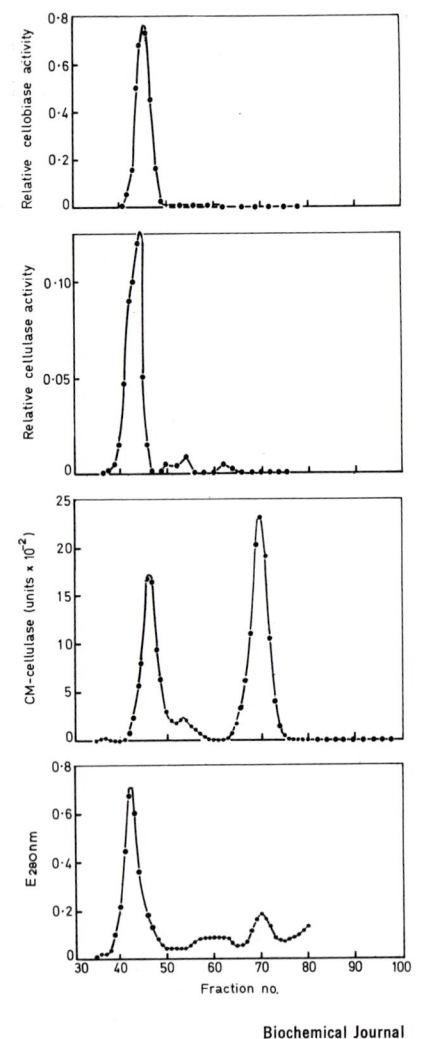

Figure 1. Fractionation of a culture filtrate from T. viride *on Sephadex G-75 (2)*

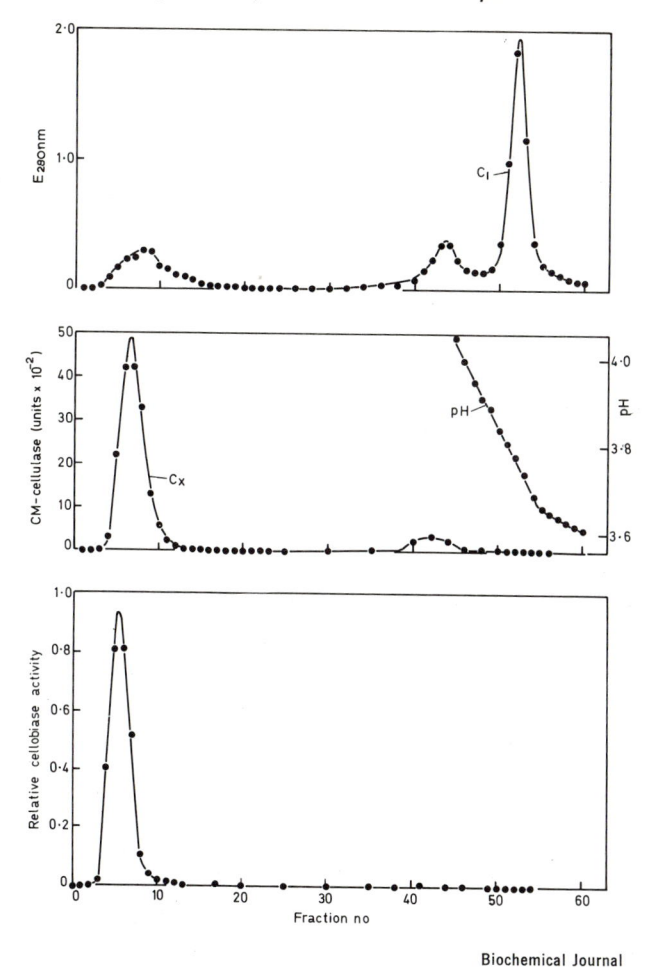

Figure 2. Separation of the C_1- and C_x-components from T. viride cellulase by chromatography on DEAE-Sephadex (2)

evidence for the existence of separate peaks of cellulase, carboxymethylcellulase, cellobiase, and material absorbing at 280 n.m. Moreover, the positions of the peaks suggested that the "cellulase" activity could result from synergism between the material absorbing at 280 n.m. and the carboxymethylcellulase and/or cellobiase. The optically dense component, believed to be C_1, was separated from cellobiase and carboxymethylcellulase by gradient elution on DEAE-Sephadex (Figure 2); the carboxymethylcellulase and cellobiase were subsequently resolved on SE-Sephadex at pH 5.1 (Figure 3).

These studies had thus yielded a carboxymethylcellulase with no activity towards cellobiose, a cellobiase with no activity towards carboxy-

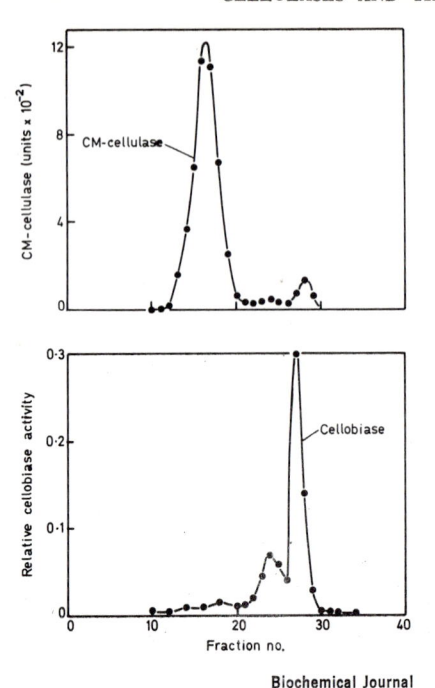

Figure 3. Separation of the carboxymethylcellulase and cellobiase of the C_x-fraction of T. viride cellulase on SE-Sephadex (2)

methylcellulose and a C_1-component, obtained from the fractionation on DEAE-Sephadex, which was inactive towards both cellobiose and carboxymethylcellulose and had virtually no activity towards cotton. However, when the C_1-component was recombined with either the unresolved C_x or appropriate mixtures of the separated cellobiase and carboxymethylcellulase, the activity towards cotton was restored completely. All possible recombinations of the three components were made in the proportions in which they were originally present in the culture filtrate. The results are shown in Table I from which the following points can be appreciated:

(1) C_1 is virtually without activity in the absence of C_x.

(2) There is no synergism between the components of C_x.

(3) Synergism between C_1 and the individual components of C_x is limited; all three must be present to account for the activity towards cotton.

Clearly the property, ascribed (2, 10) to the C_1-component of *T. viride* cellulase, of being able, on its own, to solubilize cotton and crystalline hydrocellulose, is lost when the C_1 is highly purified; this property is possessed only by mixtures of C_1 and C_x acting in synergism. In the separation on DEAE-Sephadex reported by Mandels and Reese (10), employed without prior removal of protease and low molecular-weight carboxymethylcellulase, the relative positions of the β, A, and C components were similar to those of the cellobiase, carboxymethylcellulase, and C_1-components respectively (Figure 2). The major difference was that their C-component had lost less activity towards cotton on purification than had our C_1-component, so that the synergism was less marked; from their Figures 6 and 9 the enhancement of activity on recombination would seem to be about five-fold. Subsequently, Flora (2) purified the "hydrocellulase" (C_1) from *T. viride* by adsorption on Avicel, followed by ion-exchange chromatography on a carboxylic acid resin. His "hydrocellulase" was still capable of causing extensive weight loss of native cotton fiber, however, and, although synergism was demonstrated with some of the other components separated on Avicel, the maximum effect obtained represented only a threefold enhancement of activity.

Table I

Component	Relative Cellulase Activity (%)
Original solution	100
C_1	1
C_x	5
$C_1 + C_x$	102
CMC-ase	4
Cellobiase	<1
CMC-ase + cellobiase	2
C_1 + CMC-ase	35
C_1 + cellobiase	20
C_1 + CMC-ase + cellobiase	104

Iwasaki, Hayashi, and Funatsu (5) have separated *T. viride* cellulase on hydroxyapatite at pH 6.0 into two fractions. The first, eluted at low ionic strength, contained material with a molecular weight of about 26,000, had 50 times the hydroxyethylcellulose activity of the unfractionated material, and was active on cellobiose but not on cotton. The second fraction, eluted at high ionic strength, contained material with a molecular weight of about 50,000 and was active on cotton, but much less active on cellobiose and hydroxyethylcellulose than component I. Component II, but not component I, was strongly adsorbed on cellulose. Presumably, component I contained the low molecular-weight carboxymethylcellulase and the bulk of the (other) C_x-components while component II contained C_1, together with a little C_x with which it acted synergistically on cotton.

Niwa, Okada, Ishikawa, and Nisizawa (17) have separated carboxymethylcellulase from material active towards filter paper on columns of hydroxyapatite at pH 6.0, and on DEAE-Sephadex at pH 4.0, by using buffer-concentration gradients. The low recovery of activity towards filter paper was attributed to the existence of a synergistic effect but no account of recombination experiments was given.

Koaze, Yamada, Ezawa, Goi, and Hara (7) have reported a partial separation, on Sephadex G-100 at pH 4.5, of carboxymethylcellulase from activity towards filter paper. The two components emerged in the same order as in Figure 1 but the resolution obtained was less.

Ogawa and Toyama (18) isolated a C_1-component from *T. viride* cellulase by adsorption on a column of disintegrated cotton gauze, which was then washed with buffer solutions ranging from pH 3.0 to 7.0 and finally with water. The buffer solutions removed the C_1-component, together with carboxymethylcellulase (called C_3), leaving activity towards filter paper (C_2), which was removed with water. This separation, which was similar to that achieved by Li, Flora, and King (8), would presumably have resolved C_1- and C_x-components, but no mention of recombination experiments was made. In a later paper Ogawa and Toyama (19) again used a column of disintegrated cotton gauze to separate *T. viride* cellulase. Four fractions were identified which were active, respectively, on cotton gauze (C_{1g}), Avicel (C_{1a}), filter paper (C_2), and carboxymethylcellulose (C_3). C_2 and C_3 were claimed to be different from each other and from C_1. Components C_{1a} and C_{1g}, although both considered capable of degrading native cellulose, were affected differently by adding methylcellulose; C_{1g} was stimulated but C_{1a} was inhibited. Fractionation of a preparation rich in C_2 on DEAE-Sephadex, by stepwise elution with buffers of increasing molarity and decreasing pH, showed that the C_2-component was eluted at about $0.5M$ and pH 5.0.

Niwa, Kawamura, and Nisizawa (16) have also used stepwise elution, with increasing pH and molarity, to fractionate *T. viride* cellulase into 3 major components. The middle component, containing most of the activity towards cotton and carboxymethylcellulose, but little aryl β-glucosidase, was re-fractionated on DEAE-Sephadex at pH 4.0 by stepwise elution with buffers of increasing molarity. Unfortunately, a rather short column was used (4.2 cm. $\times$ 26.5 cm.) and large (80 ml.) fractions were taken so that one column-volume was equivalent to only 4.6 fractions. Each change of ionic strength gave rise to a "peak" about $4\frac{1}{2}$ fractions later and from the results given the components in these "peaks" clearly overlapped. Nevertheless, at lower ionic strength the carboxymethylcellulase was eluted in preference to the material active towards cotton, which came off at a buffer concentration between $0.5M$ and $1.0M$. No recombination experiments were done to check for synergism, but it was demonstrated that none of the enzymes exhibited transferase activity on p-nitrophenyl-β-cellobioside. In a subsequent paper Okada, Niwa, Suzuki, and Nisizawa (20) concentrated on the fraction previously shown to be active towards cotton and modified the elution technique so that it was obtained when the molarity of the buffer was increased from 0.4 to 0.6. However, stepwise elution was still employed and rather large fractions were taken compared with the column-volume (1 column-volume was approximately 11 fractions), so that it is difficult to assess the resolution obtained. Ability to solubilize cotton was not measured in this later work; instead "filter-paper disintegrating activity" was used to estimate cellulase action. Carboxymethylcellulase, cellobiase, and S-factor were also measured. All types of activity were found in this one fraction and it was concluded therefore that the results supported the unienzyme concept. It must be remembered, however, that the fractionation technique employed had only poor resolution. The most recent paper from this school (21) reports a fractionation of *T. viride* cellulase, in which cellulase components were found with different C_1/C_x ratios but none with C_1-activity alone. Under the chromatographic conditions employed, however, the C_1-component could have been in Peak 1 of their Figure 1, which was not investigated further.

Although the existence of C_1- and C_x-components in cellulase from other microorganisms is suggested (25, 27, 32), the only positive demonstrations of the presence of C_1 and of its ability to act in synergism with C_x had come from work with *T. viride* (or *T. koningii*) cellulase. When, therefore, during a search for a more active and less variable strain of cellulase-producing microorganism, the writer's colleague, Thompson, isolated a different fungus that gave filtrates capable of solubilizing cotton completely, it was of great interest to see whether this too contained a C_1-component. The new cellulase was a mutant strain of *Penicillium funiculosum*, which produced both a powerful cellulase and a dextranase, thereby precluding direct gel-filtration on Sephadex G-75. The dextranase was removed without loss of cellulase by passage through DEAE-Sephadex at pH 3.5. Ion-exchange chromatography of the dextranase-free cellulase was less successful than with *T. viride* cellulase (Figure 4). Although a carboxymethylcellulase and a cellobiase were well separated by gradient elution, the cellobiase was still active towards cotton cellulose and this activity was enhanced synergistically by recombination with the carboxymethylcellulase, suggesting that the cellobiase contained an unresolved C_1-component. The presence of this component was demonstrated by isoelectric focusing (31); it had an isoelectric pH value of 4.05 (Figure 5) and displayed synergistic action on cotton when recombined with the carboxymethylcellulase and cellobiase. The C_1-component from *T. viride* was also examined by isoelectric focusing and had the same isoelectric pH value.

Neither ion-exchange chromatography nor the isoelectric focusing technique gave a good separation of *P. funiculosum* C_1 from the cellobiase. This was finally achieved by gel-filtration on Sephadex G-75, whereby the unresolved mixture obtained by gradient elution on DEAE-Sephadex was easily separated (Figure 6). The activity of the C_1-component towards cotton was very low and was increased more than 20-fold on recombination with the C_x-components.

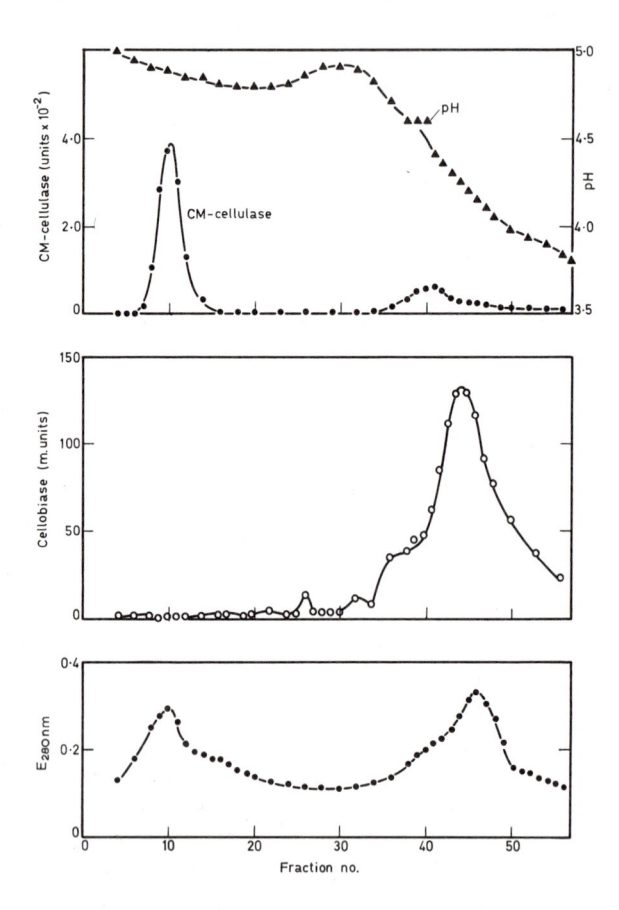

Figure 4. Separation of the carboxymethylcellulase from the cellobiase and C_1-components of P. funiculosum *cellulase on DEAE-Sephadex. The C_1-component was detected by optical density at 280 n.m. The sample was applied at pH 4.8 and eluted with a pH gradient obtained from a variable-gradient mixer with compartments containing buffer solutions (0.01M, 30 ml.) at pH 4.8, 4.5, 4.2, 3.9, 3.6, 3.6*

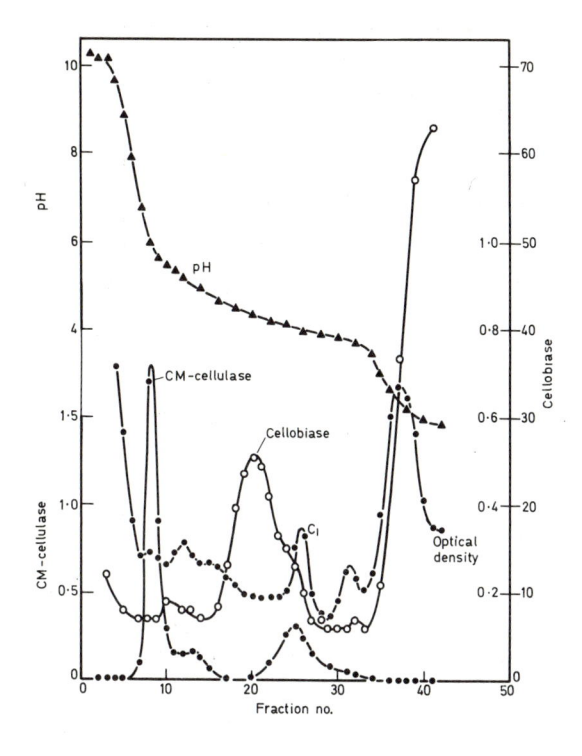

Figure 5. Existence of a C_1-component in P. funiculosum cellulase, shown by isoelectric focusing. The column had a volume of 110 ml. and the pH gradient was 3-6. The C_1-component was measured by a zone-clearance technique in a plate assay involving solubilization of cellulose suspended in agar

The C_1- and C_x-components from T. viride and P. funiculosum also exhibited cross-synergism; for example the C_1- from P. funiculosum could act almost as effectively on cotton in synergism with the C_x from T. viride as with its own C_x (Figure 7). This C_1-preparation has been used to confer the ability to solubilize highly ordered forms of cellulose on certain "cellulase" preparations, which, although apparently deficient in C_1-component, contain other useful polysaccharases.

The C_1-components appear to have all the characteristics of an enzyme, although its precise function has not been identified. Both samples are probably glycoproteins with a carbohydrate:protein ratio of about 1:1, apparent molecular weights by gel-filtration of about 57,000, and isoelectric pH values of 4.05. Their behavior on exposure to adverse conditions of pH was similar (Figure 8), but the C_1 from P. funiculosum displayed greater thermal stability than the C_1 from T. viride (Figure 9).

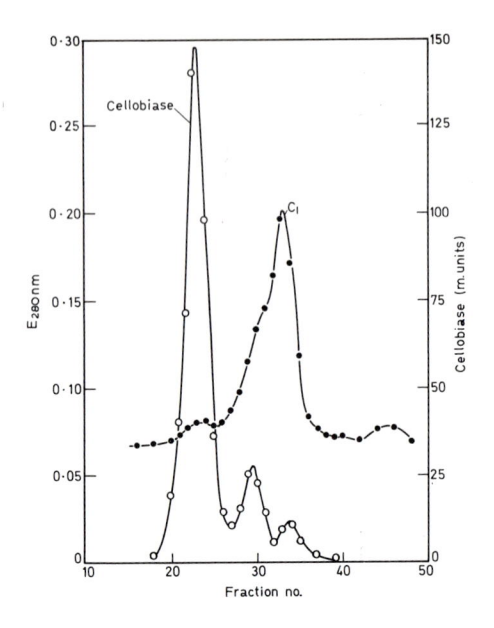

Figure 6. Separation of the C_1-component and cellobiase of P. funiculosum *cellulase on Sephadex G-75 at pH 4.5*

Although the thermal stability of the cellulase complex from *T. viride* is limited by that of the C_1-component, this is not so for *P. funiculosum*, in which the C_1-component is the most stable and the carboxymethylcellulase the most labile. Increased thermostability of the whole cellulase complex can thus be obtained by adding *T. viride* carboxymethylcellulase to the *P. funiculosum* system. The role of the C_1-component in the cellulase complex is still not clear. Opinions have been expressed from time to time concerning the possible identity of C_1 with the swelling factor discovered by Marsh and his coworkers (*11, 12, 13, 14*), for both features are presumed to be involved in the early stages of attack on cotton and swelling may reasonably be considered to be a necessary pre-requisite of hydrolytic degradation. However, Youatt (*35*), Nisizawa, Suzuki, and Nisizawa (*15*) and, very recently, Wood (*34*) have produced evidence supporting the original suggestion of Reese and Gilligan (*23*), that S-factor activity is a property of one of the C_x-components and not of C_1.

In his original concept, Siu (*29*) had suggested that rupture of hydrogen bonds might form an essential part of C_1 action. When King (*6*) reported fragmentation of hydrocellulose particles on exposure to a partially purified cellulase from *T. viride* and attributed this action to "a cellulase component not previously recognized as being involved in cellulose breakdown" it seemed that this might be a manifestation of "hydrogen bondase" action. King showed that when hydrocellulose par-

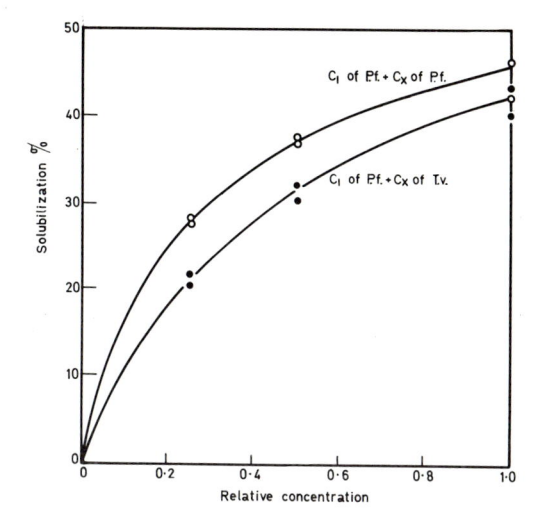

Figure 7. Cross synergism of the C_1-component of P. funiculosum cellulase with its own C_x-component and with the C_x-component of T. viride cellulase. A relative concentration of 1.0 is here used to express the concentration of components that, when recombined, produce a solubilization equivalent to that caused by a culture filtrate from P. funiculosum diluted three fold

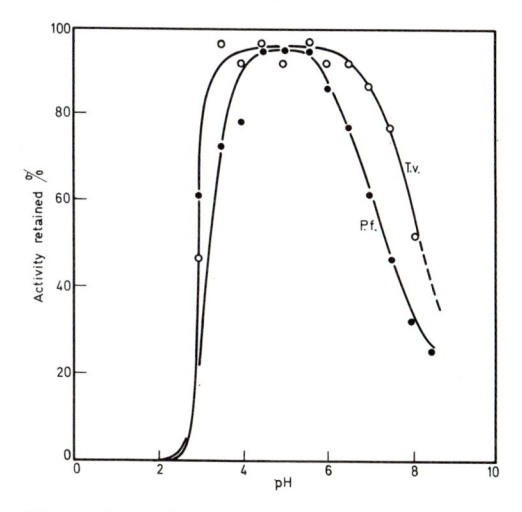

Figure 8. Effect of pH on the stability (at 40°C.) of the C_1-components of P. funiculosum and T. viride cellulases. Each sample of C_1 was kept for 4 hr. at 40°C. at the pH shown. The pH was then changed to 4.5 and residual C_1-activity measured after addition of the C_x-components

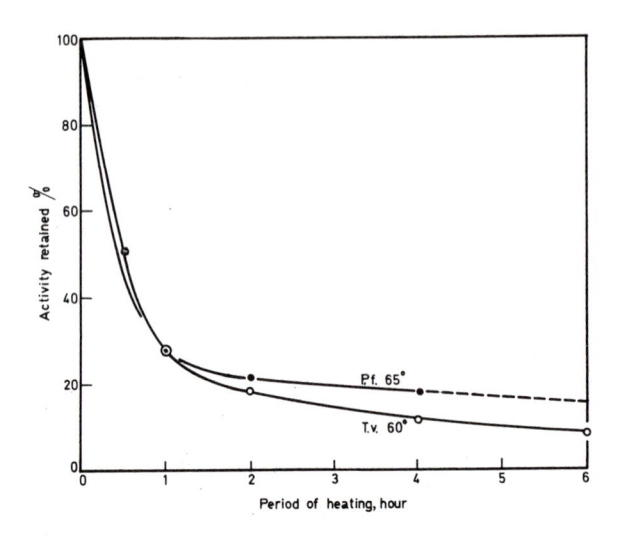

Figure 9. Thermal stability of the C_1-components of P. funiculosum *and* T. viride *cellulases at pH 4.5*

ticles were exposed to cellulase there was a very rapid fragmentation of each into at least 1500 smaller particles. It is known that acid hydrolysis of cotton produces crystallites that aggregate to form particles of about 5 µ diameter (*1*), which suggested that the fragmentation step might be regarded simply as disaggregation or peptization; particle-size measurements, reported subsequently (*9*), reinforced this view. The writer feels, however, that it is not possible from this evidence to suggest that similar disaggregation precedes attack on the cotton hair, for the fibrillar bundles in cotton cannot be held to resemble the simple aggregates used by King and would be less likely to disaggregate simply by adsorbing protein; the accessibility of the fibrillar bundles to protein molecules ought to be much lower. Nevertheless, similar particles, about the size of individual crystallites—*ca.* 60 × 600 A.—are formed when cotton is treated with whole culture filtrate. Evidence that disaggregation caused by adsorption of C_1 is not an essential prerequisite for the degradation of cotton comes from comparison of rate of adsorption with rate of degradation. The rate of adsorption of C_x-free C_1 on crystalline hydrocellulose is much more rapid than on cotton; in a typical experiment the adsorption of C_1 from solution by a sample of hydrocellulose was virtually complete in less than one minute, whereas an equal weight of cotton had not adsorbed a measureable amount of C_1 after two hours. Residual C_1 in each solution was measured by adding an excess of C_x and measuring solubilization (*28*). However, when each substrate was treated with C_1 in the presence of excess C_x, hydrocellulose was solubilized only twenty times as rapidly as cotton. The rate of adsorption of C_1 was unaffected by the presence of C_x and that of C_x was unaffected by the presence of C_1.

If, in preparation for the attack by C_x, the function of C_1 is to produce gross structural changes in the cotton hair without release of material into solution, such changes should be accompanied by a fall in strength, particularly wet strength. Losses of wet strength for given small losses of weight produced by C_1, C_x, and mixtures of C_1 and C_x, suitably diluted to reduce their activity, were compared (Figure 10) and showed that, on the contrary, rapid reduction of wet strength was more a property of C_x than C_1.

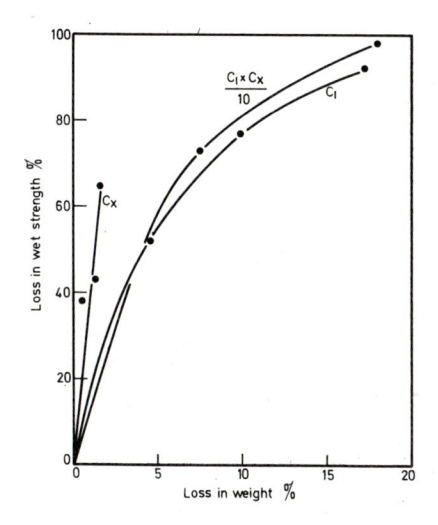

Figure 10. Comparison of losses of wet strength and of weight caused by the prolonged action on cotton of the C_1- and C_x-components of T. viride cellulase. Comparative data are given for the action of a dilute solution of the recombined components

Failure to detect any function that C_1 alone could perform, that was not equally a property of C_x or of mixtures of the enzymes led to the conclusion that both enzymes must be present on the substrate simultaneously when contributing to the solubilization of highly ordered forms of cellulose. Attempts were made, therefore, to interpret the action of C_1 in terms of its contribution to the synergism displayed by the cellulase complex.

The function of C_1 is not to aid adsorption of C_x on cellulose, since the rates of adsorption of both carboxymethylcellulase and cellobiase are unaffected by the presence of C_1. Furthermore, the low adsorption of C_1

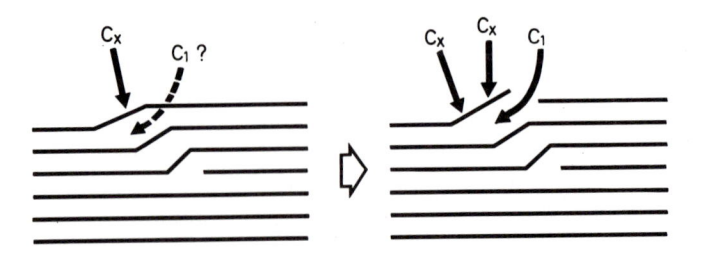

Figure 11. Possible mechanism for synergistic action of C_1 and C_x on cellulose microfibril

by cotton, already mentioned, is not increased in the presence of C_x. A further possible explanation hinges on new theories concerning the supramolecular structure of cotton (33), in which the fiber is considered to consist of completely crystalline elementary microfibrils, each containing about 100 cellulose chains. It is reasonable to assume that the regular array of molecular chains will be disturbed at intervals by the occurrence of chain ends, as shown diagrammatically in Figure 11. The accompanying disturbance in hydrogen bonding between chains in the vicinity of the chain end may be insufficient to enable C_x, acting alone, to split off soluble sugars, but when both components are present, a single bond-rupture by C_x might allow the hydrogen bonding to be further disturbed by C_1 with consequent loosening of a short length of surface chain, which might then be susceptible to more extensive attack by C_x. Chain ends occur at intervals of about 200 glucose units along the microfibril, which would explain both the occurrence of D.P. 200 fragments and the difficulty in detecting the effect of C_1 or C_x, acting alone, by measurements of swelling, loss of weight, or loss of wet strength. This theory obviously borrows features from the "hydrogen bondase" concept (29), but contains the added notion of a very limited number of sensitive sites to explain the difficulty in detecting the influence on the cotton hair of the purified components of the cellulase system when acting alone.

Acknowledgment

I wish to thank Glaxo Laboratories, Ltd. for their financial support of part of this work.

Literature Cited

(1) Battista, O. A., Smith, P. A., *Ind. Eng. Chem.* **54** (9), 20 (1962).
(2) Flora, R. M., *Ph.D. Thesis*, Virginia Polytechnic Institute; Ann Arbor, Mich.: *University Microfilms* (**65-2041**) (1965).
(3) Gilligan, W., Reese, E. T., *Can. J. Microbiol.* **1**, 90 (1954).

(4) Hashimoto, Y., Nisizawa, K., "Advances in Enzymic Hydrolysis of Cellulose and Related Materials," E. T. Reese, Ed., p. 93, Pergamon Press, London, 1963.
(5) Iwasaki, T., Hayashi, K., Funatsu, M., *J. Biochem. Tokyo* **55**, 209 (1964).
(6) King, K. W., *Biochem. Biophys. Res. Comm.* **24**, 295 (1966).
(7) Koaze, Y., Yamada, Y., Ezawa, K., Goi, H., Hara, T., "Proceedings of the Fourth Symposium on Cellulase and Related Enzymes," p. 8, Cellulase Association, Osaka Univ., Japan, 1964.
(8) Li, L. H., Flora, R. M., King, K. W., *Arch. Biochem. Biophys.* **111**, 439 (1965).
(9) Liu, T. H., King, K. W., *Arch. Biochem. Biophys.* **120**, 462 (1967).
(10) Mandels, M., Reese, E. T., "Developments in Industrial Mycology," Vol. 5, p. 5, American Institute of Biological Sciences, Washington, D. C., 1964.
(11) Marsh, P. B., *Plant Disease Reporter* **37**, 71 (1953).
(12) Marsh, P. B., Bollenbacher, K., Butler, M. L., Guthrie, L. R., *Text. Res. J.* **23**, 878 (1953).
(13) Marsh, P. B., Merola, G. V., Simpson, M. E., *Text. Res. J.* **23**, 831 (1953).
(14) Marsh, P. B., Merola, G. V., Bollenbacher, K., Butler, M. L., Simpson, M. E., *Plant Disease Reporter* **38**, 106 (1954).
(15) Nisizawa, T., Suzuki, H., Nisizawa, K., *J. Ferment. Tech.* **44**, 659 (1966).
(16) Niwa, T., Kawamura, K., Nisizawa, K., "Proceedings of the Fifth Symposium on Cellulases and Related Enzymes," p. 44, Cellulase Association, Osaka Univ., Japan, 1965.
(17) Niwa, T., Okada, G., Ishikawa, T., Nisizawa, K., "Proceedings of the Fourth Symposium on Cellulase and Related Enzymes," p. 1, Cellulase Association, Osaka Univ., Japan, 1964.
(18) Ogawa, K., Toyama, N., "Proceedings of the Fourth Symposium on Cellulase and Related Enzymes," p. 17, Cellulase Association, Osaka Univ., Japan, 1964.
(19) Ogawa, K., Toyama, N., "Proceedings of the Fifth Symposium on Cellulase and Related Enzymes," p. 85, Cellulase Association, Osaka Univ., Japan, 1965.
(20) Okada, G., Niwa, T., Suzuki, H., Nisizawa, K., *J. Ferment. Tech.* **44**, 682 (1966).
(21) Okada, G., Nisizawa, K., Suzuki, H., *J. Biochem. (Tokyo)* **63**, 591 (1968).
(22) Reese, E. T., Ed., "Advances in Enzymic Hydrolysis of Cellulose and Related Materials," Pergamon Press, London, 1963.
(23) Reese, E. T., Gilligan, W., *Text. Res. J.* **24**, 663 (1954).
(24) Reese, E. T., Siu, R. G. H., Levinson, H. S., *J. Bacteriol.* **59**, 485 (1950).
(25) Selby, K., "Advances in Enzymic Hydrolysis of Cellulose and Related Materials," E. T. Reese, Ed., p. 33, Pergamon Press, London, 1963.
(26) Selby, K., Maitland, C. C., Thompson, K. V. A., *Biochem. J.* **88**, 288 (1963).
(27) Selby, K., Maitland, C. C., *Biochem. J.* **94**, 578 (1965).
(28) *Ibid.*, **104**, 716 (1967).
(29) Siu, R. G. H., "Microbial Decomposition of Cellulose," Reinhold, New York, 1951.
(30) Siu, R. G. H., "Advances in Enzymic Hydrolysis of Cellulose and Related Materials," E. T. Reese, Ed., p. 257, Pergamon Press, London, 1963.
(31) Vesterberg, O., Svensson, H., *Acta Chem. Scand.* **20**, 820 (1966).
(32) Wakabayashi, K., Kanda, T., Nisizawa, K., *J. Ferment. Tech.* **44**, 669 (1966).
(33) Warwicker, J. O., Jeffries, R., Colbran, R. L., Robinson, R. N., *Shirley Inst. Pam.* **No. 93**, Cotton Silk and Man-Made Fibres Research Association, Manchester, 1966.

(34) Wood, T. M., *Biochem. J.* **109**, 217 (1968).
(35) Youatt, G., *Text. Res. J.* **32**, 158 (1962).

RECEIVED October 21, 1968.

Discussion

T. Cayle: "By way of confirmation of Dr. Selby's hypothesis of the opening in the fibril at which point the C_1 can enter, we have observed an increased rate of action on pulp being mechanically disrupted during the beating operation, in the presence of cellulase derived from *Aspergillus niger*. Presumably, the energy and grinding action imparted during beating provide additional openings in the fibril which permit greater entry for the C_1 component to act, thereby shortening beating time, and in some instances reducing tear strength of the finished paper."

K. Selby: "Yes, we have done a similar thing in pulp from bagasse."

K. E. Eriksson: "Have you investigated the activity of your C_1 enzyme after purification, against higher cellodextrins like cellopentose or cellohexose?"

Selby: "No, it is obvious from what Dr. King was saying earlier, that this is something that ought to be done. We haven't, in fact, done it."

J. M. Leatherwood: "I would like to ask about the adsorption of C_1. Is C_1 adsorbed to hydrocellulose in an active or in an inert form?"

Selby: "In the sense that there is little hydrolytic action going on, I consider it purely as a peptization effect."

Leatherwood: "In other words you consider it inert?"

Selby: "Yes, a breaking of secondary bonds that are holding these particles together. This was the idea."

T. K. Ghose: "I wish to refer to your data showing percent cellulase activity retained against heating period of the enzyme in hours. You know that the pH of such a system is likely to exercise considerable influence on the rate and nature of the loss taking place. I was wondering if you had done any study on this question. What was the pH of the system you worked with?"

Selby: "As to pH, we haven't done a full pH temperature optimum scan on this. And as you say, it does vary; the optimum pH stability will vary as you modify the temperature. We determine optimum pH for C_1 as about 50 and we have worked at about pH 4.5. In fact, this is not really complete until you have said, 'Let's change the pH at 60 or 65 and see what happens.'"

M. Mandels: "I would like to ask what you think your A factor from *Myrothecium* is now? (Since in your present work you equate wet strength loss with C_x) You did have that factor separated, as I remember it, from C_x?"

K. Selby: "Yes, we did separate the factor that was responsible for tensile loss, but we never saw any synergistic effect. This needs re-examining. But really it isn't such a good solubilizing system. I am now trying to say to anybody who has a good solubilizing enzyme, 'Let's have a look at it and see if we can get a C_1 out of it.'"

R. G. H. Siu: "In the past, my colleagues and I have indicated that systems of the C_1–C_x type are probably not limited to cellulase, but that they may exist wherever there is a need for the biological degradation of highly oriented polymeric materials, such as wool, walls of living cells, chitin, etc. I wish to call your attention to reactions of markedly different types where similar systems of proteins are involved (Table I). The second column here represents the C_1 (= A) and C_x (= B) story. Two proteins, each relatively inactive, together they are highly active. The nature of the activity by one of the components is known; that of the other protein (C_1) unknown.

Table I. Synergistic Action of Proteins

	Activity			
	I^a	IIa^b	IIb^b	III^c
Protein A	6	1	3.0	1.5
Protein B	0	5	6.0	0
Protein A and B	260	102	96.0	100

(2), (b) Wood (4).
^{c} = Lactose synthetase. Brew, Vanaman, and Hill (1).
^{a} = Assoc. constants. (Antibody—Antigen) Tanford (3).
^{b} = Enzymatic hydrolysis of cotton (*T. viride* C_1 and C_x) (a) Selby and Maitland

"In the third column is another enzyme system, lactose synthetase, an enzyme synthesizing lactose in mammary glands. Here again, two proteins are required. One (A) has known synthetic ability, it can synthesize N-acetyl lactosamine. The other (B) is α-lactalbumin, with no known catalytic effect in itself. Alone each is inactive; together they synthesize milk sugar. Finally, in the first column, are components A and B, derived from larger antibody molecules. The binding constants of each component with antigen is small. On re-association of A and B, the binding constant is much higher, approaching that of the undegraded antibody.

"We hope that elucidation of the action of such systems may lead to an understanding of C_1."

Literature Cited

(1) Brew, K., Vanaman, T. C., Hill, R. L., *Proc. Natl. Acad. Sci.* **59,** 491 (1968).
(2) Selby, K., Maitland, C. C., *Biochem. J.* **104,** 716 (1967).
(3) Tanford, C., *Accounts Chem. Res.* **1,** 161 (1968).
(4) Wood, T. M., *Biochem. J.* **109,** 217 (1968).

Cellulase Complex of *Ruminococcus* and a New Mechanism for Cellulose Degradation

J. M. LEATHERWOOD

Nutritional Biochemistry Section, Department of Ainmal Science,
North Carolina State University, Raleigh, N. C. 27607

> Three variants of Ruminococcus albus were designated on the basis of their clearing of the cellulose around the colonies on cellulose roll tubes. Beta-colonies do not form significant clear zones while gamma- and alpha-colonies form a sharp clear zone and a diffuse clear zone, respectively. A substance diffusing from a beta-colony interacts with a substance from either an alpha- or gamma-colony to form an enzyme complex that appears to act as a single entity. This apparent protein-protein interaction is discussed with respect to its formation, degradation, and adsorption to cellulose. A new mechanism of cellulose degradation is proposed that involves an affinity factor combined with a hydrolytic factor to form a complete cellulase.

An apparent protein-protein interaction has been observed in cultures of *Ruminococcus* that results in the formation of an enzyme complex that degrades cellulose. On the basis of these and other observations, we propose a new mechanism for cellulose degradation that involves a single cellulase complex.

Formation of a Cellulase Complex

Ruminococcus albus strain 7, (1), was grown on cellulose roll tubes according to the Hungate technique (2). Either Avicel (FMC Corporation, Marcus Hook, Pa.) or balled filter paper was used as the cellulose source. Three milliliters of melted cellulose-medium at 45°C. in 16 × 150 mm. tubes were inoculated with the bacteria under a gas phase of 95% CO_2 and 5% H_2. The tubes were subsequently stoppered and rolled in a tray of ice in order to form a film of agar medium around the inside surface of the tubes. Three variants of colony-types based on the clearing of celluose due to the cellulolytic activity of the microorganism were observed after approximately three days of incubation at 37°C. Beta-colonies do not form a significant clear zone around the colony, whereas gamma- and alpha-colonies form a sharp clear zone and a diffuse clear zone, respectively. Reversion to other colony-types occurs at approximately 1 to 5%, much too high for mutation. The reason for the three variants is unknown.

An unusual phenomenon was observed when either the alpha- or gamma-variants were grown along with the beta-variant in cellulose roll tubes at proper dilutions. A substance diffusing from a beta-colony interacts with a substance from an alpha-colony to form an enzyme complex that appears to act as a single entity (Figure 1). If a substance diffusing from a colony has a zone of equal concentration in the pattern of a circle around the colony, the theoretical pattern for clear-zone formation can be predicted for either the formation of a single complex or a two-enzyme reaction and activation of a pre-enzyme. For either a two-enzyme reaction or the activation of a pre-enzyme, the area of clearing should follow the pattern of overlapping circles with centers of the circles represented by the two colonies. However, the interaction of two components to form a single complex that has affinity to cellulose would be in the pattern of a narrow area of clearing with the outer perimeter of clearing only becoming wider as complete cellulose hydrolysis occurs. As can be seen in Figure 1 the distance of diffusion and the rate of diffusion are approximately equal for the two substances. On this basis we believe the reaction is that of two different proteins interacting to form a complete cellulase. When the complete cellulase is formed, it no longer diffuses in the presence of cellulose. The clearing of cellulose because of the complete cellulase is usually more effective than occurs around alpha-colonies. When complete cellulase is bound to cellulose, apparently it is active and is not bound in an inactive, nonspecific affinity.

A New Mechanism for Cellulose Degradation

We propose a new mechanism for cellulose degradation that is based on the formation of the cellulase complex and is compatible with the results reported in the scientific literature. An affinity factor and a hydrolytic factor are necessary for the formation of a complete cellulase which can hydrolyze native cellulose to cellobiose (Figure 2). The amount of affinity factor (xA) that combines with hydrolytic factor (yH) probably is not $x = y = 1$. This new mechanism does not require a separate enzyme for the formation of reactive cellulose. The natural formation of the complete cellulase complex should occur at the cell level. However, one should be able to separate the component factors of the complete cellulase by physical and chemical techniques. The separate hydrolytic factor can hydrolyze soluble cellulose derivatives. However, to effectively hydrolyze insoluble cellulose, the hydrolytic factor must be held in position on the insoluble cellulose by the affinity factor.

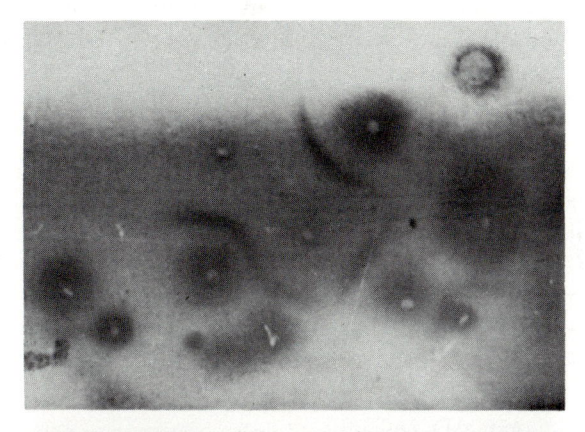

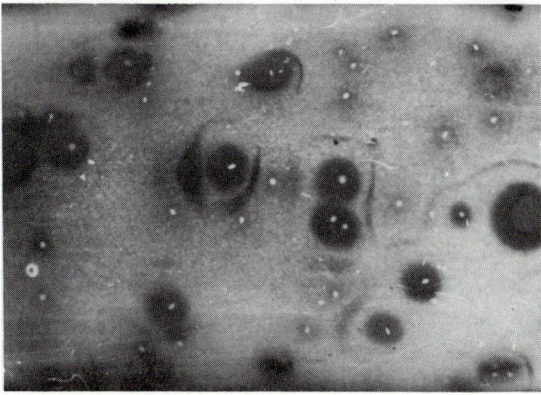

Figure 1. Cellulase complex formation as the result of substances diffusing from two colony-types on cellulose roll tubes. The arc or line of clearing occurring between the colonies is because of degradation of the cellulose

Several observations related to this phenomenon have been made on the cellulase of *Ruminococcus albus* strain 7. Crystalline cellulose is degraded by this organism. Cell-free preparations are active in decreasing the turbidity of cellulose suspensions. The amount of hydrolytic activity against soluble cellulose derivatives is not different for the three variants. Two major cellulolytic components have been separated by

G-100 Sephadex gel filtration. The small component, with molecular weight of 41,000, has been purified. The larger component, which is excluded on G-100 Sephadex, can be disassociated to an active hydrolytic enzyme of the size of the small component by treatment with β-mercaptoethanol. The large component has a strong affinity for cellulose, whereas the small component is not significantly adsorbed to cellulose. However, cellulase in these cultures prior to any purification step is adsorbed strongly to cellulose. In addition, the proportion of hydrolytic factor that is adsorbed to cellulose decreases from over 90% in a young culture to less than 50% in the late stationary phase while the hydrolytic factor continues to increase. This is attributed to a natural degradation of the complex which results in a free, non-adsorbable hydrolytic factor.

This mechanism could be unique for these anaerobic bacteria, however, the major findings reported in the literature are compatible with this proposal. "C_1 enzyme" (3, 4) requires the addition of C_x, a "hydrolytic factor," in order to effectively demonstrate activity. It may be that the C_1 component is an "affinity factor" that binds the "hydrolytic factor," C_x, to the insoluble cellulose.

AFFINITY FACTOR
(xA)

COMPLETE
CELLULASE
(AxHy)

HYDROLYTIC FACTOR
(yH)

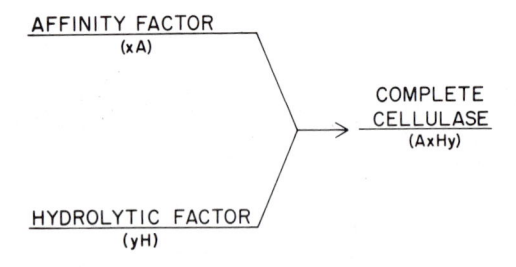

Figure 2. A new mechanism for cellulose degradation proposes the combination of affinity factor and hydrolytic factor to form a complete cellulase complex which can hydrolyze native cellulose to cellobiose

This new mechanism has been considered as an explanation for the general phenomena of resistance, extent, and nature of cellulose hydrolysis. Some speculations of these matters with respect to the new mechanism are: (A) differences in the rate of degradation of resistant celluloses may be caused by the difference in numbers and accessibility of binding sites for the affinity factor, (B) microorganisms could cause a decrease in the rate of cellulose degradation by producing proteins that bind sites in an inactive manner. This could account for large quantities of cellulosic residue remaining when reacting with certain microorganisms and enzyme preparations, and (C) the effect of fragmentation with little change in crystallinity could result if the binding sites are more prevalent in specific areas. Once bound, complete cellulase could affect complete hydrolysis of the local area. This would result in a rapid loss of fiber strength with very little weight loss and production of reducing substances.

Acknowledgment

This research was supported in part by PHS research grant UI 00572 and NSF research grant GB 7405.

Literature Cited

(1) Bryant, M. P., Small, N., Bouma, C., Robinson, I. M., *J. Bacteriol.* **76**, 529 (1958).
(2) Hungate, R. E., *Bacteriol. Rev.* **14**, 1 (1950).
(3) Mandels, M., Reese, E. T., *Develop. Ind. Microbiol.* **5**, 5 (1964).
(4) Selby, K., Maitland, C. C., *Biochem. J.* **104**, 716 (1967).

RECEIVED October 14, 1968.

Discussion

K. Selby: "We have thought of this mechanism too, and to test it we tried to see whether the addition of C_1 would modify or increase the adsorption of C_x. We did this on Avicel and we found that it didn't affect C_x adsorption. So, I am a little bit puzzled as to how the mechanism of fixation that Leatherwood is suggesting comes about. If I understand you correctly, your argument is based on the shape of the clear zone between the colonies."

J. M. Leatherwood: "That is a major consideration. In Figure 1A we have a curved line because of interaction among several colonies while Figure 1B has some lines with less curvature. If two separate activities were involved the lines would be wider and shaped like the overlap of two circles."

Selby: "The thing that is puzzling me a bit is the shape of this clearance zone and the fact that there is an uncleared zone between it and the colony producing C_1 and C_x. What you are suggesting presumably is that there is excess C_1 produced by this colony which diffuses out and that something is happening when it reaches C_x from these other colonies, which are not producing C_1. This is the basic assumption, right? Would you accept this as a possibility, that diffusing out with this C_1, and behind it is glucose, arising from the degradation of the cellulose in your plate? This glucose could inhibit solubilization, which might account for the shape of the zone. In other words, there is an area, beyond the clearance zone immediately round the colony, in which there is sufficient glucose to inhibit solubilization. From this is coming a supply of C_1 and the shape of the trailing edge may be determined by the inhibiting glucose. I submit that this hypothesis could account for the shape observed."

Leatherwood: "We put in 0.03% cellobiose with the cellulose in the media and the concentration of cellobiose decreases from this, so that I don't think simple inhibition by carbohydrate can be the explanation."

Selby: "I still am puzzled by what this mechanism implies, in view of the fact that we could not alter the adsorption of either component by adding the other."

Leatherwood: "Probably the classical protein-protein interaction is an antigen antibody reaction, in which two proteins interact to form a complex. And it is well known that excess concentration of one of these can interfere with the precipitin reaction. So, whenever you test affinity factors and hydrolytic factors, you must test at various levels. This would be one reason for overlooking the interaction. We have found an increase in adsorption of hydrolytic factor, by addition of it to solutions that had what we considered excess affinity factor."

K. E. Eriksson: "I will tell you something about the fractionation of the extra-cellular enzymes from *Chrysosporium lignorum*. We have grown this fungus on a very large scale on cellulose as the sole carbon source. The cell-free culture solution was precipitated by ammonium sulfate, dialyzed and fractionated on a polyacrylamide P 150 column. We obtained two protein peaks from the fractionation. In the first peak we had aryl-β-glucosidase activity which also contains a very slight activity against cellobiose. I will not discuss this enzyme here, but I can tell you that we have obtained about the same fractionation picture from the culture solution of *Stereum sanguinolentum*. From this fraction we were later able to get three different aryl-β-glucosidases; some of them also showed cellobiase activity. These results will soon be published in *Archives of Biochemistry and Biophysics*. But now I will discuss the *Chrysosporium lignorum* enzymes.

The peak obtained on the Sephadex G-75 column containing activity against carboxymethylcellulose was fractionated on a Sephadex A 50 column. The fractionation was obtained by a discontinuous increase of the buffer concentration. From this fractionation three CMCase peaks were obtained. The enzymes in these peaks differed in polysaccharide concentration. The first peak contained 15%, the second 10% and the third 7% of carbohydrate. We have been able to analyze the sugar component and found only mannose and traces of galactose. These certainly are not artifacts, since the organism was grown on pure cellulose. The three enzyme fractions did the same work—e.g., hydrolyzing cellodextrins. Cellopentaose was, for instance, hydrolyzed down to cellobiose and glucose. There was, however, one difference between the peaks. If a cellopentaose solution (1%) was incubated with the three different fractions, we obtained a difference in transglucosidation activity. The first peak of transglucosidase activity was so strong that insoluble cellodextrins precipitated from the solution. The second enzyme fraction did the same work but to a slighter extent, and with the third enzyme solution we did not obtain any precipitation at all.

We had found that a concentrated unfractionated culture solution hydrolyzed cotton cellulose at a reasonable rate. When a cotton fiber suspension was incubated with the different CMCase peaks (equal numbers of cellulase units) the hydrolyses of the cotton fibers were much lower than for the nonfractionated culture. The number of CMCase units were the same in both cases. The fractionation diagram from the Sephadex A 50 column contained a protein peak which had no CMCase activity. We thought that this might be the C_1 enzyme. Our next experiment was to take each of the three CMCase fractions, mix them with the same amount of the solution from the protein peak, and test their activity against cotton cellulose. By doing so we found that the rate of hydrolysis of the cotton cellulose was increased about 10 times. Well, we think this must be the C_1 enzyme. We are now trying to purify and characterize it. We have already found that when we incubate a solution of cellohexaose with the enzyme, the result of the hydrolysis is cellobiose and nothing but cellobiose. Our opinion concerning the C_1 enzyme, if this is it, of course, is that it is an enzyme acting endwise splitting off cellobiose units. In our opinion this also makes sense. We think that the reason for a synergistic effect between a random cellulase and this endwise-acting cellulase is the following. A random-acting cellulase (CMCase, C_x) hydrolyzes a β-1,4-glucosidic linkage. The chances that this bond will remain broken are not very good because of steric factors. Hydrogen bonds will put the glucose in position so that the β-1,4-glucosidic bond can be formed again. This will not happen in the presence of endwise-acting enzyme. When this enzyme has a free cellulose chain-end to work with, it will cut off cellobiose units, and the cellobiose will go into solution.

Since the enzyme which we consider the C_1 enzyme has not been completely purified yet, I will not stress our results too much, but I think the possibility for the mechanism of breakdown of cellulose I have postulated here makes sense, and I hope that it soon will be possible for us to confirm these results."

E. T. Reese: "I would like to say that we have worked with a related fungus, published under a different name. This is *Chrysosporium pruinosum* (Gilman and Abbott) Carmichael, which we have been calling *Sporotrichum pruinoides* QM 826. Its cellulase system is much like that of Dr. Erikksson's *C. lignorum*, in having relatively good C_1 activity."

6

Extracellular and Cell-bound Cellulase Components of Bacteria

H. SUZUKI, K. YAMANE, and K. NISIZAWA

Botanical Institute, Tokyo Kyoiku Univ., Otsuka, Tokyo, Japan

> Pseudomonas fluorescens *produced two extracellular (A and B) and one cell-bound (C) cellulase components, the latter being released by treatment with EDTA-lysozyme in isotonic sucrose. Culture with 0.5% glucose formed little cellulase. Cellobiose stimulated only the synthesis of C. The formation of A and B was strikingly enhanced in cultures with cellulose, sophorose, or continuous low concentration of cellobiose. The absence of extracellular cellulase synthesis in 0.5% cellobiose culture may be caused by catabolite repression. The three cellulases were purified and characterized. None of them split cellobiose, but all hydrolyzed various cellodextrins and celluloses. C easily attacked cellotriose and cellotriosyl sorbitol, but A and B had no effect. When pure B was incubated with broken spheroplasts of sophorose-grown cells, a cellulase component indistinguishable from A was formed.*

Since cellulose is insoluble high polymers under physiological conditions, cellulase which is destined to attack it has been expected to be an extracellular enzyme. In fact, most of cellulolytic microorganisms secrete some cellulase components into the culture medium, and almost all work on the cellulase have been performed using these extracellular components. In the cultures of cellulolytic bacteria, cellulases are not only found in their culture filtrates, but also are generally obtainable from the cells by treatment with autolytic agents—e.g., toluene (9, 19, 30). These facts indicate that at least certain components are existent within the bacterial cells and that their physiological function may be distinct from that of the extracellular components.

Using mixed bacterial preparations isolated from the rumen, King (18) offered the evidence for the presence of surface-localized cellulase components, and speculated that as much as 25–30% of the cellulase activity in the rumen may be normally caused by that of the bacterial cell surfaces. However, detailed studies on the formation and the localization of bacterial cellulases in the cells have been insufficient, and more critical work may be required in this field.

For the last few years, we have investigated the cellulase components of *Pseudomonas fluorescens* var. *cellulosa* in special regard to the physiological relation of their synthesis and localization to the cultural conditions. This pseudomonad is an aerobic mesophilic cellulose-decomposing bacterium isolated earlier by Ueda *et al.* (*39*) from field soil. Some of the enzymatic properties of cellulases obtained from it have previously been reported (*28*). In the present review, the results of our recent studies (*42*) are described and discussed together with the related works of other authors. [In these studies, activities of cellulase and aryl β-glucosidase were assayed by the same methods as described in a recent paper (*31*), and that of amylase by the blue value method modified by Fuwa (*10*)].

Effect of Cultural Conditions on Cellulase Formation

Cellulase Formation during Bacterial Growth. The cellulolytic *Pseudomonas* is, like *Clostridium* and *Cellulomonas* (*12*), known to produce cellulases constitutively, though they are very small in amount (0.5–5 CMC-liquefying units/ml. culture) when it was grown on individual various lower alcohols, organic acids, or amino acids as C-source (*41*).

Influences of glucose, cellobiose, sophorose, and cellulose, when they were each used as a C-source, upon the formation of both cell-bound and extracellular cellulases during the growth of this pseudomonad are shown in Figure 1. Glucose supported the bacterium for an excellent growth, but only slightly stimulated the formation of cellulases, and the enzymes produced were distributed almost equally to the cell and the culture medium. In the cellulose and sophorose cultures, the formation of cellulases, particularly that of extracellular component, was enhanced prominently (exo-type synthesis), whereas cellobiose which was a main end-product of enzymatic cellulolysis stimulated the formation of cell-bound component (endo-type synthesis). Thus, an apparent difference in the distribution of extracellular and cell-bound cellulases was noticed between the cultures with cellobiose and sophorose or cellulose.

Mandels *et al.* (*22*) found that glucose manufactured commercially by acid hydrolysis of starch contained sophorose as an impurity which is a powerful inducer of cellulase in *Trichoderma viride*. In the studies using resting cells of *Ps. fluorescens* grown on succinate medium, we have shown that sophorose is a cellulase inducer in this bacterium as well (*41*).

Electrophoretic properties of typical cellulase preparations, an extracellular cellulase from a culture on 0.5% cellulose and a cell-bound cellulase from that on 0.5% cellobiose, were compared in respect to their behavior in zone electrophoresis on cellulose acetate film. As shown in Figure 2, the former was separated into two components, A (fast moving to the cathode) and B (almost no moving). With the latter, a single component was detected under the same conditions. This fast moving component was in approximate agreement with component A in regard to its mobility, but as will be mentioned later, there was considerable difference in substrate specificity and other properties. Therefore, it seems to be a different component, and is referred to as component C.

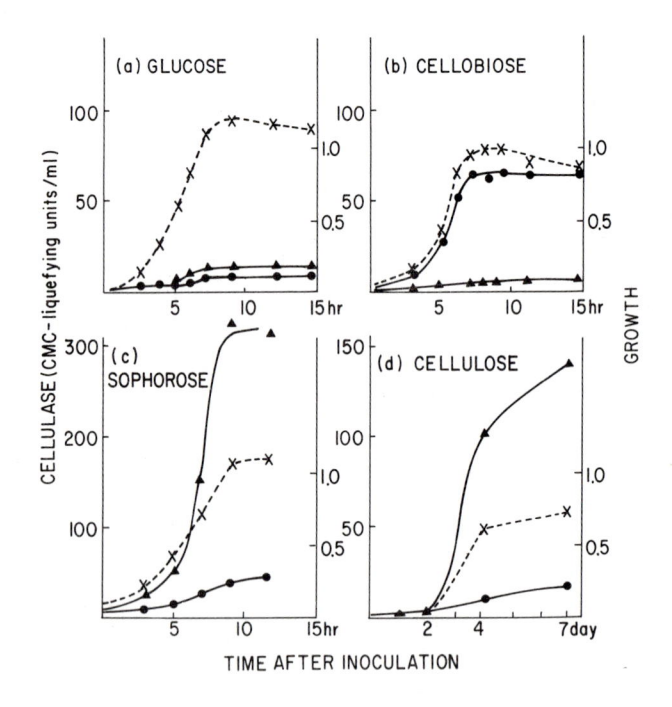

Figure 1. Formation of cellulase in cultures of Ps. fluorescens on 0.5% glucose (a), cellobiose (b), sophorose (c) and cellulose (filter paper) (d)

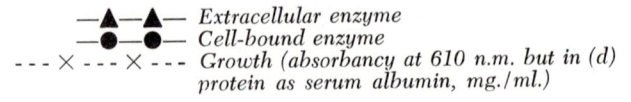

—▲—▲— *Extracellular enzyme*
—●—●— *Cell-bound enzyme*
--- × --- × --- *Growth (absorbancy at 610 n.m. but in (d) protein as serum albumin, mg./ml.)*

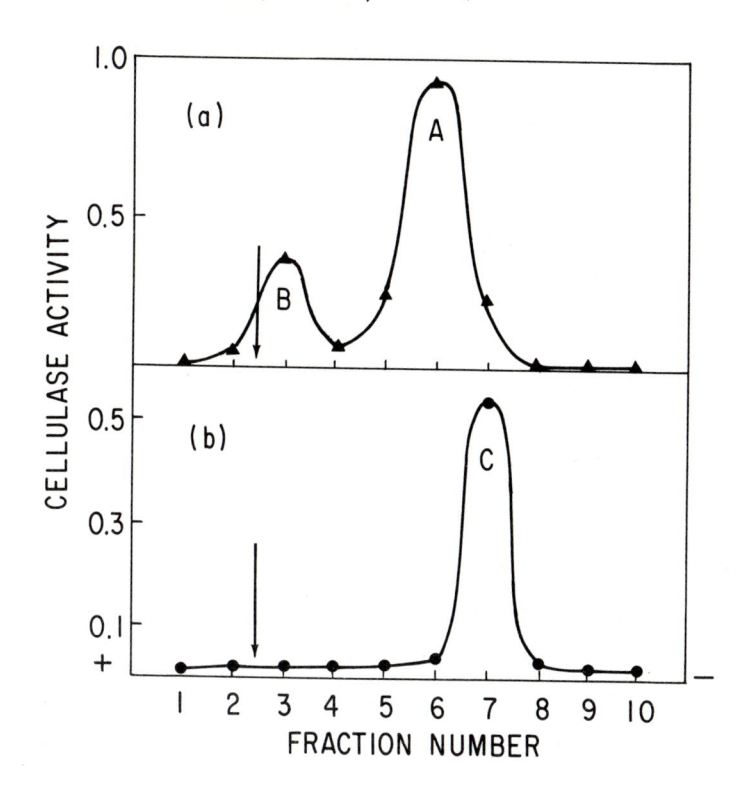

Figure 2. Zone electrophoretic patterns of (a) extracellular cellulase from 0.5% filter-paper culture and (b) cell-bound cellulase from 0.5% cellobiose culture of Ps. fluorescens

Arrow indicates the position of start. The patterns were not corrected for electroendosmosis. Electrophoretic conditions: 11 × 2 cm. cellulose acetate film, using veronal buffer at pH 8.6 ($I = 0.05$); 2-hour run at 0–2°C. and at constant current of 0.6 ma./cm. The cellulase activity was assayed by CMC-saccharification and expressed in terms of the absorbancy at 660 n.m.

Distribution of Extracellular and Cell-bound Cellulases in Cultures on Various Sugars. The formation of extracellular and cell-bound cellulases of *Ps. fluorescens* in cultures on a variety of sugars was investigated. The results are summarized in Figure 3. In cellulosic substances, as may be expected from earlier observations with other bacteria (9, 12, 19, 30), this pseudomonad produced larger quantities of cellulase, in particular extracellular component. In contrast, the cells produced intermediate quantities of cellulase as a whole in cello-oligosaccharides of different DP (degree of polymerization) including insoluble cellodextrin (DP 25), but the cell-bound cellulase was relatively larger in amount. With glucose and other sugars not related to cellulose, the cells formed a very small amount of both kinds of cellulases which seemed to be constitutive. It should also be noticed that sophorose and, though to much lesser extent, gentiobiose behaved like insoluble cellulosic substances. The possibility that sophorose would be enzymatically synthesized in culture medium appears to be very slight, and in addition cellobiose is the major end-product commonly found during enzymatic hydrolysis of cellulose and cello-oligosaccharides. Thus, it is reasonable to assume, as already supposed by Reese (33), that cellobiose rather than sophorose may be controlling the cellulase synthesis in cellulolytic microorganisms.

The synthesis of amylase in *Ps. saccharophila* (40) and *Bacillus stearothermophilus* (24) is stimulated by degradation products of starch as much as or much more than starch itself. But, such relationships could

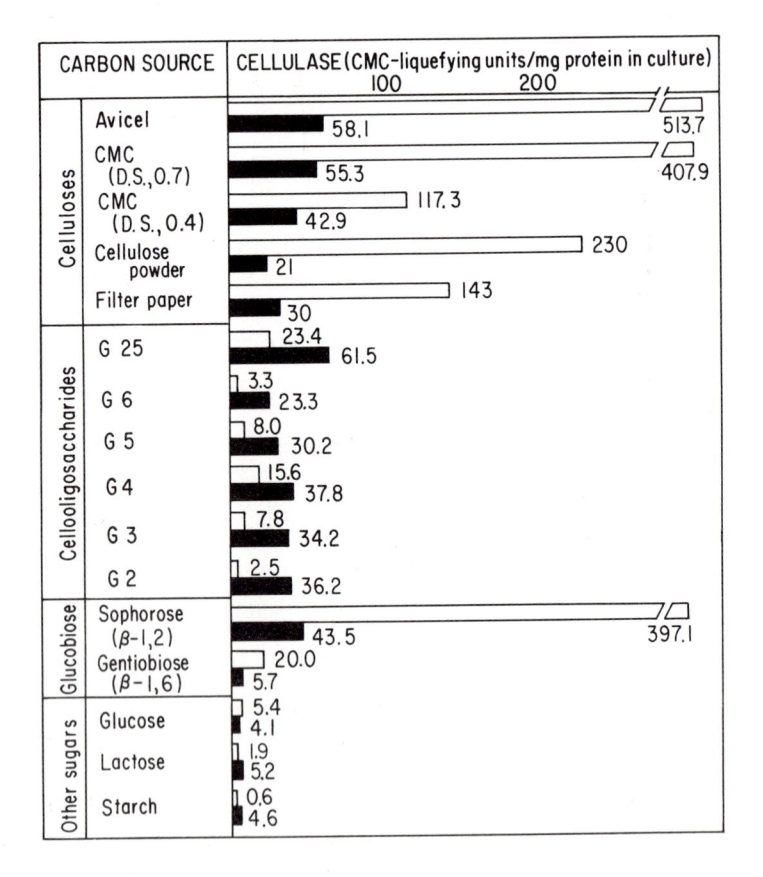

CARBON SOURCE		CELLULASE (CMC-liquefying units/mg protein in culture)
Celluloses	Avicel	58.1 / 513.7
	CMC (D.S., 0.7)	55.3 / 407.9
	CMC (D.S., 0.4)	42.9 / 117.3
	Cellulose powder	21 / 230
	Filter paper	30 / 143
Cellooligosaccharides	G 25	23.4 / 61.5
	G 6	3.3 / 23.3
	G 5	8.0 / 30.2
	G 4	15.6 / 37.8
	G 3	7.8 / 34.2
	G 2	2.5 / 36.2
Glucobiose	Sophorose (β-1,2)	43.5 / 397.1
	Gentiobiose (β-1,6)	20.0 / 5.7
Other sugars	Glucose	5.4 / 4.1
	Lactose	1.9 / 5.2
	Starch	0.6 / 4.6

Figure 3. Distribution of cellulase in cultures of Ps. fluorescens *on a variety of carbon sources of 0.5%*

Each figure is an average value of three points in the maximal region of the cellulase formation. Abbreviations: G2 to G6; cellobiose to cellohexaose, G25; cellodextrin of $DP = 25$

☐ *Extracellular cellulase*   ■ *Cell-bound celluase*

not be observed in the cellulase synthesis in *Ps. fluorescens* as well as in most cellulolytic fungi, such as *Trichoderma viride* (*23*), *Polyporus annosus* (*92*), and *Stereum sanginolentum* (*4*). As illustrated in Figure 3, the total amount of cellulase seemed to depend on the amount of extracellular cellulase, since the amount of cell-bound cellulase of *Ps. fluorescens* was nearly equal in both of the cultures on cellulose and its degradation products. In these cultures on cellulose or cello-oligosaccharides, however, the C-source available for the growth of cells must be some soluble cellulolytic products, most of which may be cellobiose. Furthermore, as seen from Figure 3, the more resistant substrates to the enzymatic cellulolysis showed the higher capacity to stimulate the cellulase synthesis. A similar observation has been reported recently for the cellulase formation in a phytopathogenic fungus, *Pyrenochaeta terrestris* (*14*). These facts led to the assumption that paucity of soluble cellulolytic products, such as cellobiose in the culture medium, would cause the cells to promote cellulase synthesis or secretion.

Culture under Controlling Supply of C-Source (C-Supply Controlling Culture). In order to prove the earlier assumption, C-supply controlling culture, which is basically similar to the technique of Clark and Marr (*8*), for slow continuous feeding of C-source was attempted. The concentration of C-source was maintained at 50–100 μg./ml. throughout the culture period, usually for 11–13 hours. Figure 4 shows the profile of cellulase synthesis in such a cellobiose-supply controlling culture of *Ps. fluorescens*. In this cultural condition, the rate of cell growth was suppressed to 1/3 of the normal growth rate in our conventional 0.5% culture, whereas the cellulase synthesis was strikingly stimulated, in particular in the formation of extracellular cellulase. Even in the cellobiose culture, the distribution of extracellular and cell-bound cellulases became similar to that in the 0.5% cellulose under such controlling concentration. Furthermore, it was found that the two electrophoretically separate components exactly corresponding to the components A and B in cellulose culture (Figure 2) were formed in this culture. Thus, the physiological conditions of the cells of "endo-type synthesis" in 0.5% cellobiose culture was completely changed into those of "exo-type synthesis," as in the cellulose culture, by controlling cellobiose-supply. However, such change in the pattern of cellulase synthesis was not specific for cellobiose, but was observed with other sugars such as D-glucose and D-xylose when they were used as C-source under similar conditions, as shown in Figure 5.

Since the inhibition of cellulase synthesis in 0.5% cellobiose culture seems to be caused by so-called catabolite repression as reported for the formation of many other hydrolases, the same results would be expected for other metabolizable sugars, which will show an enhanced cellulase synthesis depending on the supply conditions. Most of the sugars tested

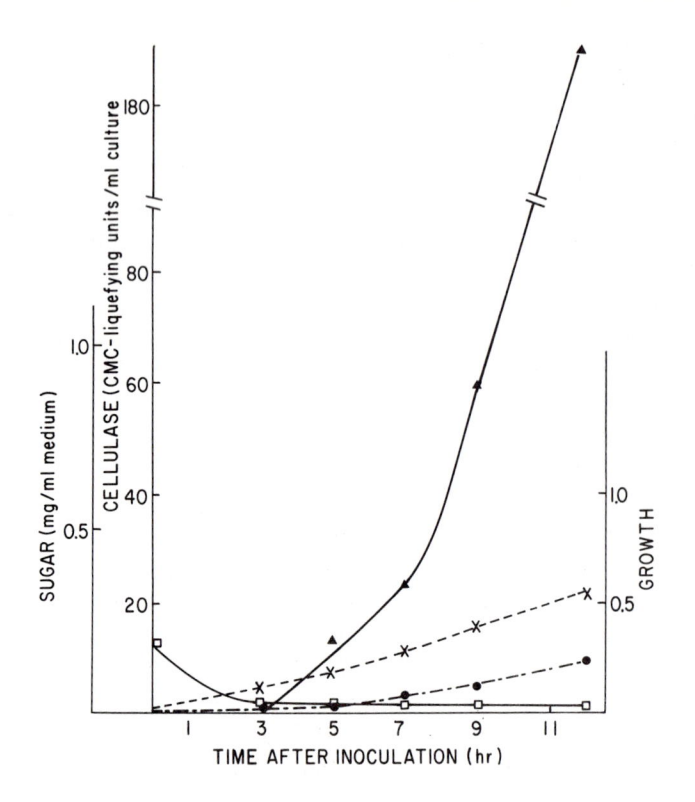

Figure 4. Cellulase formation by Ps. fluorescens *in a C-supply controlling culture using cellobiose as carbon source*

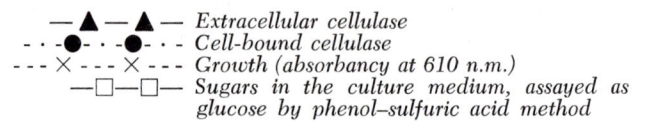

—▲—▲— *Extracellular cellulase*
..●....●... *Cell-bound cellulase*
---×---×--- *Growth (absorbancy at 610 n.m.)*
—□—□— *Sugars in the culture medium, assayed as glucose by phenol–sulfuric acid method*

behaved in this way, and therefore the catabolite repression of cellulase synthesis seemed to be partially de-repressed in the C-supply controlling culture. However, it appears rather mysterious that only sophorose behaved as an exception. At any rate, the enhanced synthesis of extra-cellular cellulase in cellulose culture of cellulolytic microorganisms seems to be caused, at least in part, by the natural control of C-supply.

In connection with the enhanced formation of cellulase by C-supply controlling culture, the possible changes in the formation of other enzymes such as amylase and β-glucosidase were investigated using some selected C-sources which had been employed for the cellulase formation. The results are shown in Figure 5. It is very interesting that amylase was enhanced in its production by C-supply controlling culture on all sugars tested, the situation being entirely similar to the cellulase formation. In contrast, the formation of β-glucosidase which was an intracellular enzyme in this pseudomonad was not influenced by the C-supply controlling culture. Moreover, the β-glucosidase activities scarcely were found in extracellular fractions in any culture. The fact may support the assumption that the enhanced formation of extracellular cellulase in C-supply controlling culture was not caused by some artifacts such as autolysis of bacterial cells during the cultural process.

The simliar C-supply controlling culture was applied to *Trichoderma viride*, but no stimulated formation of cellulase was found in both cell-bound and extracellular fractions (unpublished data in our laboratory).

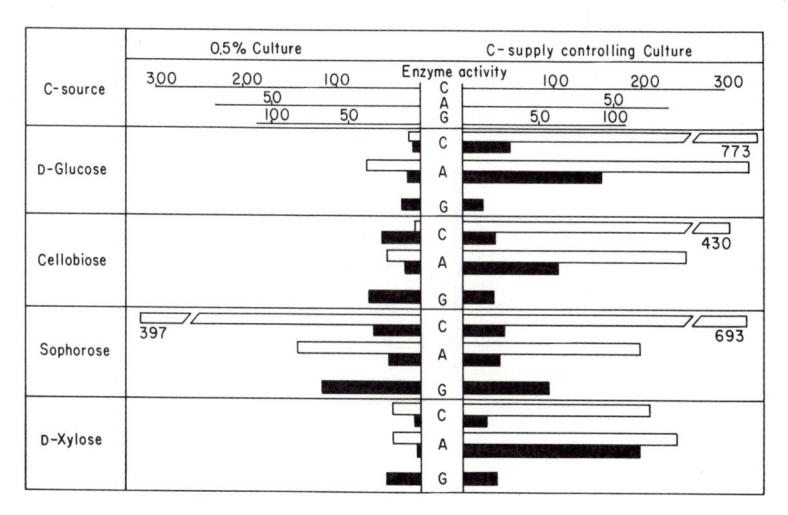

Figure 5. Patterns of synthesis of cellulase (C), amylase (A), and β-glucosidase (G; p-nitrophenyl-β-D-glucoside as substrate) of Ps. fluorescens in 0.5% and C-supply controlling cultures on some selected carbon sources

Activities are represented as units per mg. of total protein in both extracellular and cell-bound enzyme preparations; each figure is an average of values for three points selected from the maximal region of cellulase activity in the culture

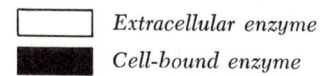

☐ Extracellular enzyme
■ Cell-bound enzyme

Localization of Cellulase and β-Glucosidase in the Cell

The multiplicity of cellulases from various organisms has been studied mainly on the basis of fractionation procedure, and their presence has often been assumed to imply the physiological significance for a complete degradation of native cellulose by utilizing each different substrate specificity. Consequently, any difference in their enzymatic properties and the process of their formation and localization in the cell must be important problems to investigate. Most microbial cellulases usually occur as a typical extracellular enzyme. It has been reported that the extracellular enzymes originated in viable cells in general and not as a result of their autolysis (32). Without such kinds of enzymes, living cells are unable to metabolize large molecules or insoluble particles that are normally found outside the cell.

Living cells are surrounded by a structure of highly selective permeability—i.e., cytoplasmic membrane. On the base of this structure, it is possible to classify the bacterial enzymes into three distinct groups, according to their location in, on, or around the cell as shown in Figure 6; (a) intracellular, (b) surface-bound, and (c) extracellular, respectively (32). The extracellular enzymes, usually separable from cells by centrifugation, are listed by Pollock (32), and the extracellular cellulases from various origins are summarized by Gascoigne and Gascoigne (11).

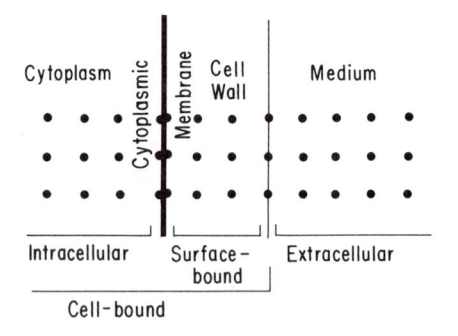

Figure 6. Diagram illustrating the location of bacterial enzymes (black dots) in relation to cell structure (20)

In recent years, Heppel (13) has reported the presence of a special group of degradative enzymes in *Escherichia coli* and related gram-negative bacteria which are confined in a surface compartment rather than free in the cytoplasm. Ten or more hydrolases, such as alkaline phosphatase, ribonuclease I, and UDPG pyrophosphatase, can be selectively set free from cells by osmotic shock or by changing them into spheroplasts. Another group of enzymes—e.g., β-galactosidase, inorganic pyrophosphatase, glucose-6-phosphate dehydrogenase, and UDPG pyrophosphorylase, are entirely associated with the cell and found in the cell sap upon fractionation of cell materials. Thus, the cell-bound enzymes are divided into surface-bound and intracellular enzymes. The former group of enzymes is obtained by treatment of gram-negative bacteria by combination of lysozyme and EDTA in an isotonic sucrose solution.

Alkaline phosphatase, for example, is almost completely released into the sucrose medium when *E. coli* cells are converted to spheroplast (21). The shock fluid obtained by the procedure of Heppel contains various surface-bound hydrolases—*e.g.*, alkaline phosphatase of *E. coli* is released into the fluid in a yield of 70% or more (26, 27). This osmotic treatment has brought about the selective release of enzymes without morphological changes in the cell surface.

Intracellular enzymes are prepared by plasmoptysis or sonic oscillation of spheroplasts or shocked cells.

It should be appreciated, however, that the enzymes of a functionally similar type may not infrequently be cell-bound in one species of bacterium and in another they may be extracellular—*e.g.*, penicillinase and alkaline phosphatase of *E. coli* are cell-bound, whereas those of *Bacillus cereus* are extracellular (6). Even in the same strain of bacterium, an enzyme may be partly cell-bound and partially extracellular; penicillinase of *Bacillus licheniformis* is an example (60). Consequently, the above classification is not so distinctive and may change, depending on the circumstances of cells or by their mutation.

In regard to the cell-bound cellulase, however, no exact locality and function in the cell has been fully studied in general. The cells of *Ps. fluorescens* grown on 0.5% cellobiose were, therefore, fractionated principally according to the method of Burrows and Wood (5). At the same time, part of the same sample of cells was subjected to direct disintegration by sonic treatment and the sonicate was centrifuged to use as a standard for enzyme assays. The results are summarized in Table I.

As clearly seen from Table I, most of the cell-bound cellulase was recovered in the fraction 4 designated as intrawall fraction. A low cellulase activity was found in cytoplasmic and membrane fractions, but it is uncertain whether these activities would be really derived from these fractions or caused by an incomplete resolution of these fractions. A similar experiment using cells grown on 0.5% sophorose indicated that about 80% of the cell-bound cellulase could be recovered in the intrawall fraction.

The cell-bound cellulase of *Ps. fluorescens* could not be removed from the cell by several washings with the basal medium. Furthermore, cellulase released in the fluid obtained by cold osmotic shock of Heppel or cellulase in a fraction corresponding to the fraction 4 obtained by treatment with EDTA in 0.32M sucrose solution only amounted to about 20% of that released in the fraction obtained by lysozyme-EDTA-0.32M sucrose-treatment (fraction 4). Therefore, most of the cell-bound cellulase of this bacterium must reside in the cell-wall region and/or on the surface of cytoplasmic membrane and possibly adsorb thereon tighter than alkaline phosphatase in E. coli (12, 21) which is known as a typical periplasmic enzyme.

On the other hand, β-glucosidase, known as a typical intracellular enzyme, was found mostly in cytoplasmic and membrane fractions. Possibly this enzyme would exist in the above two forms within the pseudomonad cell as in the yeast cell (17).

Table I. Distribution of Cellulase and β-Glucosidase among Various Fractions from the Cells of Ps. *fluorescens* Grown on 0.5% Cellobiose

Fraction	Cellulase[a]	β-Glucosidase[b]	Lysozyme[c]
1. Extracellular: supernatant (3,000 g., 20 min.) of the culture	1.4	0.08	
2. Cell-bound: supernatant (4,500 g., 30 min.) of the sonicate of cells obtained at the above step	67.3	5.05	
3. Residual: supernatant (10,000 g., 30 min.) of the sonicate of precipitate obtained at the above step	0.6	0.22	
sum	69.3	5.35	
1. Extracellular (same as above)	1.4	0.08	(14.62)[d]
4. Intrawall: supernatant (10,000 g., 30 min.) of the cell suspension after treatment to make spheroplasts with EDTA-lysozyme-0.32M sucrose	48.2	0.06	0.14
5. Cytoplasmic: supernatant (60,000 g., 30 min.) of the plasmoptysate of spheroplasts	9.9	2.02	0.60
6. Membrane: supernatant (80,000 g., 60 min.) of the sonicate of precipitate obtained at the above step	1.4	2.27	12.24
sum	60.9	4.75	12.98

[a] CMC-liquefying activity.
[b] p-Nitrophenyl-β-glucosidase activity.
[c] Ethylene glycol chitin-liquefying activity.
[d] Total units added at the beginning.

Nearly all lysozyme added to the starting material to digest cell wall was recovered together with the membrane fraction. The fact indicates that the lysozyme protein was bound with insoluble residues on the surface of spheroplasts.

Recent observations by Carpenter and Barnett (7) have shown that membrane-bound ribosomes of *Cellvibrio gilvus* contained slightly higher cellulase activity than that which occurred in cytoplasmic ribosomes while the reverse relation was seen with the β-glucosidase activities, thereby suggesting consideration of the biogenesis of these enzymes. To elucidate this problem, the exact characterization not only of extracellular and cell-bound cellulase components but also of the membrane-bound ribosomal cellulase appears to be most important, although much remains to be studied.

Purification and Properties of Extracellular and Cell-bound Cellulase Components of Ps. fluorescens

Purification and Physical and Chemical Properties. Extracellular cellulase components A and B and cell-bound cellulase component C were purified through the steps summarized in Figure 7 from the cultures of *Ps. fluorescens* on 0.5% Avicel and on 0.5% cellobiose, respectively. The purified cellulase components (Cellulases A, B, and C) thus obtained showed a single peak in zone electrophoresis on cellulose acetate film and starch bed.

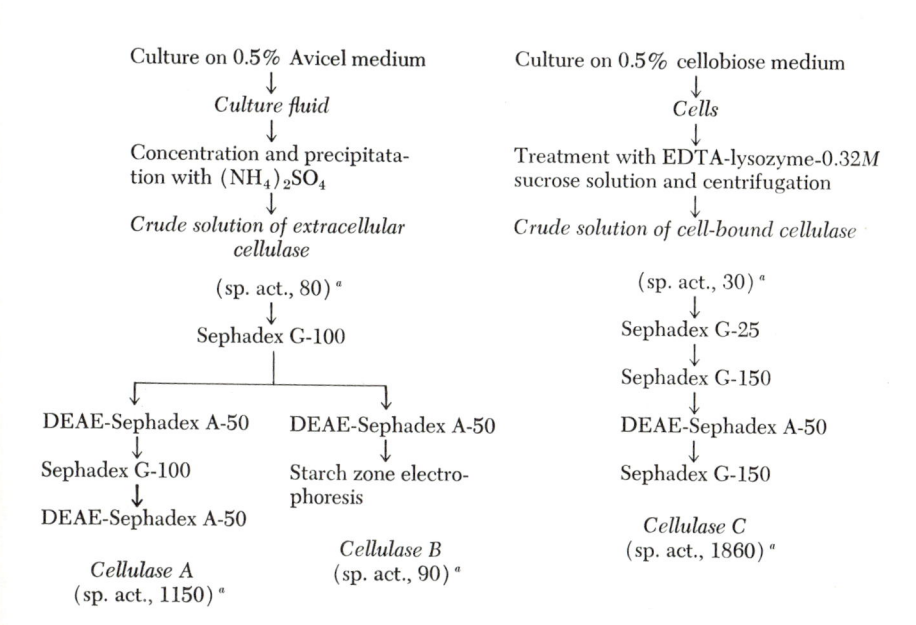

Culture on 0.5% Avicel medium	Culture on 0.5% cellobiose medium
↓	↓
Culture fluid	Cells
↓	↓
Concentration and precipitation with (NH₄)₂SO₄	Treatment with EDTA-lysozyme-0.32M sucrose solution and centrifugation
↓	↓
Crude solution of extracellular cellulase	Crude solution of cell-bound cellulase
(sp. act., 80)[a]	(sp. act., 30)[a]
↓	↓
Sephadex G-100	Sephadex G-25
	↓
	Sephadex G-150
↓ ↓	↓
DEAE-Sephadex A-50 DEAE-Sephadex A-50	DEAE-Sephadex A-50
↓ ↓	↓
Sephadex G-100 Starch zone electrophoresis	Sephadex G-150
↓	
DEAE-Sephadex A-50	Cellulase C
	(sp. act., 1860)[a]
Cellulase A Cellulase B	
(sp. act., 1150)[a] (sp. act., 90)[a]	

Figure 7. *Purification steps of extracellular (Cellulases A and B) and cell-bound (Cellulase C) cellulases of* Ps. fluorescens

[a] *Specific activity; CMC-saccharifying units per mg. protein*

The ultracentrifugal patterns of these cellulases are shown in Figure 8. The sedimentation coefficient of Cellulase A was 1.4 S, corresponding to molecular weight of about 20,000. Those of others were difficult to estimate because of their synthetic boundaries, but should be < 1.4 S. On the other hand, the exclusion chromatography on Sephadex suggested the molecular weights of Cellulases A and C to be at least 50,000. The molecular weight of Cellulase B could not be estimated, since it was difficult to elute from Sephadex to which it had been applied. This great discrepancy between the molecular weights estimated from two different methods seems to be unusual, therefore further examinations are needed.

Cellulase A   Cellulase B   Cellulase C

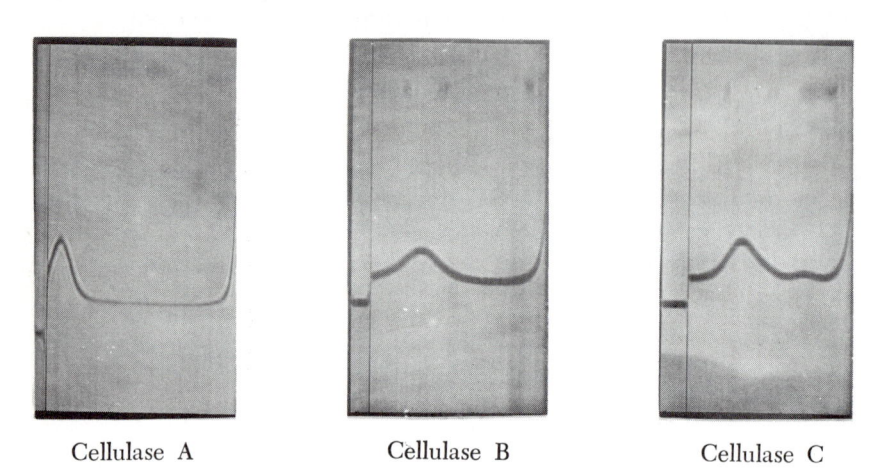

Figure 8. Ultracentrifugal patterns of Pseudomonas *cellulases*
Pictures were taken at 45 minutes after rotation had reached a maximum speed of 55,430 r.p.m. at 17°C. Protein concentration was 0.45% (Cellulase A) and 0.7% (Cellulases B and C) in 1/15M phosphate buffer of pH 7.0. The synthetic boundary cell was employed for Cellulases B and C

The optimal activities of Cellulases A and B were at pH 8.0, whereas that of Cellulase C was at pH 7.0. They were most stable at pH's 7.0 to 8.0 and were completely inactivated by heating at 60°C. for 10 minutes.

It is worthwhile to note that the three cellulases were found to contain a considerable amount of carbohydrate. Although there is no conclusive proof to show that the carbohydrate was chemically bound to the enzyme protein, it seems likely that these cellulases are glycoproteins. They were equally rich in aspartic acid, glycine, glutamic acid, and alanine, while poor in basic and sulfur-containing amino acids as are other extracellular enzymes. The constituent sugars are shown in Table II. It should be noted that glucosamine was detected in Cellulases A and B, while galactosamine was detected in Cellulase C.

Table II. Constituent Sugars Detected on Paper Chromatograms of Acid Hydrolyzate of *Pseudomonas* Cellulases A, B, and C

Cellulase	Sugar[a]					
	Galactosamine	Glucosamine	Galactose	Glucose	Mannose	Fucose
A		+	++++++	++	±	+
B		+	++	++++++	±	+
C	++		+	++++++	±	+

[a] The number of + signs indicates the relative intensity of spots; ± means a very weak spot.

Table III. Specificities of *Pseudomonas* Cellulases A, B, and C Toward Soluble and Insoluble Substrates[a]

Substrate	Cellulase		
	A	B	C
Cellobiose	0.0	0.0	0.0
Cellotriose	0.0	0.0	1010
Cellotetraose	1190	230	1850
Cellopentaose	2250	240	1370
Cellohexaose	3520	410	1810
Cellodextrin	80	7	640
Swollen cellulose	3	7	4
Avicel	0.7	0.4	0.2
CMC	1150	90	1860
DEAE-cellulose	1	1	6
Cellobiosyl sorbitol	0.0	0.0	0.0
Cellotriosyl sorbitol	0.0	0.0	230
Cellotetraosyl sorbitol	560	30	850
Cellopentaosyl sorbitol	1350	110	1100
p-Nitrophenyl-β-D-glucoside	0.0	0.0	0.0
p-Nitrophenyl-β-cellobioside	4	0.3	8

[a] Activities are expressed in terms of the absorbancy at 660 n.m. owing to reducing power per 30 minutes per mg. of protein as bovine serum albumin, except those toward p-nitrophenyl-β-D-glucoside and -cellobioside where the absorbancy at 420 n.m. owing to p-nitrophenol liberated was measured.

Substrate Specificity. Saccharifying activities of the three cellulases were examined as the measure of their cellulolytic capacities using various soluble and insoluble substrates and the results are shown in Table III. They were similarly incapable of attacking cellobiose and p-nitrophenyl β-glucoside, but hydrolyzed various substrates such as cello-oligosaccharides, amorphous celluloses, and Avicel, a highly crystalline cellulose. However, the activities of each cellulase toward these substrates were markedly different from each other. For example, Cellulase B generally showed much smaller activity toward soluble substrates including CMC and insoluble cellodextrin than Cellulases A and C, while their activity toward highly polymerized insoluble cellulose was not markedly different. The most conspicuous difference was found in the activity toward cellotriose and cellotriosyl sorbitol; only Cellulase C hydrolyzed them and this took place with ease. However, since reduced cellotriose (cellobiosyl sorbital) no longer was hydrolyzed by Cellulase C, at least three consecutive glucosyl residues seem to be necessary for the hydrolysis of cello-oligosaccharides by this cellulase component, whereas four consecutive ones were necessary for hydrolysis by Cellulases A and B. In contrast, β-cellobiosides (p-nitrophenyl β-cellobioside and vanillin β-cellobioside) which seem to correspond in appearance to cellotriose in the structure were similarly hydrolyzed by all three cellulases at a glycon bond. The degree of randomness in the hydrolysis of CMC by the three cellulases is shown in Figure 9. There is no significant difference in the slopes (ϕ_{sp}/R.P.) of the lines obtained for each individual cellulase. This fact indicates that the three cellulases hydrolyze CMC in a similar random mechanism.

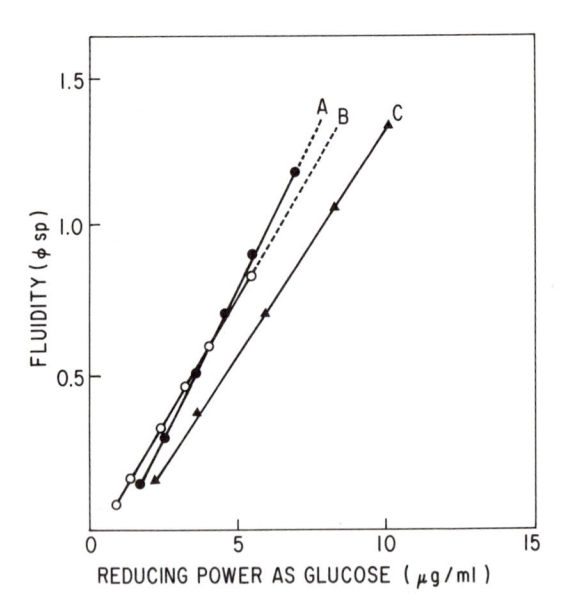

Figure 9. Relation between fluidity and reducing power during the hydrolysis of CMC by Pseudomonas Cellulases A, B, and C

From the quantitative analysis of hydrolysis products from cellotetraose and reduced cellotetraose by the three cellulases, the bond 2, when the glucosyl bonds of the cello-oligosaccharides were numbered from the nonreducing end, was found to be hydrolyzed more easily than the bond 3 by any of the three cellulases. With higher cello-oligosaccharides, all these cellulases attacked the bond 3 more easily than the bond 2 or others, the tendency being particularly distinctive for Cellulases A and B. In regard to such a mode of action, *Psuedomonas* cellulases are similar to some components of *Cellvibrio gilvus* cellulases (*37*).

Discrimination of Cellulase Components A, B, and C. The physical and chemical properties as well as substrate specificities of the highly purified cellulases of *Ps. fluorescens* have been characterized and are summarized in Table IV. Cellulase A is different from Cellulase B in the mobility on zone electrophoresis and in the pattern of Sephadex G-25 chromatography, but similar in substrate specificity toward several reduced and nonreduced cello-oligosaccharides. On the other hand, Cellulase C is different from A in the pattern of DEAE-Sephadex chromatography and the substrate specificity, and from B in all respects. These characteristics of each cellulase component are therefore different enough to be used as criteria to discriminate one from the other.

Table IV. Characteristic Properties of Purified Cellulase Components A, B, and C in the Discrimination Procedures

	Property of Cellulase Component		
Discrimination Procedure	A	B	C
Sephadex G-25 chromatography (as peak)	G-25-1st	G-25-2nd	G-25-1st
DEAE-Sephadex A-50 chromatography (eluted at)	pH 7.0, 1/15M	pH 7.0, 1/15M	pH 6.0, 1/50M
Electrophoresis (mobility toward cathode)	Rapid	Slow	Rapid
Substrate specificity (hydrolysis ability)[a]			
G2	−	−	−
G3	−	−	+
G3H	−	−	−
G4	+	+	+
G4H	−	−	+

[a] G2, G3, and G4: cellobiose, -triose and -tetraose; G3H and G4H: reduced G3 and G4.

Location of Cellulase Components in the **Pseudomonas** *Cell*

Cellulase activities of *Ps. fluorescens* were shown to be located mainly in the extracellular and intrawall fractions, to which at least three cellulase components A, B, and C were distributed as already stated (Table I). They were located for different cultures using the discrimination procedures shown in Table IV. The results are summarized in Table V.

All the extracellular cellulase preparations from various cultures contained only cellulase components A and B. On the other hand, the three cellulase components A, B, and C were found in the intrawall cellulase preparations from cultures in which a prominent production of extracellular cellulase (exo-type synthesis) occurred, while only the component C was detected in the intrawall fraction from 0.5% cellobiose culture in which a low production of cell-bound cellulase was mainly found (endo-type synthesis). These facts indicate that the presence of components A and B may be characteristic of extracellular cellulase and the presence of component C may be characteristic of cell-bound cellulase.

Table V. Identification by the Discrimination Tests of Cellulase Components in Extracellular and Intrawall Fractions from Different Cultures of *Ps. fluorescens*

	Cellulase Components Identified from	
	Extracellular Fraction	Intrawall Fraction
Culture	(Fraction 1)[a]	(Fraction 4)[a]
Avicel (0.5%)	A B	C A B
Sophorose (0.5%)	A B	C A B
Cellobiose (0.5%)	(A)[b] (B)[b] (C)[b]	C
Cellobiose (C-supply controlling)	A B	C A B
Glucose (C-supply controlling)	A B	—

[a] Procedures as described in Table I.
[b] The components in parenthesis are doubtful for their existence because of extremely small amounts.

Moreover, it is worthwhile to note that cellulase components A and B always occurred in pairs in all fractions of the cell under the cultural conditions to cause an exo-type synthesis of cellulase. The component C was, on the contrary, produced under the cultural conditions to cause both exo- and endo-type syntheses. These facts suggest not only the possible relationship between the components A and B with regard to their formation, but also the possible formation of the component C being independent of the other two.

Conversion of the Pseudomonas Cellulase Component B to A

Since the occurrence of cellulase components A and B in pairs at first suggests the possible mutual conversion between them, it was investigated if such possibility would be realized by some cytoplasmic materials. For this purpose, plasmoptysized spheroplasts were prepared from cells grown on 0.5% sophorose (S1) or 0.5% cellobiose culture (S2) and those on culture under controlling supply of cellobiose (S3). (Although, as already stated, spheroplasts retained about 20% of the cell-bound cellulase, it was presumed that they might be employed as a source of possible factor for conversion under situation minimized for the technical obstruction due to pre-existent cellulase therein.) Each plasmoptysate was mixed with an equal volume of each of the highly purified cellulase components A, B, and C dissolved in 0.05M phosphate buffer at pH 7.0. After the mixture was incubated at 30°C. for suitable periods, 0.1 ml. aliquots thereof were determined for the relative CMC-saccharifying activities.

Table VI. Changes in CMC-saccharifying Activities of *Pseudomonas* Cellulase Components During Incubation with Broken Spheroplast Preparation

Mixture (1:1)		Relative CMCase Activity[b]			
Cellulase Component	Broken[a] Spheroplast	0 hr.[c]	0.5 hr.[d]	4 hr.[d]	48 hr.[d]
A	S 1	1.00	1.19	1.33	1.55
A	S 2	1.00	1.37	1.58	1.45
A	S 3	1.00	1.55	1.48	1.78
B	S 1	1.00	3.03	4.06	5.00
B	S 2	1.00	1.29	1.60	2.06
B	S 3	1.00	3.46	4.35	4.96
C	S 1	1.00	1.33	1.49	1.62
C	S 2	1.00	1.32	1.58	1.39
C	S 3	1.00	1.49	1.63	1.69

[a] Spheroplasts obtained by treating cells from 0.5% sophorose (S 1), 0.5% cellobiose (S 2), and cellobiose-supply controlling (S 3) cultures of *Ps. fluorescens* with EDTA-lysozyme 0.32M sucrose were broken osmotically by suspending in 0.005M Tris-HCl buffer (pH 8.0) of an equal volume to the original culture.
[b] CMC-saccharifying activity was assayed by incubating aliquots of each mixture for 60 min.
[c] Activities in a mixture of an equal volume of each cellulase solution and distilled water as the substrate control (0.43, 0.45, and 0.42 for mixtures with A, B, and C, respectively) and those in a mixture of an equal volume of each preparation of broken spheroplast and distilled waster as the enzyme control (0.08, 0.01, and 0.04 for mixtures with S 1, S 2, and S 3, respectively) were separately measured. Since these activities did not change during the experimental period, the combined cellulase activity of the two was used as the whole control and taken as unity.
[d] Activities in a mixture of an equal volume of each cellulase solution and each broken spheroplast preparation after incubation at 30°C. for the given periods.

As shown in Table VI, only slight changes in the relative CMCase activity were found in the mixtures of the cellulase components A and C with any of the broken spheroplast preparations, while a considerable increase in the activity was noticed in the mixtures with the component B. The enhancement in the mixtures treated with those S1 and S3 was particularly high, which were both the broken spheroplasts from the cultures of an exo-type cellulase synthesis; the cellulase activity was enhanced five times that of the original. S2 preparation from the culture of an endo-type cellulase synthesis was much lower in the activating potency for the component B. These findings suggest that an activating factor is present in some spheroplasts and that it may be adaptively produced by the cells in response to the cultural conditions, under which an exo-type cellulase synthesis should be promoted.

Since it was assumed that the increase in the CMCase activity of Cellulase B is caused by some change in this cellulase molecule, the mixtures of this component B with S1 and S3 were separately tested for their electrophoretic behaviors on cellulose acetate film. As evidently seen from the results shown in Figure 10, a new component moving rapidly toward the cathode occurred besides the original component B in both mixtures with S1 and S3. (As expected, no definite peaks of cellulase could be detected on the electrophoretograms of mixtures containing only these broken spheroplasts.) This conversion product showed the same characteristics in every respect as those of highly purified Cellulase A. Since the CMC-saccharifying activity of Cellulase A was about 13 times as high as that of Cellulase B, this enhancement of CMCase activity in the mixtures with S1 and S3 may also account for the conversion of Cellulase B into the component A.

The conversion of cellulase component B into A may be a result of some enzymatic modification of the enzyme molecule. Similar type of *in vitro* conversion has also been reported, for example, for the extracellular cellulase of *Trichoderma viride* (38) and the cell-bound invertase of bakers' yeast (15). The occurrence of another type of conversion where the reversible association and dissociation of active subunits are operative, has been proven on the intrawall and extracellular invertases of *Neurospora crassa* (25).

Conclusion

Ps. *fluorescens* var. *cellulosa* synthesized a large amount of cellulase in 0.5% cellulose or sophorose medium, and more than 90% of the enzyme formed was released in surrounding medium (exo-type synthesis). In contrast, smaller amounts of cellulase were formed upon cultures on 0.5% cello-oligosaccharides, and more than 90% of the enzyme remained within the cells (endo-type synthesis). Since the major soluble end-products of cellulolysis by *Pseudomonas* cellulases *in vitro* were cellobiose and cellotriose (28), the direct C-source utilized by the bacterium should be these sugars even if either cellulose or cello-oligosaccharides were used as C-source. The only difference between these cultures, therefore, may be in the concentration of the end-products in each medium.

The experiments not described in this paper revealed that a relative amount of extracellular cellulase was increased with decreasing concen-

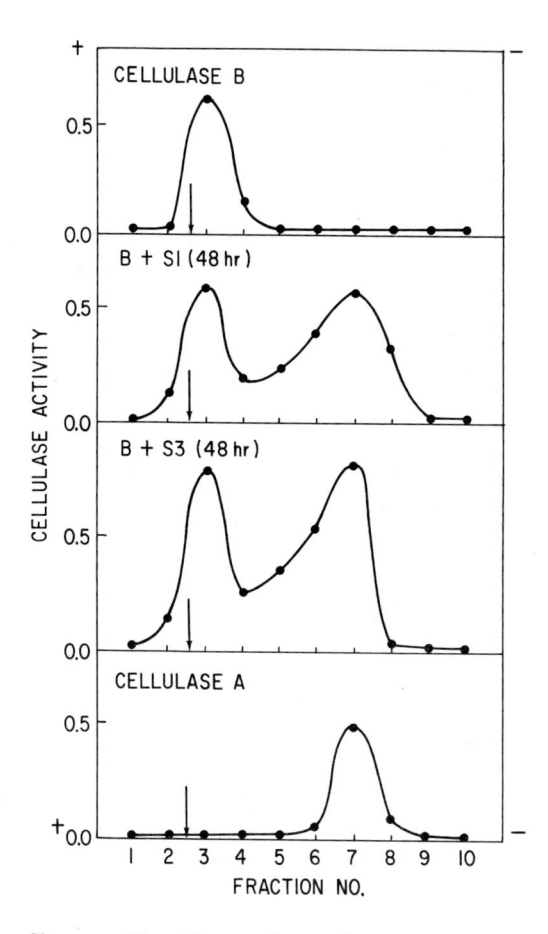

Figure 10. Zone electrophoretic patterns showing the conversion of Cellulase B into another cellulase component moving rapidly toward cathode by treatment with S1 or S3 for 48 hours. The patterns of highly purified Cellulases A and B are shown for comparison, but the size of peak in one run cannot be directly compared with that in another. Arrow indicates the position of start. Electrophoresis and enzyme assay as described in Figure 2

trations of cellobiose used as C-source. Therefore, an attempt was made to perform the culture under a controlling supply of C-source. In this culture, the cellulase synthesis was strikingly stimulated, especially in the formation of extracellular cellulase.

This bacterium produced three cellulase components A, B, and C. Of these, the components A and B occurred mainly in extracellular fraction or culture medium, and the component C was usually localized in intrawall or periplasmic fraction of the cell which was solubilized by treating with EDTA-lysozyme in isotonic sucrose solution. Furthermore, the cellulase component B seemed to be converted by the action of broken spheroplast preparations obtained from cultures which showed an exo-type synthesis of cellulase into a cellulase component indistinguishable from the component A.

All these findings led us to the assumption that components B and C would be formed by inherently different systems in the cells, and component A would be derived from B through some modification within and/or outside the cell.

The enhanced formation of cellulase in cultures on 0.5% cellulose or sophorose as well as in the C-supply controlling culture on various sugars appeared to be actually a result of the stimulation of only extracellular cellulase (components A and B); the activity level of the intrawall cellulase (component C) which occupied the largest portion of the cell-bound cellulase remained almost unchanged under these cultural conditions. This fact might be interpreted as indicating that the formation of extracellular cellulase component B was enhanced in response to the de-reperssion either by inducer such as sophorose or by release from the catabolite repression in C-supply controlling culture, while the synthesis of intrawall cellulase component C showed only poor susceptibility to such metabolic conditions, possibly being constitutive.

With regard to relationships between the substrate specificities of cellulase components A, B, and C and their localization in the cell, a very noticeable fact was observed. Since the main cellulolytic end-product by the extracellular components A and B were composed of a small amount of cellobiose in addition to a large amount of cellotriose, and no cellulase component C, which degrades cellotriose, was present outside the cell, most of sugars absorbed by the cell must be a mixture of cellobiose and cellotriose. Cellotriose must then be degraded by an intrawall cellulase component C, before it will enter the cytoplasm. Since cellobiose was not attacked even by the component C, all the sugars degraded must be absorbed by cytoplasm as the forms of cellobiose and glucose. β-Glucosidase or cellobiase has been found in the cytoplasmic and membrane fractions, and cellobiose which will enter the cytoplasm must receive the hydrolysis by this enzyme to be used as a metabolic source. It should not be overlooked, however, that the cellulolytic bacteria such as *Cellvibrio gilvus* (16), *Clostridium thermocellum* (1, 36) and *Ruminococcus flavefaciens* (2, 3) possess a cell-bound cellobiose phosphorylase, hence they seem to be able to utilize more efficiently the glucosyl-bond energy. *Clostridium thermocellum* (35) and possibly *Cellvibrio gilvus* (34) are found to have cellodextrin phosphorylase in addition to cellobiose phosphorylase, and energetically more efficient utlization of absorbed sugars may be expected. Our preliminary experiments gave an indication of the occurrence of cellobiose phosphorylase in *Ps. fluorescens*, thus, it appears that the phosphorolytic enzymes would serve as an advantageous system in these bacteria for both the absorption and utilization of extracellular cellulolytic products.

Literature Cited

(1) Alexander, J. K., *J. Bacteriol.* **81**, 903 (1961).
(2) Ayers, W. A., *J. Bacteriol.* **76**, 515 (1959).
(3) Ayers, W. A., *J. Biol. Chem.* **234**, 2819 (1959).
(4) Bucht, B., Eriksson, K.-E., *Arch. Biochem. Biophys.* **124**, 135 (1968).
(5) Burrous, S. E., Wood, W. A., *J. Bacteriol.* **84**, 364 (1962).
(6) Caohel, M., Freeze, E., *Biochem. Biophys. Res. Commun.* **16**, 541 (1964).
(7) Carpenter, S. A., Barnett, L. B., *Arch. Biochem. Biophys.* **122**, 1 (1967).
(8) Clark, D. J., Marr, A. G., *Biochim. Biophys. Acta* **92**, 85 (1964).
(9) Fåhraeus, G., "Encyclopedia of Plant Physiology," W. Ruhland, Ed., Vol. 6, p. 305, Springer-Verlag, Berlin, 1958.
(10) Fuwa, H., *J. Biochem.* **41**, 583 (1954).
(11) Gascoigne, J. A., Gascoigne, M. M., "Biological Degradation of Cellulose," p. 52, Butterworths, London, 1960.
(12) Hammerstrom, R. A., Claus, K. D., Coghlan, J. W., McBee, R. H., *Arch. Biochem. Biophys.* **56**, 123 (1955).
(13) Heppel, L. A., *Science* **156**, 1451 (1967).
(14) Horton, J. C., Keen, N. T., *Can. J. Microbiol.* **12**, 209 (1966).
(15) Hoshino, J., Momose, A., *J. Biochem.* **59**, 192 (1966).
(16) Hulcher, F. H., King, K. W., *J. Bacteriol.* **76**, 571 (1958).
(17) Kaplan, J. G., *J. Gen. Physiol.* **48**, 873 (1965).
(18) King, K. W., *J. Dairy Sci.* **42**, 1848 (1959).
(19) King, K. W., *Va. Agr. Expt. Sta. Tech. Bull.* **No. 154**, 55 (1961).
(20) Lampen, J. O., *J. Gen. Microbiol.* **48**, 249 (1962).
(21) Malamy, M., Horecker, B. L., *Biochem. Biophys. Res. Commun.* **5**, 104 (1961).
(22) Mandels, M., Parrish, F. W., Reese, E. T., *J. Bacteriol.* **83**, 400 (1962).
(23) Mandels, M., Reese, E. T., *J. Bacteriol.* **76**, 816 (1960).
(24) Markovitz, A., Klein, H. P., *J. Bacteriol.* **70**, 641 (1955).
(25) Metzenberg, R. L., *Biochim. Biophys. Acta* **89**, 291 (1964).
(26) Neu, H. C., Heppel, L. A., *Biochem. Biophys. Res. Commun.* **17**, 215 (1964).
(27) Neu, H. C., Heppel, L. A., *J. Biol. Chem.* **249**, 3685 (1965).
(28) Nisizawa, K., Hashimoto, Y., Shibata, Y., "Advances in Enzymic Hydrolysis of Cellulose and Related Materials," E. T. Reese, Ed., p. 171, Pergamon Press, New York, 1963.
(29) Norkrans, B., *Physiol. Plant.* **10**, 198 (1957).
(30) Norkrans, B., *Advan. Appl. Microbiol.* **9**, 91 (1967).
(31) Okada, G., Nisizawa, K., Suzuki, H., *J. Biochem.* **63**, 591 (1968).
(32) Pollock, M. R., "The Bacteria," I. C. Gunsalus, R. Y. Stanier, Eds., Vol. 4, p. 121, Academic Press, New York, 1962.
(33) Reese, E. T., Mandels, M., "Advances in Enzymic Hydrolysis of Cellulose and Related Materials," E. T. Reese, Ed., p. 197, Pergamon Press, New York, 1963.
(34) Schafer, M. L., King, K. W., *J. Bacteriol.* **89**, 113 (1965).
(35) Sheth, K., Alexander, J. K., *Biochim. Biophys. Acta* **148**, 808 (1967).
(36) Sih, C. J., Nelson, N. M., McBee, R. H., *Science* **126**, 1116 (1957).

(37) Storvick, W. O., Cole, F. E., King, K. W., *Biochem.* **2**, 1106 (1963).
(38) Tomita, Y., Suzuki, H., Nisizawa, K., *J. Ferment. Technol.* **46**, 701 (1968).
(39) Ueda, K., Ishikawa, S., Asai, T., *J. Agr. Chem. Soc.* **26**, 35 (1952).
(40) Welker, N. K., Campbell, L. L., *J. Bacteriol.* **86**, 681 (1963).
(41) Yamane, K., Suzuki, H., Yamaguchi, K., Tsukada, M., Nisizawa, K., *J. Ferment. Technol.* **43**, 721 (1965).
(42) Yamane, K., Suzuki, H., Nisizawa, K., *J. Biochem.* (to be published).

RECEIVED October 14, 1968.

Discussion

M. Mandels: "We are very interested in your results. In our experiments with *Trichoderma viride* we have had low yields of cellulase in 0.5–1.0% cellobiose culture, similar to your results with *P. fluorescens*. However, we could get excellent production of cellulase in these cultures when cellobiose consumption was slowed by sub-optimal temperatures, reduced aeration, or certain imbalances in the nutrients. We believe that high cellobiose levels repress cellulase formation. When the metabolic rate was decreased, however, the repression was relieved and cellobiose acted as an inducer. We believe, in fact, that cellobiose, the product of cellulase action, is the true inducer in a cellulose culture. Here the cellobiose is consumed by the fungus as rapidly as it is produced and never builds up to repressing levels.

Your experiments with carbon limited cultures seem to support this theory up to a point, but two of your findings are confusing to us. First, the enzyme induced in carbon-limited cellobiose cultures appears to be a new component. In *T. viride* we have only very low levels of constitutive enzyme, and almost no intra-cellular enzyme. It is all found in the medium. Only when the fungus is grown on an inducer do we obtain high yields of cellulase. Secondly, you also had good cellulase production on glucose and xylose in carbon-limited cultures. This seems to rule out cellobiose as a specific inducer in your system. We can consider sophorose and lactose as cellobiose analogues capable of inducing cellulase in some organisms. Simple sugars are another matter. We did not find them to be inducers in our tests."

K. Yamane: "Yes, we do have cellulase production in carbon-limited cultures on glucose and xylose, and also on mannose and arabinose."

7

New Methods for the Investigation of Cellulases

KARL-ERIK ERIKSSON

Swedish Forest Products Research Laboratory, Paper Technology Department, Stockholm, Sweden

> *New methods for the investigation of cellulases are described and discussed. The methods used for the concentration of large volumes of cell-free culture solutions are critically examined. A new viscometric method making possible the determination of units of activity of polymer degrading enzymes in absolute terms is recommended for the assay of the activity of randomly acting cellulases. Of the new methods presented for separation and purification of cellulases the isoelectric focusing method has been found very convenient for detecting heterogeneities in the protein material. For the characterization of cellulases and studies of their structure and function, most of the new methods in protein chemistry have been used. The application of these methods to cellulases is summarized.*

The cellulolytic enzymes are defined as the enzymes which hydrolyze cellulose, thereby ultimately yielding soluble sugars small enough to pass through cell walls. From this definition it follows that these enzymes are extracellular. This paper will only deal with methods used for the study of cellulolytic enzymes of fungal and bacterial origin. Reviews of works on such cellulolytic enzymes are given by Siu (47), Gascoigne and Gascoigne (14), Reese (41), Norkrans (33), Jurazek et al. (26), and Norkrans (34).

Anyone who has read this literature attentively must have noticed that conflicting reports flourish concerning the composition of the cellulase system and the multiplicity and mechanism of action of the cellulolytic enzymes.

It has been frequently reported that some cellulolytic microorganisms can degrade cellulosic materials only after some kind of modification of the cellulose has occurred. Other organisms degrade cellulose in the native form, such as cotton fibers. Only the latter organisms have been considered as truly cellulolytic.

In 1950 Reese and his co-workers (*43*) introduced their "C_1, C_x" hypothesis. It was postulated that "truly cellulolytic microorganisms" are equipped with both C_1 and C_x enzymes, while microorganisms able to hydrolyze only modified cellulose lack the C_1 enzyme.

However, cell-free enzyme preparations from many, according to the above definition, truly cellulolytic microorganisms fail to solubilize native cotton completely unless this is first converted into a reactive form by swelling in acidic or basic milieu (*14, 41*). Thus, these organisms do not seem to produce a C_1 enzyme. The most well-known example is culture filtrate from *Myrothecium verrucaria* (*26, 55*).

The C_1, C_x concept has recently been extensively studied by a number of workers (*16, 23, 24, 27, 31, 32, 46*), for two species of *Trichoderma*, namely *T. viride* and *T. koningi*. It was shown that cell-free culture solutions of these fungi were able to solubilize cotton fibers. The solubilization of cotton fibers is also a way of measuring the activity of the C_1 enzyme. In the most recent of these publications (*46*) Selby and Maitland were able to isolate the C_1 enzyme from culture filtrates of *T. viride*. The enzyme was shown not to act upon cellobiose or carboxymethyl cellulose and to lose its ability to solubilize cotton in the absence of the C_x component. The mechanism of action of the C_1 enzyme is thus still obscure although many different hypotheses have been presented (*34*).

The term cellulase will in this survey be used for endoenzymes which will hydrolyze β-1.4-glucosidic linkages between anhydroglucose units in cellulose, cellulose derivatives, or cellodextrins. Cellulase is thus the trivial name for enzymes with the systematic name β-1,4-glucan glucanohydrolases (*13*). Though the C_1- enzyme with the present knowledge cannot be characterized as a cellulase in the sense defined above, it definitely belongs to the cellulase system. The new methods used for the study of the C_1 enzyme will thus also be dealt with in this paper.

Concentration of Culture Solutions

The cellulolytic enzymes of fungi are mostly present in the liquid in which the organism has been grown. The fungus is normally cultured in a liquid medium containing cellulose and different mineral salts. The culturing process is followed by filtration or centrifugation to remove the cells and residual cellulose.

To obtain protein material enough for allowing a careful study of the enzymes, large volumes of culture solution are necessary. To facilitate handling and to obtain satisfactory concentrates of enzyme for further purification work a convenient method for concentration of the bulk of liquid is essential. Various methods have been used satisfactorily based on the following procedures or combinations of these procedures:

(a) precipitation methods,
(b) freeze-drying,
(c) vacuum evaporation.

Literature information on concentration methods is given in References 14 and 41.

For the concentration of enzyme solutions it is important to use methods giving the highest possible yield of enzymic activity—*i.e.*, loss of activity should be brought to a minimum. None of the above methods will in all cases fulfill these requirements since denaturation and consequently low yield of enzymic activity are reported for all these methods. To avoid denaturation owing to precipitation as well as owing to freeze-drying and evaporation, Petterson *et al.* (39) introduced the dextran gel concentration method of Flodin *et al.* (12) into cellulase research. This method was found to give essentially quantitative recovery of cellulase activity from culture solutions of *Polyporus versicolor*.

However, even this method has its drawbacks and limitations. It is time consuming and the large quantities of gel required to concentrate larger volumes makes it also expensive. Also this method causes loss of activity probably owing to retention of the enzymes to the gel matrix. Low yields of cellulases fractionated on Sephadex columns have been reported by Eriksson and Pettersson (8). These authors (8) used a modification of the method of Björling (6) for the concentration of culture solution of *Stereum sanguinolentum*. Concentration was obtained by suspending DEAE Sephadex A-50 in the culture solution. The suspension was stirred and the ion exchanger filtered off and eluted. Even with this method large losses of enzyme material were obtained. A yield of 50% was obtained at best.

Studies of the possibility to use a Diaflo Ultrafiltration apparatus obtained from Amicon Corporation, Cambridge, Mass., U.S.A., for the concentration of cellulolytic enzymes from the culture solution of the fungus *Chrysosporium lignorum* have been carried out by Eriksson and Rzedowski (10). No loss of enzymic activities could be detected during the concentration. The speed of permeation of the solution through the membrane was, however, not more than 240 ml. hour^{-1}, which means that also this method is rather time consuming. It also has the drawback that the flow through the membrane will cease.

In conclusion, it can be said that none of the used methods for the concentration of large enzyme solutions is perfect. A useful method should be very valuable in the field of cellulase research. It seems reasonable to assume that such a method should be based upon the principles of electrodialysis.

Assay of Cellulase Activity

A wide variety of substrates have been used for the assay of cellulase activity.

The primary requirement for such a substrate is that it must discriminate between cellulases and other related enzymes. The most commonly used substrates fulfilling this requirement are native cellulose, regenerated cellulose, and soluble cellulose derivatives.

The assay methods are normally based on the following procedures:

(a) loss in weight of insoluble substrates

(b) decrease in mechanical properties of fibers or films

(c) change in turbidity of cellulose suspensions

(d) increase in reducing end groups

(e) decrease of viscosity of cellulose derivatives

(f) colorimetric determination of dissolved decomposed products of cellulose

(g) measurements of clearance zones in cellulose agar.

Lists on assay methods of cellulase activity are given in most of the review literature concerning cellulose degradation (14, 26, 34, 41, 47).

For the assay of cellulase activity it is desirable to use a method by which the number of cellulase units can be directly determined. A number of quantities are defined in terms of units of enzyme, such as the concentration of an enzyme, normally given as units/ml., the specific activity defined as units/mg. of protein, and the molecular activity, defined as units/μmole of enzyme. The determination of units of activity does not present any difficulties for enzymes where the change of the substrate can be expressed in absolute terms. However, for enzymes whose activity is not measured in terms of a chemical reaction but in terms of some physical change, such as a decrease in viscosity of the substrate, complications arise and it has not been possible to express the activity in the units mentioned above.

If the recommendations of the Commission on Enzymes are applied to cellulases, a cellulase unit should be defined as that amount of enzyme which, under defined conditions, hydrolyzes β-1,4-glucosidic linkages at the initial rate of 1 μ equiv./min. (*13*).

Theoretically the determination of the rate of formation of reducing groups directly gives the required information. Too many drawbacks are, however, involved in this method. For example, the measurements must be carried out at the leveling-off stage of the enzyme reaction where complications, such as competitive inhibition and transglucosidation reactions, arise.

The Commission on Enzymes states that, wherever possible, enzyme assays should be based on measurements of initial rates of reaction. A viscometric assay method should be especially suitable for this purpose since the mean molecular weight of a polymer can be rapidly determined by such methods. A useful viscometric method has been worked out by Almin *et al.* (*3, 4, 5*).

The enzymic activity, A, on a polymer substrate has been expressed as the number of bonds (equivalents) broken per unit time in the initial stage of the degradation process—*i.e.*,

$$A = \left[\frac{d}{dt}\left(\frac{Cs}{M_n}\right)\right] t = 0 \qquad (1)$$

where Cs is substrate concentration and M_n is the number average molecular weight.

An expression for the calculation of units of enzymic activity in absolute terms from viscosity measurements has been derived by the use of:

(1) A modified Staudinger equation

$$[\eta] = C_m \cdot M_v^x \qquad (2)$$

where $[\eta]$ is intrinsic viscosity, C_m and x are constants and M_v is viscosity average molecular weight.

(2) A law giving the relation between η_{sp} and the substrate concentration Cs. For the substrates used in the experiments in References 3, 4, and 5 the Hess-Philippoff modification of the Baker relation

$$\eta_{sp} + 1 = (1 + [\eta] Cs/8)^8 \qquad (3)$$

has been used.

(3) An empirical relation describing the change in $[\eta]$ with time

$$[\eta]^{-\alpha} = Bt/Cs + [\eta]_o - \alpha \qquad (4)$$

where B and α are constants and $[\eta]_o$ is the intrinsic viscosity at the start of the reaction. Methods for the estimation of B and α are given in Reference 3.

On the basis of the above equations the following expression for the enzymic activity has been derived

$$A = \kappa \cdot \frac{B}{x \cdot d} \cdot \frac{[\eta]_o^\alpha}{[M_v]_o} \qquad (5)$$

where κ is the ratio between the viscosity average and number average molecular weights and the index o refers to the start of the reaction. Equation 5 should be applied to solutions where the enzymic activity is independent of the substrate concentration which thus has to be appreciably higher than the one indicated by the Michaelis-Menten constant.

An experimental method based on the above theoretical concepts has been worked out for the determination of enzymic activity towards polymer substrates. The applicability of this method has been demonstrated in Reference 5 where the molecular activity of a cellulase has been determined.

In the third paper by Almin et al. (4) cellulase activities of different cellulases towards several CMC preparations were determined and compared. Cellulases of different fungal origin and purity were selected for these studies. The reasons for the choice of CMC as substrate were manifold. Previously, Reese et al. (30, 42, 43) had shown that the DS of cellulose derivatives considerably influences the rate of enzymic hydrolysis so that the higher the DS the lower the rate of hydrolysis. One substituent per anhydroglucose unit is considered to render a cellulose derivative resistant to enzymic hydrolysis. It was also shown by these authors that the rate of hydrolysis was independent of DP. However, the DP range covered, 125–200, was too narrow to allow a definite conclusion. Since DS and DP could be expected to be the primary parameters controlling the rate of enzymic hydrolysis of cellulose derivatives, it was necessary to select a derivative where samples covering a vast DS and DP interval were easily available. Both CMC and hydroxyethylcellulose (HEC) could fulfill this requirement. However, Husemann (22) has found that the smaller the substituents in a cellulose derivative are, the less is the resistance to enzymic hydrolysis of the β-1,4-glucosidic bonds. In HEC the substituents can be present as individual units or as polyethyleneoxide chains of varying sizes. This implies that direct comparisons of cellulase activities or the possibility of deriving conversion factors for activities obtained with different HEC samples should be more complicated than with CMC samples.

It is, however, to be noticed that the viscosity of CMC solutions is affected by the pH value and the ionic strength of the solvent as demonstrated by Iwasaki et al. (25). Thus, the pH value and ionic strength of CMC solutions must be fixed when comparisons of cellulase activity are carried out with different CMC solutions.

The experimental results (4) give support to the findings reported by many investigators (14, 41) that the resistance of CMC samples to enzymic hydrolysis increases with increasing DS. This is no doubt true as far as the extent of the reaction is concerned. However, the result (4) suggests that the primary influence upon the activity, measured in bonds broken per unit time, is not exerted by the DS provided this is > 0.5. At DS values < 0.5 the DP must also be low to obtain solubility of the derivative. The influence of the low DP upon the rate of hydrolysis cannot be excluded but an appreciable scatter and too few measurements with samples of this type obscured a unique relationship.

At high DS values the reaction rate cannot be accurately determined owing to the low concentration of β-1,4-glucosidic bonds available for hydrolysis. As far as can be judged from the results the rate of hydrolysis is not influenced by the DS even if this is > 1.0.

In order to avoid the complications arising from the use of CMC samples of too low and too high DS values, samples in the range 0.8 < DS < 1.0 were recommended for comparative studies of cellulase activity. To make comparisons of absolute activity determinations meaningful it was necessary to determine the numerical values of the quotient κ and the Staudinger exponent x for the CMC samples studied. With CMC samples in the mentioned DS range the following conclusions can be drawn from measurements with four different cellulases:

(1) The enzymes are ranked in the same order with respect to their activities no matter which CMC sample is used as substrate.

(2) The DP values influence the A values so that normally the lower the DP the higher the A values.

The influence of DP is evident from Figure 1 where the conversion factor interrelating activity, obtained with different CMC preparations— *cf.* Reference 4, Table II—is plotted as a function of the molecular weight of the CMC samples.

The influence of DS and DP was not found to be so clear-cut that conversion factors could be predicted from these parameters alone. In addition to DS and DP there is some other structural characteristic of the CMC samples influencing the enzyme-substrate reaction. These conversion factors interrelating activity, obtained with different CMC preparations, must be empirically determined. The accuracy obtained in the A determinations by the viscometric method outlined above is reflected by the coefficient of variation which was found to be approximately 0.2.

Separation and Purification of Cellulolytic Enzymes

The latest summary of separation and purification of cellulases and related enzymes is written by Hashimoto and Nisizawa (17). Their list of methods used in the study of these enzymes comprises the following:

(a) separation owing to differences in stability to heat
(b) purification by fractional precipitation
(c) different electrophoretic techniques for purification
(d) paper chromatography and column chromatography for separation of cellulolytic enzymes.

The modern technique for separation and purification of proteins has in all essentials been applied to cellulases. The methods which will be further discussed here are the following:

(a) gel filtration technique
(b) ion exchange on DEAE Sephadex columns
(c) isoelectric focusing
(d) specific adsorption on different cellulose columns.

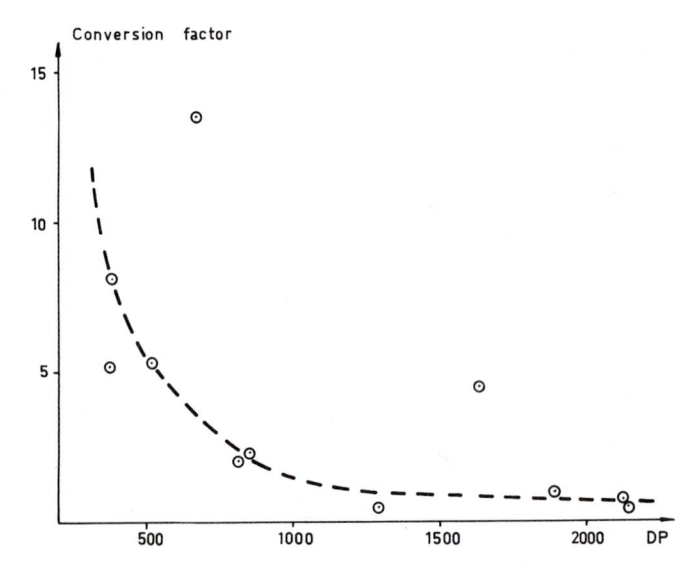

Figure 1. Cellulase activity (in the diagram given as "conversion factor") plotted as a function of the molecular weight of the CMC samples (11a)

Gel Filtration Technique. Cellulases were among the first proteins to be separated by the gel filtration technique. Cellulolytic enzymes from *Polyporus versicolor* were thus separated by Pettersson et al. (39) on Sephadex G-75 (Pharmacia Fine Chemicals, Uppsala, Sweden) and simultaneously Whitaker et al. (56) applied the same technique in studies on the cellulolytic enzymes from *Myrothecium verrucaria*. The gel filtration method has since been so frequently used in separation and purification that it would be difficult to refer to all these works here. For works on Sephadex gels the reader is referred to "Literature References" continuously distributed by Pharmacia Fine Chemicals, Uppsala, Sweden.

In addition to Sephadex gels the polyacrylamide gels (Bio-Rad Laboratories, Richmond, California) have been used for separation of cellulolytic enzymes. The cellulases from *Sterum sanguinolentum* have thus been separated on a polyacrylamide P-150 column by Eriksson and Pettersson (8) and the cellulolytic enzymes from the fungus *Chrysosporium lignorum* have been separated on the same type of gel by Eriksson and Rzedowski (11).

In Reference 8 it is mentioned that the enzyme yield is low upon fractionation on Sephadex columns. We have found the yield on both Sephadex gels and polyacrylamide gels to be low when the cellulases from *C. lignorum* are fractionated. In a system of ammonium acetate buffer pH 5.0, 0.1M the enzyme yield was 31% on a Sephadex column and 37% on a polyacrylamide gel column. Evidence exists that the yield is dependent upon the buffer system and the pH value of the buffer system. Different buffer systems for gel filtration are now studied in our laboratory in order to find out if the yield can be increased. One of the reasons why the low yield of cellulases upon gel filtration has not been noticed before is that the activity of these enzymes has usually been assayed viscometrically using only relative enzyme units as a basis.

Separation of cellulolytic enzymes from fungi by Pettersson *et al.* (35, 39) on gel filtration columns indicated that cellulases, capable of hydrolyzing high molecular weight substrates, are small molecules compared with the β-glucosidases that hydrolyze low molecular weight substrates. These findings were confirmed by Ahlgren *et al.* (1, 2) to be valid for enzymes from another four fungi, all potent wood destroyers, namely *S. sanguinolentum* and *Fomus annosus* belonging to *Basidiomycetes*, *C. lignorum*, probably belonging to *Fungi imperfecti* and one representing the genus *Aspergillus* belonging to the *Ascomycetes*. In addition to the confirmation of the results from separation of β-glucosidases from cellulase it was found that mannanase and xylanase isolated from the four fungi were smaller molecules than their corresponding glycosidases, mannosidase, and xylosidase. In Figure 2 a diagram of one representative gel filtration experiment on *Aspergillus* enzymes is given. These reports make us believe that molecular sieving on gel columns probably is the best method to separate endo-enzymes of this kind from their corresponding glycosidases.

Ion Exchange on DEAE Sephadex Columns. The use of various types of DEAE Sephadex ion-exchangers has been reported lately. Pettersson (36) found DEAE Sephadex A-25 to be a convenient material for the purification of cellulase from *Penicillium notatum*. Eriksson and Pettersson (8) separated two cellulase peaks, P_1 and P_2 from culture solutions of *S. sanguinolentum* on DEAE Sephadex A-50. In the latter case it was found that the P_2 enzyme peak was a carbohydrate complex of enzyme P_1 and that the fungus only excretes one cellulase enzyme in a culture solution with powdered cellulose as carbon source. In a paper by Eriksson and Rzedowski (11) three cellulase peaks were separated on a DEAE Sephadex A-50 column from culture solution of the fungus *C. lignorum*.

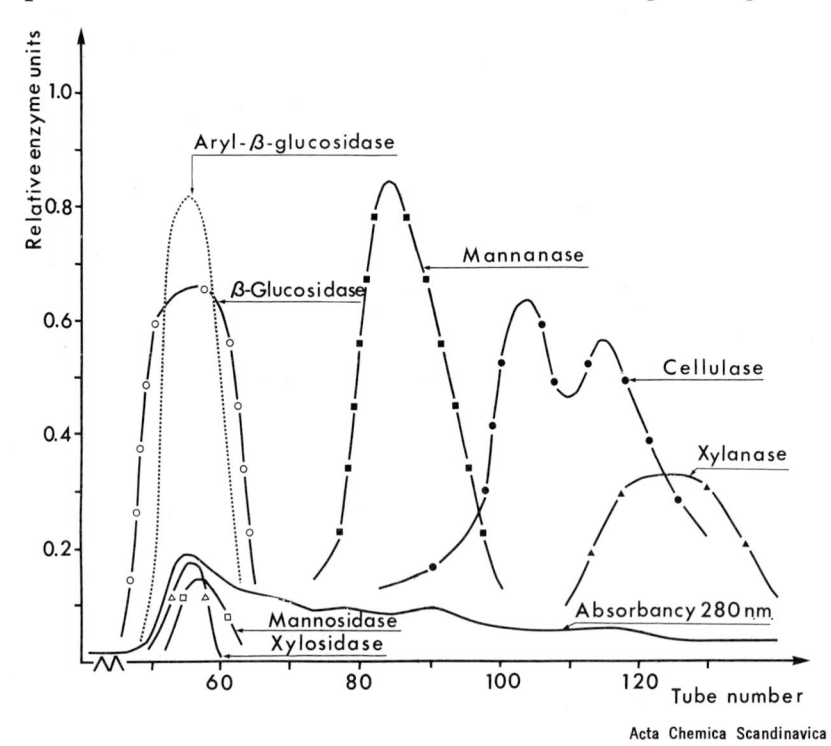

Figure 2. Distribution of protein and enzymic activities after gel filtration of a partly purified commercial enzyme powder (Cellulase 36, Rohm & Haas Co., Philadelphia) on two connected Sephadex G-75 columns. The void volume of the system was 1170 ml. The fraction volume was 27 ml. (2)

In the work on *P. natatum* cellulase (36) relatively high yield of enzyme is reported from the separation on the DEAE Sephadex A-25 column. In the works from *S. sanguinolentum* (8) and *C. lignorum* (11) low yields from the separations on DEAE Sephadex A-50 are noticed. In Reference 8 the total sugar content of the cellulases was found to increase after this fractionation.

Isoelectric Focusing. The principle of the isoelectric focusing method is to focus proteins at their respective pI's (isoelectric points) in a stable pH gradient created by the electrolysis of suitable ampholytes, *viz.* carrier ampholytes (*50, 51*). Detailed presentations of the isoelectric focusing method are given in References 15 and 53 and in a periodical called "Acta Ampholinae" (LKB-Produkter AB, Fack, 161 25 Bromma 1, Stockholm) a literature list of works where the isoelectric focusing method has been used is continuously given.

With the isoelectric focusing method it has been possible to separate proteins differing in pI by only 0.06 pH units. The resolving power of the method has been mathematically calculated in Reference 54. In addition to the separation thus obtained, the method permits a direct determination of the pI of proteins by measuring the pH at their point of focusing in the pH gradient. These pI values are obtained with a high degree of reproducibility, and are thus valuable for characterization and identification of proteins.

The isoelectric focusing method has been applied to characterization of cellulases and related enzymes from the fungi *S. sanguinolentum, F. annosus, C. lignorum* and an *Aspergillus* specie by Ahlgren *et al.* (*1, 2*). In Figure 3 a diagram of one representative isoelectric focusing experiment on *Aspergillus* enzymes is given. The pI values of the enzymes can be read off from this figure. At least three of the enzymes seem to be heterogeneous. Thus, three peaks with cellulase activity, two peaks with xylanase activity, and two peaks with aryl-β-glucosidase activity were obtained.

The investigated enzymes from the four fungi reported above were found to be stable at their isoelectric points. In the experiments referred to above, the density gradient was built up of ethanediol since the common compounds normally used for preparing density gradients, *viz.* sucrose, and glycerol, could not be employed in this case since the enzyme preparations contained sucrase activity which disturbed the enzyme analyses. Glycerol was found to act as a strong inhibitor of some of the enzymes being studied. As a result of the search for another compound ethanediol was selected since it proved to be the best one available in spite of its inhibiting effect on some of the investigated exoenzymes (*1*).

The enzymes cellulase, mannanase, and aryl-β-glucosidase showed no significant decrease in activity as determined in the presence of up to 60% of ethanediol. Xylanase, β-glucosidase, and mannosidase lost approximately half of their initial activity at this ethanediol concentration while xylosidase lost most of its activity. The inhibition caused by ethanediol was found to be reversible.

In References 1 and 2 a comparison of the resolving power of gel filtration, isoelectric focusing, and zone electrophoresis has been undertaken. It was found that the resolving power of the different methods shows such variations from one fungus to another that none of the methods can be specially recommended for the purification of the enzymes. However, in order to discover charge heterogeneities of a protein, the isoelectric focusing method seems to be the most convenient as it separates entirely according to the electric charge possessed by the protein. This is not the case in an electrophoretic column where the mobility, in addition to the charge, most likely also depends upon the size and shape of the protein molecule, the nature of the carrier medium and the buffer solution.

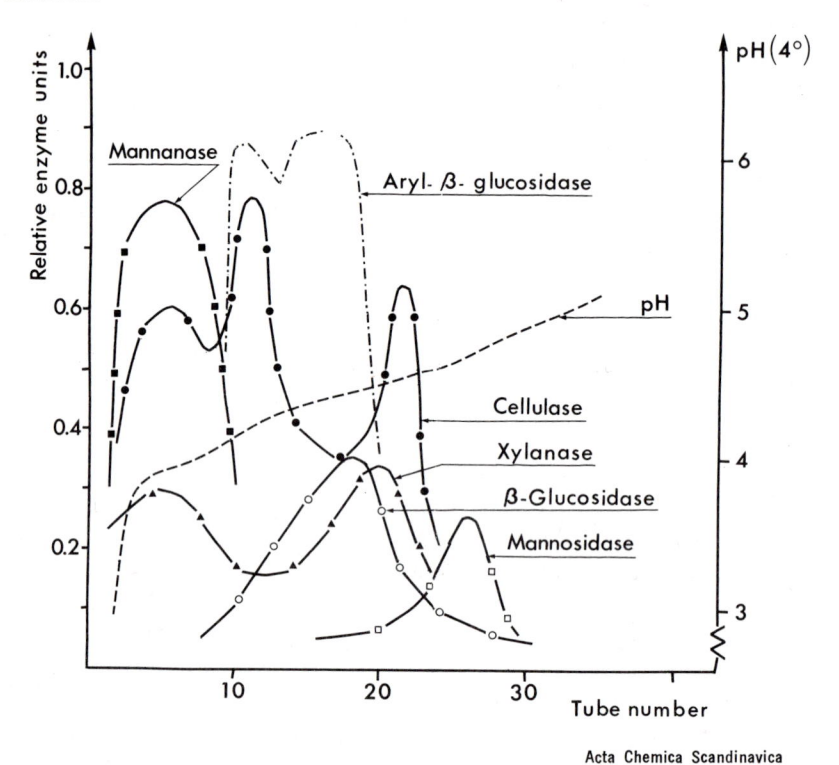

Figure 3. pH diagram with distribution of enzymic activities from isoelectric separation. The fraction volume was 2.3 ml. (2)

In Reference 7 it has been investigated by the aid of isoelectric focusing if prolonged cultivation of *S. sanguinolentum* caused an increasing heterogeneity of the cellulolytic and related enzymes in the culture solution. If the heterogeneities in a protein are not large enough to allow resolution of the protein into separate peaks, the broader the peaks, the more heterogeneous is the material. The difference in pH value between the two points on a curve which lie on half maximum activity was considered to give a value of the width of the curves. The results (7) imply that prolonged cultivation gives rise to somewhat broader peaks and thus to a slight increase in the heterogeneities of the enzymes, the pH ranges being greater after the long cultivation period. The cellulase, however, seemed to be the least affected enzyme and therefore suggested a greater stability than the other studied enzymes.

The influence of prolonged cultivation upon *C. lignorum* enzymes in culture solution has been investigated by isoelectric focusing in Reference 10. It was found that the isoelectric points of the studied enzymes were shifted to higher pH values during the course of cultivation. In Reference 11 three cellulase fractions from culture solution of *C. lignorum* were obtained by fractionation on a DEAE Sephadex A-50 column. These fractions were subject to isoelectric focusing and found to have distinctly different isoelectric points. The studies clearly demonstrate the high resolution obtained by the isoelectric focusing method.

Specific Adsorption on Different Cellulose Columns. Purification of cellulolytic enzymes, including the C_1 enzyme, has been carried out by Li *et al.* (*32*) by specific adsorption. By passing the crude enzyme through a column of Avicel the C_1 fraction was retained by the column while 95% of the cellulase activity could be eluted. The cellulase fraction was then further purified by passing a column of alkali-swollen cellulose where the cellulase was retained. The enzyme could, however, be eluted by buffer in a high yield from the column. The adsorption on an alkali-swollen cellulose column was then repeated. The yield of enzyme obtained with the specific adsorption method is surprisingly high if compared with the yield on Sephadex and on polyacrylamide gel columns.

Characterization of Cellulases

In an excellent chapter under the headline "Criteria for Characterization of Cellulases," Whitaker (*55*) has reviewed the results of work on this subject up to 1961. This chapter dealt with works comprising characterization of cellulases both as enzymes and as proteins. As far as known to the present author no essentially new methods have been presented since Whitaker's review concerning characterization of cellulases as enzymes. Characterization of the C_1 enzyme has failed so far though a considerable amount of work has been carried out lately on the C_1 enzyme from two species of *Trichoderma*, namely *T. viride* and *T. koningi*.

On the other hand a lot of work with new methods has been carried out concerning characterization of cellulases as proteins. The isoelectric focusing works by Ahlgren *et al.* (*1, 2*) have already been described. The cellulase, the protein structure of which is best characterized, is the cellulase from *P. notatum*. The work carried out by Pettersson *et al.* (*9,*

38, 40) introduces a lot of new methods into the structure investigation of cellulases. These methods are in some cases new also in protein research as such.

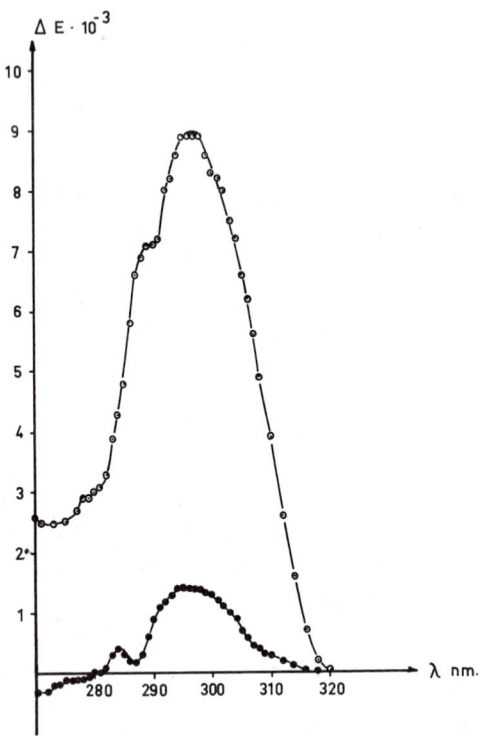

Figure 4. Difference spectra from the addition of mercuric ions to Penicillium notatum *cellulase (○) and tryptophan (●). The concentrations of cellulase, tryptophan, and mercuric acetate were 2.4×10^{-6}M, 4.7×10^{-5}M, and 2.3×10^{-4}M, respectively* (9)

Reproduced by permission of Academic Press, copyright (1968).

Difference Spectrophotometric Studies. In Reference 9, Hg^{2+} was found very strongly to inhibit the cellulase from *P. notatum*. To investigate the influence of Hg^{2+} upon the spectral properties of the cellulase, difference spectrophotometric studies were carried out. The spectra were recorded at room temperature in a Beckman DB spectrophotometer half an hour after mixing the enzyme solution with the mercuric acetate solution. The difference spectrum of cellulase in the presence of Hg^{2+} was compared with that of tryptophan under similar conditions. The similarity between the two difference spectra for cellulase and tryptophan in the presence of Hg^{2+} is evident—*cf.*, Figure 4. These studies showed that the mercuric ions interact with tryptophyl groups in the enzyme. It is not, however, possible to judge from these studies whether the inhibition caused by the mercuric ions is owing only to the interaction with the tryptophyl groups or if other groups are involved as well.

In Reference 37, cellobiose, which together with glucose is an end product of the reaction of the cellulase investigated in this paper, was tried as a neutral perturbant using the difference spectrophotometry method according to Herskowitz and Laskowski (*18*). It was found that cellobiose gave rise to a typical tryptophan spectrum and that the perturbation effect was much more pronounced than with sucrose as a perturbant (Figure 5). It is concluded in Reference 37 that these results do not necessarily mean that there is a specific interaction between cellobiose and the tryptophyl residues in the protein.

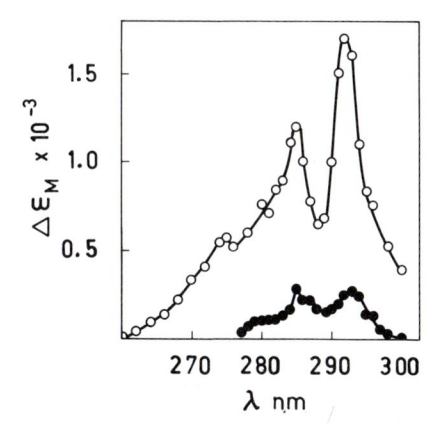

Archives of Biochemistry and Biophysics

Figure 5. Perturbation spectra of cellulase. Media: 10% cellobiose, open circles, and 10% sucrose, solid circles. Protein concentration: 0.5 mg./ml., pH 4.1 (37)

Reproduced by permission of Academic Press, copyright (1968).

The difference spectrophotometric technique (*38*) has been applied to *P. notatum* cellulase at pH 1.9 and 2.4 relative to native cellulase. Only a minor acid difference spectrum was obtained, while after heat denaturation a much more pronounced spectrum was observed. It was possible to evaluate from these measurements that a striking structural change has occurred in the environment of the tryptophyl groups upon heat denaturation. It may be concluded from these data that tryptophyl groups are buried in the interior of the native protein.

Electrolytic Reduction. Amino acid analyses of the *P. notatum* cellulase have given evidence for the presence of two half-cysteines. Investigations of the primary structure of the enzyme have shown that the enzyme was composed of a single peptide chain which was internally cross-linked by the two cysteine residues forming a disulfide bridge (*40*). To examine how this disulfide bridge influenced the enzymic activity, the enzyme was reduced electrolytically (*9*). This reduction was carried out essentially as described by Leach *et al.* (*29*). The current, initially 20 ma., decreased rapidly (*see* Figure 6). Aliquots (25 µl.) were removed from the cathode compartment and tested for cellulase activity after dilution with buffer. As the figure shows, the enzymic activity decreased concomitantly with the current.

The results of the electrolytic reduction showed that the disulfide bridge is necessary for the enzymic activity. This does not imply, however, that it should be directly involved in the catalytic mechanism. The disulfide bond most likely only helps to keep the native protein structure intact.

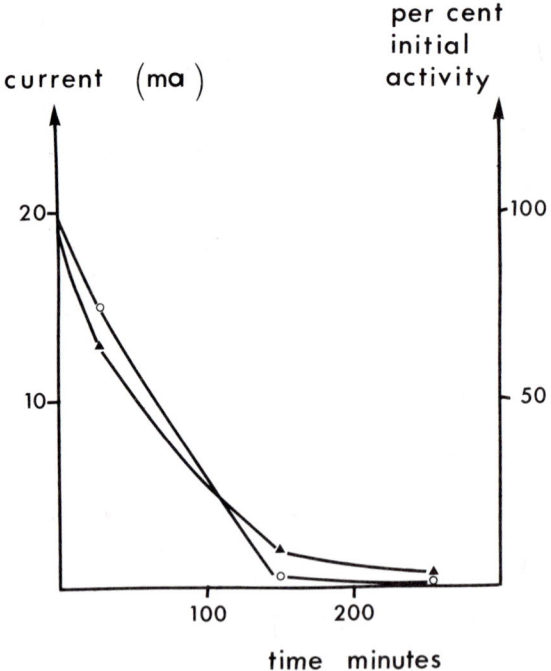

Archives of Biochemistry and Biophysics

Figure 6. Electrolytic reduction of a cellulase from the fungus Penicillium notatum *at pH 7.6 and room temperature. Current in ma. (▲); cellulase activity (% of initial activity)* (○) (*9*)

Reproduced by permission of Academic Press, copyright (1968).

Studies by Proteolytic Enzymes. The influence of proteolytic enzymes upon the cellulase from the *P. notatum* fungus has been investigated (9). When the cellulase was incubated with carboxypeptidase and leucine aminopeptidase at pH 7.7, 30°C., no decrease in cellulase activity could be detected nor a split off of amino acids even after an incubation time of 24 hours.

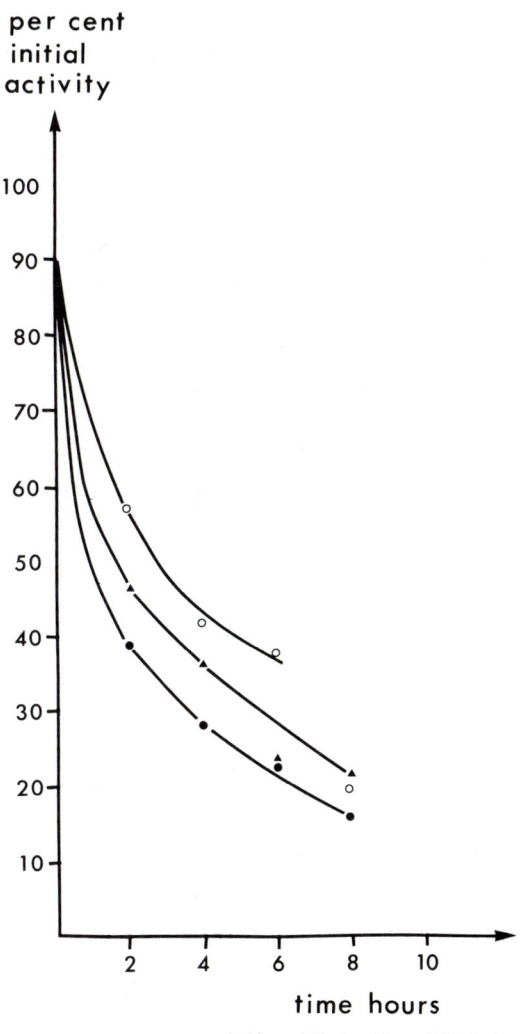

Figure 7. Proteolytic degradation of a cellulase from the fungus Penicillium notatum *at pH 7.7 and 30°C. by the endopeptidases, trypsin (○), subtilisin (▲), and pronase (●) (9)*

Reproduced by permission of Academic Press, copyright (1968).

When the cellulase under the same conditions as described above was incubated with the endopeptidases subtilisin, pronase, and trypsin, the cellulase activity decreased 50% within four hours. The decrease in cellulolytic activity caused by the different proteolytic enzymes is seen in Figure 7. The action of the endopeptidases upon the cellulase was found to increase in the order trypsin, subtilisin, and pronase. The degradation velocity of the cellulase with the endopeptidases is half that of serum albumin. It is obvious that the native cellulase is very resistant to proteolytic enzymes. The possibility of obtaining an active fragment of the cellulase from *Penicillium notatum* was also investigated in Reference 9. The cellulase was incubated with subtilisin in the protein ratio 100:1. After 4 hours incubation at 30°C., when the cellulolytic activity had decreased by approximately 50%, the solution was fractionated on a Sephadex G-75 column. The elution pattern is shown in Figure 8 which shows that the cellulolytic activity appears only with the main A_{280} peak. (This peak has the same elution volume as the intact cellulolytic enzyme tested under identical conditions.) It was not possible to detect any differences in amino acid composition between the active enzyme treated with subtilisin and the original one.

Free-zone Electrophoresis. The *P. notatum* cellulase has been the object for free-zone electrophoresis (40). This method has been worked

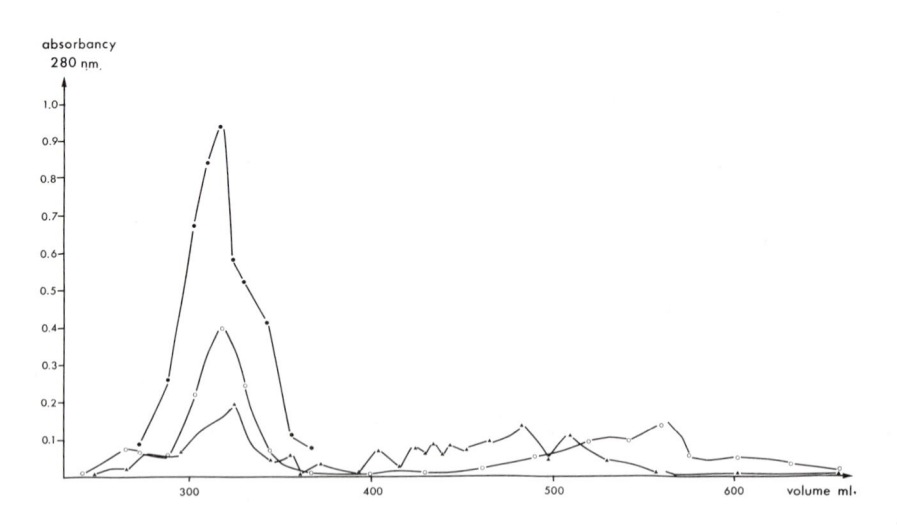

Archives of Biochemistry and Biophysics

Figure 8. Gel filtration on Sephadex G-75 of a subtilisin-digested cellulase from the fungus Penicillium notatum. *Void volume, 150 ml.; absorbancy at 280 n.m. (○); absorbancy at 570 n.m., ninhydrin reaction (▲); cellulase activity (●) (9)*

Reproduced by permission of Academic Press, copyright (1968).

out by Hjertén (19) and offers in addition to isoelectric focusing one of the most convenient and sensitive methods to discover heterogeneities in a protein material.

Potentiometric Titration, Spectrophotometric Titration, and ORD Studies. The *P. notatum* cellulase (38) has been the object of potentiometric titration and ORD studies. With potentiometric titration the relation between the hydrogen ion activity and the state of particular titratable groups can be determined as well as the mean net charge on the protein molecules. The shape of the titration curve is dependent on the exposure of titratable groups accompanying any unfolding reactions during the titration. This effect can be minimized by performing the titration quickly. The results gained with this method are normally rather uncertain since certain approximations must be done. The theoretical titration curve is thus computed assuming that all the members belonging to the same kind of titratable group have the same intrinsic pK value.

The spectrophotometric titrations were carried out at 245 n.m. and 295 n.m. in the pH interval 7.0–14. This titration showed that all the tyrosyl groups titrate as if they were equivalent. The value of $DE_{245} : DE_{295}$ was found to be 5.5, which is higher than the ratio found for most proteins studied so far (44). High values are ascribed to proteins rich in tryptophan.

The optical rotation and rotatory dispersion measurements were carried out with a Model 220 Rudolph Automatic Recording Spectropolarimeter equipped with a 450 W Xenon lamp. Rotatory dispersion data in the 350–600 n.m. region were treated according to the equation of Moffitt (52):

$$[m'] = a_o \frac{\lambda_o^2}{\lambda^2 - \lambda_o^2} + b_o \left(\frac{\lambda_o^2}{\lambda^2 - \lambda_o^2}\right)^2$$

Determinations of some of these parameters makes it possible to calculate at least a semi-quantitative measure of alpha helix structure in proteins and polypeptides (52). The alpha helix content for the cellulase was found to be about 30%.

Studies of Structure and Function by Chemical Modification

Investigation of the Tryptophyl Groups. The structure and function of the cellulase from *P. notatum* has been studied by chemical modification and solvent accessibility (37).

The tryptophyl residues were modified by oxidation with N-bromosuccinimide (NBS) according to Spande and Witkop (48) and by reaction with 2-hydroxy-5-nitrobenzyl bromide according to Horton and Koshland (21). Oxidation of tryptophyl residues with NBS has been shown to be a specific and convenient method for determining the tryptophan content in proteins (48). The method is based upon the decrease in absorbance at 280 n.m. accompanying the transformation of the indole to the oxindole chromophore. If the titration with NBS is carried out in a denaturing medium (8M urea) almost all the tryptophyl residues are titrated. In the cellulase from P. *notatum* twelve of the earlier found thirteen residues were oxidized.

2-Hydroxy-5-nitrobenzyl bromide is also a reagent of high degree of specificity towards tryptophyl residues. In addition to tryptophyl only cystein reacts with this reagent. The results from the titration with ϕ Br indicated that in the studied cellulase only a few of the thirteen tryptophyl residues seem to be exposed.

The accessibility of the tryptophyl residues was further investigated by using difference spectrophotometry (*c.f.* above) and fluorescence studies (49). For a discussion of the interesting conclusions that can be drawn from studies of the tryptophyl residues in the P. *notatum* cellulase —*c.f.*, Reference 37.

Investigation of the Tyrosyl Groups. Tetranitromethane (TNM) has been shown to be a mild and specific reagent for the nitration of tyrosyl groups in proteins (20). From spectrophotometric titrations of the P. *notatum* cellulase it was established that the tyrosyl groups titrated freely and with the same pH (38). With TNM only 5 tyrosines out of 15 were nitrated. When the cellulase was heat denatured before the TNM titration, 11–12 tyrosines were nitrated. Amino acid analysis on the other hand revealed that all tyrosyl groups were nitrated. The author assigns these discrepancies to be attributed to the fact that a conformation change occurs before the spectrophotometric titration. It might thus well be that some tyrosyl groups might be unexposed in the native molecule. The difference between the methods is ascribed to the limited precision in the spectrophotometric method.

Determination of Histidyl Groups. Horinishi *et al.* (20) introduced the reagent diazonium-1-H-tetrazole (DHT) for spectrophotometric determination of histidyl residues in proteins. The method has later been modified by Sokolovsky and Vallee (45). When the P. *notatum* cellulase (37) reacted with DHT a complete loss of activity was observed at a very low excess of reagent. The DHT is, however, a rather unspecific reagent and it is difficult to draw any definite conclusions about the reasons for the inhibition.

In addition to the rather specific reagent reported above the P. *notatum* cellulase also reacted with succinic anhydride which reacts with amino groups but under specific conditions also with hydroxyl groups (28). Riboflavin-5-phosphate and histamine were also used as reagents. They were both shown to be weak inhibitors of the cellulase activity.

Literature Cited

(1) Ahlgren, E., Eriksson, K-E., *Acta Chem. Scand.* **21**, 1193 (1967).
(2) Ahlgren, E., Eriksson, K-E., Vesterberg, O., *Acta Chem. Scand.* **21**, 937 (1967).
(3) Almin, K. E., Eriksson, K-E., *Biochim. Biophys. Acta* **139**, 238 (1967).
(4) Almin, K. E., Eriksson, K-E., *Arch. Biochem. Biophys.* **124**, 129 (1968).
(5) Almin, K. E., Eriksson, K-E., Jansson, C., *Biochim. Biophys. Acta* **139**, 248 (1967).
(6) Björling, H., *Vox Sang* **8**, 641 (1963).
(7) Bucht, B., Eriksson, K-E., *Arch. Biochem. Biophys.* **124**, 135 (1968).
(8) Eriksson, K.-E., Pettersson, B., *Arch. Biochem. Biophys.* **124**, 142 (1968).
(9) *Ibid.*, p. 160.
(10) Eriksson, K.-E., Rzedowski, W. I., *Arch. Biochem. Biophys.* **129**, 683 (1969).
(11) *Ibid.*, p. 689.
(11a) Eriksson, K.-E., *Svensk Kem. Tidskr.* **79**, 660 (1967).
(12) Flodin, P. E., Gelotte, B., Porath, J., *Nature* **188**, 493 (1961).
(13) Florkin, M., Stotz, E. H., "Comprehensive Biochemistry," Vol. 13, 2nd ed., Elsevier, Amsterdam, 1965.
(14) Gascoigne, J. A., Gascoigne, M. M., "Biological Degradation of Cellulose," Butterworths, London, 1960.
(15) Haglund, H., *Science Tools* **14**, 2 (August 1967).
(16) Halliwell, G., *Biochem. J.* **100**, 315 (1966).
(17) Hashimoto, Y., Nisizawa, K., "Advances in Enzymic Hydrolysis of Cellulose and Related Materials," E. T. Reese, Ed., Pergamon Press, London, 1963.
(18) Herskowitz, T. T., Laskowski, M., Jr., *J. Biol. Chem.* **237**, 2481 (1962).
(19) Hjertén, S., Dissertation, Uppsala (1967).
(20) Horinishi, H., Hashimori, Y., Kurihara, K., Shibata, K., *Biochim. Biophys. Acta* **86**, 477 (1964).
(21) Horton, H. R., Koshland, E. E., Jr., *J. Am. Chem. Soc.* **87**, 1126 (1965).
(22) Husemann, E., *Das Papier* **8**, 157 (1954).
(23) Iwasaki, T., Havashi, K., Funatsu, M., *J. Biochem. (Tokyo)* **57**, 467 (1965).
(24) Iwasaki, T., Ikeda, R., Havashi, K., Funatsu, M., *J. Biochem. (Tokyo)* **57**, 478 (1965).
(25) Iwasaki, T., Kokuyasu, K., Funatsu, M., *J. Biochem. (Tokyo)* **55**, 30 (1964).
(26) Jurazek, L., Colvin, J. R., Whitaker, D. R., *Adv. Appl. Microbiol.* **9**, 131 (1962).
(27) King, K. V., *J. Fermentation Technol. (Japan)* **43**, 79 (1965).
(28) Klotz, J. M., "Methods in Enzymology XI," p. 576, New York-London, 1967.
(29) Leach, S. J., Meschers, A., Swanepool, O. A., *Biochemistry* **4**, 23 (1965).
(30) Levinson, H. S., Reese, E. T., *J. Gen. Physiol.* **33**, 601 (1950).
(31) Li, L. H., Flora, R. M., King, K. V., *J. Fermentation Technol. (Japan)* **41**, 98 (1963).
(32) Li, L. H., Flora, R. M., King, K. V., *Arch. Biochem. Biophys.* **111**, 439 (1965).
(33) Norkrans, B., *Ann. Rev. Phytopathol.* **1**, 325 (1963).
(34) Norkrans, B., *Adv. Appl. Microbiol.* **9**, 91 (1967).
(35) Pettersson, G., *Biochim. Biophys. Acta* **77**, 665 (1963).

(36) Pettersson, G., *Arch. Biochem. Biophys.* **123**, 307 (1968).
(37) Pettersson, G., *Arch. Biochem. Biophys.* **126**, 776 (1968).
(38) Pettersson, G., Andersson, L., *Arch. Biochem. Biophys.* **124**, 497 (1968).
(39) Pettersson, G., Cowling, E. B., Porath, J., *Biochim. Biophys. Acta* **67**, 1 (1963).
(40) Pettersson, G., Eaker, D. L., *Arch. Biochem. Biophys.* **124**, 154 (1968).
(41) Reese, E. T., "Advances in Enzymic Hydrolysis of Cellulose and Related Materials," Pergamon Press, London, 1963.
(42) Reese, E. T., *Ind. Eng. Chem.* **49**, 89 (1957).
(43) Reese, E. T., Siu, R. G. H., Levinson, H. S., *J. Bact.* **59**, 485 (1950).
(44) Riddiford, L. M., Stellwagen, R. H., Achta, S., Edvall, J. T., *J. Biol. Chem.* **240**, 3305 (1965).
(45) Scholowsky, M., Vallee, B. L., *Biochemistry* **5**, 3574 (1966).
(46) Selby, K., Maitland, C. C., *Arch. Biochem. Biophys.* **118**, 254 (1967).
(47) Siu, R. G. H., "Microbial Decomposition of Cellulose," Reinhold, New York, 1951.
(48) Spande, T. F., Witkop, B., "Methods in Enzymology," XI, p. 498, Academic Press, New York-London, 1967.
(49) Steiner, R. F., Sippoldt, R., Edelhoch, H., Frattali, V., *Biopolymers Symp.* **Part 1**, 355 (1964).
(50) Svensson, H., *Acta Chem. Scand.* **15**, 325 (1961).
(51) Svensson, H., *Arch. Biochem. Biophys.* **Suppl. 1**, 132 (1962).
(52) Urnes, P., Dety, P., *Advan. Protein Chem.* **16**, 401 (1961).
(53) Vesterberg, O., Dissertation, Stockholm, 1968; *Sv. Kem. Tidskr.* (in press).
(54) Vesterberg, O., Svensson, H., *Acta Chem. Scand.* **20**, 820 (1966).
(55) Whitaker, D. R., "Advances in Enzymic Hydrolysis of Cellulose and Related Materials," E. T. Reese, Ed., Pergamon Press, London, 1963.
(56) Whitaker, D. R., Hansson, K. R., Datta, P. K., *Can. J. Biochem. Biophys.* **41**, 671 (1963).

RECEIVED October 14, 1968.

8

Enzyme-Substrate Relationships Among β-Glucan Hydrolases

D. R. BARRAS,[1] A. E. MOORE, and B. A. STONE

Russell Grimwade School of Biochemistry, University of Melbourne, Parkville, Victoria, Australia

The β-glucan exo- and endo-hydrolases are discussed with reference to newer techniques for the investigation of their specificity and action pattern. Those exo-hydrolases which have been well characterized are described individually. The endo-hydrolases are examined from the point of view of their linkage specificity, action on substituted glucans and their specificity for various monomer units. The significance of "more random" and "less random" endo-action patterns is considered in relation to single or multiple attack mechanisms. Certain features of β-glucan endo-hydrolase catalyzed reactions are discussed in relation to current views on the three-dimensional structure and mechanism of action of lysozyme.

The three dimensional structure of an enzyme-substrate analogue complex has recently been described for lysozyme (24), an enzyme which hydrolyzes a polysaccharide substrate. By inspection of this structure, data have been obtained on the dimensions of the substrate binding site, the nature of the amino acids involved in substrate binding and the identity of the amino acids involved in the cleavage of glycosidic linkages in the polysaccharide substrate (see *Comparison of β-Glucan Endo-hydrolases with Lysozyme*).

This information must now form the basis of any reaction mechanism proposed to describe the action of lysozyme (126), although certain dynamic aspects of glycosidic cleavage are yet to be explained. It remains to be seen how closely this model system fits other enzymes

---

[1] Present address: Department of Biochemistry and Molecular Biology, National Research Council of Canada, Ottawa, Ontario, Canada.

hydrolyzing polysaccharides, since there is only isolated information about the mechanism of action of other glycan hydrolases (*101, 102, 105*), and nothing is known about the three dimensional structure of their enzyme-substrate complexes. However, much has been learned indirectly about the dimensions of the substrate binding site, its specificity and the events occurring there during the hydrolysis of glycosidic linkages in glycosides, oligosaccharides, and polysaccharides. Of these recent investigations, one of the most significant has been the discovery by Legler (*119, 120*) that conduritol B-epoxide inhibits *Aspergillus wentii* β-glucosidase by forming a covalent compound with the enzyme. This opens the way for the identification of the amino acids in the immediate proximity of the active site and suggests that a carboxyl group may be involved in glycoside cleavage.

One of the most fruitful methods of exploring enzyme-substrate relationships is to test a variety of compounds chemically related to the substrate, for their ability to react with or to inhibit the action of the enzyme. This approach has been fully exploited for β-glucosidases (*73, 86, 220*), and similar studies relating to β-glucan hydrolases will be discussed in this chapter.

In such investigations, considerable use has been made of the wide range of structurally different β-glucans of natural origin (*see* Table I). In addition, certain soluble derivatives such as CM-cellulose (carboxymethylcellulose), HE-cellulose (hydroxyethylcellulose), and CM-pachyman (carboxymethylpachyman) (*43*) and chemically modified forms, such as the oxidized-reduced laminarin introduced by Smith and coworkers (*151*) and oxidized-reduced paramylon (*17*), have been of great value in deciding action patterns of glucan hydrolases. Oligosaccharide substrates have also been widely used in these investigations and methods are now available for preparing specifically labelled oligosaccharide substrates by enzymic syntheses (*209, 233*). In the case of β-glucan hydrolases, these methods have not yet been fully exploited, but their usefulness has been demonstrated in the case of α-glucan hydrolases by the work of French and colleagues (*3, 66, 67*).

Among methods of following the progress of the enzymic depolymerization of β-glucans, it is unfortunate that there is no simple colorimetric test, comparable to the starch-I_2 reaction, which can be applied to measure changes in chain length (CL). However, measurements of the viscosity of solutions of CM-cellulose (*232*), CM-pachyman (*43*) and of the soluble, mixed-linked glucans from oat and barley (*179*), have provided a ready means of detecting changes in CL of these substrates during enzymic attack. Viscosity data can also be used to calculate the activity of cellulases in absolute terms, as recently described (*5, 6*). The release of reducing residues has been the main procedure for following the action of polysaccharide hydrolases, and this method has been made more sensitive by the use of tritiated sodium borohydride for labelling newly formed reducing groups (147, 219). In addition, glucose liberated during enzymic hydrolysis may be specifically estimated by methods involving the use of glucose oxidase, thus enabling the distinction to be made between glucose and other reducing sugars liberated. It should be noted that the widely used dinitrosalicylic acid method for reducing sugar determination can lead to over-estimation, when substrates of the 1,3-glucan type are involved, because of the "alkaline-peeling" reaction (58, 194).

The identification of oligosaccharide hydrolysis products, which has been made possible by the availability of oligosaccharides of various linkage series as markers (14), has enabled a more detailed investigation of the finer points of linkage specificity. Newer chromatographic techniques which have been applied to the separation and identification of soluble oligosaccharides in enzymic hydrolysates, include gel filtration (156) and thin layer (38) and gas-liquid systems (30, 141, 166). Gas-liquid chromatography has also enabled the rapid determination of the anomeric configuration of sugars released during enzymic hydrolysis (164). These methods, together with the improvements in paper (215) and charcoal (68) chromatography which allow isolation of soluble, long chain oligosaccharides, will enable fuller analysis of the course of enzymic depolymerization.

The identification of oligosaccharide products, which relies on chemical methods such as periodate and lead tetraacetate oxidation, partial acid hydrolysis before and after reduction, electrophoretic mobility in borate and molybdate buffers and chromatography, has been described by Bouveng and Lindberg (29). The alkaline degradation products of oligosaccharides are characteristic of the linkage present and a TLC technique for their separation and detection has recently been described (16).

Classification of Enzymes Hydrolyzing β-Glucosidic Linkages

There are a large number of enzymes which may be classified as hydrolases for β-glucosidic linkages. These hydrolases can be grouped according to characteristics which may depend on differences in the mechanism of cleavage of the glycosidic linkage or on differences in the substrate, inhibitor and, in some cases, acceptor binding sites on the enzyme. A tentative classification based on these criteria is set out in Table II.

Table I. Structure and Structural Linkages

Glucan	Source	Structure	Structural Linkages
Cellulose		L	1,4
Amyloids	Dicotyledon seed cell walls	SCB	1,4
Luteic acid	*Penicillium luteum* cultures	L	1,6
Islandic acid	*Penicillium islandicum* cultures	L	1,6
Pustulan	*Umbilicaria pustulata* fronds	L	1,6
Crown gall polysaccharide	*Agrobacterium* spp. cultures	L	1,2
Mycoplasma glucan	Bovine mycoplasma strains	L?	1,2
Oat glucan	*Avena sativa* endosperm walls	L	1,4 (65%); 1,3 (35%)
Barley glucan	*Hordeum vulgare* endosperm walls	L	1,4 (70%); 1,3 (30%)
Lichenin	*Cetraria icelandica* fronds	L	1,4 (70%); 1,3 (30%)
Monodus glucan	*Monodus subterraneus* cell walls	?	1,3 (<30%); 1,4
Pachyman	*Poria cocus* sclerotia	L	1,3; 1,6 (2%)
Paramylon	Cells of—		
	Euglena gracilis	L	1,3
	Peranema trichophorum	L	1,3
	Astasia ocellata	L	1,3
Curdlan	*Alcaligenes faecalis* var. *myxogenes*	L	1,3; 1,6 (0.5%)
Callose	*Vitis vinifera* sieve tubes	L	1,3
	Pinus mugo pollen	L	1,3
Aureobasidium cell wall glucan	*Aureobasidium* (*Pullularia*) *pullulans* cell walls	L	1,3
Laminarins	*Laminaria digitata*	B	1,3 (~90%); 1,6
	L. hyperborea (*L. cloustoni*)	L	1,3 (~95%); 1,6
	L. saccharina	B	1,3 (~90%); 1,6
	Eisenia bicyclis	L	1,3 (~76%); 1,6 (~24%)
Leucosin	*Ochromonas malhamensis*	?	1,3 (~90%); 1,6
	Mixture of diatoms (Chrysolaminarin)	?	1,3 (90%); 1,6
	Phaeodactylum tricornutum	?	1,3; 1,6

Occurrence of β-Glucans

Other Monomers	DP	CL	Reference
	1000-10,000		(108, 109)
xylose, galactose	74		(115, 116)
esterified with malonic acid	84		(56) (see also (125, 148)
esterified with malonic acid (1 mole per mole glucose)		80	(11, 56)
			(53, 87, 123)
	22		(72, 134, 180)
			(174)
	190-920		(44)
	100		(44, 130)
	52-410		(44)
			(63)
	255	40	(199, 225)
	150		(42)
	80		(9)
	50-55	43	(140)
	~135-455	125	(82, 83, 199)
uronic acid (2%)	>90		(10)
arabinose (2%)			(27)
			(33)
mannitol (3%)	25.5	9.8	(44, 60)
	26-31	7-10	(61)
mannitol (2%)	24	19	(7, 8, 44)
mannitol (2.5-2.7%)	28	11	(7, 8)
	22		(79, 135)
	34		(9)
	21	12	(20)
			(64)

Table I.

Glucan	Source	Structure	Structural Linkages
Claviceps-type glucans	*Claviceps purpurea*	SCB	1,3 (75%); 1,6
	Claviceps fusiformis	SCB	1,3 (73-80%); 1,6
	Sclerotium rolfsii	SCB	1,3 (75%); 1,6
	Aureobasidium (Pullularia) pullulans	SCB	1,3 (60%); 1,6
	Plectania occidentalis	SCB?	1,3 (60-70%); 1,6
	Helotium sp.	SCB?	1,3 (45%); 1,6 (25%); 1,2 or 1,4 (30%)
Sclerotan	*Sclerotinia libertiana*	SCB	1,3; 1,6
Isosclerotan	*Sclerotinia libertiana*	B (12.5%)	1,3 (48%); 1,4 (16%); 1,6 (36%)
Yeast glucans	*Saccharomyces cerevisiae* cell walls	B	1,3 (80-90%); 1,6
	Candida albicans cell walls	B	1,3 (~28%); 1,6
	Candida parapsilosis cell walls	B	1,3; 1,4; 1,6

Abbreviations used:
DP   Degree of polymerization   L   Linear
                                        SCB   Side chain branched
CL   Chain length   B   Branched

A distinction between aryl β-glucosidases and other β-glucosidases was made by Youatt and Jermyn (236) for enzymes from *Stachybotrys atra*, on the basis of types of glycosides hydrolyzed and has also been made for enzymes from *Myrothecium verrucaria* (85) and *Neurospora crassa* (55, 136). An examination of the anomeric configuration of glucose released by the aryl β-glucosidases and of the point of cleavage of the glycosidic linkage has yet to be made, but will be of interest in relation to their mechanism of action.

The possibility has been considered for some time that glucosidases acting on oligosaccharides with an action pattern which can formally be described as "exo," may either be non-specific β-glucoside hydrolases or specific glucan hydrolases (206). Reese, Maguire, and Parrish (184) have made a comprehensive, comparative study of such glucosidases and have clearly differentiated them on the basis of their glycosidic linkage specificity, oligosaccharide CL specificity, their relative inhibition by 1,5-gluconolactone (and methyl β-glucoside) and the anomeric configuration of sugars released. The differences are indicated in Table II.

The distinction between exo-hydrolases and endo-hydrolases based on their action patterns during the hydrolysis of polymeric and oligomeric glucans is well established. As shown in Table II, there are also differences in the anomeric configuration of reducing groups released, although

Continued

Other Monomers	DP	CL	Reference
			(170)
	~110	90	(34) (106)
uronic acid (3%)			(28)
			(223)
			(223)
			(111, 112, 113, 212)
			(157)
	410		(44, 139, 144)
	30 ± 2		(22, 237)
	33		(237)

the number of endo-enzymes for which this information is available is small. Among the endo- and exo-hydrolases, various sub-groups can be identified and a discussion of these will be the subject of the remainder of this chapter.

β-Glucan Exo-hydrolases

General Properties. Enzymes in the broad class of exo-glucan hydrolases can be subclassified into those yielding glucose as the sole initial product from polymeric substrates and those producing glucose disaccharides. Thus, among the α-1,4-glucan hydrolases, are the glucoamylase (or amyloglucosidase) types which produce glucose and the β-amylase types which produce maltose and they have their counterparts among the β-glucan exo-hydrolases. The specificity for the linkage position of the glucosidic linkage cleaved is high, but may not be absolute, and this is also the case with the α-1,4- (2) and α-1,6-glucan hydrolases (15, 239). Characteristically, the polarity of attack by exo-enzymes is from the nonreducing end of the polymer chain and there is an inversion of the configuration of the glucose or diglucose released (21, 78, 164). However, the recent report that pullulanase is an exo-enzyme releasing α-maltotriose from the α-glucan, pullulan, may provide an exception to this rule (224).

Table II. Tentative Classification of β-glucoside hydrolases

	Aryl β-glucoside hydrolases[a]	β-glucoside hydrolases
Specificity for position of linkage i.e. 1,3; 1,4 etc.	Not applicable	Low
Specificity for glycosyl residue	High. Preference for β-glucopyranosyl	High. Preference for β-glucopyranosyl, but probably others, e.g. β-gal p, β-xyl p
Specificity for "aglycone"	High. Aryl alcohols	Low. Alkyl, aryl and oligosaccharides (not polysaccharides)
Rate of hydrolysis of oligomers	Not applicable	di > tri > tetra >>> penta
Inhibition by 1,5-gluconolactone	?	K_i low
Configuration of anomeric-OH released	?	β i.e., config. retained
Transfer activity	+ config. retained	+ config. retained
Specificity for transfer acceptors	Low	Variable

[a] A further subdivision of glycoside hydrolases into *taka* and *emulsin* types is based corresponding unsubstituted phenylglycosides (93, 100, 105).

Whether a single enzyme-substrate encounter produces more than one successful glycosidic cleavage (13, 65)—*i.e.*, whether multiple or single chain attack occurs—has only been determined for β-amylase amongst enzymes of this class. In this case the measurements of Bailey and French (12) suggest 4,3 attacks per successful encounter for β-amylase acting on a linear α-1,4-glucan of CL 44. This evidence suggests that following the cleavage of a glycosidic linkage, a new and productive enzyme-substrate "complex" is formed by rearrangement of the substrate on the active site of the enzyme. The mechanics of the relative movement of enzyme and substrate to allow multiple attack are at present obscure.

Hanson (80) has given a comprehensive mathematical treatment of the kinetics of action of the exo-enzymes, taking into account situations where multiple attack and self-competitive effects occur. He has proposed a kinetic test for multiple action, as an alternative to the statistical treatment of Bailey and French (12). Hiromi, Ono, and coworkers (88, 89,

Enzymes Hydrolyzing β-Glucosidic Linkages

β-glucan exo-hydrolases	β-glucan endo-hydrolases
High or absolute	High or absolute
High. Glucopyranosyl residues preferred, but accepts 6-substituted and ? xylopyranosyl residues	Absolute. *Must* involve β-glucopyranosyl residue, or residues? Varying linkage types (*see* Table IV)
High. Glucans preferred but oligosaccharides cleaved. May be substituted or ? an aryl residue	High (but variable). Must involve β-glucopyranosyl residue or residues. May be substituted
tri < tetra < penta <<< polymer	tri < tetra < penta < hexa << polymer
K_i high	Not inhibited
α *i.e.*, config. inverted ± config. retained ?	β *i.e.*, config. retained ± config. retained High?

on the differences in attack on o-substituted phenyl glycosides compared with the

90) have also published kinetic treatments of exo-enzyme attack on polymeric substrates.

β-1,4-Glucan Glucosyl Hydrolases. None of the enzymes in this group have been investigated in depth, so that their inclusion must be regarded as tentative. It is interesting to note that while exo-hydrolases for β-1,3- and α-1,4-glucans have been reported from a wide variety of sources, few β-1,4-glucan exo-hydrolases are known. Whether this is because of difficulties in their detection or because of a real scarcity in nature remains to be seen.

Trichoderma viride (*122*). This enzyme was prepared from culture filtrates of *T. viride* by fractionation on alkali-swollen cellulose, followed by Sephadex chromatography. The final preparation had a low specificity, hydrolyzing aryl glucosides, several β-glucose disaccharides and polymeric substrates including CM-cellulose and β-1,3-glucans, but not crystalline cellulose. Glucose was released from the terminal non-reducing ends of reduced β-1,4-oligoglucosides and was shown by optical rotation measurements to be predominantly in the α-configuration. Whether this is an exo-enzyme [as classified by Reese (*184*)] with a rather wide specificity range, or is a mixture of β-glucoside hydrolases and glucan exohydrolases remains to be determined.

Nisizawa has recently reported the detection of an exo-enzyme in a fraction from a *T. viride* cellulase preparation (*159*).

Aspergillus niger (*48, 110, 121*). A crude extra-cellular enzyme preparation from *A. niger* has been separated by ion-exchange and adsorption chromatography into a number of β-glucoside hydrolase fractions. One of these (fraction I), thought to be an exo-enzyme, had a low activity on CM-cellulose, both in reducing sugar and viscosity assays. Quantitation of the individual products of the hydrolysis of cellopentaitol showed that the predominant cleavage was at the first glycosidic linkage from the nonreducing end, releasing glucose, but that the second and third linkages were also hydrolyzed at low frequencies. The possibility of the presence of traces of other enzymes with different modes of action was not excluded. The preparation hydrolyzed p-nitrophenyl-β-O-glucoside.

Stachybotrys atra (*235*). An intracellular enzyme from this fungus, which released glucose as the sole product from β-1,4-oligoglucosides up to CL 11 and was not inhibited by 1,5-gluconolactone (*236*), is apparently an exo-hydrolase. The β-1,4-glucosyl linkage in p-nitrophenyl-α-cellobioside was also hydrolyzed.

β-1,4-Glucan Cellobiosyl Hydrolases. *Cellvibrio gilvus* (*210*). Enzymes from the culture filtrates of *Cellvibrio gilvus* have been described, which hydrolyze alkali-swollen cellulose and cellulose dextrins with an exo-action pattern. One of the four fractions from an electrophoretic separation was investigated in detail (*209*). Its optimum chain length for substrates was six glucose units or greater, and no glycosyl transferase activity was detected. Using cellopentaitol, it was shown that the primary attack was at the second glycosidic linkage from the nonreducing end (93%), while 7% of the cleavages occurred at the third linkage. When cellohexaitol was substrate, 50% of the attacks occurred at the second and 48% at the third glycosidic linkage. These results were confirmed using a ^{14}C-oligosaccharide substrate. The other three fractions also hydrolyzed the second and third linkages, but with different frequencies (*48*).

These enzymes differ from other exo-hydrolases by not hydrolyzing their substrates at unique sites, at least with the β-1,4-oligoglucosides of CL up to 6. When swollen cellulose was the substrate, cellobiose was an important product for each of the enzyme fractions, but there is no information as to whether cellotriose was also a product. The polarity of attack and the inversion of the configuration of the oligosaccharides released, distinguish this enzyme from endo-hydrolases.

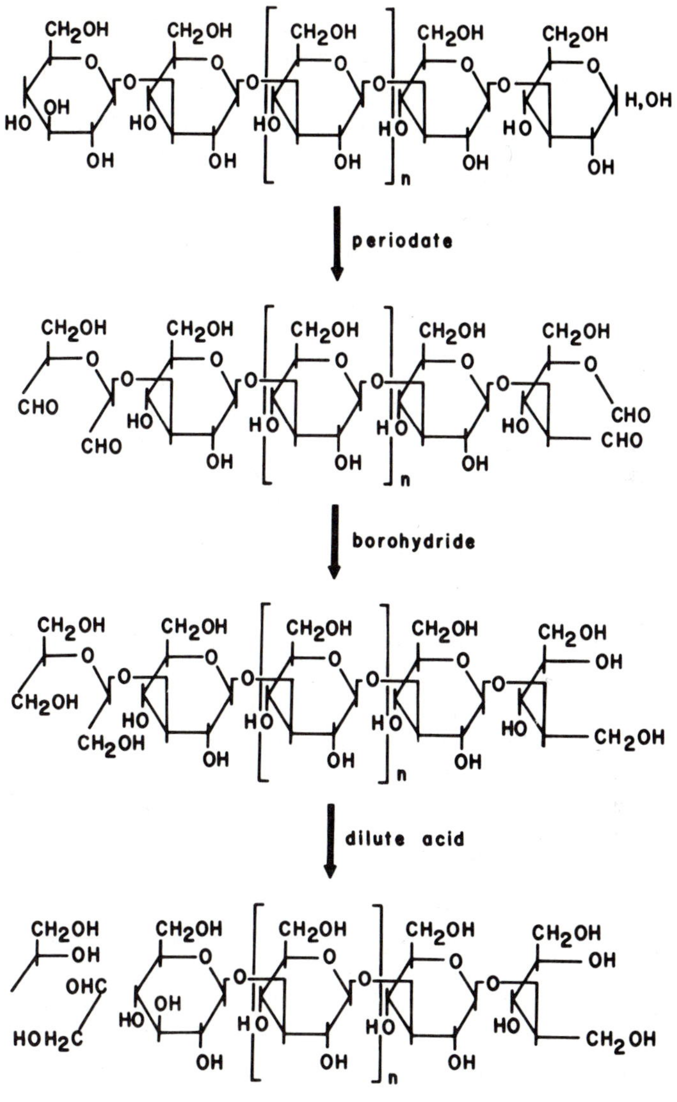

Figure 1. Structural modifications of the nonreducing and reducing terminal glucose units of a linear β-1,3-glucan, as a result of oxidation with periodate and reduction with borohydride (151)

β-1,3-**Glucan Glucosyl Hydrolases.** A large number of microorganisms (*35, 185, 211*), invertebrates (*208*) and plants (*43, 186*) have been screened for their β-1,3-glucan hydrolase activities and exo-types have been found in some. A few preparations have been examined in sufficient detail to enable useful comments on their enzyme-substrate relationships to be made and are described in the following sections.

Basidiomycete QM 806 (185). The *Basidiomycete* enzyme, first studied by Reese and coworkers, has been used in a number of investigations on the structure of β-glucans. It has been purified by ammonium sulfate fractionation, DEAE-cellulose chromatography, and preparative acrylamide gel electrophoresis, yielding a preparation homogeneous as judged by disc electrophoresis on acrylamide gel, its sedimentation characteristics in the ultracentrifuge and the absence of contaminating enzymes. Two isozymes of the enzyme have been separated (95).

This enzyme produces only glucose during the hydrolysis of β-1,3-glucans (*185*). It does not hydrolyze laminarin which has been modified at both nonreducing and reducing ends by periodate oxidation followed by borohydride reduction (Figure 1) (*151*). However, following mild acid hydrolysis which removes the modified residues at the nonreducing ends of the laminarin molecules, the glucan is hydrolyzed by the enzyme. These findings clearly indicate that the enzyme is a typical exo-enzyme. The glucose is released in the α-configuration (*164*).

The enzyme shows a preference for longer chains of β-1,3-glucose oligosaccharides and it seems likely that binding and rate of attack have not reached their maximum with substrates of CL 15–20 (95). The enzyme is highly specific, attacking only β-1,3-linked glucans or the homomorphous β-1,3-xylan (*150*).

The *Basidiomycete* enzyme releases both glucose and gentiobiose from *Claviceps*-type glucans (*see* Figure 2), indicating that 6-glucosyl substitution of the β-1,3-glucosyl residues is not a barrier to hydrolysis (*106, 170*). Extensive enzymic hydrolysis of laminarin produced glucose (70%), laminaribiose, laminaritriose, gentiobiose, and oligosaccharides containing mannitol (*151*), indicating that the inter-chain β-1,6-linkages are bypassed and that the hydrolysis ceases as the chain length of the oligosaccharides reaches 2 or 3 or as the glycosyl-mannitol residues are approached.

It has been briefly reported (77) that the *Basidiomycete* enzyme releases glucose from the cell wall glucan of *Saccharomyces cerevisiae* var. *ellipsoideus* and that in the later stages of the hydrolysis, gentiosaccharides (CL 2–5) are detectable. The glucose evidently arises from the side chains of β-1,3-linked glucosaccharides and the gentiodextrins from the main backbone (*see* Figure 2). If this structure is correct, then it appears that the enzyme can cleave 1,3 linkages when these are flanked by runs of 1,6-linked residues (*see* also *Euglena gracilis* enzyme).

Sclerotinia libertiana (54, 200). An enzyme which produces glucose and gentiobiose from laminarin and sclerotan has been crystallized from the culture filtrates of *S. libertiana*. It is almost certainly an exo-enzyme, but no information is available on the configuration of the glucose liberated, substrate chain length specificity, or behavior with inhibitors.

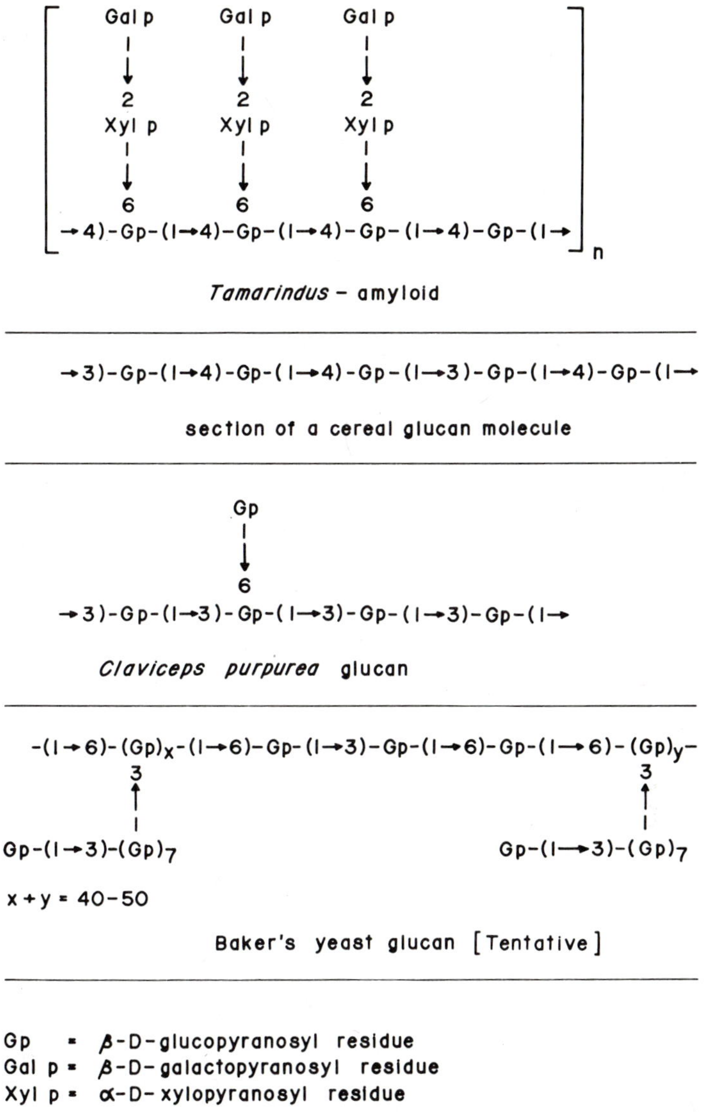

8. BARRAS ET AL.   *β-Glucan Hydrolases*   117

$$\begin{bmatrix}
\text{Gal p} & \text{Gal p} & \text{Gal p} \\
\downarrow & \downarrow & \downarrow \\
2 & 2 & 2 \\
\text{Xyl p} & \text{Xyl p} & \text{Xyl p} \\
\downarrow & \downarrow & \downarrow \\
6 & 6 & 6 \\
\rightarrow 4)\text{-Gp-}(1\rightarrow 4)\text{-Gp-}(1\rightarrow 4)\text{-Gp-}(1\rightarrow 4)\text{-Gp-}(1\rightarrow
\end{bmatrix}_n$$

Tamarindus – amyloid

→3)-Gp-(1→4)-Gp-(1→4)-Gp-(1→3)-Gp-(1→4)-Gp-(1→

section of a cereal glucan molecule

Gp
|
↓
6
→3)-Gp-(1→3)-Gp-(1→3)-Gp-(1→3)-Gp-(1→

Claviceps purpurea glucan

-(1→6)-(Gp)$_x$-(1→6)-Gp-(1→3)-Gp-(1→6)-Gp-(1→6)-(Gp)$_y$-
    3   3
    ↑   ↑
    | |
Gp-(1→3)-(Gp)$_7$ Gp-(1→3)-(Gp)$_7$

x + y = 40–50

Baker's yeast glucan [Tentative]

Gp   = β-D-glucopyranosyl residue
Gal p = β-D-galactopyranosyl residue
Xyl p = α-D-xylopyranosyl residue

Figure 2. Structures of glucans from Tamarindus *(115),* cereal endosperm *(98, 163),* Claviceps purpurea *(170) and* Saccharomyces cerevisiae *(144)*

Sporotrichum pruinosum QM 826 (185). This organism produces an exo-enzyme with properties similar to *Basidiomycete QM 806*. Its relative rate of action on the trisaccharides shown in Table III *(184)* indicates that the relative susceptibility of the terminal glycosidic linkage to hydrolysis is determined both by its nature and by the nature of the

Table III. Hydrolysis of Glucose Trimers by Exo-β-1,3-Glucanase of *Sporotrichum pruinosum* QM 826 (184)

β-Trimer of Glucose	Relative Susceptibility
G 3 G 4 G[a]	1.2
G 4 G 3 G	0.0
G 3 G 3 G	100.0
G 4 G 4 G	0.7
G 6 G 4 G	0.0

[a] G 3 G 4 G = O-β-D-glucopyranosyl (1 → 3)-O-β-D-glucopyranosyl (1 → 4)-D-glucose. Other oligosaccharides and sequences are represented in a similar way.

adjacent linkage joining the aglycone residue. Similar specificity characteristics are shown by the A. niger amyloglucosidase (2).

Euglena gracilis. Enzymes from *E. gracilis* (59) and other euglenoid flagellates (142) which hydrolyze laminarin have been described. An intracellular β-1,3-glucan exo-hydrolase from *E. gracilis* has been purified by CM-cellulose ion exchange chromatography, which yielded two apparently interconvertible components (17, 18). One of these has been examined in detail and releases glucose as the sole product during the hydrolysis of paramylon and pachyman. The glucose is released in the α-configuration. Oxidized-reduced paramylon (17) and laminarin (151) are not hydrolyzed, but following mild acid hydrolysis they become substrates, indicating that the attack occurs at the nonreducing ends. The enzyme does not hydrolyze laminaribiose and the rate of reaction on the β-1,3-oligoglucoside series increases as the chain length of the substrate increases. 1,5-Gluconolactone does not inhibit the reaction at lactone-substrate ratios which are effective in inhibiting β-glucoside hydrolases (184). The enzyme is highly specific for β-1,3-glucans and does not hydrolyze a β-1,3-xylan from *Caulerpa brownii*. Only small amounts of glucose are released from CM-pachyman and from the mixed-linked glucans, lichenin, and oat and barley glucan. The enzyme attacks 1,3-linkages joining the 6-glucosyl substituted glucose residues in a *Claviceps*-type glucan (34) (Figure 2) and releases glucose and oligosaccharides from *Eisenia bicyclis* laminarin (see Table I), where 3-β-glucosyl residues are flanked by runs of β-1,6-glucosyl residues. Yeast glucan is attacked, with the liberation of both glucose and oligosaccharides. No transfer reactions were observed when ^{14}C-glucose at a final concentration of 20% was present during the hydrolysis of a β-1,3-glucan, or when ^{14}C-glucose (20%) was incubated alone with the enzyme. This contrasts with the behavior of the purified α-1,4-glucan exo-hydrolases from *Aspergillus niger* and *Rhizopus delamar*, which form disaccharides on incubation with 60% glucose at 55°C. (165).

Hansenula and Saccharomyces Enzymes. Brock (*31, 32*) has provided evidence for the occurrence of an intracellular enzyme in *Hansenula wingei*, which hydrolyzes laminarin to yield glucose, pustulan to give gentiobiose and gentiotriose and is also able to hydrolyze *p*-nitrophenyl β-glucoside. However, cellulose, lichenin, salicin, methyl β-glucoside, and phenyl β-glucoside are not hydrolyzed. Similar enzymes which hydrolyze pustulan and laminarin with an exo-action pattern have been demonstrated in *Saccharomyces cerevisiae*, *Fabospora* (*Saccharomyces*) *fragilis*, and *Hansenula anomala* (*1*). Since they have less restricted linkage specificities than the enzymes discussed under *β-1,3-Glucan Glucosyl Hydrolases*, they may represent a further class of exo-enzymes.

β-Glucan Endo-hydrolases

Linkage Specificity Amongst Endo-hydrolases. Ploetz and Pogacar (*175*) in 1949 investigated the possibility that specific enzymes for the hydrolysis of cellulose, yeast glucan, and lutean were present in a preparation from *Aspergillus oryzae*. From kinetic data they concluded that nonspecific oligosaccharases and polysaccharases were present. It is now clear, however, that a number of different β-glucan endo-hydrolases exist, with relatively narrow linkage specificities. Thus, endo-hydrolases for β-glucans containing 1,2- (*189*), 1,3- (*186*), 1,4- (*186*) and 1,6- (*149, 190*) linkages have been identified and shown to be separate and specific enzymes.

Knowledge of the specificity requirements of endo-hydrolases has been considerably widened by the use of glucan substrates containing more than one type of glucosidic linkage. For example, *Rhizopus arrhizus* endo-hydrolase preparations, which depolymerize β-1,3-glucans to yield laminaribiose and laminaritriose as major products (*185*), also hydrolyze β-1,3; 1,4-glucans (*49, 163, 169*). The nature of the hydrolysis products (*see* Table IV) indicates that the glycosyl residue of all linkages hydrolyzed must be 3-substituted. Further, it has been concluded (*162*) that in addition to 1,3-linkages, 1,4-linkages may also be split, since the oligosaccharide hydrolysis products accounted for most of the 1,3-linkages present in the original substrate. β-1,3-glucan endo-hydrolases with this type of substrate specificity have also been found in preparations from *Schizophyllum commune* (*201*) and *Bacillus circulans* (*96*).

Although many β-1,4-glucan endo-hydrolases have been examined, the ability to hydrolyze the 1,3; 1,4 mixed-linked glucans has been investigated in detail only for a *Streptomyces* preparation (*163, 169, 193*), for a purified *Aspergillus niger* enzyme (*45, 46*), and for a fraction from a *Trichoderma viride* preparation (*97*). In each case the nature of the products (*see* Table IV) makes it clear that 4-β-glucosyl residues are specifically required in the glycosyl portion of the linkage hydrolyzed. It is not yet known whether linkages other than 1,4 can be split.

The use of heterogeneously linked glucans in addition to the homogeneous types has also shown that endo-hydrolases with at least two other types of specific substrate requirements can be distinguished: those splitting only mixed linked glucans and those splitting only homogeneously linked glucans.

Luchsinger has presented evidence that among the enzymes from germinating barley there are components which hydrolyze barley glucan, but not β-1,3- or β-1,4-glucans (*37, 129, 130, 132*). The nature of the hydrolysis products (*see* Table IV) indicates that there is a specific requirement for —G 4 G 3 G— in the glycosyl residue of the linkage cleaved. Enzymes from *Bacillus subtilis* have a similarly restricted substrate requirement, acting only on barley glucan or lichenin (*146, 188, 191*).

In contrast, several β-1,3-glucan endo-hydrolases have now been described whose action seems to be confined to linear β-1,3-glucans or to substrates containing long runs of β-1,3-glucosyl residues. These enzymes are found in germinating barley (*131*), callus tissues originating from pepper (*137*) and roots of wheat and barley (*69*), and in *Nicotiana glutinosa* leaves (*43*).

The *Nicotiana* enzyme (*145*) has been extensively purified by DEAE-cellulose and exclusion chromatography. The enzyme hydrolyzes pachyman, paramylon, and laminarin yielding a series of β-1,3-oligoglucosides of CL 2–7 and rapidly lowers the viscosity of CM-pachyman with an accompanying slow release of reducing groups. These properties are characteristic of an endo-hydrolase. The purified preparation showed no action on β-1,4-, β-1,6-, β-1,2-glucans, a β-1,3-xylan, or *Claviceps* glucan, but yeast glucan was degraded to a small extent. The viscosity of both oat and barley glucans was very slowly reduced on incubation with the enzyme.

This group of enzymes seems to require long runs of 1,3-linked glucose residues in their substrates and these may be found in certain linear β-1,3-glucans (*see* Table I), and in the β-1,3-glucan side chains of yeast glucan (Figure 2). The enzymes from callus cultures (*69, 137*) and germinating barley (*131*) were reported to have no action on samples of mixed-linked glucans, as detected by either reductometric or chromatographic assays and, in the case of barley, also by viscometric assay. The different results for the various enzymes with mixed-linked glucans may reflect differences in sensitivity of the viscometric and reductometric or chromatographic assays. On the other hand, they could be caused either by differences in certain aspects of enzyme specificity or by differences in the linkage arrangement of the samples of glucans tested.

There appear to be variations among samples of cereal glucans in the proportions of adjacent 1,3-linked glucose residues. For example, application of the Smith degradation procedure (71) to certain oat and barley (130) glucans showed runs of up to four adjacent 1,3-linkages, but other samples of these glucans analyzed in the same way showed no evidence of adjacent 1,3-linkages (98, 130).

Endo-enzymes with a similar specificity for runs of β-1,4-glucosyl residues may also occur and a fraction, designated F-I, from a *Trametes sanguinea* culture (97), which was without action on a barley glucan, but hydrolyzed CM-cellulose to give polymeric products, probably belongs to this class. The use of mixed-linked glucans as test substrates, may therefore provide additional information on specificity differences between the β-1,4-glucan endo-hydrolase components, which have been described in many preparations—e.g., *T. viride* (159, 202), *I. lacteus* (222), etc.

The β-glucan endo-hydrolases listed in Table IV show the wide range of structural requirements of the glycosyl component of the glucosidic linkage split. No precise information is available on the structural requirements of the glucan chain forming the aglycone component of the glucosidic linkage, but it is likely to be as specific as for the glycosyl portion. For enzymes of group 4 (Table IV), this requirement is likely to be at least G 3 G.

It is worth noting here that among the α-glucan endo-hydrolases, there exist enzymes exhibiting a range of substrate specificities comparable with those of the β-glucan endo-hydrolases. For example, the α-amylases hydrolyze only α-1,4-glucans (228), and the dextranases only α-1,6-glucans (26), and may be compared with the enzymes of group 4 in Table IV. On the other hand, mycodextranase (154, 155, 187, 218, 219) is apparently without action on α-1,4- or α-1,3-glucans, but cleaves the α-1,3; α-1,4-glucan, nigeran, at the α-1,4-linkages as shown
$$\downarrow \quad\quad \downarrow$$
—G 3 G 4 G 3 G 4 G—. Mycodextranase thus has a specificity like the enzymes in group 2, Table II.

Action of Endo-enzymes on Substituted Glucans. The action of a number of β-1,4-glucan endo-hydrolases on substituted celluloses has led to the general conclusions that as the degree of substitution (D.S.) of the substrate is raised, the hydrolysis rate is lowered and that the identity of the substituent is not of prime significance in determining the rate of attack (74, 183, 192, 204, 205). Certain derivatives—e.g., sulfoethyl—however, were resistant to attack at very low D.S. (0.28 or 0.078), while in other instances, highly substituted derivatives (D.S. 1.2–1.3) were hydrolyzed (114). The latter observation has been attributed to the non-homogeneity of substitution of the substrate (183).

Table IV. Classification of β-1,3⁻, β-1,4⁻,

Probable Minimum Glycosyl Requirement	Linkage Split	β-Glucan Substrate
1. ...G 4 G...	1,4 or ?1,3	1,4
		1,4; 1,3
2. ...G 4 G 3 G...	1,4	1,4; 1,3
3. ...G 3 G...	1,3 or 1,4	1,3
		1,4; 1,3
4. ...G 3 G 3 G...	1,3	1,3

ᵃ Not from *T. viride*.

Holden and Tracey (92) pointed out a relationship between the D.S. of substituted cellulose ethers and their susceptibility to hydrolysis by *Helix pomatia* enzymes. Assuming that only linkages between unsubstituted residues were split, they found that the calculated degree of hydrolysis was in agreement with the experimental value. This agreement was better for CM-cellulose than for EM-cellulose (ethylmethylcellulose). Subsequently, studies with other cellulase preparations were made (*36, 124, 133*), which confirmed and extended these findings. Klop and Kooiman (*114*) found that the identity of the isolated products of the hydrolysis of HE- and CM-cellulose by a *Myrothecium verrucaria* preparation was consistent with the assumption that the glucosidic linkage of an unsubstituted glucosyl residue is susceptible to enzymic attack, provided the aglycone portion is either unsubstituted or is a 6-substituted glycosyl residue.

Klop and Kooiman (*114*) have also explored the effect of xylosyl and galactosyl-xylosyl substituents on the hydrolysis of the β-1,4-glucan backbone of *Tamarindus* amyloid (*see* Figure 2). Again the products of hydrolysis are consistent with the assumption that linkages between unsubstituted glucosyl residues and substituted glucosyl residues may be broken. Similar results were obtained with the Luizym preparation,

and β-1,3⁻; 1,4⁻-Glucan Endo-hydrolases

Products	Examples	Reference
(G) G 4 G G 4 G 4 G G 4 G G 3 G 4 G G 4 G 3 G 4 Gᵃ G 3 G 4 G 4 G	Streptomyces QM B814 Aspergillus niger Trichoderma viride	(163, 169, 193) (45, 46) (97)
G 4 G 3 G G 4 G 4 G 3 G	Bacillus subtilis germinating barley	(146, 191) (37, 129, 130, 132)
(G) G 3 G G 3 G 3 G G 3 G G 4 G 3 G G 4 G 4 G 3 G	Rhizopus arrhizus Schizophyllum commune Bacillus circulans	(49, 163, 169, 185) (201) (94, 96, 213)
(G) G 3 G G 3 G 3 G G 3 G 3 G 3 G etc.	Pepper callus Nicotiana glutinosa leaves germinating barley wheat and barley root callus	(137) (145) (131) (69)

except that with this enzyme the *Tamarindus* amyloid was hydrolyzed more extensively. This difference could be explained if the Luizym preparation contained enzymes capable of removing the xylosyl-galactose side branches of the amyloid.

Hydrolysis of linear β-1,3-glucans by *Rhizopus arrhizus* endo-enzyme is markedly reduced by substitution of glucose residues, such as that present in *Claviceps* glucan (Figure 2). Perlin and Taber (*170*) suggested that the 6-substituted glucosyl residues hindered the approach and alignment of the enzyme into a productive enzyme-substrate complex. The total lack of hydrolysis of the *Claviceps fusiformis* polysaccharide by the *Nicotiana* enzyme (*145*) could be attributed to the inability of the substrate to meet the specific binding-site requirements of the enzyme (*see* Table IV), owing to the frequency of 6-glucosyl substitution. It will be noted that among the glucans with the *Claviceps* glucan-type structures listed in Table I, there is a range of degree of substitution and the *C. fusiformis* polysaccharide has the lowest recorded.

Fleming, Manners, and Masson (*62*) found that a β-1,3-glucan endo-hydrolase from a bacterial source was able to hydrolyze laminarin to glucose, laminaribiose, laminaritriose, and other oligosaccharides. The identity of the oligosaccharides indicated that the enzyme could bypass β-1,6-interchain linkages, but could not hydrolyze certain β-1,3-linkages adjacent to the interchain linkage or to a nonreducing mannitol end group.

Specificity for Monomer Type in Substrate. The availability of linear β-mannans, β-glucomannans, β-galactans, and β-xylans with various types of linkages (216, 217) makes it possible to test the effect of alteration of the configuration or size of the monomer unit on the action of various β-glucan hydrolases. To date, few purified enzymes have been tested in this way, but the results for several β-glucan hydrolases indicate that a high specificity for glucose polymers is the general rule. Thus, a β-1,4-glucan hydrolase from *Myrothecium verrucaria* did not produce dialyzable products from the β-1,4-glucomannan from jack pine (167).

A purified β-1,4-glucan endo-hydrolase from *A. niger* (45) was unable to hydrolyze a β-1,4-galactan, a β-1,4-xylan, pneumococcal type III polysaccharide, chitin, or hyaluronic acid. These substrates contain β-1,4-glycosidic linkages and therefore it is apparent that the enzyme cannot accommodate changes in glucosyl residues to galactosyl, xylosyl, glucuronosyl, or N-acetyl aminoglucosyl residues.

Significance of "More Random" and "Less Random" Action Patterns. Gilligan and Reese (70) showed that the different "cellulase" fractions from a *Trichoderma viride* culture filtrate, obtained by calcium phosphate gel chromatography, could be distinguished by the slope of the curves obtained by plotting the change in fluidity (ϕ) of CM-cellulose in the presence of each fraction, against the corresponding increase in reducing sugars. Nisizawa and colleagues (152) determined the $d\phi/dRS$ relationships for a series of β-1,4-glucan endo-hydrolase fractions from *Irpex lacteus* and described those enzymes which gave steep curves (similar to that obtained for HCl hydrolysis) as "more random" and the flatter curves as "less random." The members of groups of β-1,4-glucan hydrolases from *Pseudomonas fluorescens* (152), *Trichoderma viride* (159), *Trichoderma koningi* (99) and the hepatopancreas of *Dolabella* sp. (158), a marine mollusc, showed similar variations in the slope of the curves.

The varying relationships between changes in fluidity and accompanying release of reducing sugars suggest differences in the pattern of attack of the enzyme fractions on the CM-cellulose substrate. The explanation for these differences must lie in the nature of events occurring at the active sites of the enzymes. It is possible that following the initial hydrolytic event at an internal linkage, the residual portions of the polysaccharide may diffuse away, to be later replaced by another substrate molecule. On the other hand, one of the fragments of the polymer substrate could be retained at the surface of the enzyme and become involved in another enzyme-substrate association, which could lead to a further catalytic event. By repetition of this process, multiple attack on fragments derived from the one molecule could occur.

In the first case, the random attack on different molecules could lead to the steep $d\phi/dRS$ curves described as "more random" by Nisizawa. In the second case, where more than one attack occurs on successive fragments of a single molecule, $d\phi/dRS$ would be lower and would give the curves described as "less random." In other words, the difference in the pattern of action of the enzymes, as measured by the slope of the $d\phi/dRS$ curves, would depend upon the numbers of linkages cleaved per encounter between enzyme and substrate.

At present, there is no experimental evidence available to define the nature of the events at the active site, during and after the cleavage of the glycosidic link by β-glucan endo-hydrolases. However, French has briefly reported (3) the situation for pig pancreatic α-amylase cleaving an amylose substrate and has presented evidence that this enzyme has an action which involves more than one attack per encounter. On the other hand, the action of the α-amylase from *Aspergillus oryzae* involves only one attack per encounter. French and Robyt (67) have computed the number of attacks per encounter for pig pancreatic amylase as eight and have shown that the principal points of attack are the second and third glycosidic bonds from the reducing end of the amylose substrate. Thus, for pig pancreatic amylase, although the initial attack may be random, the continued attack is not and the polarity of the attack is towards the nonreducing end of the substrate. This would indicate that the portion of the original substrate bearing the nonreducing end is translocated on the enzyme surface and repeatedly attacked.

The mechanism by which "multiple" attack takes place is not at all clear, although the finding that pig pancreatic amylase has two apparently independent binding sites for substrates and forms multimolecular enzyme-substrate complexes (127) suggests a way in which enzyme and substrate could be constrained to continue the association necessary for "multiple" attack.

It is possible that the formation of multimolecular enzyme-substrate complexes could also explain the results of Perlin's interesting experiments (168) with the *Streptomyces* "cellulase" and the *Rhizopus arrhizus* "laminarinase." These enzymes were incubated with lichenin, in varying proportions (cellulase:laminarinase = 0.5:1, 1:1, and 1:2) and the oligosaccharide products were analyzed. The results indicated that only the cellulase was attacking the lichenin, although independently the two enzymes showed little difference in ability to hydrolyze lichenin. When the ratio of cellulase to laminarinase was reduced to 1:4 or 5, the *Rhizopus* enzyme functioned in the mixed system. Control experiments appear to rule out irreversible inactivation of the *Rhizopus* enzyme by the *Streptomyces* enzyme, the presence of selective inhibitors in the cellulase preparation, or inhibition of the *Rhizopus* enzyme by products of cellulase hydrolysis. The suppression of the laminarinase activity in the presence of the cellulase could be explained by the formation of cellulase-lichenin complexes of the type described for the amylose-amylase system, which would effectively remove the lichenin substrate from the *Rhizopus* enzyme.

Analysis of Oligosaccharide End Products

The analysis of the oligosaccharides found in enzymic hydrolysates of polymeric substrates during the course of the reaction and after prolonged incubation, provide useful evidence concerning the action pattern of the hydrolases. The interpretation of these results may be complicated by the presence of traces of other enzymes, especially where the hydrolyses are prolonged.

This may be illustrated by the results of investigations using a fractionated *Aspergillus niger* β-1,4-glucan hydrolase acting on mixed-linked glucans (46), where the final hydrolysis products were glucose and three oligosaccharides. When trace amounts of a contaminating β-1,3-glucan exo-hydrolase were removed by treatment with laminarin, several additional oligosaccharides of longer chain length were found as products of prolonged hydrolysis. This example shows the importance of examining preparations for other enzymes with activities which could alter the final pattern of oligosaccharide products and thus lead to errors in the interpretation of the specificity of the enzyme.

A further example (17, 18, 19) is provided by a *Euglena gracilis* enzyme preparation acting on β-1,3-glucan substrates, where glucose was the sole chromatographically detectable product, suggesting the presence of an exo-hydrolase. However, independent experiments with CM-pachyman as a viscometric substrate and with the use of both the specific glucose oxidase method and of the Somogyi-Nelson method to follow the progress of the reaction, indicated that some endo-hydrolase action was also occurring. This endo-hydrolase could be inactivated without altering the exo-hydrolase activity, by preparing the *Euglena* extracts at a lower pH. This situation, where the presence of one enzyme may be masked by the action of a second, also emphasizes the necessity for examining the specificity properties of an enzyme preparation by more than one method.

The patterns of end products of hydrolysis of homogeneous glucans by various endo-hydrolases of known purity show differences which are dependent on the specificity of the enzymes. For example, purified β-1,3-

glucan endo-hydrolases from *Rhizopus* and from *Nicotiana* produce different end products in the hydrolysis of paramylon (see Table IV). These results are indicative of different binding site specificities for CL of substrate and this conclusion is supported by evidence, previously discussed, for their action on mixed-linked glucans.

The action of purified glucan hydrolases on the lower members of the oligoglucoside series (*143*, *229*), their reduced analogues (*230*) and methyl- (*81*) or *p*-nitrophenyl-β-oligoglucoside derivatives (*152*) has provided information on the susceptibility to cleavage of the various linkages in these oligosaccharides. Thus, the *M. verrucaria* (*81*, *229*, *230*), *I. lacteus* (*152*, *222*), *T. viride* (*122*, *153*, *160*), *A. niger* (*45*), and *Stereum sanguinolentum* (*23*) β-1,4-endo-hydrolases have been examined in this way and it has been generally concluded that they cleave the internal linkages more readily than terminal linkages. The extension of such studies to longer members (CL 7 and upwards) of the homologous series is limited by the availability of substrates in the case of insoluble oligosaccharides, owing to the difficulties of fractionation. However, two investigations using cellodextrins (average CL 24) confirm the results for the short chain oligosaccharides for *M. verrucaria* (*231*) and fractions from *T. viride* (*153*, *160*).

In the case of certain other endo-hydrolases [α-amylases (*76*), lysozyme (*195*)], it is relevant that as the CL of the oligosaccharide substrate increases, the hydrolytic attack occurs preferentially at linkages nearer the reducing end, indicating some asymmetry of the specificity requirements of the glycosyl and aglycone components of the glucosidic linkage attacked. It is likely then, that the use of the lower members of the homologous series could give an incomplete picture of substrate specificity requirements. The change in linkage preference as the substrate chain length increases has been explained for lysozyme (*197*) (see *Comparison of β-Glucan Endo-hydrolases with Lysozyme*).

Enzymic transglycosylation, which may accompany the action of glycoside hydrolases (*102*, *184*), and in some circumstances exo- (*165*) and endo-hydrolase action (*23*, *91*, *176*, *226*), could produce oligosaccharide patterns during the hydrolysis of glucan substrates, which would lead to possible misinterpretation of enzyme specificity.

Transglycosylation by glycoside hydrolases (*102*) is well established and various hydroxyl compounds including mono- and oligosaccharides are known to be acceptors. Jermyn (*104*) has shown that there is not only high specificity for the substrate, but also for the acceptor and that competition between hydroxyl compounds for the acceptor site can occur.

The two exo-hydrolases for β-glucans which have so far been examined (*17*, *18*, *209*) showed no transfer activity. However, purified α-1,4-glucan gluco-hydrolases from *A. niger* and *R. delamar* (*165*) when tested in the presence of a high glucose concentration, were shown to produce isomaltose. Occurrence of transferase activity among β-glucan endo-hydrolases is variable. Enzymes from A. *niger* (*45*) and *Irpex lacteus* (*152*) have been tested and showed no ability to transfer from glucan substrates to glucose acceptors. On the other hand, a preparation from *Pseudomonas fluorescens* (*152*) showed transglucosylase activity from cellodextrin or cellotetraose to glucose and a *Stereum sanguinolentum* preparation (*23*) catalyzes transglycosylation from 2% cellotetraose to yield insoluble oligosaccharides. Some other endo-hydrolases, lysozyme (*118, 176*) and hyaluronidase (*91, 226*), have been shown to exhibit transferase activity.

Comparison of β-Glucan Endo-hydrolases with Lysozyme

Lysozyme is unique among the polysaccharide hydrolases in that its chemical (*107*) and physical (*24, 25, 173, 197*) structure is accurately known and the mechanism of its hydrolytic action explained in considerable detail (*182, 195, 221*). Comparison of the characteristics of lysozyme action (Table V) with those of the β-glucan endo-hydrolases (Table II) shows that these enzymes share many common features.

From x-ray crystallographic investigations of lysozyme (*24, 25, 173, 197*), Phillips and coworkers deduced the detailed three-dimensional structure of the lysozyme molecule and of its complex with substrate analogues (*24, 84*). By these techniques the substrate binding site was shown to be a deep cleft in the surface of the molecule, long enough to accommodate a portion of the polysaccharide substrate consisting of six hexopyranosyl residues. Each residue has its individual binding sub-site associated with particular amino acids lining the cleft. The catalytic amino acids have also been identified and are situated in the active site so that cleavage of a bound hexasaccharide liberates a disaccharide from its reducing end.

Data on the binding of molecules to lysozyme in solution, obtained from studies based on fluorescence spectroscopy (*39*), equilibrium dialysis (*39*), nuclear magnetic resonance (*47, 51, 52, 181*), difference spectra (*50, 196*), difference spectropolarimetry (*4*), and temperature jump relaxation methods (*41*) have led to conclusions about the characteristics of the individual sub-sites, which are generally consistent with and support the concepts of lysozyme action developed on the basis of crystallographic studies. It follows that the structure of the active site of the lysozyme molecule in solution and in the crystal are very similar. This conclusion is further supported by tritium exchange data (*178*) which suggest that the organization of the molecule as a whole is unaltered by the change from the crystalline state to solution.

Table V. Features of Lysozyme-catalyzed Reactions

Site of glycosidic bond cleavage	Between C_1–0 (195)
Structural requirements of substrates	Substrates [... NAMβ(1 → 4) NAGβ(1 → 4) ...]$_n$ [... NAGβ(1 → 4) NAGβ(1 → 4) ...]$_n$ chitin (177) Not substrates [... GAβ(1 → 4) GAβ(1 → 4) ...]$_n$ chitosan (227) [... Gβ(1 → 4) Gβ(1 → 4) ...]$_n$ cellulose (227)
Rate of hydrolysis of NAG-oligomers	tri < tetra <<<< penta << hexa (40, 197)
Transfer activity	+ (40, 161, 176, 197, 198, 203, 238) configuration retained (40, 182, 197)
Inhibitors	NAG, di-NAG, tetra-NAG, di-NAG-NAM and higher oligosaccharides, but not cellobiose or glucosamine (24)
Abbreviations:	NAM, N-acetyl muramic acid NAG, N-acetyl glucosamine GA, glucosamine G, glucose

Certain properties of lysozyme, which are common to β-glucan endo-hydrolases, have been explained following the elucidation of the structure of the active site. It is therefore relevant to discuss the current concepts of lysozyme structure and action and to compare these with the rather fragmentary observations so far made on the action of the various β-glucan endo-hydrolases.

(1) The binding site of lysozyme consists of six independent sub-sites, each binding a hexopyranosyl residue. Thermodynamic binding studies show that three of these sub-sites interact very strongly with the three residues at the reducing end of a hexaccharide substrate, while the remaining sites contribute only slightly to substrate binding (196). In contrast, kinetic studies indicate that more than three residues must interact with the active site during the reaction, since the relative rates of hydrolysis of trimer, tetramer, pentamer, and hexamer are 1:8:4000:30,000 (40, 197).

Binding constants for the substrates of β-glucan endo-hydrolases are not known, but the kinetics of the splitting of β-1,4-oligoglucosides are similarly dependent on CL, suggesting that multiple sub-site binding is a general property of these endo-hydrolases (81, 122, 229, 230).

(2) The sub-sites involved in binding of the lysozyme substrates do not have equivalent affinities for the two types of hexopyranosyl residues present. Thus, the binding of N-acetyl glucosamine or N-acetyl muramic acid residues is favored at certain sites and forbidden at others and since

the catalytic amino acids are adjacent to a compulsory N-acetyl muramic acid binding site, a muramyl glycosidic linkage is always hydrolyzed (*24, 39*). In the same way, differences in sub-site affinities could account for the varied specificities of β-1,3-, β-1,4-, and β-1,3;1,4-glucan hydrolases (*see* Table IV).

(3) The asymmetric placement of the catalytic amino acids with respect to the binding sub-sites of lysozyme explains why the hydrolysis of a hexasaccharide substrate occurs predominantly at the second glycosidic linkage from the reducing end (*192*). The patterns of products of β-1,4-glucan endo-hydrolases acting on β-1,4-oligoglucosides similarly indicate that certain linkages are cleaved in preference to others (*81, 122, 152, 229*). The hydrolysis of specific linkages must, as in the case of lysozyme, reflect the positioning of the catalytic amino acids with respect to the binding sub-sites in the active centers of these enzymes.

In the case of lysozyme hydrolyzing oligosaccharide substrates at high concentrations, linkages are cleaved at positions other than those normally preferred. This has been interpreted as the result of the simultaneous binding of parts of two oligosaccharide molecules in the active site cleft (*197*). The dependence of cleavage pattern on the concentration of oligosaccharide has not been examined for β-glucan endo-hydrolases.

(4) The specific structural requirements for lysozyme substrates are clearly explained by the precise relationships of the amino acid residues lining the active site to the substituents on the hexopyranosyl units of the bound substrate. By inspection of the three-dimensional enzyme-substrate structure it can be seen that N-acetyl groups are required for binding so that chitin is a substrate (*177*) but chitosan and cellulose are not (*227*).

At present none of the β-glucan endo-hydrolases have been examined crystallographically, nor have their amino acid sequences been determined, so that their three-dimensional structures cannot yet be deduced. Without this information it is not possible to relate substrate or inhibitor specificity to amino acid orientation at the active site.

(5) Examination of the lysozyme-substrate model (*24*) suggests that the catalytic amino acids at the active site are glutamate (residue 35) and aspartate (residue 52). These carboxylic amino acid residues are situated close to the glycosidic linkage hydrolyzed and it has been proposed (*25*) that they participate in an acid-base catalyzed hydrolysis mechanism. On the basis of transglycosylation experiments with synthetic lysozyme substrates, Raftery and Rand-Meir (*182*) were able to eliminate a number of possible mechanistic pathways and concluded that the most likely was either a glycosyl carbonium ion mechanism (*25, 198, 221*), giving rise stereospecifically to β-anomeric products, or a double displacement mechanism, also resulting in retention of configuration (*117*).

An observation basic to the elucidation of the mechanism of hydrolysis of lysozyme substrates is that the glycosidic linkages are cleaved at the C_1—O bond (*195*). No similar information is available for the β-glucan endo-hydrolases. Another key to the description of the hydrolytic mechanism of lysozyme is the observed retention of the configuration of the substrate by the products of transfer (*40, 182, 197*). Such retention of configuration is a feature of all β-glucan endo-hydrolases so far investigated in this way, and hence the suggestions made for the lysozyme mechanism have direct relevance to their mode of action.

(6) The formation of higher oligosaccharides during lysozyme action on oligosaccharide and glycoside substrates is well documented (*40, 118, 161, 176, 197, 198, 203, 238*). Analysis of these reactions shows that transglycosylation is occurring, possibly *via* a metastable glycosyl intermediate rather than by the reversal of hydrolysis (*40*). β-Glucan endo-hydrolases do not universally exhibit transferase action (*see* Table II, and the last part of *Analysis of Oligosaccharide End Products*), although a few apparently resemble lysozyme in this respect (*23, 152*).

The transfer activity observed for individual hydrolases would, in terms of the glycosyl intermediate mechanism, depend on the relative rates at which these intermediates are attacked by water or by acceptor oligosaccharides and glycosides (*40*) and on the affinity of the acceptors for the remaining binding sites on the enzyme (*40, 103*).

(7) Competitive inhibition of lysozyme by potential oligosaccharide substrates has been explained as being caused by their non-productive binding at sites adjacent to, but not across, the bond-breaking site (*24*). The inhibition of certain β-1,4-glucan endo-hydrolases by cellobiose and lactose (*138*) may have an analogous physical explanation.

N-acetyl glucosamine is also an inhibitor of lysozyme, but in contrast to the oligosaccharide inhibitors, its action is due to binding across, rather than in, the cleft of the active site (*24*). Monosaccharides are known to inhibit some β-glucan endo-hydrolases (*202*) and a physical explanation of this kind is possible, but would be difficult to confirm without recourse to crystallographic methods.

This brief comparative resume shows clearly that a number of facets of the action of β-glucan endo-hydrolases may be explained by reference to an enzyme model based on the structure found for lysozyme. The crystallographic data which provide the basis for the three-dimensional model of lysozyme have been supplemented by information on enzyme-substrate interactions in solution (*4, 19, 39, 41, 47, 51, 52, 181, 196*). Since it is not yet possible to examine the β-glucan endo-hydrolases crystallographically, the methods which have been used to study lysozyme-substrate interactions in solution will be of primary importance in the elucidation of the action of these enzymes. Already certain β-1,4-glucan endo-hydrolases have been examined using some of these techniques (*57, 171, 172*) but wider application of such methods is needed.

None of the exo-hydrolases, for α- or β-glucans, have yet been studied crystallographically so that the physico-chemical basis for enzyme-substrate interactions of these enzymes is unknown. For the present, investigations will be guided by schematic mechanisms such as that proposed by Thoma and Koshland for the exo-hydrolase, β-amylase, which explains certain features of its substrate and inhibitor behavior (*214*). Koshland has also suggested a reaction mechanism to account for the inversion of configuration of the glycosidic linkage of the substrate (*117*), which has been observed for β-amylase and is a general property of other glucan exo-hydrolases (*see* "General Properties" in *β-Glucan Exo-hydrolases*).

Conclusion

The foregoing summary of enzyme-substrate relationships among β-glucan hydrolases makes it apparent that, except in a few instances, our knowledge of enzyme properties including substrate specificities, is incomplete. Nevertheless, it is possible to classify the known hydrolases into groups of apparently similar types. The main division of glucan hydrolases into exo- and endo-types is firmly established, but within these sub-groups, the evidence suggests that a wide variety of specificities exists. Certain enzymes, such as the exohydrolases *Cellvibrio gilvus* (*48, 209, 210*) and *T. viride* (*122*) do not fit easily into the classification scheme and such exceptions suggest that, as further knowledge is gained using new techniques, this tentative classification scheme will need revision.

For the enzymologist, information on enzyme specificity is all important for an accurate description of enzyme action. However, biologists are interested in the functional aspects of enzymes in living systems, as for example their role in (a) germination processes in cereal grains (*128*), (b) mobilization of storage glucans such as paramylon or laminarin (*19*), (c) changes in the yeast cell wall during the budding process (*207*), (d) morphogenetic changes during the development of plants (*207*), (e) the digestion of plant cell wall polysaccharides in the rumen or caecum of herbivores and in the soil (*207*), and (f) enabling parasites to gain access to cells (*234*). In such situations, the patterns of specificity which emerge from the studies outlined in this chapter assume their fullest meaning only when considered in relation, firstly to other enzymes present, secondly to their changes during the life cycle of the organism or in response to environmental variations, thirdly to their localization within cells, on cell surfaces or extracellularly, and finally in relation to their selective advantage for the cell or organism in its natural environment.

Acknowledgment

The authors wish to acknowledge financial support from the Australian Research Grants Committee.

Literature Cited

(1) Abd-El-Al, A. T. H., Phaff, H. J., *Biochem. J.* **109**, 347 (1968).
(2) Abdullah, M., Fleming, I. D., Taylor, P. M., Whelan, W. J., *Biochem. J.* **89**, 35P (1963).
(3) Abdullah, M., French, D., Robyt, J. F., *Arch. Biochem. Biophys.* **114**, 595 (1966).
(4) Adkins, B. J., Yang, J. T., *Biochemistry* **7**, 266 (1968).

(5) Almin, K. E., Eriksson, K. E., *Biochim. Biophys. Acta* **139**, 238 (1967).
(6) Almin, K. E., Eriksson, K. E., *Arch. Biochem. Biophys.* **124**, 129 (1968).
(7) Annan, W. D., Hirst, E., Manners, D. J., *J. Chem. Soc.* **1965**, 220.
(8) *Ibid.*, **1965**, 885.
(9) Archibald, A. R., Cunningham, W. L., Manners, D. J., Stark, J. R., Ryley, J. F., *Biochem. J.* **88**, 444 (1963).
(10) Aspinall, G. O., Kessler, G., *Chem. Ind. (London)* **1957**, 1296.
(11) Baddiley, J., Buchanan, J. G., Thain, E. M., *J. Chem. Soc.* **1953**, 1944.
(12) Bailey, J. M., French, D., *J. Biol. Chem.* **226**, 1 (1957).
(13) Bailey, J. M., Whelan, W. J., *Biochem. J.* **67**, 540 (1957).
(14) Bailey, R. W., "Oligosaccharides," Pergamon Press, Oxford, 1965.
(15) Bailey, R. W., Roberton, A. M., *Biochem. J.* **82**, 272 (1962).
(16) Barker, S. A., Edwards, J. M., Somers, P. J., Repas, A., *Carbohydrate Res.* **6**, 341 (1968).
(17) Barras, D. R., *Ph.D. Thesis*, University of Melbourne (1968).
(18) Barras, D. R., Stone, B. A., *Proc. Australian Biochemical Soc.* **1968**, 9.
(19) Barras, D. R., Stone, B. A., "The Biology of Euglena," Vol. 2, Chapter 7, p. 149, D. E. Buetow, Ed., Academic Press, New York, 1968.
(20) Beattie, A., Hirst, E. L., Percival, E., *Biochem. J.* **79**, 531 (1961).
(21) Ben-Gershom, E., Leibowitz, J., *Enzymologia* **20**, 148 (1958).
(22) Bishop, C. T., Blank, F., Gardiner, P. E., *Can. J. Chem.* **38**, 869 (1960).
(23) Björndal, H., Eriksson, K. E., *Arch. Biochem. Biophys.* **124**, 149 (1968).
(24) Blake, C. C. F., Johnson, L. N., Mair, G. A., North, A. C. T., Phillips, D. C., Sarma, V. R., *Proc. Roy. Soc. London, Series B* **167**, 378 (1967).
(25) Blake, C. C. F., Mair, G. A., North, A. C. T., Phillips, D. C., Sarma, V. R., *Proc. Roy. Soc. London, Series B* **167**, 365 (1967).
(26) Bourne, E. J., Hutson, D. H., Weigel, H., *Biochem. J.* **85**, 158 (1962).
(27) Bouveng, H. O., *Phytochem.* **2**, 341 (1963).
(28) Bouveng, H. O., Kiessling, H., Lindberg, B., McKay, J., *Acta Chem. Scand.* **17**, 1351 (1963).
(29) Bouveng, H. O., Lindberg, B., *Advan. Carbohydrate Chem.* **15**, 53 (1960).
(30) Brobst, K. M., Lott, C. E., *Cereal Chem.* **43**, 35 (1966).
(31) Brock, T. D., *J. Cell Biol.* **23**, 15A (1964).
(32) Brock, T. D., *Biochem. Biophys. Res. Comm.* **19**, 623 (1965).
(33) Brown, R. G., Lindberg, B., *Acta Chem. Scand.* **21**, 2379 (1967).
(34) Buck, K. W., Chen, A. W., Dickerson, A. G., Chain, E. B., *J. Gen. Microbiol.* **51**, 337 (1968).
(35) Bull, A. T., Chesters, C. G. C., *Advan. Enzymology* **28**, 325 (1966).
(36) Cayle, T., *Wallerstein Lab. Commun.* **25**, 349 (1962).
(37) Chen, S. C., Luchsinger, W. W., *Arch. Biochem. Biophys.* **106**, 71 (1964).
(38) Chiba, S., Shimomura, T., *Agr. Biol. Chem.* **29**, 486 (1965).
(39) Chipman, D. M., Grisaro, V., Sharon, N., *J. Biol. Chem.* **242**, 4388 (1967).
(40) Chipman, D. M., Pollock, J. J., Sharon, N., *J. Biol. Chem.* **243**, 487 (1968).
(41) Chipman, D. M., Schimmel, P. R., *J. Biol. Chem.* **243**, 3771 (1968).
(42) Clarke, A. E., Stone, B. A., *Biochim. Biophys. Acta* **44**, 161 (1960).
(43) Clarke, A. E., Stone, B. A., *Phytochem.* **1**, 175 (1962).
(44) Clarke, A. E., Stone, B. A., *Rev. Pure Appl. Chem.* **13**, 134 (1963).
(45) Clarke, A. E., Stone, B. A., *Biochem. J.* **96**, 802 (1965).
(46) *Ibid.*, **99**, 582 (1966).
(47) Cohen, J. S., Jardetzky, O., *Proc. Natl. Acad. Sci.* **60**, 92 (1968).
(48) Cole, F. E., King, K. W., *Biochim. Biophys. Acta* **81**, 122 (1964).
(49) Cunningham, W. L., Manners, D. J., *Biochem. J.* **90**, 596 (1964).

(50) Dahlquist, F. W., Jao, L., Raftery, M., *Proc. Natl. Acad. Sci.* **56**, 26 (1966).
(51) Dahlquist, F. W., Raftery, M. A., *Biochemistry* **7**, 3269 (1968).
(52) *Ibid.*, **7**, 3277 (1968).
(53) Drake, B., *Biochem. Z.* **313**, 388 (1943).
(54) Ebata, J., Satomura, Y., *Agr. Biol. Chem.* **27**, 478 (1963).
(55) Eberhart, B., Cross, D. F., Chase, L. R., *J. Bact.* **87**, 761 (1964).
(56) Ebert, E., Zenk, M. H., *Phytochem.* **6**, 309 (1967).
(57) Eriksson, K. E., Pettersson, G., *Arch. Biochem. Biophys.* **124**, 160 (1968).
(58) Eveleigh, D. E., *Can. J. Microbiol.* **13**, 727 (1967).
(59) Fellig, J., *Science* **131**, 832 (1960).
(60) Fleming, M., Hirst, E., Manners, D. J., "Proceedings of the Fifth International Seaweed Symposium," p. 255, Pergamon Press, Oxford, 1966.
(61) Fleming, M., Manners, D. J., *Biochem. J.* **94**, 17P (1965).
(62) Fleming, M., Manners, D. J., Masson, A. J., *Biochem. J.* **104**, 32P (1967).
(63) Ford, C. W., Percival, E., *J. Chem. Soc.* **1965**, 3014.
(64) *Ibid.*, **1965**, 7035.
(65) French, D., "The Enzymes," 2nd ed., Vol. 4, p. 345, P. D. Boyer, H. Lardy, K. Myrbäck, Eds., Academic Press, New York, 1960.
(66) French, D., Abdullah, M., *Cereal Chem.* **43**, 555 (1966).
(67) French, D., Robyt, J. F., *7th Intern. Congr. of Biochem. (Tokyo)*, p. 795, F-188 (1967).
(68) French, D., Robyt, J. F., Weintraub, M., Knock, P., *J. Chromatog.* **24**, 68 (1966).
(69) Gamborg, O. L., Eveleigh, D. E., *Can. J. Biochem.* **46**, 417 (1968).
(70) Gilligan, W., Reese, E. T., *Can. J. Microbiol.* **1**, 90 (1954).
(71) Goldstein, I. J., Hay, G. W., Lewis, B. A., Smith, F., "Methods of Carbohydrate Chemistry," Vol. 5, p. 361, R. L. Whistler, Ed., Academic Press, New York, 1965.
(72) Gorin, P. A. J., Spencer, J. F. T., Westlake, D. W. S., *Can. J. Chem.* **39**, 1067 (1961).
(73) Gottschalk, A., *Advan. Carbohydrate Chem.* **5**, 49 (1950).
(74) Greathouse, G. A., *Textile Res. J.* **20**, 227 (1950).
(75) Greenwood, C. T., Milne, E. A., *Die Stärke* **20**, 101 (1968).
(76) *Ibid.*, **20**, 139 (1968).
(77) Grimes, J. P., Nelson, T. E., Kirkwood, S., Scaletti, J. V., *Bacteriol. Proc.* **1966**, 84.
(78) Hamauzu, Z. I., Hiromi, K., Ono, S., *J. Biochem. (Tokyo)* **57**, 39 (1965).
(79) Handa, N., Nisizawa, K., *Nature* **192**, 1078 (1961).
(80) Hanson, K. R., *Biochemistry* **1**, 723 (1962).
(81) Hanstein, F. G., Whitaker, D. R., *Can. J. Biochem. Physiol.* **41**, 707 (1963).
(82) Harada, T., *Arch. Biochem. Biophys.* **112**, 65 (1965).
(83) Harada, T., Misaki, A., Saito, H., *Arch. Biochem. Biophys.* **124**, 292 (1968).
(84) Harte, R. A., Rupley, J. A., *J. Biol. Chem.* **243**, 1663 (1968).
(85) Hash, J. H., King, K. W., *J. Biol. Chem.* **232**, 381 (1958).
(86) Helferich, B., *Ergeb. Enzymforsch.* **7**, 83 (1938).
(87) Hellerqvist, C. G., Lindberg, B., Samuelsson, K., *Acta Chem. Scand.* **22**, 2763 (1968).
(88) Hiromi, K., Ogawa, K., Nakanishi, N., Ono, S., *J. Biochem. (Tokyo)* **60**, 439 (1966).
(89) Hiromi, K., Ono, S., *J. Biochem. (Tokyo)* **53**, 164 (1963).
(90) *Ibid.*, **61**, 654 (1967).
(91) Hoffman, P., Meyer, K., Linker, A., *J. Biol. Chem.* **219**, 653 (1956).
(92) Holden, M., Tracey, M. V., *Biochem. J.* **47**, 407 (1950).
(93) Horikoshi, K., *J. Biochem. (Tokyo)* **35**, 39 (1942).

(94) Horikoshi, K., Koffler, H., Arima, K., *Biochim. Biophys. Acta* **73**, 267 (1963).
(95) Huotari, F. I., Nelson, T. E., Smith, F., Kirkwood, S., *J. Biol. Chem.* **243**, 952 (1968).
(96) Igarashi, O., Igoshi, M., Sakurai, Y., *Agr. Biol. Chem.* **30**, 1254 (1966).
(97) Igarashi, O., Noguchi, M., Fujimaki, M., *Agr. Biol. Chem.* **32**, 272 (1968).
(98) Igarashi, O., Sakurai, Y., *Agr. Biol. Chem.* **30**, 642 (1966).
(99) Iwasaki, T., Hayashi, K., Funatsu, M., *J. Biochem. (Tokyo)* **57**, 467 (1965).
(100) Jermyn, M. A., *Australian J. Biol. Sci.* **8**, 577 (1955).
(101) Jermyn, M. A., *Science* **125**, 12 (1957).
(102) Jermyn, M. A., *Rev. Pure Appl. Chem.* **11**, 92 (1961).
(103) Jermyn, M. A., *Australian J. Biol. Sci.* **19**, 903 (1966).
(104) *Ibid.*, **20**, 1227 (1967).
(105) Jermyn, M. A. (to be published).
(106) Johnson, J., Kirkwood, S., Misaki, A., Nelson, T. E., Scaletti, J. V., Smith, F., *Chem. Ind. (London)* **1963**, 820.
(107) Jolles, P., *Proc. Roy. Soc. London, Series B* **167**, 350 (1967).
(108) Jones, D. M., *Advan. Carbohydrate Chem.* **19**, 219 (1964).
(109) King, K. W., *Virginia Agr. Expt. Sta. Tech. Bull.* **1961**, 154.
(110) King, K. W., Smibert, R. M., *Appl. Microbiol.* **11**, 315 (1963).
(111) Kitahara, M., Takeuchi, Y., *Gifu Daigaku Nogakubu Kenkyu Hokoku (Res. Bull. Fac. Agr. Gifu Univ.)* **11**, 127 (1959).
(112) Kitahara, M., Takeuchi, Y., *J. Agr. Chem. Soc. (Japan)* **35**, 468 (1961).
(113) *Ibid.*, **35**, 474 (1961).
(114) Klop, W., Kooiman, P., *Biochim. Biophys. Acta* **99**, 102 (1965).
(115) Kooiman, P., *Rec. Trav. Chim.* **80**, 849 (1961).
(116) Kooiman, P., *Phytochem.* **6**, 1665 (1967).
(117) Koshland, D. E., *Biol. Rev. Cambridge Phil. Soc.* **28**, 416 (1953).
(118) Kravchenko, N. A., *Proc. Roy. Soc. London, Series B* **167**, 429 (1967).
(119) Legler, G., *Z. Physiol. Chem.* **349**, 767 (1968).
(120) Legler, G., *Biochim. Biophys. Acta* **151**, 728 (1968).
(121) Li, L. H., Flora, R. M., King, K. W., *J. Ferment. Technol.* **41**, 98 (1963).
(122) Li, L. H., Flora, R. M., King, K. W., *Arch. Biochem. Biophys.* **111**, 439 (1965).
(123) Lindberg, B., McPherson, J., *Acta Chem. Scand.* **8**, 985 (1954).
(124) Lindenfors, S., *Acta Chem. Scand.* **16**, 1111 (1962).
(125) Lloyd, P. F., Pon, G., Stacey, M., *Chem. Ind. (London)* **1956**, 172.
(126) Lowe, G., Sheppard, G., Sinnott, M. L., Williams, A., *Biochem. J.* **104**, 893 (1967).
(127) Loyter, A., Schramm, M., *J. Biol. Chem.* **241**, 2611 (1966).
(128) Luchsinger, W. W., *Cereal Science Today* **11**, No. 2 (1966).
(129) Luchsinger, W. W., Chen, S. C., Richards, A. W., *Arch. Biochem. Biophys.* **112**, 524 (1965).
(130) Luchsinger, W. W., Chen, S. C., Richards, A. W., *Arch. Biochem Biophys.* **112**, 531 (1965).
(131) Luchsinger, W. W., Ferrell, W. J., Schneberger, G. L., *Cereal Chem.* **40**, 554 (1963).
(132) Luchsinger, W. W., Richards, A. W., *Arch. Biochem. Biophys.* **106**, 65 (1964).
(133) McClendon, J. H., *Biochim. Biophys. Acta* **48**, 398 (1961).
(134) Madsen, N. B., *Anal. Biochem.* **4**, 509 (1962).
(135) Maeda, M., Nisizawa, K., *J. Biochem. (Tokyo)* **63**, 199 (1968).
(136) Mahadevan, P. R., Eberhart, B., *Biochim. Biophys. Acta* **90**, 214 (1964).
(137) Mandels, M., Parrish, F. W., Reese, E. T., *Phytochem.* **6**, 1097 (1967).

(138) Mandels, M., Reese, E. T., "Advances in Enzymic Hydrolysis of Cellulose and Related Materials," E. T. Reese, Ed., p. 115, Pergamon Press, Oxford, 1963.
(139) Manners, D. J., Patterson, J. C., *Biochem. J.* **98**, 19C (1966).
(140) Manners, D. J., Ryley, J. F., Stark, J. R., *Biochem. J.* **101**, 323 (1966).
(141) Marinelli, L., Whitney, D., *J. Inst. Brew.* **73**, 35 (1967).
(142) Meeuse, B. J. D., *Basteria* **28**, 67 (1964).
(143) Miller, G. L., *Anal. Biochem.* **1**, 133 (1960).
(144) Misaki, A., Johnson, J., Kirkwood, S., Scaletti, J. V., Smith, F., *Carbohydrate Res.* **6**, 150 (1968).
(145) Moore, A. E., Stone, B. A., *Abstr. of Ann. Gen. Meeting Aust. Biochem. Soc.*, p. 117 (1967).
(146) Moscatelli, E. A., Ham, E. A., Rickes, E. L., *J. Biol. Chem.* **236**, 2858 (1961).
(147) Murakami, M., Winzler, R. J., *J. Chromatog.* **28**, 344 (1967).
(148) Nakamura, N., Ooyama, J., Tanabe, O., *J. Agr. Chem. Soc. (Japan)* **35**, 949 (1961).
(149) Nakamura, N., Tanabe, O., *Nature* **196**, 774 (1962).
(150) Nelson, T. E., Scaletti, J. V., Smith, F., Kirkwood, S., *Fed. Proc.* **22**, 297 (1963).
(151) Nelson, T. E., Scaletti, J. V., Smith, F., Kirkwood, S., *Can. J. Chem.* **41**, 1671 (1963).
(152) Nisizawa, K., Hashimoto, Y., Shibata, Y., "Advances in Enzymic Hydrolysis of Cellulose and Related Materials," E. T. Reese, Ed., p. 171, Pergamon Press, Oxford, 1963.
(153) Niwa, T., Kawamura, K., Nisizawa, K., *J. Ferment. Technol.* **43**, 286 (1965).
(154) Nordin, J. H., Hasegawa, S., Smith, F., Kirkwood, S., *Nature* **210**, 303 (1966).
(155) Nordin, J. H., Kirkwood, S., *Ann. Rev. Plant Physiol.* **16**, 393 (1965).
(156) Nordin, P., *Arch. Biochem. Biophys.* **99**, 101 (1962).
(157) Oi, S., Konishi, I., Satomura, Y., *Agr. Biol. Chem.* **30**, 266 (1966).
(158) Okada, G., Nisizawa, T., Nisizawa, K., *Biochem. J.* **99**, 214 (1966).
(159) Okada, G., Nisizawa, K., Suzuki, H., *J. Biochem. (Tokyo)* **63**, 591 (1968).
(160) Okada, G., Niwa, T., Suzuki, H., Nisizawa, K., *J. Ferment. Technol.* **44**, 682 (1966).
(161) Osawa, T., *Carbohydrate Res.* **7**, 217 (1968).
(162) Parrish, F. W., Perlin, A. S., *Nature* **187**, 1110 (1960).
(163) Parrish, F. W., Perlin, A. S., Reese, E. T., *Can. J. Chem.* **38**, 2094 (1960).
(164) Parrish, F. W., Reese, E. T., *Carbohydrate Res.* **3**, 424 (1967).
(165) Pazur, J. H., Okada, S., *Carbohydrate Res.* **4**, 371 (1967).
(166) Percival, E., *Carbohydrate Res.* **4**, 441 (1967).
(167) Perila, O., Bishop, C. T., *Can. J. Chem.* **39**, 815 (1961).
(168) Perlin, A. S., *Biochem. Biophys. Res. Comm.* **18**, 538 (1965).
(169) Perlin, A. S., Suzuki, S., *Can. J. Chem.* **40**, 50 (1962).
(170) Perlin, A. S., Taber, W. A., *Can. J. Chem.* **41**, 2278 (1963).
(171) Pettersson, G., Andersson, L., *Arch. Biochem. Biophys.* **124**, 497 (1968).
(172) Pettersson, G., Eaker, D. L., *Arch. Biochem. Biophys.* **124**, 154 (1968).
(173) Phillips, D. C., *Proc. Natl. Acad. Sci.* **57**, 484 (1967).
(174) Plackett, P., Buttery, S. H., Cottew, G. S., *Recent Progress in Microbiology, Symposia Volume of the 8th Internat. Congr. for Microbiology, Montreal*, pp. 535-547 (1962).
(175) Ploetz, T., Pogacar, P., *Leibig's Ann. Chem.* **562**, 36 (1949).
(176) Pollock, J. J., Chipman, D. M., Sharon, N., *Arch. Biochem. Biophys.* **120**, 235 (1967).
(177) Powning, R. F., Irzykiewicz, H., *Biochim. Biophys. Acta* **124**, 218 (1966).

(178) Praissman, M., Rupley, J. A., *Biochemistry* **7**, 2446 (1968).
(179) Preece, I. A., Hoggan, J., *J. Inst. Brewing* **62**, 486 (1956).
(180) Putman, E. W., Potter, A. L., Hodgson, R., Hassid, W. Z., *J. Am. Chem. Soc.* **72**, 5024 (1950).
(181) Raftery, M. A., Dahlquist, F. W., Chan, S. I., Parsons, S. M., *J. Biol. Chem.* **243**, 4175 (1968).
(182) Raftery, M. A., Rand-Meir, T., *Biochemistry* **7**, 3281 (1968).
(183) Reese, E. T., *Ind. Eng. Chem.* **49**, 89 (1957).
(184) Reese, E. T., Maguire, A. H., Parrish, F. W., *Can. J. Biochem.* **46**, 25 (1968).
(185) Reese, E. T., Mandels, M., *Can. J. Microbiol.* **5**, 173 (1959).
(186) Reese, E. T., Mandels, M., "Advances in Enzymic Hydrolysis of Cellulose and Related Materials," E. T. Reese, Ed., p. 197, Pergamon Press, Oxford, 1963.
(187) Reese, E. T., Mandels, M., *Can. J. Microbiol.* **10**, 103 (1964).
(188) Reese, E. T., Mandels, M., "Methods in Enzymology," E. F. Neufeld, V. Ginsburg, Eds., Volume 8, p. 607, Academic Press, N. Y., 1966.
(189) Reese, E. T., Parrish, F. W., Mandels, M., *Can. J. Microbiol.* **7**, 309 (1961).
(190) *Ibid.*, **8**, 327 (1962).
(191) Reese, E. T., Perlin, A. S., *Biochem. Biophys. Res. Comm.* **12**, 194 (1963).
(192) Reese, E. T., Siu, R. G. H., Levinson, H. S., *J. Bacteriol.* **59**, 485 (1950).
(193) Reese, E. T., Smakula, E., Perlin, A. S., *Arch. Biochem. Biophys.* **85**, 171 (1959).
(194) Robyt, J. F., Whelan, W. J., *Biochem. J.* **95**, 10P (1965).
(195) Rupley, J. A., *Proc. Roy. Soc. London, Series B* **167**, 416 (1967).
(196) Rupley, J. A., Butler, L., Gerring, M., Hartdegen, F. J., Pecoraro, R., *Proc. Natl. Acad. Sci.* **57**, 1088 (1967).
(197) Rupley, J. A., Gates, V., *Proc. Natl. Acad. Sci.* **57**, 496 (1967).
(198) Rupley, J. A., Gates, V., Bilbrey, R., *J. Am. Chem. Soc.* **90**, 5633 (1968).
(199) Saito, H., Misaki, A., Harada, T., *Agr. Biol. Chem.* **32**, 1261 (1968).
(200) Satomura, Y., Ebata, J., *Hakko Kyokaishi* **22**, 162 (1964).
(201) Schneberger, G. L., Luchsinger, W. W., *Can. J. Microbiol.* **13**, 969 (1967).
(202) Selby, K., Maitland, C. C., *Biochem. J.* **104**, 716 (1967).
(203) Sharon, N., *Proc. Roy. Soc. London, Series B* **167**, 402 (1967).
(204) Siu, R. G. H., "Microbial Decomposition of Cellulose," Reinhold, N. Y., 1951.
(205) Siu, R. G. H., Darby, R. T., Burkholder, P. R., Barghoorn, E. S., *Textile Res. J.* **19**, 484 (1949).
(206) Stone, B. A., *Nature* **182**, 687 (1958).
(207) Stone, B. A., *J. Ferment. Technol.* **45**, 687 (1967).
(208) Stone, B. A., Morton, J. E., *Proc. Malacological Soc. Lond.* **33**, 127 (1958).
(209) Storvick, W. O., Cole, F. E., King, K. W., *Biochemistry* **2**, 1106 (1963).
(210) Storvick, W. O., King, K. W., *J. Biol. Chem.* **235**, 303 (1960).
(211) Tabata, S., Hirao, K., Terui, G., *J. Ferment. Technol.* **43**, 772 (1965).
(212) Takeuchi, Y., Kitahara, M., *Agr. Biol. Chem.* **30**, 523 (1966).
(213) Tanaka, H., Phaff, H. J., *J. Bact.* **89**, 1570 (1965).
(214) Thoma, J. A., Koshland, D. E., *J. Am. Chem. Soc.* **82**, 3329 (1960).
(215) Thoma, J. A., Wright, H. B., French, D., *Arch. Biochem. Biophys.* **85**, 452 (1959).
(216) Timell, T. E., *Advan. Carbohydrate Chem.* **19**, 247 (1964).
(217) *Ibid.*, **20**, 409 (1965).
(218) Tung, K. K., Nordin, J. H., *Biochem. Biophys. Res. Comm.* **28**, 519 (1967).

(219) Tung, K. K., Nordin, J. H., *Biochim. Biophys. Acta* **158**, 154 (1968).
(220) Veibel, S., "The Enzymes," Vol. 1, Part 1, p. 583, J. B. Sumner, K. Myrbäck, Eds., Academic Press, N. Y., 1950.
(221) Vernon, C. A., *Proc. Roy. Soc. London, Series B* **167**, 389 (1967).
(222) Wakabayashi, K., Kanda, T., Nisizawa, K., *J. Ferment. Technol.* **44**, 669 (1966).
(223) Wallen, L. L., Rhodes, R. A., Shulke, H. R., *Appl. Microbiol.* **13**, 272 (1965).
(224) Wallenfels, K., Rached, I. R., Hucho, F., *Euro. J. Biochem.* **7**, 231 (1969).
(225) Warsi, S. A., Whelan, W. J., *Chem. Ind. (London)* **1957**, 1573.
(226) Weissmann, B., *J. Biol. Chem.* **216**, 783 (1955).
(227) Wenzel, M., Lenk, H. P., Schutte, E., *Z. Physiol. Chem.* **322**, 13 (1962).
(228) Whelan, W. J., "Methods in Carbohydrate Chemistry," Vol. 4, p. 252, R. L. Whistler, Ed., Academic Press, N. Y., 1964.
(229) Whitaker, D. R., *Arch. Biochem. Biophys.* **53**, 439 (1954).
(230) Whitaker, D. R., *Can. J. Biochem. Physiol.* **34**, 102 (1956).
(231) *Ibid.*, **34**, 488 (1956).
(232) Whitaker, D. R., "Advances in Enzymic Hydrolysis of Cellulose and Related Materials," E. T. Reese, Ed., p. 51, Pergamon Press, Oxford, 1963.
(233) Whitaker, D. R., Merler, E., *Can. J. Biochem. Physiol.* **34**, 83 (1956).
(234) Wood, R. K. S., *Ann. Rev. Plant Physiol.* **11**, 299 (1960).
(235) Youatt, G., *Australian J. Biol. Sci.* **11**, 209 (1958).
(236) Youatt, G., Jermyn, M. A., "Marine Boring and Fouling Organisms," Dixy Lee Ray, Ed., p. 397, Friday Harbor Symposia, University of Washington Press, Seattle, 1959.
(237) Yu, R. J., Bishop, C. T., Cooper, F. P., Blank, F., Hasenclever, H. F., *Can. J. Chem.* **45**, 2264 (1967).
(238) Zehavi, U., Jeanloz, R. W., *Carbohydrate Res.* **6**, 129 (1968).
(239) Zevenhuizen, L. P. T. M., *Carbohydrate Res.* **6**, 310 (1968).

RECEIVED October 4, 1968.

Discussion

Query: "What is the origin of these enzymes? Are these just the enzymes that you get from the fermentation of the organisms?"

B. A. Stone: "The enzymes referred to (Table IV) are either from the culture filtrates of micro-organisms or from plant sources. In most cases they have been subjected to fractionation prior to examination for specificity."

Query: "Does the pH have any influence at all on the specificity? How do you choose the pH at which you carry out your experiments?"

B. A. Stone: "We have not studied the effect of pH on the specificity of either the *Euglena* or *Nicotiana* enzymes. The pH chosen for tests on specificity was that determined for optimal activity on a β-1,3-glucan substrate."

# Recent Progress on the Structure and Morphology of Cellulose

BENGT RÅNBY[1]

The Royal Institute of Technology, Stockholm, Sweden

Recent progress on the structure and morphology of cellulose is described—e.g., the cellulose chain conformation with its intrachain hydrogen bonds, the fibrillar morphology of native cellulose containing very thin elementary microfibrils (30–40 A. wide), the acid hydrolysis of cellulose fibers to microcrystalline cellulose giving rodlike particles (micelles) of the same width as the fibrils, and the new but controversial helical model of the cellulose microfibrils. Penetration measurements of native cellulose fiber walls using aqueous polymer solutions have given average pore dimensions of 5 A. (cotton fibers) to 10 A. (delignified wood fibers). Accessibility measurements using deuterium and tritium oxide exchange have shown that native cellulose probably is deposited in crystalline form. Holocellulose fibers (delignified wood) are shown to be less accessible than purified wood pulp fibers, and tentative interpretations are given.

The chemical structure of cellulose chains was established by Haworth and Hibbert more than 40 years ago. Native cellulose occurs in solid state and in partly crystalline form. A schematic model for the native cellulose lattice was worked out about 1930 by Meyer, Mark, and Misch. The basic morphology of native cellulose could not be resolved, however, before electron microscopy with high resolution was developed and applied. During the last two decades, it has been amply shown and is now generally accepted that native cellulose basically is composed of microfibrils of a width 100 A. or less as was reviewed in 1956 (*12*).

[1] Visiting Professor of Polymer Chemistry, North Carolina State University at Raleigh, North Carolina and Camille Dreyfus Laboratory, Research Triangle Institute, Durham, North Carolina.

No interwoven fringe micelles, no membranes, and no networks of cellulose chains seem to occur except when formed from microfibrils, laid down as units in the morphological structure. Thicker fibrils, lamellae of largely parallel microfibrils, and layers and walls of crossing fibrillar units all contain microfibrils of a certain rather well defined thickness and indefinite length.

This is not attempted to be a comprehensive review of recent progress on the structure and morphology of native cellulose and related substances. It is limited to aspects supposed to have more direct bearing on the enzymatic degradation. The discussions of the various implications are of necessity brief.

Conformation of Cellulose Chains

Infrared absorption measurements have shown that the chains in solid cellulose are to a large extent tied together with hydrogen bonds between hydroxyl groups and oxygen atoms in the structure (10). In dry native cellulose, there are no measurable amounts of free hydroxyl groups—i.e., OH groups not engaged in hydrogen bonds. Cellulose chains in solution are not considered here, as they represent a state not found in nature.

The cellulose chains in native crystalline cellulose seem to contain intrachain hydrogen bonds (Figure 1) between OH_3 in one glucose unit and O_5 in the following unit as suggested by Hermans in the 1940's and later supported by infrared measurements of Liang and Marchessault (7). For mercerized cellulose similar intrachain hydrogen bonds are proposed. It is suggested but not proved that these bonds also occur in non-crystalline celluloses.

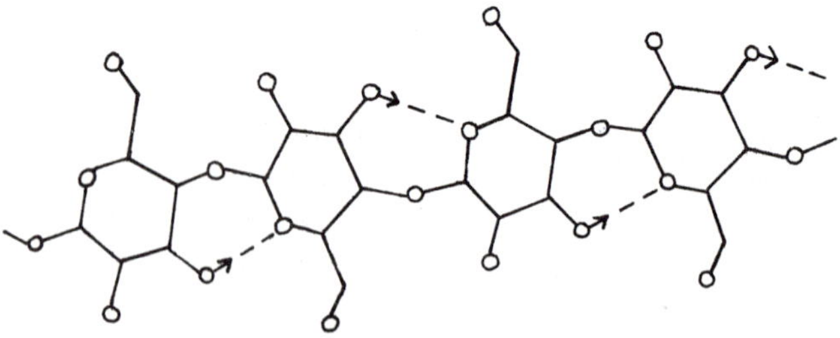

Figure 1. Cellulose molecule with intrachain hydrogen bonds between OH_3 and O_5

Part of an infrared absorption spectrum from a membrane of bacterial cellulose is shown in Figure 2 (16). The absorption bands refer to the stretching frequencies of the hydrogen-bonded OH groups. Free OH groups would give an absorption band at 3640 cm.$^{-1}$. All OH groups are apparently hydrogen-bonded in different but rather specific ways. Two of the absorption bands are assigned to interchain H-bonds in different lattice planes (101⁻ and 101, respectively). The most intense band is assigned to the intrachain H-bond OH_3-O_5, in the 002-planes as previously discussed. The infrared band at 3245 cm.⁻¹ with the largest shift from 3640 cm.⁻¹ (free OH groups)—*i.e.*, due the strongest H-bonds—is not yet assigned to a specific hydrogen bond.

When native cellulose is wetted with water, an accessible surface layer of the microfibrils is assumed to react. The cellulose lattice, however, is not penetrated or otherwise affected by the water in any way measurable with x-ray diffraction.

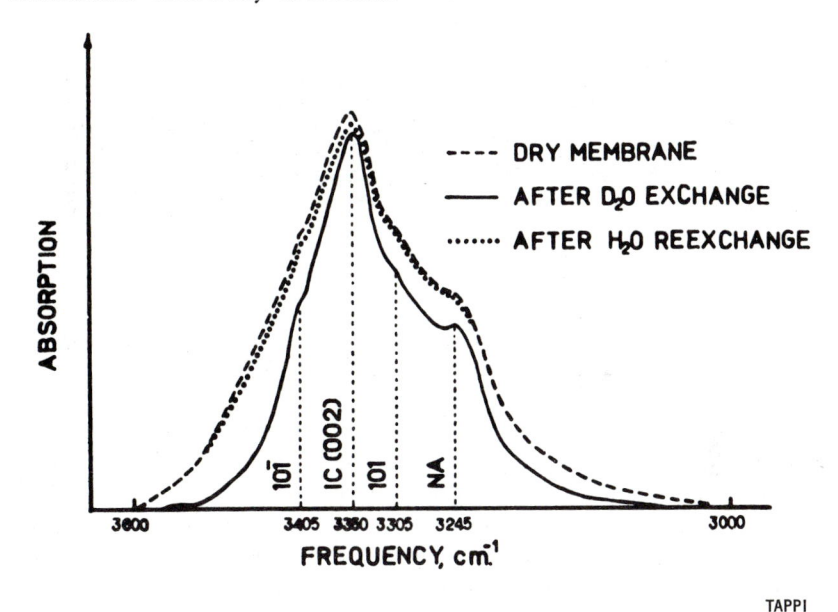

Figure 2. The OH stretching frequency bands in the infrared absorption spectra of bacterial cellulose (16)

Dimensions and Structure of Native Cellulose Microfibrils

In electron micrographs, the native cellulose microfibrils are usually seen as bundles of lamellae containing an indefinite number of fibrillar units. A schematic representation of the cross section of a small lamellae of microfibrils is shown in Figure 3 (*14*). Two structural features should be pointed out. The cellulose lattice extends through the whole cross section of the microfibrils but the surface layers are supposed to be disordered to some extent because they represent a discontinuity. The microfibrils show a preferred orientation with their lattice planes 101 well aligned parallel with the flat surface of the lamellae—*i.e.*, the tangential plane of the cell wall or the membrane where they occur. It seems therefore likely that the cellulose microfibrils have a flat cross section. This orientation effect was first found by x-ray diffraction and later confirmed by infrared absorption measurements. Owing to the prevailing orientation of the lattice planes and the parallel alignment of the microfibrils, it is likely that the cellulose lattice from one microfibril frequently extends into and through adjacent microfibrils as indicated in Figure 3. But no bundles of microfibrils of preferred size—e.g., 250–300 A. as previously reported (5), seem to occur in native cellulose.

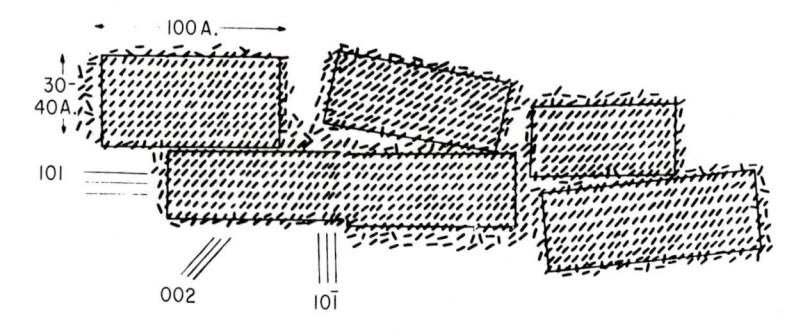

Figure 3. Schematic cross section of a lamellae of cellulose microfibrils from the secondary wall of a plant cell (14)

Along the microfibrils, there are areas of sufficient disorder to allow disintegration by hydrolysis into rodlike particles (micelles) with aqueous, non-swelling strong acid. After washing out the acid with distilled water, aqueous colloidal solutions of the cellulose micelles can be prepared (11). The disordered areas of the microfibrils may be native or formed by mechanical forces, giving deformation beyond the limit of elastic recovery of the microfibrils. The length of the resulting particles after acid hydrolysis (micelles or microcrystals) varies with the pretreatment of the native cellulose and corresponds to the leveling-off degree of polymerization (LO-DP) of the hydrocellulose (2). Repeated cycles of drying and swelling the native cellulose with water or dilute aqueous caustic soda tend to decrease the length of the particles obtained in subsequent hydrolysis. At the same time, these treatments are found to increase the crystallinity and decrease the chemical accessibility of the cellulose material (13).

The preparation of microcrystalline cellulose by acid hydrolysis of native and regenerated fibers has been studied extensively (3) and developed into a commercial process by Battista *et al.* (4). The resulting products are used as aqueous gels with high water-bonding capacity, inert food and drug additives, viscosity regulators, and stabilizers in colloidal solutions. Microcrystalline celluloses are also used as substrates for enzymatic degradation by fungi and microorganisms.

In recent years a new model for the molecular morphology of native microfibrils has been developed by Manley (9). He has found that the microfibrils have an average width of about 35 A. (or rather 30–40 A.) and not 60–100 A. as reported 15–20 years ago. The resolving power of the electron microscope has increased in recent years, and it seems likely that the dimensions reported now do represent some finite morphological unit (Figure 4). Particles observed along the microfibrils on the electron micrographs are taken as evidence for a detailed periodic structure of the microfibrils. This is ambiguous because the specimens were stained —e.g., with uranyl acetate. Both the supporting membrane and the cellulose specimen show grain of the same order of magnitude. Manley has worked out a detailed helical model of folded cellulose chains for the structure of cellulose microfibrils (Figure 5). The helix is assumed to be built as a spiral ribbon of folded chains. The helix could be unwound to a flat ribbon by mechanical disruption. A serious objection to this model is that it does not give a preferred orientation of the (101) lattice planes parallel with the fiber wall, as previously reported from x-ray diffraction and infrared absorption work.

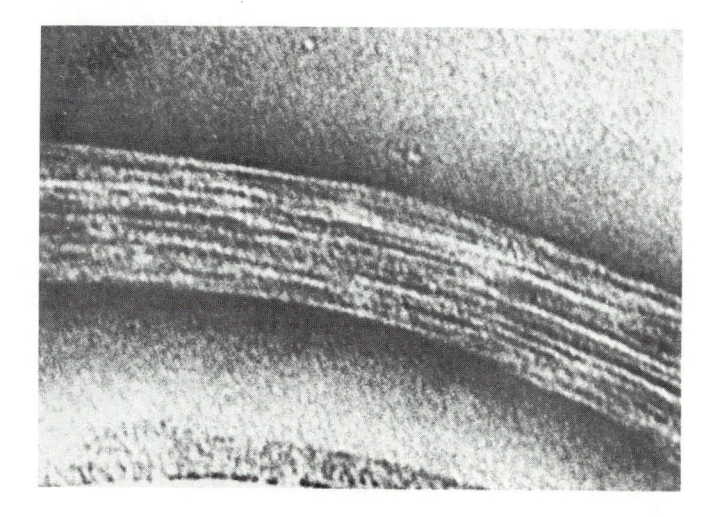

Figure 4. Cellulose microfibrils in a fragment of a ramie fiber wall (Manley)

The circular shape of the cross section of the fibril according to Manley, and the cylindrical hole in the middle of the fibril are difficult to reconcile with the measured density of the fiber wall, known to be as high as about 1.5. For pure crystalline cellulose a density of 1.59 is calculated from the lattice parameters. To meet these objections, Manley has indicated that the cylindrical fibrils may be deformed (flattened) to an oval or elliptical cross section in the fiber walls. Such a deformation seems unlikely, as it would require strong forces and also mean a very extensive deformation of the cellulose lattice. Further evidence is required for the construction of an acceptable model for the native cellulose microfibrils.

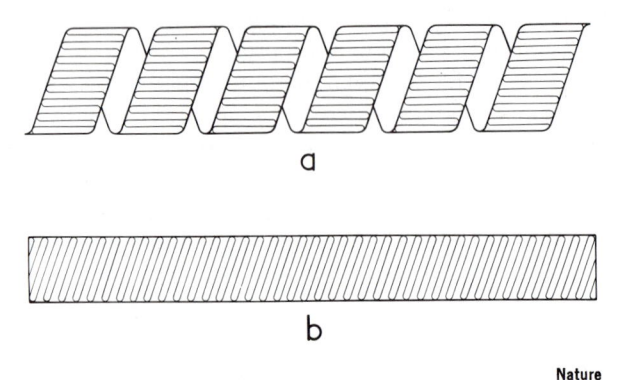

Figure 5. Structural model of a cellulose microfibril containing folded cellulose chains (9)

Porosity of the Cell Walls of the Native Cellulose Fibers

Measurements of the pore size distribution of cellulose fibers have been made using a penetration method related to the principle of gel permeation chromatography (*1*). A cellulose fiber is known to imbibe a considerable amount of water—e.g., for purified native cotton fibers about 0.4 grams of water per gram of dry cellulose at the fiber saturation point. The excess of water is removed by centrifugation in a field of about $100 \times$ gravity. If the dry fibers are immersed in aqueous solutions of a polymer with a certain molecular weight, the polymer molecules can penetrate only a fraction of the volume for water penetration. This can be derived from chemical analysis of the polymer concentration in the solution before and after immersion of the dry fibers in the solution. The excluded volume for the polymer molecules increases with increasing molecular weight. If a series of polymers with increasing molecular weights are used, the pore size distribution can then be derived from the calculated molecular dimensions of the polymer molecules in solution (the radius of gyration of the polymer molecules). The measurements of Aggebrandt and Samuelson (*1*), using polyethyleneglycol as polymer, have shown that the most frequent pore diameter in cotton fibers is 5 A. Seventy-five percent of the total pore volume (corresponds to 0.3 ml. per gram of dry fiber) was found in pores of less than 20 A. diameter. Delignified wood fibers show two to four times larger pore diameters (*15*).

These measurements give an indication of the possibilities for enzymes to penetrate a fiber wall as will be described in later papers during this symposium.

Accessibility of Native Cellulose

Two powerful methods have recently been applied to measure the accessibility of native cellulose and wood specimens.

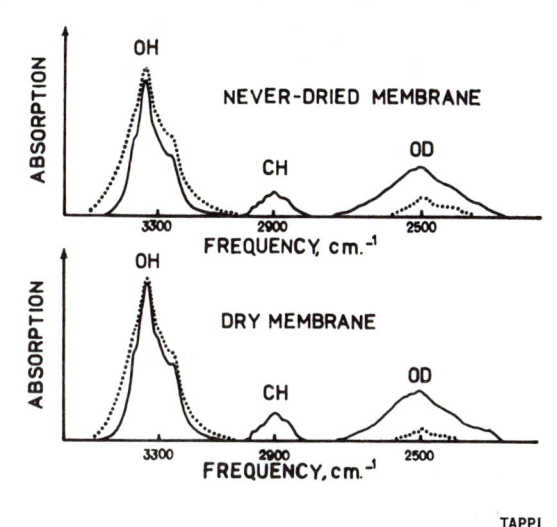

Figure 6. Infrared spectra of deuterated bacterial cellulose (solid line) and rehydrogenated bacterial cellulose (dashed line) (16)

Deuterium Oxide. Exchange with deuterium oxide (heavy water) and subsequent infrared analysis can be applied to transparent samples (6, 10, 16). Thin membranes of bacterial cellulose are ideal as model samples. When the dry membranes are exposed to heavy water as vapor or liquid, the accessible OH groups are exchanged to OD groups. Infrared spectra of a D_2O-exchanged and rehydrogenated membrane are given in Figure 6. Deuterium exchange is recorded as a shift from OH stretching frequencies in the band at about 3300 cm.$^{-1}$ to the OD stretching frequency band at about 2500 cm.$^{-1}$. The non-accessible OH groups are retained in the 3300 cm.$^{-1}$ band. Upon rehydrogenation, some OD groups are retained (not exchanged to OH groups), which indicates recrystallization and trapping in the wetting and drying processes. In one series of experiments, the virgin bacterial cellulose gel containing about 0.5% cellulose was purified completely in heavy water by treatments with dilute NaOD solutions in D_2O (16). The subsequent infrared measurements showed that about 55% of the OH groups in the gel cellulose membranes were inaccessible to deuterium exchange. This is interpreted to mean that the bacterial cellulose microfibrils most probably are deposited directly in crystalline form. The data are collected in Table I, and they show further that wet (never-dried) membranes are accessible to 33% and dried membranes to 27%. Prolonged treatment with D_2O gives only a minor increase in accessibility. These results for bacterial cellulose are taken to be applicable also to cotton and ramie cellulose fibers which have very closely the same degree of crystalline order as the bacterial cellulose. The microfibrils of cotton cellulose are oriented as spirals around the fiber axis while those of ramie are closely parallel with the fiber axis.

Table I. The Total Accessibility of Bacterial Cellulose Membranes in Heavy Water by Infrared Analysis

	A Total accessibility, %	B Total amount of resistant OD, %	A − B [a]
Membranes purified in heavy water	44.7	12.2	32.5
Never-dried membranes	39.9	6.9	33.0
Dry membranes	31.4	4.4	27.0
Dry membranes (deuteration for 15 hr.)	32.2	—	—

[a] (A − B) represents amount (%) of reexchangeable OD groups.

Various attempts to interpret the accessibility data as related to surface area of microfibrils of measured dimensions have been made. If 10% of the microfibrils are accessible because they contain a disordered cellulose lattice, indicated by rapid hydrolysis in acid, the remaining 20–25% of the accessible OH groups should be located on microfibril surfaces.

Tritiated Water. Exchange with tritiated water and subsequent scintillation analysis of the tritium content is a very useful method for accessibility measurements when the samples are not transparent to IR and visible light. Thin shives of sapwood from black spruce (a common softwood), and white birch (a common hardwood) where studied before and after delignification by treatment with peracetic acid (17). Also wood cambium from the same softwood was studied. The cambium is the recently formed fiber layer, located close to the bark and not yet lignified. Spruce and birch wood contain about the same amounts of cellulose, 42 and 44%, respectively. Spruce wood has more lignin (28 vs. 18%), while birch has more hemicellulose (35 vs. 28%), in particular more pentosans than spruce (24 vs. 14%).

Table II. Percentage Accessibility

Sample		Accessibility, % Spruce	Birch
Sapwood	Never-dried	60.6	53.4
	Dried	60.5	52.3
Holocellulose (yield, 83%)	Never-dried	67.5	54.7
	Dried	64.6	53.8
Holocellulose (yield, 77%)	Never-dried	—	55.7
	Dried	—	55.5
Cambium	Never-dried	63.5	—

The accessibility of the OH groups in spruce wood is about 60% while the corresponding value for birch wood is only 53% (Table II). Removal of lignin results in an increase in the number of accessible OH groups. The OH groups in lignin are rather few and they are accessible to at least 40% but probably not 100%. Calculations of the carbohydrate accessibility is dependent on the estimates of lignin accessibility (Table III). Pure wood cellulose (the hemicellulose removed) has an accessibility of 58 to 60%. It can be calculated from the data that hemicellulose in wood, although x-ray amorphous, is only partially accessible to tritium exchange. This is an indication of molecular order in the cellulose-hemicellulose arrangement in the woods—*e.g.*, by ordered hydrogen-bonding or some other form of ordered association which makes OH groups inaccessible. This result is well in line with the finding from infrared measurements on fibers that the hemicellulose chains in the wood fibers are arranged largely parallel with the cellulose microfibrils (8).

Table III. Accessibility of Carbohydrates

		Carbohydrate accessibility, %			
		Lignin accessibility, 40%		Lignin accessibility, 100%	
Sample		Spruce	Birch	Spruce	Birch
Sapwood	Never-dried	63.8	54.7	54.7	49.3
	Dried	63.7	52.4	54.6	48.0
Holocellulose (yield 83%)	Never-dried	69.0	54.7	65.5	54.3
	Dried	66.0	53.8	62.5	53.3
Holocellulose (yield 77%)	Never-dried	—	55.7	—	55.7
	Dried	—	55.5	—	55.5
Cambium	Never-dried	63.5	—	63.5	—

Birch wood and birch holocellulose have particularly low accessibility (Table II and III) compared to purified cellulose from birch with an accessibility close to 60%. This is at least to some extent related to the relative inaccessibility of the xylan portion. This could partly be a result of strong internal hydrogen bonding between acetyl groups and neighboring OH groups. It could also be partly attributed to regularly arranged hydrogen bonds between cellulose microfibrils and xylan chains. Taken together, the data for OH group accessibility of wood, holocellulose, and purified wood cellulose fibers indicate that the carbohydrate part of the wood is regularly organized on the molecular level—*e.g.*, by hydrogen bonding between cellulose and hemicellulose. This molecular order reduces the accessibility of the x-ray amorphous hemicellulose portion to tritium exchange and would be of importance also for the accessibility to enzymatic degradation. Native wood is known to be less accessible to enzymatic degradation than purified wood cellulose fibers.

Acknowledgment

The author is indebted to R. St. J. Manley for mailing a review paper (including Fig. 4) available before publication.

Literature Cited

(1) Aggebrandt, L., Samuelson, O., *J. Appl. Polymer Sci.* **8,** 2813 (1964).
(2) Battista, O. A., *Ind. Eng. Chem.* **42,** 502 (1950).
(3) Battista, O. A., Coppick, S., Howsmon, J. A., Morehead, F. F., Sisson, W. A., *Ind. Eng. Chem.* **48,** 333 (1956).
(4) Battista, O. A., Smith, P. A., *U. S. Patent* **2,978,446** April 4, 1961).
(5) Frey-Wyssling, A., Mühlethaler, K., *Fortschr. Chem. Org. Naturstoffe* **8,** 1 (1951).
(6) Lang, A. R. G., Mason, S. G., *Can. J. Chem.* **38,** 373 (1960).
(7) Liang, C. Y., Marchessault, R. H., *J. Polymer Sci.* **37,** 385 (1959).
(8) Liang, C. Y., Bassett, K. H., McGinnes, E. A., Marchessault, R. H., *TAPPI* **43,** 1017 (1960).
(9) Manley, R. St. J., *Nature* **204,** 1155 (1964).
(10) Mann, J., Marrinan, H. J., *Trans. Faraday Soc.* **52,** 492 (1956).
(11) Rånby, B., *Acta Chem. Scand.* **3,** 649 (1949).
(12) Rånby, B., Rydholm, S. A., " 'Polymer Processes,' High Polymers," C. E. Schildknecht, Ed., Vol. 10, Chapt. IX, p. 351, Interscience, N. Y., 1956.
(13) Rånby, B., "Fundamentals of Papermaking Fibers," F. Bolam, Ed., p. 55, Cambridge, 1957.
(14) Rånby, B., *Das Papier* **18,** 593 (October 1964).
(15) Stone, J. E., Scallan, A. M., *Pulp Paper Mag. Can.* **69,** 288 (1968).
(16) Sumi, Y., Hale, R. D., Rånby, B. G., *TAPPI* **46,** 126 (1963).
(17) Sumi, Y., Hale, R. D., Meyer, J. A., Leopold, B., Rånby, B. G., *TAPPI* **47,** 621 (1964).

RECEIVED October 28, 1968.

Discussion

K. Selby: "We are talking now about accessibility, and crystallinity in cotton, so I thought you might like to see some results we got on the course of solubilization of cotton by cellulases. We were rather surprised to find this follows a first order reaction with respect to substrate. This cotton sample was solubilized by *Trichoderma viride* cellulase up to about 35% solubilization (Figure A). The next one (Figure B) is a similar curve from the *P. funiculosum* cellulase that I was talking about this morning; and the final one (Figure C) contains those two curves and also another obtained with a concentrate of *Trichoderma viride* cellulase which is producing up to 97% solubilization. Now the point is that that reaction is either first order all the way, or, in some cases, it departs slightly from it as solubilization proceeds. We were a little bit wide-eyed about this; what this is all about, we don't really know. But what is of considerable interest in this context is that we have established very very carefully that this reaction is first order over this critical range up to 35%

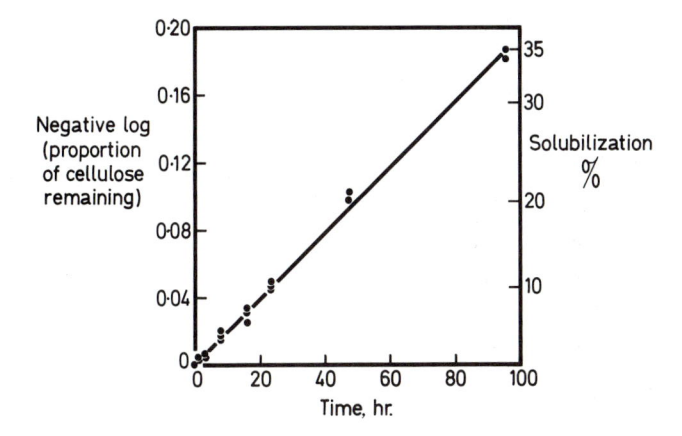

Figure A. Solubilization of cotton by T. viride *cellulase*

solubilization. We have done a lot of determinations here because this is a log plot. What we are saying is that, if there were any departure from uniform accessibility and reactivity as suggested in the old theories involving the idea of amorphous regions, we must surely have a break somewhere in this curve. We are interested in the fact that this is a straight line."

B. Rånby: "I have not seen these data reported before. In the hydrolysis by acid we know that it is an initial phase which takes something like 10% away from the cellulose very quickly and then the rate of reaction is much slower. Therefore, the reaction is not first order throughout. Referring to Dr. Selby's report apparently the enzyme is

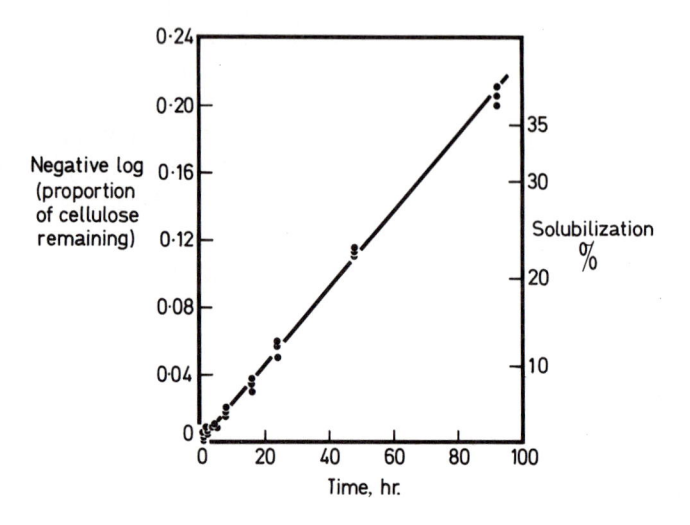

Figure B. Solubilization of cotton by P. funiculosum cellulase

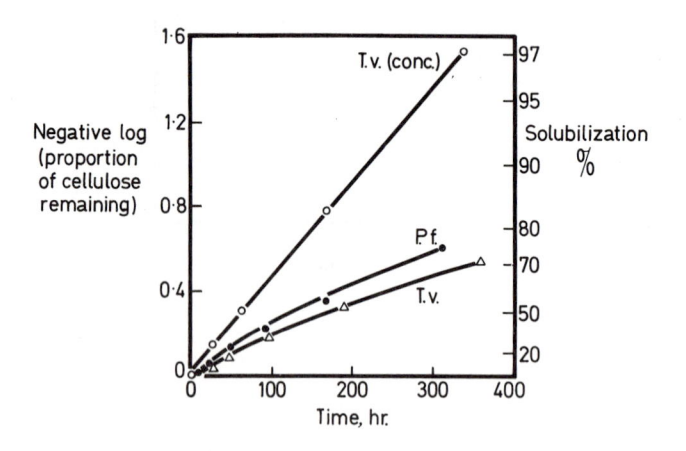

Figure C. Solubilization of cotton by cellulase

attacking the cellulose microfibrils evenly whether they are disordered or not in some way. The material is appearing as homogeneous to the enzyme. That is my preliminary conclusion."

K. Selby: "The attack of mineral acid on cotton does not always follow the same course—i.e., a rapid solubilization of the first 10%). Jeffries, Roberts, and Robinson (1) have recently shown that the size of the rapidly hydrolyzed fraction varies with acid concentration and deduce that "two clearly defined fractions are not present in the structure" but that "reaction is at an accessible surface, the accessibility depending on the concentration of the hydrolyzing acid."

The conclusion would therefore seem to be that the structure has very low accessibility to cellulase (as to the weakest mineral acid). The interesting fact still remains, however, that as the degradation proceeds there is no sudden change in accessibility or reactivity that would suggest the early disappearance of a more accessible fraction."

Literature Cited

(1) Jeffries, Roberts, and Robinson, *Text. Res. J.* **38,** 234 (1968).

10

Structural Features of Cellulosic Materials in Relation to Enzymatic Hydrolysis

ELLIS B. COWLING and WYNFORD BROWN

Departments of Plant Pathology and Wood and Paper Science, North Carolina State University, Raleigh, N. C. 27607

The susceptibility of cellulose to enzymatic hydrolysis is determined largely by its accessibility to extracellular enzymes secreted by or bound on the surface of cellulolytic microorganisms. Direct physical contact between these enzymes and the cellulosic substrate molecules is an essential prerequisite to hydrolysis. Since cellulose is an insoluble and structurally complex substrate, this contact can be achieved only by diffusion of the enzymes from the organism into the complex structural matrix of the cellulose. Any structural feature that limits the accessibility of the cellulose to enzymes by diffusion within the fiber will diminish the susceptibility of the cellulose of that fiber to enzymatic degradation. In this review, the influence of eight such structural features have been discussed in detail.

In this symposium it is important to recognize the importance of cellulolytic enzymes and their substrates both in nature and in commerce. Cellulose forms the bulk of the cell-wall material of all higher plants. It is thus by far the most abundant organic material on earth. It has been one of man's most generally available and useful natural resources since the beginning of civilization. It is unique among modern industrial raw materials in being available from renewable and therefore potentially inexhaustible sources of supply. In common with all other natural products it is susceptible to enzymatic degradation by a wide range of organisms. As discussed in the next four paragraphs, this biodegradability is both a biological necessity and a serious limitation to the usefulness of many cellulosic commodities.

Cellulolytic enzymes play an essential role in the carbon cycle of the earth. Photosynthetic organisms transform atmospheric carbon dioxide into plant materials. These materials are the primary source of food for all forms of animal life and the fossil fuels that drive man's industries and warm his homes. Since there is a finite amount of carbon on the earth, this element must be recycled if life is to be sustained. The organisms that degrade cellulose recycle the bulk of that carbon—about 85 billion metric tons of it annually. If these organisms were to cease their activity while photosynthesis proceeded at its usual pace, life as we know it on earth would stagnate for lack of atmospheric carbon dioxide in about 20 years.

As the major component of wood, flax, jute, cotton, pulp, paper, and more recently of rayon and certain plastic products, cellulose has always been one of man's most abundant and generally useful natural resources. In certain cellulosic commodities such as wooden building materials and many cordage and textile products, a high degree of resistance to microbiological deterioration is essential to their utility. In others, such as disposable packaging materials and animal feeds for example, rapid biodegradability is an essential or at least a desirable attribute. Unfortunately, cellulolytic microorganisms do not distinguish between these two categories of useful materials and the plant debris they evolved to destroy. Man has therefore found it necessary to invest huge sums to prevent degradation of certain products and to increase the biodegradability of others. These burdensome taxes will be reduced as more economical and effective means of inhibiting or increasing the activity of cellulolytic enzymes and microorganisms become available.

In view of the high consumption of disposable cellulosic materials in developed societies, it is not surprising that more than half the total solid wastes of major urban centers consists of cellulosic materials of various types. These materials must be disposed of by reutilization, composting, or by incineration, of which the latter method is the most common. The resulting dumping grounds, and the smoke and particulate materials that enter the atmosphere from incinerators, contribute substantially to urban environmental pollution. Alternative microbiological or enzymatic methods for treatment and disposal of waste materials will become ever more important future applications for cellulolytic enzymes.

As the human population of the earth continues to grow and the resulting crisis in world food supplies becomes more critical, utilization of wood cellulose and cellulosic waste materials as sources of food for animals or even for human consumption may become imperative. Food processors already are experimenting with fermentation methods and treatments with cellulolytic enzymes as means to increase the nutritive value of wood and other refractory food materials or waste products. It is conceivable that cellulolytic enzymes produced in microbiological cultures could be ingested by man or other animals together with foods containing cellulose to give useful sources of energy. Applications for cellulases in the realm of animal and human nutrition are certain to increase in the future.

In view of the great biological and economic importance of cellulose degradation, it is surprising that present knowledge of the enzymatic processes involved remains so relatively incomplete. A major challenge for man in the decades immediately ahead is to learn enough about these enzymes to increase the extent of human benefit both from their applications and their control.

The purpose of this paper is to summarize present knowledge of the structural features of cellulose and its associated substances and to describe how these features determine the susceptibility and resistance of cellulose fibers to enzymatic hydrolysis. In 1963 a similar review was prepared by one of us (9) in connection with a symposium on "Enzymatic Hydrolysis of Cellulose and Related Materials." Since that time a number of advances has been made in knowledge of the structure of cellulose and the materials with which it is associated in nature. Progress also has been made in understanding the size and other molecular properties of cellulolytic enzymes. Synthesis of cellulose *in vitro* has been accomplished recently. More complete understanding of the elementary building unit of cellulose, the microfibril, has been achieved in the past five years. Also a growing uncertainty has arisen about the adequacy of the fringe micellar theory of cellulose structure. Many of these advances have provided greater insight into the influence of structure on the susceptibility of cellulose to enzymatic hydrolysis. Thus, in the context of a new status report concerning "Cellulases and Their Applications" the present paper has been prepared in an attempt to bring the subject up-to-date once again.

Since the specific features of structure that were recognized to be important in 1963 are still important today, the organization of this contribution follows that of the original review. We will first summarize recent advances in the study of cellulose fibers, then consider the relationship between the fiber and cellulolytic organisms and their enzymes, and finally discuss the specific structural features of cellulose that influence the rate and mechanism by which cellulolytic enzymes act to cause degradation. Where little or no advance in knowledge has been made, an abbreviated discussion is presented. Where new insight or perspectives are available, they have been incorporated. Readers of the earlier review will recognize that it was prepared by a biologist with limited physicochemical perspective. We hope that the present paper shows that two heads are better than one.

Structure of Cellulose Fibers

The natural fibers of cotton and wood are the most important commercial sources of cellulose. Thus, knowledge of their structure is essential to any discussion of its enzymatic degradation. Several excellent reviews of cellulose structure have been published recently; those on cotton by Hamby (22) and Warwicker *et al.* (76), and on wood in the volumes edited by Zimmerman (79) and Côté (7), as well as the discussions of the ultrastructure of plant cell walls by Mühlethaler (47) and Rollins (59) make more than a general summary here unnecessary.

Cellulose exists in various states of purity in plant cell walls. It makes up about 90% of cotton fibers but only about 45% of typical wood cell walls. The cellulose in cotton and wood is very similar in molecular structure. But the two types of cells differ both in gross structure (as discussed below) and in the nature and amounts of substances with which the cellulose is associated.

Cotton fibers are single-celled hairs that cover the surface of the cotton seed. They are formed independently and thus contain no intercellular substance (Figure 1). Wood fibers, on the other hand, form a cohesive three-dimensional structure (Figure 2) whose integrity is assured by large amounts of intercellular substance (Figure 3). The outermost layer of cotton is called the cuticle. The equivalent layer in the multicellular structure of wood is the middle lamella.

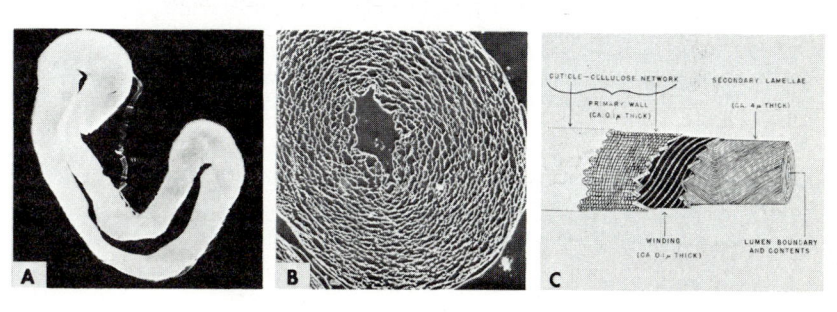

Forest Products Journal

Figure 1. Structure of cotton fibers showing: (A) cross section of a single fiber; (B) cell wall expanded to separate the concentric growth layers within the secondary wall; and (C) diagrammatic sketch of the organization of a single cotton fiber and the orientation of microfibrils within each layer (59)

Both cotton and wood fibers have a thin primary wall that consists of a loose, random fibrillar network and surrounds the relatively thick secondary wall. The primary wall and adjacent intercellular substance between contiguous cells in wood is referred to as the compound middle lamella. In both wood and cotton the secondary wall usually consists of three layers designated S1, S2, and S3. The S1 and S3 layers usually are

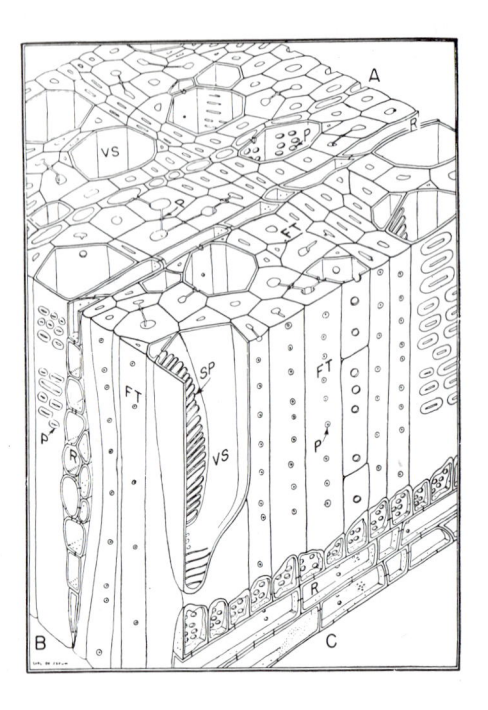

Figure 2. Minute anatomy of the typical angiosperm Liquidambar styraciflua *L. Each type of wood cell is depicted in transverse (A), tangential (B), and radial (C) view; the walls of the fiber tracheid cells (FT), vessel segments (VS), and ray cells (R) comprise the bulk of the wood substance; sclariform plates (SP) and pit pairs (P) between adjacent cells permit movement of fluids from one cell to another*

From "Textbook of Wood Technology" by Brown et al., copyright 1949. Used with permission of McGraw-Hill Book Company

quite thin; the S2 layer is of variable thickness but usually forms the bulk of the cell-wall substance. As shown in Figure 3 the cellulose molecules in the S1 and S3 layers are deposited in a flat helix with respect to the fiber axis where as those in the S2 layer are deposited nearly parallel to the fiber axis. The innermost, S3 layer lining the lumen is prominent in wood cells but is not always evident in cotton fibers (Figure 1). In cotton the S2 layer is deposited in a series of concentric zones, the number of which has been correlated with the growth of the fiber (Figure 1).

Within each layer of the secondary wall, the cellulose and other cell-wall constituents are aggregated into long slender bundles called microfibrils. The microfibrils are distinct entities in that few cellulose molecules, if any, cross over from one microfibril to another. Within each microfibril, the linear molecules of cellulose are bound laterally by hydrogen bonds and are associated in various degrees of parallelism—regions that contain highly oriented molecules are called crystalline whereas those of lesser order are called paracrystalline or amorphous regions. The specific dimensions of microfibrils and the spatial relationship between the crystalline and the paracrystalline regions is a matter of some controversy.

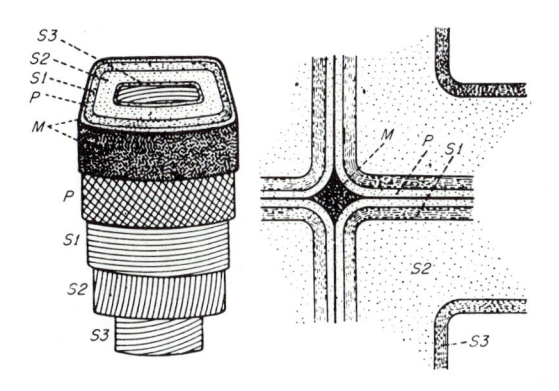

Figure 3. Diagrammatic sketch showing the various layers of wood cell walls. The true intercellular substance or middle lamella (M) and adjacent primary walls (P) of contiguous cells comprise the compound middle lamella. The secondary walls are composed of outer (S1), middle (S2), and inner (S3) layers which can be distinguished by the orientation of microfibrils within each layer

From "Chemistry of Wood" by Hagglund, copyright 1949. Used by permission of Academic Press

According to the traditional fringe micellar concept the microfibrils are 150 to 250 A. in diameter and contain about 600 molecules. The molecules are arranged in part in rod-shaped, highly ordered, micelles 50–60 A. in diameter and about 600 A. long, and in part in less ordered regions. These regions are interspersed with one another—the molecules grading gradually from crystalline to the amorphous phase.

Three more recent models of the microfibril are shown in Figure 4 (47). According to the concept shown at a in Figure 4, the microfibril is about 50 × 100 A. in cross section and consists of a "crystalline core" of highly ordered cellulose molecules arranged in a flat ribbon, rectangular in cross section. This crystalline core is surrounded by a "paracrystalline sheath" that in cotton contains mainly cellulose molecules but in wood also contains hemicellulose and lignin molecules. The crystalline core may (24) or may not (53) be continuous. In the model with periodic discontinuities in the crystalline core, the cellulose molecules are believed to be much longer than individual crystallites and therefore to traverse several crystalline and amorphous regions. A related concept (shown at *b* in Figure 4) suggests that the microfibril is made up of a number of "elementary fibrils" which contain 15 to 40 cellulose molecules and show a minimum cross-sectional dimension of about 35 A. According to this model the association of cellulose molecules is less well ordered at certain points along the length of the microfibril. The very novel concept shown at *c* in Figure 4 suggests that the cellulose molecules exist in a folded chain lattice formed as a ribbon which in turn is wound in a tight helix.

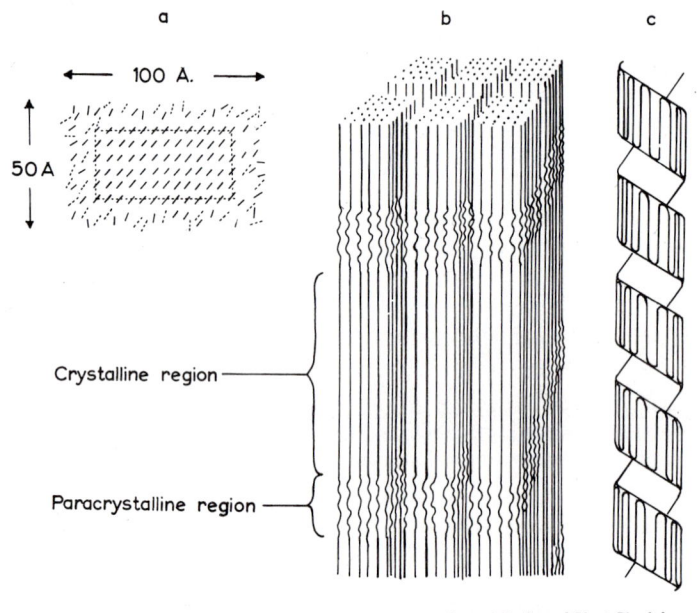

Figure 4. Recent concepts of the structure of microfibrils (47). (a) According to Preston and Cronshaw (53) the microfibril consists of a rectangular crystalline core surrounded by a paracrystalline cortex. The solid lines represent the planes of glucose residues; the broken lines represent the orientation of hemicellulose molecules. (b) According to Hess, Mahl, and Gütter (25) the microfibril contains several elementary fibrils which contain 15 to 40 cellulose molecules and are segmented into crystalline and paracrystalline regions. (c) According to Manley (42) the cellulose molecule is first folded into a flat ribbon which in turn is wound in a tight helix

Since neither elementary fibrils, microfibrils, or larger structures are visible in untreated cotton or wood fibers, and because similar plant fibers treated in different ways and different fibers treated in similar ways have shown aggregates of variable dimensions in the range between 100 and 1000 A., some controversy about which are real and which are artifactual aggregates is understandable. A more detailed discussion about these conflicting concepts and points of view is given by Sullivan (70). Since entities with minimum dimensions of 35 A. have been demonstrated in such diverse cellulose preparations as bacterial cellulose, algal cell walls, cotton, wood cell walls, and regenerated cellulose, the existence of the so-called elementary fibril is assured.

Chemical Constituents of Cellulose Fibers

The chemical constituents of wood and cotton fibers include cellulose, several hemicelluloses, lignin, a wide variety of extraneous materials including certain nitrogenous substances, and a small amount of inorganic matter. Unfortunately, sharp lines of demarcation among these constituents cannot be drawn. But the amounts indicated in Table I show the gross composition of cotton and typical angiospermous and gymnospermous species of wood. The non-cellulosic materials exert a significant influence on the susceptibility of cellulose in natural fibers to enzymatic hydrolysis. Thus, their structure and distribution will be discussed briefly.

Table I. Composition of Cotton and Typical Angiospermous (Birch) and Gymnospermous (Spruce) Fibers

Constituent	Cotton % by weight	Birch % by weight	Spruce % by weight
Holocellulose	94.0	77.6	70.7
Cellulose	89.0	44.9	46.1
Noncellulosic polysaccharides	5.0	32.7	24.6
Lignin	0.0	19.3	26.3
Protein ($N \times 6.25$)	1.3	0.5	0.2
Extractable extraneous materials	2.5	2.3	2.5
Ash	1.2	0.3	0.3

The celluloses of wood and cotton are very similar in molecular structure. Both are linear polymers of β-D-glucopyranose units linked by B-($1 \rightarrow 4$)-glycosidic bonds. As isolated in a form amenable to molecular weight determinations, cellulose molecules are of indefinite length; the number of glucose units per molecule (degree of polymerization) ranges from as few as 15 or less to as high as 10,000–14,000 with an average of about 3,000. Marx-Figini and Schulz (44) provide evidence, however, suggesting that the cellulose in the secondary walls of developing cotton fibers may be monodisperse with a degree of polymerization of about 14,000.

For many years the question of whether adjacent cellulose molecules run in the same or in opposite directions has been unresolved experimentally. The anti-parallel arrangement was generally assumed, however. Recently, Colvin (6) demonstrated that bacterial cellulose is formed in the anti-parallel arrangement. This follows logically from his earlier observations that microfibrils are synthesized by endwise growth, rather than by aggregation of preformed cellulose molecules and that the rate of growth is similar at both ends of the microfibril.

The most abundant hemicelluloses in cotton are pectic substances. Those in wood consist of relatively short heteropolymers of glucose, xylose, galactose, mannose, and arabinose, as well as uronic acids of glucose and galactose. These monosaccharide units and derivatives are linked together by $1 \to 3$, $1 \to 6$, and $1 \to 4$ glycosidic bonds. Most are linear molecules but some possess short branches. The hemicelluloses are much lower in molecular weight than cellulose; their degree of polymerization seldom exceeds 200. An appreciable number of acetyl groups are present as substituents, particularly on the glucomannans of gymnosperms and the xylans of angiosperms.

Lignin is a complex three-dimensional polymer formed from *p*-hydroxycinnamyl alcohols. These phenylpropane type groups include coniferyl units in the case of gymnosperms and both coniferyl and syringyl units in the case of angiosperms. A composite structure proposed by Freudenberg (16), illustrating most of the presently accepted types of linkages in spruce lignin, is shown in Figure 5. Although this is not a structural formula in the usual sense, it well illustrates the great complexity of lignin and provides a basis for understanding some of its essential features. A portion of the lignin is believed to be linked by covalent bonds to certain of the hemicelluloses. See Harkin (23) for a recent review of lignin chemistry.

The extraneous materials are a heterogeneous group of non-structural constituents most of which are organic compounds extractable in neutral solvents such as ether, acetone, ethyl alcohol, ethyl alcohol-benzene, and water. They include waxes, fats, essential oils, tannins, resin and fatty acids, terpenes, alkaloids, starch, soluble saccharides (gums), and various cytoplasmic constituents such as amino acids, proteins, and nucleic acids. An excellent review on the general subject of wood extractives is that of Hillis (26). The distribution of cytoplasmic constituents in wood has also been reviewed recently (11, 46).

The mineral constituents of cotton and wood include all the inorganic elements needed by plants (and fiber-degrading microorganisms) for growth. The concentration of these elements varies greatly with the environment in which the fiber is grown.

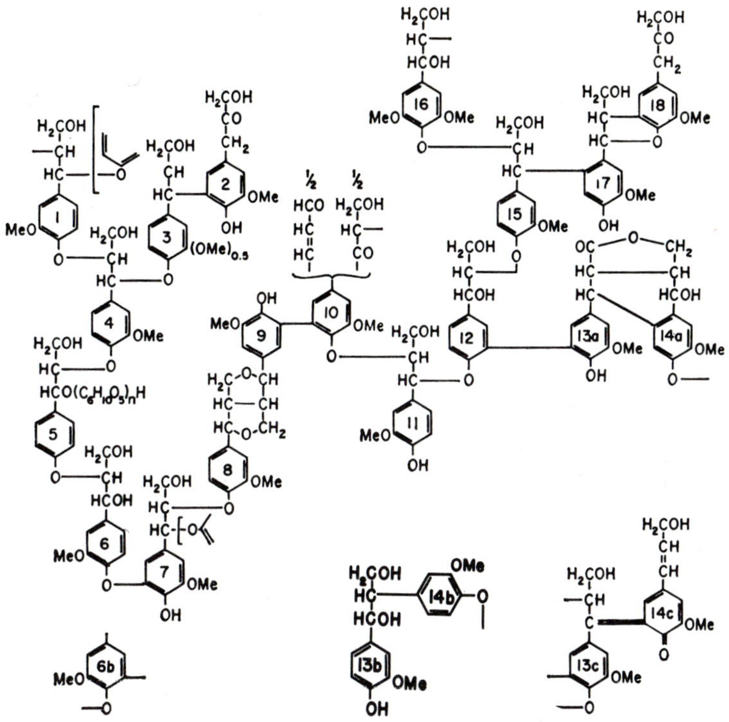

Figure 5. Schematic model showing the various types of monomer units and intermonomer bonds known to exist in spruce lignin (16)

Copyright 1965 by the American Association for the Advancement of Science

Distribution of Constituents Within Cellulose Fibers

The accessibility of cellulose to the extracellular enzymes of cellulolytic microorganisms is determined in part by its distribution within the cell wall and the nature of the structural relationships among the various cell-wall constituents. These relationships are reviewed briefly.

The distribution of constituents is quite simple in cotton. The secondary walls of cotton fibers consist almost entirely of highly crystalline cellulose. Almost all the hemicelluloses and extraneous materials (waxes, pectins, and certain nitrogenous substances) are contained in the cuticle and primary wall layers. In wood, on the other hand, the non-cellulosic materials are deposited in all regions of the cell walls from the lumen through the compound middle lamella.

As Meier (45) demonstrated, cellulose is in highest concentration in the secondary wall and diminishes toward the middle lamella (Table II). Hemicelluloses predominate among the polysaccharides in the compound middle lamella and decrease toward the lumen. The minimum concentration of hemicelluloses and maximum concentration of cellulose occurs in the S3 layer of angiosperms and in the S2 layer of gymnosperms (Table II). The S3 layer of gymnosperms contains appreciable lignin and hemicelluloses and thus is closer in composition to the compound middle lamella than the other layers of the secondary wall.

The extraneous materials are deposited in part in the lumina of wood cells and in part within the cell walls. Mineral constituents are distributed in all cell-wall layers of both cotton and wood fibers.

Table II. Percentages of Polysaccharides in Various Layers of Wood Fibers According to Meier (45)

Polysaccharide	Compound Middle Lamella	S1	S2	S3
Birch				
Cellulose	41.4	49.8	48.0	60.0
Total hemicelluloses	58.6	50.2	52.0	40.2
Galactan	16.9	1.2	0.7	0.0
Glucomannan	3.1	2.8	2.1	5.1
Glucuconoxylan	25.2	44.1	47.7	35.1
Arabinan	13.4	1.9	1.5	0.0
Spruce				
Cellulose	35.5	61.5	66.5	47.5
Total hemicelluloses	64.5	38.5	33.5	52.5
Galactan	20.1	5.2	1.6	3.2
Glucomannan	7.7	16.9	24.6	27.2
Glucuconoxylan	7.3	15.7	7.4	19.4
Arabinan	29.4	0.6	0.0	2.4

Within the various layers of wood cell walls, the hemicelluloses, lignin, extraneous and mineral constituents are concentrated in the spaces between microfibrils or elementary fibrils. The hemicelluloses and lignin form a matrix surrounding the cellulose. Within a given microfibril, lignin and the hemicelluloses may penetrate the spaces between cellulose molecules in the amorphous regions.

Covalent chemical bonds between lignin and certain of the polysaccharides in wood cell walls have been suggested for many years. Evidence for its existence remains inconclusive largely because the possibility of physical entanglement cannot be excluded. Freudenberg and Harkin (17) have suggested that free-radicals formed during the biogenesis of lignin will combine with hydroxyl groups of the polysaccharides as well as those of other lignin precursors. Whether or not a true covalent linkage does occur, it is certain that lignin and wood polysaccharides form a mutually interpenetrating system of polymers. As will be discussed later, this intimate physical (and possibly chemical) association with lignin accounts for much of the resistance of wood to many organisms that can readily degrade cotton fibers. The lignin apparently prevents the cellulases and hemicellulases of these organisms from contacting a sufficient number of glycosidic bonds to permit significant hydrolysis.

Organism-Substrate Relationships

To appreciate fully the influence of the structural features of natural fibers on their susceptibility and resistance to enzymatic degradation, it is necessary to understand the relationship between cellulolytic microorganisms, their extracellular enzymes, and the fiber substrate itself.

The organisms that degrade natural fibers live either on the exterior surface of the fibers, as most often is the case with cotton, or in the fiber lumina as is necessary in the case of wood. Here they secrete extracellular enzymes that catalyze the dissolution of the high-polymeric constituents of the fiber to soluble products that can be assimilated and metabolized by the organisms.

These extracellular enzymes are either: (1) bound on the surface of the organism and act on fiber surfaces with which the organisms are in contact, or (2) secreted into the exterior environment, diffusing some distance away from the organism, and acting on the accessible microfibrillar or molecular surfaces within the fine capillary structure of the fiber. Enzymes bound on the surfaces of cellulolytic microorganisms give rise to three typical degradation patterns that are visible in the light microscope: (1) surface pitting of cotton fibers caused by cellulolytic bacteria and fungi (64); (2) cylindrical cavities formed by soft-rot fungi as they grow parallel to the orientation of microfibrils in the S2 layer of wood cells as shown in Figure 6 (5, 13); (3) formation of bore holes as fungi pass from cell to cell in wood as shown in Figures 7 and 8 (10, 54). Figure 8 also shows hydrolysis at a distance from the hyphae.

For many years it was assumed, in part because no success had been achieved in attempts to obtain cell-free enzyme systems active on natural fibers, that degradation of these fibers occurred only in close proximity to cellulolytic microorganisms. For example, Gascoigne and Gascoigne (18) concluded as recently as 1960 that "In the case of fungal and bacterial attack, the digestion of cellulose with the aid of enzymes is highly localized, occurring only very close to or at points of contact of the fiber with the organism." More recently, several investigators have succeeded in isolating cell-free systems active on native cotton (43, 61) and isolated holocellulose of wood (33). Wilcox (77) showed that at least one hypha of the white-rot and brown-rot types of wood-destroying fungi must be present in wood cells in which degradation occurred. He also observed that the dissolution of cell-wall substance resulting in a general thinning of cell walls by white-rot fungi occurred more or less uniformly in the cells rather than just at points of contact between the hyphae and the lumen surfaces. This type of dissolution of cell wall substance is shown in Figures 8 and 9.

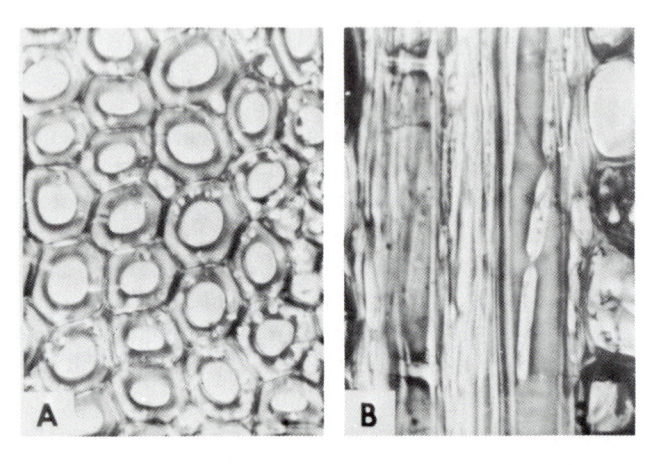

Department of Agriculture, Food Products Lab., Madison, Wisc.

Figure 6. Cylindrical cavities formed in the secondary walls of wood fibers by soft-rot fungi. Note hyphae of these organisms developing within the secondary walls (A) and causing dissolution of cell-wall substance in very close proximity to the hyphae (B) (13)

Enzyme-Substrate Relationships

As indicated earlier in this review, most of the constituents of natural fibers are high molecular weight, insoluble polymers of considerable structural complexity. Before these substances can be utilized by fiber-degrading organisms, they must be depolymerized to soluble products that can be assimilated. This is accomplished by usually very highly substrate-specific extracellular enzymes. The more complex the substrate, the greater will be the number of specific enzyme proteins that will be required to degrade the substrate completely. Each enzyme in its turn must achieve direct physical contact with its specific substrate before catalysis can occur. Thus, any chemical or physical feature of a natural fiber or its constituents that limits the ability of an organism to synthesize the necessary enzyme proteins will profoundly influence the susceptibility or resistance of the fiber to degradation by that organism.

A proposed sequence of enzymatic reactions involved in the degradation of native cellulose is shown in Figure 10 (58). One component (C_1)

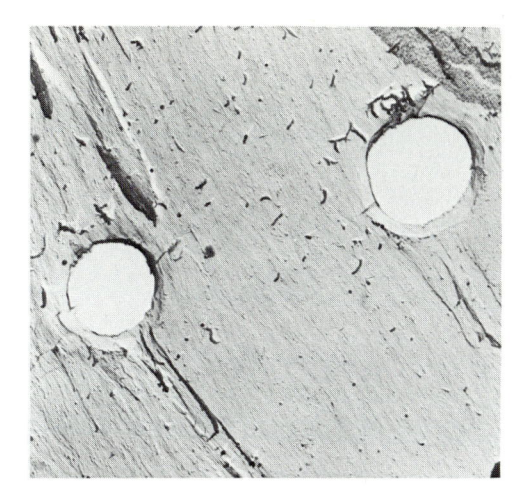

Figure 7. Replica of a split surface of sitka spruce sapwood (Picea sitchensis *(Bong.) Carr.) decayed by* Polyporus versicolor *L. Note bore holes formed as the fungus passed through the walls of contiguous tracheids and the "ring" around each hole indicating localized alteration of the wall material in the immediate vicinity of the hypha that formed the hole*

From "Cellular Ultra-Ultrastructure of Plant Cell Walls" by Coté. Copyright 1965. Used by permission of Syracuse Univ. Press.

is believed to separate the cellulose molecules and other components (C_x's) to catalyze their hydrolytic cleavage. These partial degradation products are then further cleaved as necessary to reduce them to cellobiose, glucose, or other soluble products that diffuse back to the organism and are assimilated. Although much less is known about the enzymes involved in depolymerization of hemicelluloses and lignin, it may be assumed that a similar sequence of enzymes probably is required for depolymerization of each of these other cell-wall constituents.

As shown by the following general sequence (which applies to all enzymatic reactions) direct physical contact between each particular enzyme and its specific substrate is essential for any enzymatic reaction:

Enzyme + Substrate ⇌ Enzyme-substrate complex → Enzyme + Products

Since the cellulose and other major constituents of natural fibers are insoluble molecules and are deposited within the cell walls in an intimate physical mixture of great structural complexity, formation of this requisite physical association can be achieved only by diffusion of these enzymes to susceptible sites on the gross surfaces of the fiber or the microfibrillar and molecular surfaces within the fiber wall. Thus, any structural feature

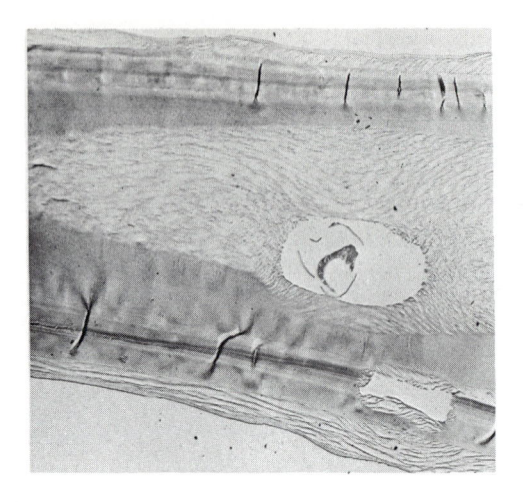

Figure 8. Longitudinal section of sweetgum sapwood (Liquidambar styraciflua L.) *decayed by* Polyporus versicolor L. *Note evidence for enzymatic disintegration of the cell walls between contiguous fiber tracheids in the immediate vicinity of the bore holes formed by hyphae in passing from one cell to another. Note similar "loosening" of the cell wall material also along lumen surfaces at some distance from the bore holes. The hypha in the largest bore hole apparently is in the process of autolysis*

From "Cellular Ultra-Ultrastructure of Plant Cell Walls" by Coté. Copyright 1965. Used by permission of Syracuse Univ. Press

of the fiber or its constituents that limits its accessibility or the diffusion of cellulolytic enzymes in close proximity to that fiber will exert a profound influence on the susceptibility or resistance of the fiber to enzymatic degradation.

Influence of Fiber Structure on Its Susceptibility to Enzymatic Degradation

Structural features of cellulosic materials that determine their susceptibility to enzymatic degradation include: (1) the moisture content of the fiber; (2) the size and diffusibility of the enzyme molecules involved in relation to the size and surface properties of the gross capillaries, and the spaces between microfibrils and the cellulose molecules in the amorphous regions; (3) the degree of crystallinity of the cellulose; (4) its unit-cell dimensions; (5) the conformation and steric rigidity of the anhydroglucose units; (6) the degree of polymerization of the cellulose;

(7) the nature of the substances with which the cellulose is associated; and (8) the nature, concentration, and distribution of substituent groups. The influence of each of these factors will be discussed in turn.

Moisture Content of the Fiber. Cellulosic materials are effectively protected from deterioration by microorganisms so long as their moisture content is maintained below some critical level that is characteristic of the material and organisms involved. In general, this critical level for wood is slightly above the fiber saturation point (*15*). The fiber saturation point of most temperate-zone wood species is approximately 40% of its oven-dry weight. The fiber saturation point is defined as that moisture content at which all sorption sites within the cell walls are fully saturated but no free water is present in the lumina. Recent research by Stone and Scallan (*65*) has provided much improved insight into the relationship between relative humidity and the moisture content of wood, cotton and pulps. So long as wood is in equilibrium with an atmosphere that is somewhat below 100% relative humidity, it will not pick up enough moisture for decay to be initiated. For reasons yet unknown, as little as 10% moisture is adequate to allow degradation of cotton fibers (*64*). Ten percent moisture is achieved in equilibrium with 80% relative humidity and is significantly below the fiber saturation point.

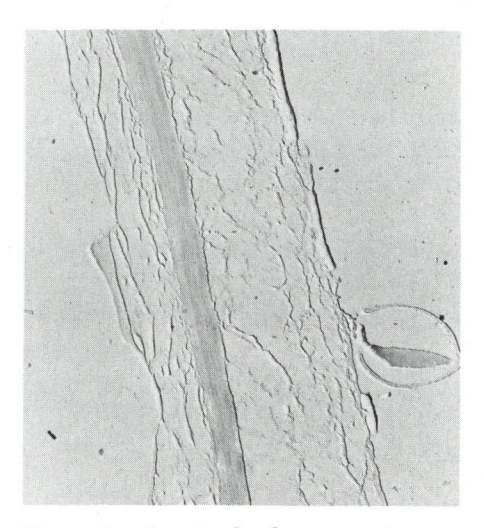

Figure 9. Longitudinal section of sweet-gum sapwood (Liquidambar styraciflua L.) decayed by Polyporus versicolor L. Note essentially complete dissolution of cell-wall substance both near the hypha and at some distance (in terms of hyphal diameters)

From "Cellular Ultra-Ultrastructure of Plant Cell Walls" by Coté. Copyright 1965. Used by permission of Syracuse Univ. Press

ENZYMATIC DEGRADATION OF CELLULOSE
(PROPOSED BY REESE AND ASSOCIATES)

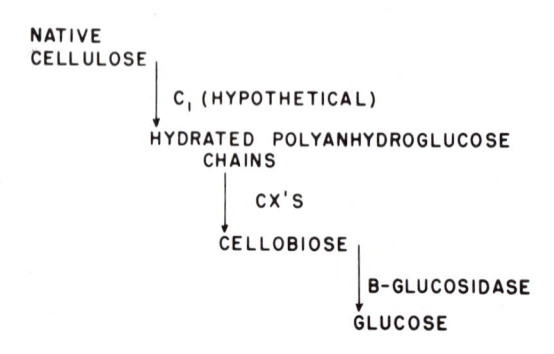

NATIVE CELLULOSE
↓ C_1 (HYPOTHETICAL)
HYDRATED POLYANHYDROGLUCOSE CHAINS
↓ C_x's
CELLOBIOSE
↓ β-GLUCOSIDASE
GLUCOSE

Journal of Bacteriology

Figure 10. Sequence of enzymatic reactions involved in the degradation of native cellulose as proposed by Reese, Siu, and Levinson (58)

Moisture plays three major roles in the degradation of cellulose: (1) it swells the fiber by hydrating the cellulose molecules, thus opening up the fine structure in such a way that the substrate is more accessible to cellulolytic enzymes; (2) it provides, between the organism and the fiber, a medium of diffusion for the extracellular enzymes of the organism and the partial degradation products of the fiber from which the organism derives its nourishment; and (3) the elements of water are added to the cellulose during hydrolytic cleavage of each glucosidic link in the molecule. The ability of cotton-destroying fungi to degrade cotton fabrics below their fiber saturation point may be related to their usual habit of degrading the fiber mainly at points of contact with the organism (64).

Size and Diffusibility of Cellulolytic Enzymes in Relation to the Capillary Structure of Cellulose. As discussed earlier, enzymatic degradation of cellulose requires that the cellulolytic and other extracellular enzymes of the organisms diffuse from the organism producing them to accessible surfaces on or in the walls of the fiber. This accessible surface is defined by the size, shape, and surface properties of the microscopic and submicroscopic capillaries within the fiber in relation to the size, shape, and diffusibility of the enzyme molecules themselves. The influence of these relationships on the susceptibility and resistance of cellulose to enzymatic hydrolysis has not been verified experimentally in natural fibers but the validity of the concepts that follow is demonstrated by the work of Stone, Scallan, Donefer, and Ahlgren (69).

CAPILLARY STRUCTURE OF CELLULOSE FIBERS. The capillary voids in wood and cotton fibers fall into two main categories: (1) gross capillaries such as the cell lumina, pit apertures, and pit-membrane pores that are visible in the light microscope and thus are in the range between 2000 A. and 10 or more microns in diameter; (2) cell-wall capillaries such as the spaces between microfibrils and the cellulose molecules in the amorphous regions. Most of the cell-wall capillaries are closed when the walls are free of water but open again when moisture is adsorbed. When wood and cotton are fully saturated with water, the cell-wall capillaries attain their maximum dimensions. Some expand to about 200 A. in diameter but most are substantially smaller in diameter.

The total surface area exposed in the gross capillaries is quite large—approximately one square meter per gram of wood or cotton—but is several orders of magnitude smaller than the total surface area potentially available to a small molecule within the water-swollen cell wall—approximately 300 square meters per gram. Thus, if cellulolytic enzymes can penetrate the cell-wall capillaries, substantially greater rates of dissolution of cell-wall constituents can be expected than if the enzymes are so large that they are confined to the surfaces of the gross capillaries. The area exposed on the gross capillary surfaces of one gram of wood or cotton is sufficient to accommodate about 3×10^{15} randomly oriented enzyme molecules 200×35 A. in size. That is equivalent to approximately 3 mg. of enzyme protein per gram of wood or cotton.

As discussed earlier in this review most of the hemicelluloses, lignin, and extraneous and mineral constituents of cotton and wood occupy the spaces between microfibrils and the cellulose molecules in the amorphous regions. When moisture free, these regions contain a very small amount of void space. The density of dry wood, wood pulp, cotton, and ramie fibers was shown by measurements of mercury intrusion to be very close to that of the cell wall components themselves (*68*). When moisture is adsorbed, however, the structure swells and an appreciable amount of void space is created within the cell walls. The dimensions of these so-called transient cell-wall capillaries (solid-solution structures) have been determined recently by using a series of dextran or polyethylene glycol polymers of known molecular size "go-no-go-gauges" to measure the dimensions of cell-wall capillaries. This method, which has been called the "inaccessible water" technique, was developed by Aggebrandt and Samuelson (*1*) and adapted to wood by Stone and Scallan (*65, 66, 67*) and Tarkow, Feist, and Southerland (*72*). Figure 11 shows the results of such an analysis plotted as a frequency distribution of capillary dimensions in water-swollen wood, cotton, and wood pulp. The approximate median and maximum dimensions of cell-wall capillaries in these three materials were, respectively, about 10 and 35, 5 and 75, and 25 and 150 A. It should be emphasized that the dextran and polyethylene glycol molecules used to obtain these estimates do not have the specific affinity for the substrate that a cellulolytic enzyme has. The substantial increase in median and maximum pore size during pulping results in part by removal of lignin and hemicelluloses from the cell-wall capillaries. How do these dimensions compare with the size of the cellulolytic enzymes?

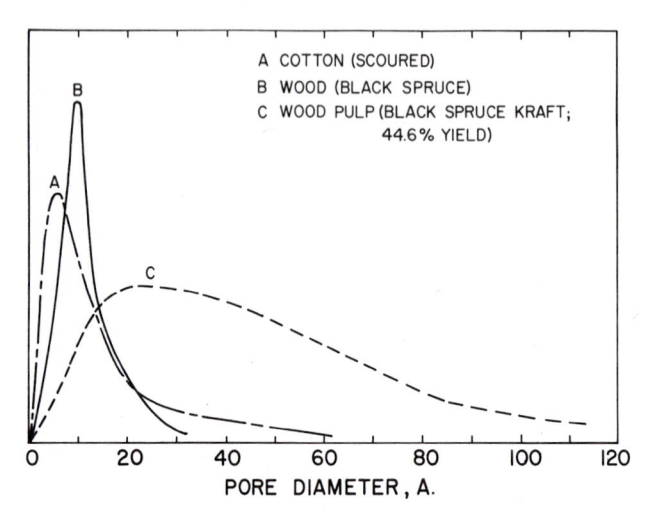

Figure 11. Frequency distribution for cell wall capillaries in (A) scoured cotton, (B) black spruce wood (Picea mariana (Mill.) B.S.P.) and (C) kraft pulp from black spruce wood as determined by the "inaccessible water" technique by Aggebrandt and Samuelson (1) and Stone and Scallan (66)

SIZE AND SHAPE OF CELLULASE MOLECULES. The cellulolytic enzymes of microorganisms, in common with other enzymes, are water-soluble protein molecules of high particle weight. The most complete physicochemical characterization of a cellulolytic enzyme molecule to date remains the original work of Whitaker, Colvin, and Cook (78) on the cellulase of *Myrothecium verrucaria*. Their data indicate that this particular enzyme is an asymmetric globular protein with a molecular weight of about 63,000. In their words, "it corresponds to a cigar-shaped molecule roughly 200 A. long and 33 A. broad at the widest point." The validity of their judgment about the shape of this cellulase molecule is supported by comparison of hydrodynamic parameters (Table III) and estimates of size and shape derived from available viscosity and diffusion data for a well-characterized globular protein, serum albumin, and an elongated rod-shaped protein, tropomyosin (Table IV). The properties of the cellulase molecule are much closer to those of serum albumin than to tropomyosin.

Estimates of the dimensions of the cellulolytic enzymes of other microorganisms are given in Table V. In some cases (marked GF) the dimensions given were calculated by the procedure of Laurent and Killander (*36*) from available gel-filtration data on the enzymes. In other cases (marked MW) the dimensions were obtained by interpolating in a

Table III. Comparison of Hydrodynamic Parameters for a Cellulolytic Enzyme from *Myrothecium verrucaria* (*78*), a Globular Protein, Serum Albumin, and an Elongated Rod-shaped Protein, Tropomyosin

Protein	Molecular Weight $\overline{M_{sD}}$[a]	Sedimentation Coefficient $S_o \times 10^{13}$	Diffusion Coefficient $D_o \times 10^7$	Frictional Ratio[b] f/f_{min}	Intrinsic Viscosity $[n]$ ml./g.
Cellulase	63,000	3.72	5.61	1.44	8.70
Serum albumin	65,000	4.31	5.94	1.35	3.70
Tropomyosin	93,000	—	2.24	3.22	52

[a] Based on sedimentation diffusion data.
[a] f_{min} refers to the frictional factor for the hydrodynamically equivalent sphere.

Table IV. Estimates of Size and Shape of a Cellulolytic Enzyme from *Myrothecium verrucaria* (*78*), a Globular Protein, Serum Albumin, and an Elongated Rod-shaped Protein, Tropomyosin

	Fully Solvated Equivalent Sphere		Unsolvated Equivalent Ellipsoid	Partially Solvated Equivalent Ellipsoid
	Water of Solvation g/g	Diameter A.	Length × Width A.	Length × Width A.
Derived from viscosity data				
Cellulase	2.8	90	380 × 40	300 × 50
Serum albumin	0.8	70	250 × 40	230 × 45
Tropomyosin	20.0	180	1700 × 60	—
Derived from diffusion data				
Cellulase	1.5	80	300 × 40	250 × 40
Serum albumin	1.1	70	250 × 40	210 × 40
Tropomyosin	23.0	190	3000 × 50	—

log-log plot of molecular radius *vs.* molecular weight for the cellulase of *Myrothecium verrucaria* and other well-characterized globular proteins (Figure 12) using data summarized by Tanford (*71*). A considerable range of dimensions is shown for the various cellulase preparations included in Table V. If the enzymes are spherical, they range from about 25 to 80 A. in diameter with an average of 60 A. If the enzymes are ellipsoids with an axial ratio of about six, as demonstrated for the enzyme

Table V. Estimated Size of Cellulases of Various Fungi Calculated from Given Are for Hydrodynamically Equivalent Spheres or Ellipsoids with verrucaria by Whitaker,

Organism	Molecular Weight	Equivalent Sphere Diameter, A.
Aspergillus niger	—	60
Aspergillus niger (Rohm and Haas Cellulose 36)	—	30
		25
Chrysosporium lignorum	—	35
Fomes annosus	—	30
Myrothecium verrucaria	63,000	40
		40
Myrothecium verrucaria	49,000	80
Myrothecium verrucaria	55,000	65
	30,000	70
	30,000	50
	5,300	55
Penicillium notatum	35,000	25
Penicillium notatum	35,000	65
Polyporus versicolor	51,000	55
	11,400	65
Stereum sanguinolentum	20,500	35
	—	60
Stereum sanguinolentum	—	45
Trichoderma koningi	50,000	35
	26,000	65
Trichoderma viride	76,000	45
	49,000	75
Trichoderma viride	61,000	65
	12,600	70
	39,000	35
Average for all enzymes taken together		50

[a] Dimensions for the equivalent ellipsoid include water of solvation assumed to be 0.2 gram water per gram protein.
[b] Estimates were calculated from either: gel filtration (GF) data using the correlation between elution volume and molecular radius according to the relationship de-

studied by Whitaker, Colvin, and Cook (78), they range from about 15 × 80 A. to 40 × 250 A. in width and length, respectively, with an average of 35 × 200 A.

PENETRATION OF CAPILLARIES IN NATURAL FIBERS. The maximum dimensions of cellulolytic enzyme molecules listed in Table V are all smaller than the gross capillaries of both wood and cotton fibers. Thus,

Available Gel Filtration or Molecular Weight Data. The Dimensions an Axial Ratio of Six as Demonstrated for the Cellulase of *Myrothecium* Colvin, and Cook (78)

Equivalent Ellipsoid[a] W × L, A.	Method of Estimation	Investigator
30 × 190	GF	Pettersson (50)
15 × 100	GF	Ahlgren, Eriksson, and
15 × 85	GF	Vesterberg (3)
20 × 120	GF	Ahlgren, Eriksson, and
15 × 90	GF	Vesterberg (3)
25 × 140	GF	Ahlgren and Eriksson (2)
20 × 130	GF	
40 × 250	MW	Whitaker, Colvin, and Cook (78)
35 × 210	MW	Datta, Hanson, and Whitaker (12)
35 × 220	MW	Selby and Maitland (62)
30 × 170	MW	
30 × 190	GF	
15 × 80	MW	
35 × 210	GF	Pettersson (50)
30 × 180	MW	Pettersson and Eaker (52)
35 × 210	MW	Pettersson, Cowling, and Porath (51)
20 × 110	MW	
25 × 140	GF	Eriksson and Pettersson (14)
25 × 140	MW	
20 × 120	GF	Ahlgren and Eriksson (2)
35 × 210	MW	Iwaski, Hayashi, and Funatsu (28)
25 × 150	MW	
40 × 250	MW	Li, Flora, and King (39)
35 × 210	MW	
40 × 230	MW	Selby and Maitland (63)
20 × 110	MW	
30 × 170		

veloped by Laurent and Killander (36); or molecular weight (MW) data using a log-log plot of molecular radius of the equivalent hydrodynamic sphere against molecular weight constructed from experimentally determined diffusion constants and molecular weights of globular proteins as given by Tanford (71), see Figure 12.

from consideration of size and shape alone, these molecules would be expected to diffuse readily within the gross capillaries and act on cellulose molecules exposed on the surface of these capillaries. By the same token, however, only a small fraction of the cell-wall voids in water-swollen wood and cotton fibers are sufficiently large to permit penetration of most of the cellulolytic enzyme molecules described in Table V. The pulp prepared from spruce wood contains cell-wall capillaries that are adequate to admit many of the enzyme molecules as spheres or by endwise penetration if they are ellipsoids (Figure 11). But since both rotational and translational modes of motion are generally assumed during diffusion of dissolved solute molecules, it is most realistic to assume that pores as large or larger than the largest (rather than the smallest) dimension of a given enzyme molecule would be required for unimpeded accessibility. For this reason, it is likely that cellulolytic enzyme molecules are physically excluded from all but the largest capillaries. Consequently, they must gain access to the interior regions of the cell wall by enlarging the existing capillaries.

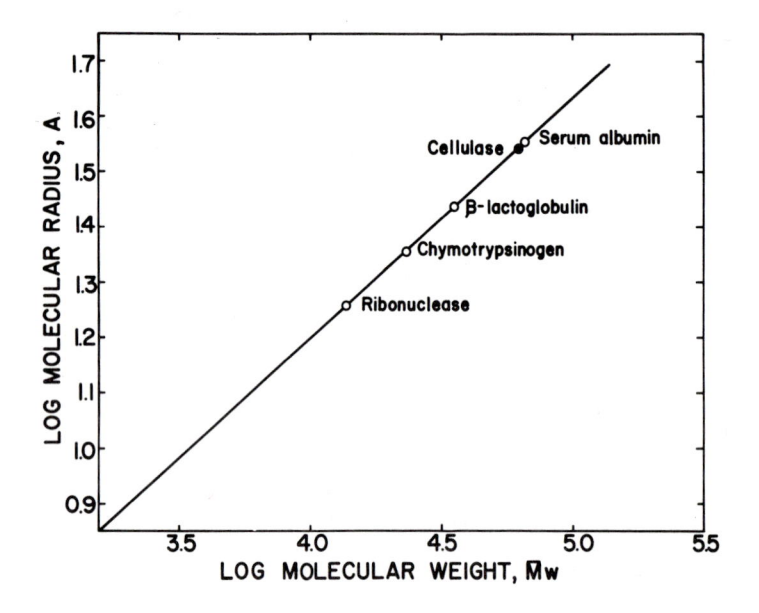

Figure 12. Logarithmic plot of the molecular radius of the equivalent hydrodynamic sphere against molecular weight for cellulase of Myrothecium verrucaria *(77) and other globular proteins. This relationship was used to estimate the probable size of cellulolytic enzymes of other fungi as shown in Table V*

In addition to considerations of size and shape, the surface properties of the fiber capillaries and the diffusibility of the cellulolytic enzyme molecules within them can profoundly influence the susceptibility of cellulose to enzymatic hydrolysis. In contrast to inorganic catalysts, enzymes have a very strong and specific affinity for their specific substrate molecules. This affinity accounts for their susceptibility to competitive inhibitors. When the substrate exists as an insoluble polymer in a complex structural matrix, this specific affinity drastically reduces the rate of diffusion of the enzyme in the presence of the substrate. This retarded diffusion has been shown experimentally by Reese and Norkrans in their attempts to purify cellulolytic culture filtrates. It probably also accounts for the observation by Wilcox (77) that at least one hyphal strand must be present in each wood cell that is degraded by wood-destroying fungi; it suggests that these extracellular enzymes may not diffuse outside the cell in which they are secreted.

Although no doubt exists that extensive degradation takes place on the gross capillary surfaces of wood and cotton fibers in close proximity to microbial cells, Cowling (8) has presented evidence that the submicroscopic capillary surfaces also are involved in cellulose degradation at least in the presence of microorganisms. Microscopic examination of decayed wood samples showed that the weight loss due to decay cannot be accounted for by formation of bore holes (Figures 7 and 8). Also, measurements of progressive changes in alpha-, beta-, and gamma-cellulose, average degree of polymerization of holocellulose (Figure 13), and hygroscopicity of wood during decay by the brown-rot fungus, *Poria monticola,* all indicate penetration of the entire secondary wall structure. During the initial stages of decay the attack apparently occurred primarily in the amorphous regions; in later stages the residual crystalline material also was degraded. These observations strongly suggest that the hydrolytic catalysts of this organism are able readily to penetrate and act within (although possibly after enlargement of) the cell-wall capillaries.

These conclusions for the brown-rot fungus, *Poria monticola,* are in distinct contrast to those drawn from the same analyses for the white-rot fungus, *Polyporus versicolor.* In this case the cellulases appeared to depolymerize the wood polysaccharides in such a way that the degree of polymerization of the residual cellulose remained high in all stages of decay (Figure 13). In this case, both crystalline and amorphous cellulose were degraded simultaneously at rates that were essentially constant in all stages of decay and proportional to the amounts present in undecayed wood (8).

One possible explanation for these different modes of cellulose depolymerization in the same species of wood is that the cellulolytic enzyme molecules of *Poria monticola* are smaller than those of *Polyporus versicolor* and for that reason would be able to penetrate and act in regions of the fine structure of the fibers that are not accessible to those of the latter fungus. This hypothesis has led to efforts (as yet incomplete) to determine the molecular size of the cellulolytic enzyme proteins of these two organisms. Another possible explanation is that the initial dissolution of cellulose and other cell-wall polysaccharides is accomplished by catalysts that are not enzyme proteins and therefore could be substantially smaller in molecular size. Halliwell (21) has described experiments on the

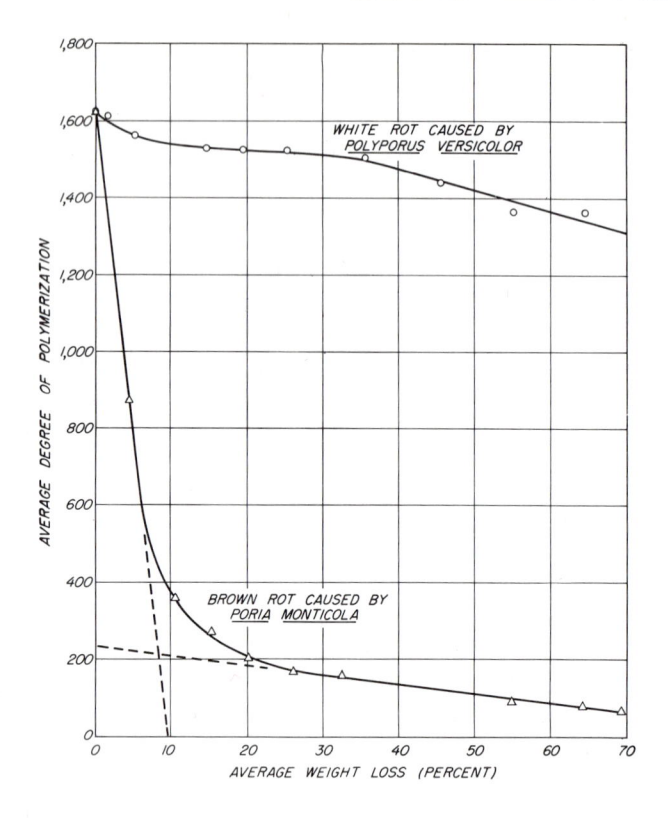

Figure 13. Average degree of polymerization (DP) of the holocellulose of sweetgum sapwood (Liquidambar styraciflua L.) *in progressive stages of decay by* Polyporus versicolor L. *and* Poria monticola *Murr. Extrapolation of the rapidly descending portion of the lower curve to zero DP indicates that about 10% of the wood is easily degraded by* Poria monticola. *Extrapolation of the gradually descending portion of the same curve to zero weight loss indicates that the average DP of the resistant (presumably crystalline) fraction of the holocellulose is about 230 glucose units. The curves indicate that the dissolution of carbohydrates by the enzymes (or other lytic catalysts) of* Poria monticola *enter and act within the cell wall capillaries whereas those of* Polyporus versicolor *act mainly on the gross capillary surfaces of the wood*

catalytic decomposition of cellulose under physiological conditions with hydrogen peroxide and ferrous salts.

Degree of Crystallinity. The influence of degree of crystallinity on the susceptibility of cellulose to enzymatic hydrolysis has been studied by Norkrans (48), Walseth (74, 75) and Reese, Segal, and Tripp (57), among others. Using several cellulose samples precipitated after swelling in phosphoric acid, Walseth (74) showed that the preparations with lower moisture regain values (higher crystallinity) were more resistant to hydrolysis by isolated enzymes of *Aspergillus niger* than those with higher regain values. Using x-ray diffraction data Norkrans (49) drew the same conclusion for cellulose regenerated from cuprammonium solution. She concluded that the cellulolytic enzymes were degrading the more readily accessible amorphous portions of her regenerated cellulose and were unable to attack the less accessible crystalline material. Complete degradation of the resistant residues was possible only by successive swelling and enzyme treatments.

Such swelling pretreatments do not appear necessary for cellulases to degrade native cellulosic materials in the presence of the organism. Preferential attack of amorphous cellulose also is indicated in some organisms but not in all. Figure 13 shows the rate of change in degree of polymerization (DP) of holocellulose obtained from sweetgum wood after progressive stages of decay by *Poria monticola* and *Polyporus versicolor*. Extrapolation of the initial linear portion of the lower curve to zero DP indicates that the fraction of sweetgum sapwood that was readily hydrolyzed by *P. monticola* constituted about 10% of the original wood substance. This estimate is in substantial agreement with estimates of the proportion of amorphous cellulose in several native cellulosic materials as determined by acid hydrolysis. Extrapolation of the second linear portion of the lower curve to zero weight loss indicates that the more resistant fraction of the holocellulose consisted of entities with an average length of about 230 DP units. This value is in substantial agreement with estimates of the length of crystallites in other cellulosic materials as determined by acid hydrolysis.

The slow, gradual decrease in DP of holocellulose in the case of the white-rot fungus, *Polyporus versicolor,* indicates that the organism degraded the crystalline and amorphous cellulose simultaneously and at rates proportional to the amounts originally present. This conclusion was confirmed by moisture regain and x-ray diffraction analyses (8).

In Norkrans' work (49), the rate of hydrolysis also was followed by degree of polymerization measurements. The rates were initially very rapid but much slower in later stages to the point of complete inhibition at a "leveling off DP" analogous to that observed during acid hydrolysis of cellulose. As in the case of acid hydrolysis, there also is observed during hydrolysis with enzymes, a significant increase in crystallinity during the reaction. As the more accessible portions of the cellulose molecules are hydrolyzed, they became shorter but more free to aggregate with one another and thus become more resistant to further enzymatic action.

From these observations it follows that any treatment which will alter the proportion of crystalline material or the degree of perfection (parallelism) of the crystallites present may modify the susceptibility of the material to enzymatic attack. Treatments that would increase susceptibility include: reprecipitation from solution, mechanical disruption such as in a vibratory ball mill, or ionizing radiation.

The Unit Cell Dimensions of the Crystallites Present. Cellulose occurs in four recognized crystal structures designated Cellulose I, II, III, and IV (27). These can be distinguished by their characteristic x-ray diffraction patterns. Cellulose I is the crystal form in native cellulosic materials. Cellulose II is found in regenerated materials such as viscose filaments, cellophane, and mercerized cotton. Cellulose III and IV are formed by treatment with anhydrous ethylamine and certain high temperatures, respectively. These four crystal forms differ in unit cell dimensions—i.e., the repeating three-dimensional unit within the crystalline regions. These dimensions are shown in Table VI for the four crystal forms.

Since it is essentially impossible to prepare Celluloses II, III, and IV without also altering the degree of crystallinity of the material, it is difficult to determine whether their modified susceptibility to enzymatic degradation is owing to changes in the unit cell dimensions of the crystallites alone or also to differences in the amount of amorphous material present.

Table VI. The Unit Cell Structures of the Various Crystal Forms of Cellulose

Axis	Cellulose I	Cellulose II	Cellulose III	Cellulose IV
a	8.35 A.	8.1 A.	7.7 A.	8.1 A.
b	10.3 A.	10.3 A.	10.3 A.	10.3 A.
c	7.9 A.	9.1 A.	9.9 A.	7.9 A.
Angle (B) between a and c axes	84°	62°	58°	90°

Recently Rautela (55) cultivated *Trichoderma viride* in separate synthetic media containing as sole source of carbon, cellulose preparations of all four crystal forms and then determined the activation energies for the reaction of the enzymes produced on each separate substrate during its activity on all four substrates. The results show that the organism produced enzymes that required minimal activation energy for catalysis of the specific substrate on which it was cultivated. These results are remarkable because they indicate that the enzyme-synthesizing machinery of this fungus can adapt the structure of the active site of the enzyme so as to accommodate a specific crystal lattice structure in its substrate. This observation also reinforces the generalizations in *Enzyme-Substrate Relationships*.

Conformation and Steric Rigidity of the Anhydroglucose Units. King (*31*) suggested that the greater resistance of crystalline as compared with amorphous cellulose may not be due just to its physical inaccessibility to enzyme molecules but also to the conformation and steric rigidity of the anhydroglucose units within the crystalline regions. In these regions the glucose units occur in a so-called Cl (chair) conformation with alternate glucopyranose units oriented in opposite directions within the lattice. It is possible, although not necessarily likely, that this particular conformation and planar orientation of the glucose units is the one of best fit in the active site of the enzyme. It is also possible, but still more unlikely in terms of modern understanding of enzyme specificity, that the active site of the enzyme is designed in such a way that the cellulose chains on all sides of a crystallite are in a steric orientation equally ideal for complex formation and subsequent substrate cleavage.

These considerations strongly suggest the possibility that an induced fit mechanism as described by Thoma and Koshland (*73*) for beta-amylases may be involved in cellulose hydrolysis by enzymes. The possibility for adjustments in conformation and respective orientations of the glucose rings in cellulose increases as the degree of crystallinity decreases and the degree of hydration increases. It is in exactly this way that susceptibility of cellulose to enzymatic degradation also increases. Thus, it seems highly probable that conformation and steric rigidity as well as accessibility may have a bearing on the resistance of crystalline cellulose to enzymatic hydrolysis. Experimental verification of these speculations provides a very challenging problem in the stereospecificity of enzyme reactions.

Degree of Polymerization of the Cellulose. The length of cellulose molecules in a fiber varies over a wide range from gamma-celluloses containing less than 15 glucose units to alpha-cellulose with as many as 10,000 to 14,000 glucose units per molecules. This variation would be expected to affect the rate of hydrolysis very considerably, particularly by enzymes that cleave the cellulose molecules by an endwise mechanism. King (*32*) has discussed evidence for endwise hydrolysis of cellulose. But most isolated cellulases studied to date apparently hydrolyze cellulose at random along the length of the molecules. Only slight differences have been noted in the resistance of cellulose preparations of different average DP to a wide variety of fabric-destroying fungi (*20, 64*).

Gilligan and Reese (*19*) have shown that chromatographically separated cellulolytic components of *Trichoderma viride* showed as high as twofold differences in rate of hydrolysis of regenerated cellulose preparations of DP 500 and DP 50, respectively. They suggested that cellulolytic enzymes may vary in affinity for chains of different DP, but their results may in part have been the result of undetermined differences in accessibility of their particular substrates.

When the DP of cellulose is reduced during acid hydrolysis, the disaggregated cellulose chain ends in the amorphous regions have a tendency to recrystallize and make the residue more resistant to enzymatic hydrolysis. This has been observed by Norkrans (49). When the DP of cellulose is reduced to such an extent (DP 6 to 10) that the molecules are soluble and thus no longer maintain their structural relationships one with another, quite obviously there will be a great increase in susceptibility to enzymatic hydrolysis. But this change is more the result of solubilization than it is the direct result of chain shortening.

Thus, DP of itself probably is of limited significance in determining the susceptibility of cellulose to enzymatic hydrolysis except in the relatively rare case of enzymes that act by an endwise mechanism.

The Nature of the Substances with Which the Cellulose Is Associated and the Nature of That Association. As discussed earlier in this review, cellulose in cotton and wood is associated with a variety of other materials. These substances frequently have an influence on the susceptibility of the cellulose to enzymatic hydrolysis. The influence of several different types of material will be discussed in turn.

MINERAL CONSTITUENTS. Cellulose fibers usually contain about 1% ash. The mineral elements contained in the ash include all those essential for the growth and development of cellulolytic microorganisms. This fact has led to speculations and certain experimental treatments by Baechler (4) and others showing that chelating agents applied to cotton or wood can so effectively bind these essential mineral elements that the growth of cellulolytic microorganisms is prevented.

In untreated fibers, however, the essential elements are readily available to the organisms and these play their characteristic metabolic roles. Mandels and Reese (40) have studied the induction of cellulase by various metal ions and have shown that cobalt is particularly stimulatory to cellulase production. Gascoigne and Gascoigne (18) have summarized recent literature on the inhibition and activation of cellulases by various inorganic substances. They concluded that mercury, silver, copper, chromium, and zinc salts are generally inhibitory whereas manganese, cobalt, magnesium, and calcium in the presence of phosphate have been reported as stimulatory, though exceptions to these generalizations also have been reported.

EXTRANEOUS MATERIALS. The extraneous materials in cotton and wood include a wide variety of organic substances that are soluble in such neutral solvents as acetone, ether, alcohol, benzene, and water. The water-insoluble substances usually are deposited within the gross capillary structure of the fibers; the water-soluble materials, at least in part, are deposited in the fine capillary structure within the fiber walls.

These materials influence the susceptibility of cellulose fibers to enzymatic degradation in five major ways: (1) Growth promoting substances such as vitamins (particularly thiamin) and certain soluble carbohydrates provide substrates for rapid growth and development of cellulolytic microorganisms within cellulose fibers. (2) Poisonous substances, particularly toxic phenolic materials, inhibit the normal growth and development of cellulolytic organisms. (3) Various substances are deposited within the fine capillary structure of the cell wall and thus reduce the accessibility of the cellulose to extracellular enzymes. (4) Certain specific enzyme inhibitors act directly to reduce the rate or extent of enzymatic hydrolysis of the cellulose. (5) The very small amounts of nitrogen and phosphorus present make both wood and cotton comparatively deficient substrates for microorganisms that are not specifically adapted to cope with these deficiencies.

Jennison (29) has shown that thiamin is required for growth of many wood-destroying fungi. Kennedy (30) has shown that the arabino-galactan of larch wood provides a substrate which the fungi use in becoming initially established within this wood.

A recent review by Scheffer and Cowling (60) summarizes extensive evidence that the natural durability of many wood species is because of the toxicity of certain phenolic substances that are deposited in the process of heartwood formation. These substances act as poisons to the cellulolytic microorganism rather than by direct action on the enzymatic process of deterioration.

The deposition of extraneous materials within the fine structure of wood fibers has been shown to reduce the accessibility of the cellulose in the fiber wall to the hydrolyzing action of strong acids and bases and also to water as shown by the reduced moisture regain capacity of certain woods high in extractives content, such as redwood. Although it has not been demonstrated experimentally, it is logical to suspect that these substances also would reduce the accessibility of cellulose to cellulases.

The presence of specific enzyme inhibitors in a wide variety of plant materials including the heartwood of certain wood species has been reviewed recently by Mandels and Reese (41). Many of these materials are phenols. They usually are more effective against the C_1 than the Cx components (Figure 11).

The deficiency of nitrogen and phosphorus in wood fibers increases their natural resistance to microbiological degradation (11). The efficiency of wood-destroying fungi in utilizing wood cell wall substance despite this deficiency probably is a result of preferential allocation of available nitrogen to cellulolytic (and presumably other) enzymes essential to the utilization of cell-wall constituents (37) and to a dynamic and continuous system of autolysis of older less active cells and reuse of their cytoplasmic constituents by younger more active cells (38). A hypha in the process of autolysis is shown in Figure 9.

LIGNIN. The combination of lignin with partially crystalline cellulose that occurs in wood forms one of the natural materials most resistant to enzymatic hydrolysis. Many virulently cellulolytic microorganisms, including many cotton-fabric-destroying fungi and rumen bacteria, are prevented from degrading the cellulose in wood by this association with lignin. Although the nature of the detailed chemical structure of lignin has finally become reasonably clear through the work of Freudenberg (16) and others, the nature of the association between lignin and the wood polysaccharides is not well understood. To degrade the cellulose in wood, organisms must possess not only the ability to degrade cellulose but also to degrade the lignin or at least break down its association with cellulose. The alkyl-phenyl carbon-to-carbon bonds in certain lignin model compounds can be cleaved by an extracellular phenoloxidase of the laccase type. This mechanism of cleavage may be important in the enzymatic degradation of lignin by wood-destroying fungi of the white-rot type (34, 35).

A chemical bond between lignin and cellulose has been postulated by many investigators (17). Present evidence suggests, however, that the association is largely physical in nature—the lignin and amorphous cellulose forming a mutually interpenetrating system of high polymers. Lignin apparently decreases the accessibility of wood cellulose to enzyme molecules diffusing within the fine structure of wood fibers. Interestingly, this decreased accessibility to enzymes is achieved in wood without a corresponding decrease in moisture regain capacity as compared to cotton.

The Nature, Concentration, and Distribution of Substituent Groups. Because of the great resistance of native cellulosic materials to hydrolysis by isolated enzymes, soluble derivatives of cellulose, particularly sodium carboxymethylcellulose (CMC), have been widely used as test substrates in studies of cellulolytic enzymes. These materials circumvent the prehydrolytic phase of cellulose degradation by providing isolated anhydroglucose chain molecules in solution.

Cellulose derivatives are formed by replacing the hydrogen of the primary and secondary hydroxyl groups of cellulose with such groups as methyl, ethyl, hydroxyethyl, carboxymethyl, etc. The resulting cellulose ethers are very stable; the substituent groups are not readily removed by enzymatic means.

The addition of these groups usually makes cellulose non-crystalline and more soluble in water in proportion to its degree of substitution (DS) and the solvating capacity of the substituent groups. (DS refers to the average number of substituent groups attached to each glucose unit in the cellulose.) The DS at which complete solubility is attained ranges from 0.5 to 0.7 depending on the solvating capacity of the substituents (58).

The susceptibility of substituted cellulose derivatives to enzymatic hydrolysis increases as they become more water soluble and less crystalline up to the point of complete solubility (56). After this point, susceptibility decreases with increasing DS until complete immunity to enzymatic action results, usually at a DS somewhat greater than 1.0. Apparently a substituent group on each glucose unit protects a cellulose sample from hydrolysis. Actually a substituent group on every other glucose unit insures resistance to enzymatic degradation. Selective substitution of alternate glucose units cannot be achieved practically, however. Substituent groups of large molecular dimensions are more effective in contributing resistance to enzymatic degradation than small groups (56).

The average DS for a particular sample does not indicate that that particular number of groups has been added to each glucose unit. This is of particular importance when cotton and wood are substituted to provide decay resistance. The substituting reagents must penetrate the fine structure of the fibers and will react most readily with the most accessible molecules. Thus, all of the structural features of cellulose that influence its accessibility to enzymes, as discussed in this review, also will influence its accessibility to the reagents of substitution. Since these reagents are considerably smaller in molecular size than enzymes, the distribution of substituent groups will be most high, and therefore most effective, at the very sites at which enzyme molecules would be most likely to act. A determination of average DS of a fiber does not indicate the distribution of substituents within it. Thus, the proportion of unsubstituted glucose units in the amorphous regions would, if measurable, be a better index of resistance than DS alone.

All of the above generalizations can be interpreted in terms of the accessibility of the cellulose to enzyme molecules. Direct physical contact with the hydroxyl groups of the cellulose chain itself probably is necessary for hydrolysis. Substituent groups can effectively prevent this contact by blocking the hydroxyl groups themselves. The larger the substituent group, the further away the enzyme will be held; the more substituent groups present per glucose unit, the more certainly the enzymes will be held away from all potential sites for contact; the more hydrophilic the substituent, the more highly swollen or more soluble the cellulose will be in water, and the greater will be the probability of contact between the enzyme in solution and adjacent unsubstituted groups in the cellulose chain molecules.

Summary

In summary it is appropriate to reiterate the refrain of this entire review. The susceptibility of cellulose to enzymatic hydrolysis is determined largely by its accessibility to extracellular enzymes secreted by or bound on the surface of cellulolytic microorganisms. Direct physical contact between these enzymes and the cellulosic substrate molecules is an essential prerequisite to hydrolysis. Since cellulose is an insoluble and structurally complex substrate, this contact can be achieved only by diffusion of the enzymes from the organism into the complex structural matrix of the cellulose. Any structural feature that limits the accessibility of the cellulose to enzymes by diffusion within the fiber will diminish the susceptibility of the cellulose of that fiber to enzymatic degradation. In this review, eight such structural features have been discussed in detail. Present knowledge based on experimental evidence, and speculations based on logical inferences by us and others cited, are presented in the hope that further thought and experimentation about these subjects eventually will provide more complete understanding of the enzymatic processes by which cellulose is degraded by microorganisms.

Literature Cited

(1) Aggebrandt, L. G., Samuelson, O., *J. Appl. Polymer Sci.* **8**, 2801 (1964).
(2) Ahlgren, E., Eriksson, K. E., *Acta Chem. Scand.* **21**, 1193 (1967).
(3) Ahlgren, E., Eriksson, K. E., Vesterberg, O., *Acta Chem. Scand.* **21**, 937 (1967).
(4) Baechler, R. H., *Appl. Microbiol.* **4**, 229 (1956).
(5) Bailey, I. W., Vestal, M. R., *J. Arnold Arboretum* **18**, 196 (1937).
(6) Colvin, J. R., *Polymer Letters* **4**, 747 (1966).
(7) Coté, W. A., Jr., Ed., "Cellular Ultrastructure of Woody Plants," p. 603, Syracuse University Press, Syracuse, New York, 1965.
(8) Cowling, E. B., *U. S. Dept. Agr. Tech. Bull.* No. **1258**, p. 79 (1961).
(9) Cowling, E. B., "Biological Degradation of Cellulose and Related Materials," E. T. Reese, Ed., p. 1, Pergamon Press, London, 1963.
(10) Cowling, E. B., "Cellular Ultrastructure of Woody Plants," W. A. Cote, Ed., p. 341, Syracuse University Press, Syracuse, New York, 1965.
(11) Cowling, E. B., Merrill, W., *Can. J. Bot.* **44**, 1539 (1966).
(12) Datta, P. K., Hanson, K. R., Whitaker, D. R., *Can. J. Biochem. Phys.* **41**, 697 (1963).
(13) Duncan, C., *U. S. For. Prod. Lab. Rept.* No. **2173**, p. 28 (1960).
(14) Eriksson, K. E., Pettersson, G., *Arch. Biochem. Biophys.* **124**, 142 (1968).
(15) Etheridge, D. E., *Can. J. Bot.* **35**, 615 (1957).
(16) Freudenberg, K., *Science* **148**, 595 (1965).
(17) Freudenberg, K., Harkin, J. M., *Chem. Ber.* **93**, 2814 (1960).
(18) Gascoigne, J. A., Gascoigne, M. M., "Biological Degradation of Cellulose," p. 264, Butterworths, London, 1960.
(19) Gilligan, W., Reese, E. T., *Can. J. Microbiol.* **1**, 90 (1954).
(20) Greathouse, G., *Text. Res. J.* **20**, 227 (1950).
(21) Halliwell, G., *Biochem. J.* **95**, 35 (1964).

(22) Hamby, D. S., "The American Cotton Handbook," Interscience, N. Y., 1965.
(23) Harkin, J. M., *Fortschr. Chem. Forsch.* **6**, 101 (1966).
(24) Hearle, J. W. S., Peters, R. H., Eds., "Fibre Structure," p. 667, Butterworths, Manchester, England, 1963.
(25) Hess, K., Mahl, H., Gütter, E., *Kolloid Z.* **155**, 1 (1957).
(26) Hillis, W. E., "Wood Extractives and Their Significance to the Pulp and Paper Industries," p. 513, Academic Press, New York, 1962.
(27) Honeyman, J., "Recent Advances in the Chemistry of Cellulose and Starch," p. 496, Interscience Publishers, Inc., New York, 1959.
(28) Iwasaki, T., Hayashi, K., Funatsu, M., *J. Biochem. (Tokyo)* **55**, 209 (1964).
(29) Jennison, M. W., "Physiology of Wood-rotting Fungi," Final report No. 8, p. 151, Office of Naval Research, Microbiology Branch, Syracuse University, Syracuse, N. Y., 1952.
(30) Kennedy, R. W., *For. Prod. J.* **8**, 308 (1958).
(31) King, K. W., *Va. Agr. Expt. Sta. Tech. Bull.* No. **154**, 55 (1961).
(32) King, K. W., "Advances in Enzymic Hydrolysis of Cellulose and Related Materials," E. T. Reese, Ed., p. 150, Pergamon Press, London, 1963.
(33) King, N. J., *Nature* **218**, 1173 (1968).
(34) Kirk, T. K., Harkin, J. M., Cowling, E. B., *Biochim. Biophys. Acta* **165**, 134 (1968).
(35) Kirk, T. K., Harkin, J. M., Cowling, E. B., *Biochim. Biophys. Acta* **165**, 145 (1968).
(36) Laurent, T. C., Killander, J., *J. Chromatog.* **14**, 317 (1964).
(37) Levi, M. P., Cowling, E. B., *Phytopathology* **59**, (in press).
(38) Levi, M. P., Merrill, W., Cowling, E. B., *Phytopathology* **58**, 626 (1968).
(39) Li, L. H., Flora, R. M., King, K. W., *Arch. Biochem. Biophys.* **111**, 439 (1965).
(40) Mandels, M., Reese, E. T., *J. Bacteriol.* **73**, 269 (1957).
(41) Mandels, M., Reese, E. T., *Ann. Rev. Phytopathology* **3**, 85 (1965).
(42) Manley, R. St. J., *Nature* **204**, 1155 (1964).
(43) Marsh, P. B., Simpson, M. E., *Phytopathology* **55**, 52 (1965).
(44) Marx-Figini, M., Schulz, G. V., *Biochim. Biophys. Acta* **112**, 81 (1966).
(45) Meier, H., "The Formation of Wood in Forest Trees," M. H. Zimmerman, Ed., p. 137, Academic Press, N. Y., 1964.
(46) Merrill, W., Cowling, E. B., *Can. J. Bot.* **44**, 1555 (1966).
(47) Mühlethaler, K., *Ann. Rev. Plant Phys.* **18**, 1 (1967).
(48) Norkrans, B., *Symbolae Botanicae, Uppsala, Sweden*, No. **11**, p. 126 (1950).
(49) Norkrans, B., *Physiol. Plant.* **3**, 75 (1950).
(50) Pettersson, G., *Biochim. Biophys. Acta* **77**, 665 (1963).
(51) Pettersson, G., Cowling, E. B., Porath, J., *Biochim. Biophys. Acta* **67**, 1 (1963).
(52) Pettersson, G., Eaker, D. L., *Arch. Biochem. Biophys.* **124**, 154 (1968).
(53) Preston, R. D., Cronshaw, J., *Nature* **181**, 248 (1958).
(54) Proctor, P., *School of Forestry Bull., Yale Univ.*, No. **47**, p. 31 (1941).
(55) Rautela, G. S., PhD Thesis, Virginia Polytech. Inst., Blacksburg, Va., p. 108 (1967).
(56) Reese, E. T., *Ind. and Eng. Chem.* **49**, 89 (1957).
(57) Reese, E. T., Segal, L., Tripp, V. W., *Text. Res. J.* **27**, 626 (1957).
(58) Reese, E. T., Siu, R. G. H., Levinson, H. S., *J. Bact.* **59**, 485 (1950).
(59) Rollins, M. L., *Forest Prod. J.* **18**, 91 (1968).
(60) Scheffer, T. C., Cowling, E. B., *Ann. Rev. Phytopath.* **4**, 147 (1965).
(61) Selby, K., "Advances in Enzymic Hydrolysis of Cellulose and Related Materials," E. T. Reese, Ed., p. 33, Pergamon Press, London, 1963.
(62) Selby, K., Maitland, C. C., *Biochem. J.* **94**, 578 (1965).

(63) Selby, K., Maitland, C. C., *Biochem. J.* **104,** 716 (1967).
(64) Siu, R. G. H., "Microbial Decomposition of Cellulose," p. 531, Reinhold Publishing Co., N. Y., 1951.
(65) Stone, J. A., Scallan, A. M., *TAPPI* **50,** 496 (1967).
(66) Stone, J. A., Scallan, A. M., *Cellulose Chem. Tech.* (in press).
(67) Stone, J. E., Scallan, A. M., *Pulp and Paper Mag. Can.* **69,** 288 (1968).
(68) Stone, J. A., Scallan, A. M., Aberson, G. M. A., *Pulp and Paper Mag. Can.* **67,** 263 (1966).
(69) Stone, J. A., Scallan, A. M., Donefer, E., Ahlgren, E., ADVAN. CHEM. SER. **95,** 105 (1969).
(70) Sullivan, J. D., *TAPPI* **51,** 501 (1968).
(71) Tanford, C., "Physical Chemistry of Macromolecules," p. 710, Wiley Pub. Corp., N. Y., 1961.
(72) Tarkow, H., Feist, W. C., Southerland, C. F., *Forest Prod. J.* **16,** 61 (1966).
(73) Thoma, J. A., Koshland, Jr., D. E., *J. Am. Chem. Soc.* **82,** 2511 (1960).
(74) Walseth, C. S., *TAPPI* **35,** 228 (1952).
(75) *Ibid.*, **35,** 233 (1952).
(76) Warwicker, J. O., Jeffries, R., Colbran, R. L., Robinson, R. N., *Shirley Inst. Pamphlet* **No. 93,** p. 247, Manchester, England (1966).
(77) Wilcox, W. W., *Forest Prod. J.* **7,** 255 (1965).
(78) Whitaker, D. R., Colvin, J. R., Cook, W. H., *Arch. Biochem. Biophys.* **49,** 247 (1954).
(79) Zimmerman, M. H., Ed., "The Formation of Wood in Forest Trees," p. 562, Academic Press, N. Y., 1964.

RECEIVED January 10, 1969.

Discussion

Swenson: "Would you please comment on the similarity in partial specific volume between cellulolytic enzymes and other proteins."

Brown: "It is commonly observed that the partial specific volume of globular proteins is a relatively invariable parameter. It is approximately 0.70 to 0.75 for most proteins irrespective of their shape."

Reese: "Does Cowling now withdraw his original suggestion that the enzymes of wood-destroying fungi act at a distance? Does he believe that non-enzymatic factors may be involved?"

Brown: "The determinations of size of the cell-wall capillaries and the cellulases of fungi indicate that the enzymes that have been isolated to date are so large that they probably can penetrate only a few cell-wall capillaries in wood and cotton. This conclusion is supported by Cowling's DP data for the action of the white-rot fungus, *Polyporus versicolor*, on wood (Figure 13). But it is also contradicted by the same type of data for the effect of both the brown-rot fungus, *Poria monticola*, on wood (Figure 13) and of *Myrothecium verrucaria* on mercerized cotton as shown by Selby (*60*). Thus we believe that the catalysts responsible for the initial depolymerization of cellulose in wood and cotton by these two organisms must be able to act within the cell-wall capillaries and therefore at a distance from the hyphae. It is easier to believe that a small non-enzymatic catalyst can penetrate these capillaries than to believe that there is an active enzyme protein that is small enough to do so. In this connection we are very much interested in the observations of Halliwell (21) on the degradation of cellulose under physiological conditions by hydrogen peroxide."

11

Constitutive Cellulolytic Enzymes of *Diplodia zeae*

J. N. BEMILLER, DIANE O. TEGTMEIER, and A. J. PAPPELIS

Departments of Chemistry and Botany, Southern Illinois University, Carbondale, Ill. 62901

> *Diplodia zeae produces cellulolytic activity even when grown in media containing D-glucose as the only carbohydrate. The cellulolytic enzymes are believed to be attached to the surface of the fungal hyphae and released to the culture medium after the growth period, when most of the available carbohydrate is used up. It is suggested that not the synthesis of the C_x enzyme(s) but its release is controlled by the presence of D-glucose.*

It has been reported that fungal cellulases are induced enzymes and that cellulose preparations induce cellulolytic activity while easily assimilable carbon sources give the best fungal growth but less production of enzyme activity (*9, 12, 14*). For example, Horton and Keen (*10*) found that $7.5 \times 10^{-3} M$ D-glucose repressed the synthesis of cellulase to a basal level in *Pyrenochaeta terrestris* and suggested that cellulase synthesis was regulated by an induction-repression mechanism.

In this laboratory, we have studied stalk rot of corn (*Zea mays* L.), a disease caused in part by *Diplodia zeae* (Schw.) Lev. The organism produces cellulolytic activity, although it is of a very low level (*3*), which destroys cell wall structures as the pathogen spreads through parenchyma tissue, hollowing out the corn stalk. The parenchyma tissue in which the fungus grows, however, contains relatively large amounts of sugars. [Approximate ranges are reducing sugars 0.1–$0.5M$, sucrose 0.1–$0.35M$, total sugars 0.15–$0.5M$ (*4*)]. Hence, we became interested in the effects of cellulose and D-glucose on the induction and repression, respectively, of the cellulolytic enzymes of *D. zeae*.

Experimental

D. zeae Culture. *Diplodia zeae* (Schw.) Lev. was grown in liquid shake culture as described by BeMiller *et al.* (*3*) with variations in the source and amount of carbohydrate.

Assay Methods. D-Glucose in the culture media was determined with Glucostat reagent (Worthington Biochemical Corp., Freehold, N. J.). The assay medium contained 0.67% carboxymethylcellulose (CMC 7LF, Hercules Inc., Wilmington, Del.) in pH 4.5 acetate buffer (0.2M). Total cellulolytic (C_x) activity was determined by measurement of the total reducing power in the assay medium with sodium 3,5-dinitrosalicylate (DNSA) reagent. β-Glucosidase activity was determined by measurement, with Glucostat reagent, of the D-glucose produced in the assay medium. Details of the DNSA and Glucostat methods have been described previously (*3*).

Cellulolytic Enzymes in Culture Media as a Function of Growth Time. *D. zeae* was grown in two kinds of media, one containing 2.0 grams/liter of D-glucose (0.01M) and 7.5 grams/liter of cellulose and the other containing 10.0 grams/liter of D-glucose (0.05M) and 7.5 grams/liter of cellulose as the carbohydrate source. Other components were 1 gram of casein hydrolysate, 0.25 gram of yeast extract, 2.5 grams of NH_4NO_3, 1.25 grams of KH_2PO_4, 0.625 grams of $MgSO_4 \cdot 7H_2O$, and 0.005 grams $FeCl_3$ per liter as previously reported (*3*). Aliquots were taken at 12-hour intervals and determinations of D-glucose concentration, C_x activity, and β-glucosidase activity were made. Results of three determinations of each were averaged.

Release of Cellulolytic Enzymes. Culture medium was filtered from 4-day-old cultures of *D. zeae*. The mycelial mats were rinsed with distilled water, then stripped from the filter paper and fragmented in a minimum of distilled water in a semi-micro Waring Blendor cup. Portions of the mycelial suspension were pipeted into solutions of the compounds being tested, and the flasks were shaken in the dark for 20 minutes. The suspensions were then filtered, and the filtrates were assayed for enzyme activity.

To determine the amount of enzyme activity remaining on the mycelium after treatment with lignosulfonate solutions, cultures were filtered and mycelial mats recovered. One-fourth of the mats were incubated for 20 minutes with a 2.0% lignosulfonate solution pH (4.5); the mixture was then filtered, the mycelial mats recovered, and the filtrates assayed. All of the treated mats and one-half of the untreated mats were placed in an acetone-dry ice bath for 15 minutes, a treatment which killed the organism. Enzyme activity remaining on these mycelial mats was determined by placing them directly into the assay medium. Also, to determine the amount of the reducing power due to sugars being leached from the mycelium rather than enzyme activity, one-fourth of the mycelium was placed in buffer alone, the buffer was assayed for reducing power and D-glucose, and these values were subtracted from the values found in the enzyme assays. The assay mixtures were filtered and aliquots of the filtrates were assayed for enzyme activity to confirm that active enzyme was present and was released during assay of the mycelium. Since C_x

enzymes are not inhibited by Merthiolate, the same procedure was repeated using 1% Merthiolate to kill the organism.

Sonification. Mycelium was fragmented in buffer using a Waring Blendor. This suspension was subjected to sonification with a 125-watt Branson Sonifer in a 4°C. bath for 20 minutes. No cell structures could be observed under the microscope.

Results and Discussion

The results of experiments set up to determine if the concentration of D-glucose in culture media containing both D-glucose and cellulose reaches a steady state level owing to an induction-repression mechanism are presented in Figures 1 and 2. Some observations can be made from these data. No steady state concentration of D-glucose was found. Solid cellulose particles disappear from the media before more than traces of enzyme activity can be measured in the media. Since each enzyme molecule must come into direct physical contact with its specific substrate molecule before catalysis can occur and since only traces of enzyme activity can be found in the media before and during the disappearance of cellulose, it is suggested that the cellulolytic enzymes of *D. zeae* occur on the surface of the hyphae.

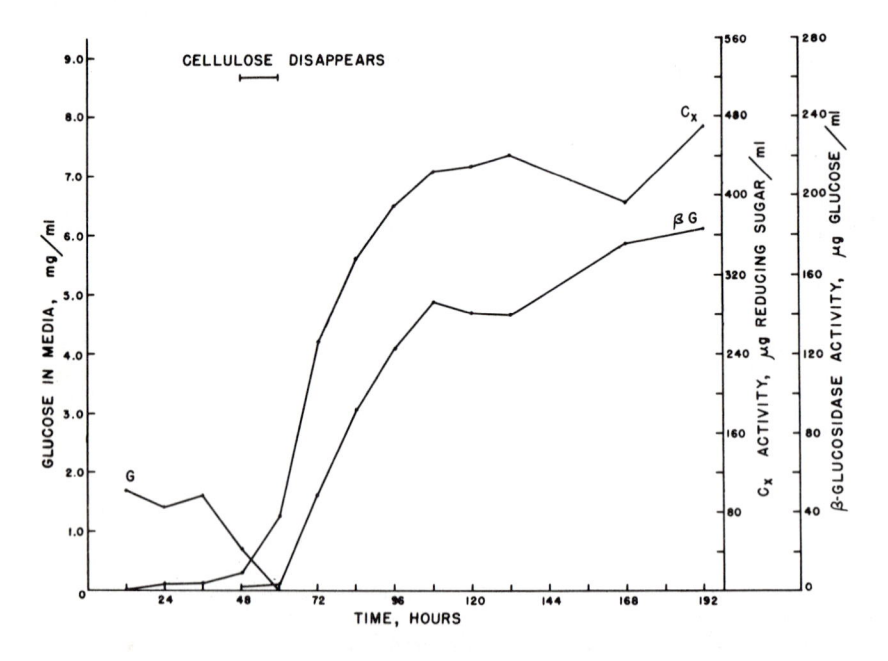

Figure 1. Concentration of D-glucose and activity of C_x and β-glucosidase enzymes in D. zeae culture media as a function of time. Original D-glucose concentration. 2.0 grams/liter; cellulose concentration, 7.5 grams/liter

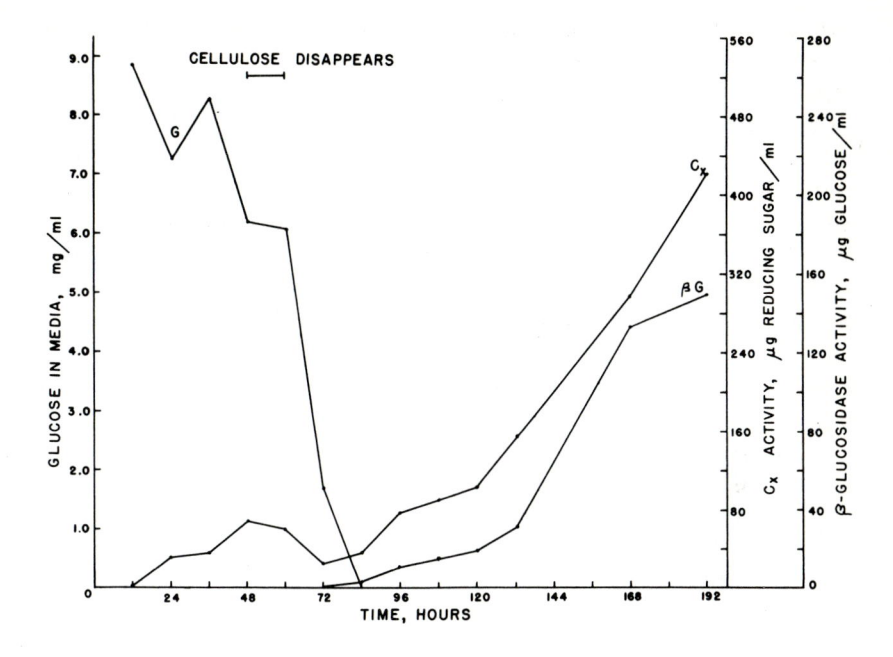

Figure 2. Concentration of D-glucose and activity of C_x and β-glucosidase enzymes in D. zeae culture media as a function of time. Original D-glucose concentration, 10.0 grams/liter; cellulose concentration, 7.5 grams/liter

Cowling and Kelman (7) reported that, in cultures of dikaryotic decayed tissue isolates of *Fomes annosus* with relatively high cellulase activity, up to 34% of total activity was bound to the mycelium (14%) and cellulose particles (21%); and Tashpulatov (15) reported that part of the cellulolytic enzymes remained in the mycelium during growth of *Aspergillus fumigatus*.

It can also be observed from Figures 1 and 2 that little C_x activity and no β-glucosidase activity could be detected in the culture filtrates until the carbohydrate (D-glucose) had almost completely disappeared; enzyme activity appeared sooner in cultures containing the smaller amount of D-glucose. It is suggested that easily assimilable carbohydrate such as D-glucose regulates, not the synthesis of superficial cellulolytic enzymes, but their release into the culture medium. This release is a function of the age or condition of the culture and occurs, for the most part, after the growth period. Deshpande and Deshpande (8) report that cellulase is liberated 48 hours after innoculation for *Aspergillus fonsecaeus* grown on 0.05M D-glucose, and Moreau and Trique (13) found a continued liberation of cellulase into the culture medium in the absence of substrate and after the growth period of *Penicillium cyclopium* and *Aspergillus versicolor*. Isaac (*11*) found cellulase activity around certain short, branched hyphae in older parts of mycelium of *Rhizoctonia solani*, but none in either young hyphal tips or older trunk hyphae bearing active side branches.

It is obvious that conclusions about the amount of enzyme synthesis could not be based on the amount of cellulolytic activity found in the culture medium since the amount of activity found depended on the time at which measurements were made. On the basis of the hypothesis that the cellulolytic enzymes are bound to the surface of the hyphae and are released to the culture medium only slowly after the growth period, we undertook to investigate the release of the enzymes; for, in order to determine if an induction-repression mechanism operates, it is necessary to measure total enzyme activity.

Coles and Gross (*5, 6*) have found that some of the penicillinase of *Staphylococcus aureus* is bound ionically to the surface of the cell wall and can be liberated by low concentrations of anions. Polyanions that do not penetrate the cell wall and tricarboxylic acids were particularly effective as release agents. Weimberg and Orton (*16*) had found earlier that acid phosphatase could be released from the cell surface of *Saccharomyces mellis* by chloride ions in combination with 2-mercaptoethanol and concluded that this enzyme is held on the cell surface by a combination of electrostatic forces and disulfide bonds. Likewise, polygalacturonase is released from the surface of spores of *Geotrichum candidum* with polyanions and 2-mercaptoethanol (*2*).

To investigate the release of cellulolytic enzymes from *D. zeae*, harvested mycelium was treated with various reagents. No appreciable activity was released by pH 4.5 acetate buffer (0.1–0.8M) or the following anionic reagents: 0.5M KCl, 2M KCl, 0.15M Na$_2$HPO$_4$ (pH 6.0, 7.5), 0.05M succinic acid (pH 4.5), 0.04M malic acid (pH 4.5), 0.03M citric acid (pH 4.5), 0.4M oxalic acid (pH 4.5), 0.5% water-soluble polyacrylamide, 1.0% Reten A-1 (Hercules Inc., Wilmington, Del.).

Activity was not released by 1.0% Reten 205M (a cationic polymer) (Hercules Inc., Wilmington, Del.), 0.13–0.38M 2-mercaptoethanol, liquid phenol, incubation with 0.1% lysozyme, osmotic shock (*1*), or the following surfactants: 0.4% Triton X-100 (Rohm and Haas Co., Philadelphia, Pa.), 0.4% sodium lauryl sulfate, 0.5% digitonin in 0.05M tris buffer (trishydroxymethyl-aminomethane).

More C_x activity was found in the wash solution when the harvested mycelium was treated with 2.2–4.0% polygalacturonic acid (pH 4.5); this activity was about 15 times that found in washes of pH 4.5 acetate buffer. Lignosulfonic acid (2.0%, pH 4.5) also either released or activated C_x activity but inhibited the glucostat reagent so that the β-glucosidase activity could not be determined; the C_x activity was about five times that found in washes of pH 4.5 acetate buffer. A solution of 1.0% lignosulfonic acid and 2.0% polygalacturonic acid (pH 4.5) released no more activity than solutions of either polyanion alone. About the same amount of enzyme activity was released from mycelium of *D. zeae* grown in a medium containing only D-glucose as carbohydrate (no cellulose or CMC) by solutions of either polygalacturonate or lignosulfonate. Since lignosulfonate was effective as a release agent, it was added to the growth medium; no significant increase in C_x activity of the medium was found. Some C_x and β-glucosidase activity (two to three times that of the controls) was released from harvested mycelium by solutions of highly substituted carboxymethylcellulose CMC 12M31P (Hercules Inc., Wilmington, Del.).

Table I. Enzyme Activity Remaining on Mycelium After Treatment with Lignosulfonate

Treatment	C_x	β-Glucosidase
Acetone-dry ice		
Non-frozen mats in assay medium	280	100
Acetone-dry ice treated mats in assay medium	160	80
Acetone-dry ice treated mats previously washed with lignosulfonate in assay medium	195	40
Merthiolate		
Untreated mats in assay medium	580	130
Merthiolate treated mats in assay medium	310	60
Merthiolate treated mats previously washed with lignosulfonate in assay medium	320	30

Relative Enzyme Activity (Reducing Power as D-Glucose per mg. Fresh wt.)

An investigation was then carried out to determine the amount of enzyme activity washed from the mycelium with lignosulfonate. Cultures were grown and mycelium was collected by filtration. Part of the mycelium was washed with lignosulfonate; then the washed mycelium and part of the unwashed mycelium was killed with either acetone-dry ice or Merthiolate. The mycelium was then placed directly into the assay medium and the reducing power and D-glucose were determined. Results of this investigation, given in Table I, indicate that some of the enzyme activity is lost during the killing process but that little of the C_x activity is removed and/or the C_x enzyme(s) is activated by the lignosulfonate wash. The latter hypothesis was proved to be true by the addition of lignosulfonic acid to active culture filtrates. The β-glucosidase activity is either removed or enough lignosulfonate remains to partially inhibit the glucostat reagent.

To get a better idea of the portion of the C_x activity released by treatment with the lignosulfonate polyanion, mycelium was disrupted by sonification after a lignosulfonate washing. The results in Table II again show that there was some greater removal and/or activation of the C_x enzyme(s) by the lignosulfonate wash than by the acetate buffer wash. However, no more than half of the total activity was solubilized even by very extensive sonification. Treatment of remaining pellets with the polyanions, surfactants, and enzymes used before removed no additional enzyme.

Table II. C_x Enzyme Distribution After Sonification (20 min.) and Centrifugation

Percentage of Total C_x Activity

	Mycelium Unwashed Before Sonification	Mycelium Washed with 0.2M Acetate Buffer (pH 4.5) Before Sonification	Mycelium Washed with 0.4% Lignosulfonate Before Sonification
Cultrate filtrate	35%	29%	28%
Wash	—	15%	22%
Supernatant after sonification and centrifugation	32%	28%	24%
Pellet after sonification and centrifugation	33%	27%	25%

This is a preliminary report of our work. Our results to date indicate that cellulolytic enzymes may be found extracellularly, both in the culture medium and attached to the surface of hyphae, and, perhaps, intracellularly. In addition, they may be found in active and inactive forms in all places. The binding of the enzymes to the mycelium and the natural release mechanism is unknown. In order to investigate induction and repression mechanisms and the influence of chemical and environmental factors on the synthesis of cellulolytic enzymes, total enzyme activity must be measured. Further work along these lines is in progress.

Literature Cited

(1) Anraku, Y., Heppel, L. A., *J. Biol. Chem.* **242**, 2561 (1967).
(2) Barash, I., Klein, L., *Phytopathology* **59**, 319 (1969).
(3) BeMiller, J. N., Tegtmeier, D. O., Pappelis, A. J., *Phytopathology* **58**, 1336 (1968).

(4) Betterton, H. O., M.S. Thesis, Southern Illinois University, Carbondale, Illinois (1963).
(5) Coles, N. W., Gross, R., *Biochem. J.* **102,** 742 (1967).
(6) *Ibid.,* **102,** 748 (1967).
(7) Cowling, E. B., Kelman, A., *Phytopathology* **53,** 873 (1963).
(8) Deshpande, K. B., Deshpande, D. S., *Enzymologia* **30,** 206 (1966).
(9) Horton, J. C., Keen, N. T., *Phytopathology* **56,** 908 (1966).
(10) Horton, J. C., Keen, N. T., *Can. J. Microbiol.* **12,** 209 (1966).
(11) Isaac, P. K., *Can. J. Microbiol.* **10,** 621 (1964).
(12) Mandels, M., Reese, E. T., *Ann. Rev. Phytopathology* **3,** 85 (1965).
(13) Moreau, M., Trique, B., *Compt. Rend.* **263,** 239 (1966).
(14) Noviello, C., *Ann. Fac. Sci. Agrar. Univ. Studii Napoli Portici* **30,** 461 (1965); *Chem. Abstr.* **65,** 2679 (1966).
(15) Tashpulatov, Zh. *Mikrobiologiya* **35,** 27 (1966); *Chem. Abstr.* **64,** 16322 (1966).
(16) Weimberg, R., Orton, W. L., *J. Bacteriol.* **90,** 82 (1965).

RECEIVED October 14, 1968. A contribution of Interdisciplinary Research in Senescence, a Cooperative Research Project of Southern Illinois University.

Discussion

J. K. Alexander (Philadelphia): "Where you have increase of enzyme activity, I wondered if you have looked at the possibility of a cellulase inhibitor being released, therefore, freeing your activity?"

J. N. BeMiller: "Yes. We are relatively confident that we can rule out inhibition as a factor. We have considered this and can find no evidence of inhibitors being present and then being destroyed, by the various criteria that we have used. We have also ruled out, at least in our own minds, the idea of a precursor being present which becomes activated to form the true enzyme. However, we are trying to keep an open mind on these subjects and will continue to consider them."

E. T. Reese: "We have found that nearly all cellulolytic organisms, where cellulase is induced, do produce small amounts of constitutive enzyme, very very low levels. Your measurements here indicate similar very low levels of enzymatic activity. Under the conditions that you use I notice that the glucose did not disappear very rapidly. In our work on cellulase induction, we actually try to reduce the rate at which the substrate is consumed. I am wondering whether your conditions, where it is obvious that the organism wasn't very happy with its environment, were actually limiting growth and so stimulating induction?"

J. N. BeMiller: "It may be. I do believe that stress is a necessary prerequisite to induction and that unfavorable growth conditions usually produce higher amounts of enzyme. However, our growth conditions are actually quite good for this organism, which is considered to be a very poor pathogen by the plant pathologists. It doesn't grow very well even in the corn stalk; it grows well enough to do considerable damage, but it is not what would be considered virulent.

12

A Mechanism for Improving the Digestibility of Lignocellulosic Materials with Dilute Alkali and Liquid Ammonia

HAROLD TARKOW and WILLIAM C. FEIST

Forest Products Laboratory, Forest Service, U. S. Department of Agriculture

> *The fiber saturation point of hardwoods is doubled following treatment with 1% NaOH or liquid ammonia. The swelling capacity of the wood substance is approximately doubled. This, together with the indication that the upper molecular weight limit for penetration of diffusing materials is increased somewhat, accounts for the improved digestibility of such treated materials by cellulolytic micro-organisms. The increase in swelling capacity results from the saponification or ammonolysis of esters of 4-O-methylglucuronic acid attached to xylan chains. In the natural condition, these esters act as cross-links, limiting the swelling or dispersion of polymer segments in water. Such treatments may provide a means for converting unused coarse plant materials, via ruminants, into protein rich foods.*

An important problem is the increasing gap between world food requirement and world food production. Several reports originating under U. S. or international auspices (6, 14, 27, 45) forewarn of Malthusian conditions. Adding to the seriousness of the situation is the fact that the quality of foods for many people is substandard although caloric intake meets minimum standards. Quality is determined by many factors, one of which is the content of protein derived from animals.

Ruminants derive their energy, carbon and nitrogen, from grains and forages. The forages and hays also provide roughage. With the increasing use of nonprotein nitrogen (urea), the importance of plant nitrogen in feeds will probably diminish (47). The important function of the plant-derived feed component will be as of a source of energy and carbon. Because of the increasing importance of large feedlots in localities removed from sources of forages and hay, other sources of energy-producing feeds must be found.

The total land area of the world is about 35 billion acres. By far, the largest crop harvested is wood, a lignocellulosic material, growing on 10 billion acres (*44*). In contrast 1,200 million acres are devoted to cereal production. Associated with the "wood agriculture" are large quantities of low-quality wood and non-utilized residues, which in the U. S. can amount to much more than 100 million tons per year (*46*). Where concentrated, these forms of wood offer an attractive inducement for economic exploitation. An increasing amount of these materials is being chipped and sold to pulp companies, yet a considerable amount remains unused.

Efforts have been expended to develop processes for hydrolyzing the carbohydrates in these wood residues, to recover the solubles, and to concentrate them to a "molasses" for supplementary animal feed (*16*). Generally, these processes are not economical. A large producer of dense fiberboard, faced with the problem of disposing the 20% solubles resulting from the board production, learned how to convert these pollution-producing solubles into a highly acceptable and profitable cattle feed (*43*).

Another source of unused lignocellulose is straw derived from rice, wheat, and other cereals. For example, one third of the population of the world lives on rice. The great bulk of the rice straw, however, is unused. A small amount is pulped (Egypt).

These various lignocellulosic materials, such as wood and straw have not been used successfully as "forage" supplements because of the poor digestibility of their carbohydrate components. An economical means of increasing the digestibility of these materials would provide a source of large quantities of new "forage" material containing more than 75% carbohydrate (*39, 40*).

Brief Review of Past Work

A comprehensive review of "Utilization of Lignocellulosic Materials in Animal Feeds" is given by Baumgardt *et al.* (*4*). During World War II, Scandinavian countries used chemical pulps (delignified wood) as a feed component for ruminants (*20, 24*). The practice was discontinued after the war because of the high costs. The success of this earlier application strengthened the view that the lignin in lignocellulosic materials prevents the digestion of the carbohydrate fraction. Yet, indications are that slight modifications of these materials, with no removal of lignin, increase their digestibility.

Pew (26) found digestibility of hardwoods by cellulases improved markedly by pretreatment with aqueous NaOH; the effect was considerably more pronounced with hardwoods than with softwoods. Stranks (35) reported that the *in vitro* rate of digestion of hardwoods by rumen-inhabiting bacteria, measured by succinic acid production, was markedly increased by pretreatment with 1–5% NaOH, whereas softwoods were unaffected by the treatment. The effect of NaOH in improving the digestibility of straw has long been known (5).

An example of this effect of NaOH treatment (49) is shown in Figure 1 relating the *in vitro* digestibility by rumen-inhabiting organisms of two lignocellulosic materials treated with increasing amounts of NaOH. The upper curve for the digestibility of wheat straw shows that the improvement is aproximately linear with NaOH content up to about seven grams NaOH per 100 grams straw. At higher NaOH contents, the additional gain is small. The lower curve is for poplar, a hardwood of low density. Digestibility of the poplar also increases with NaOH pretreatment at the same rate and also levels off at about 7% NaOH (based on the wood).

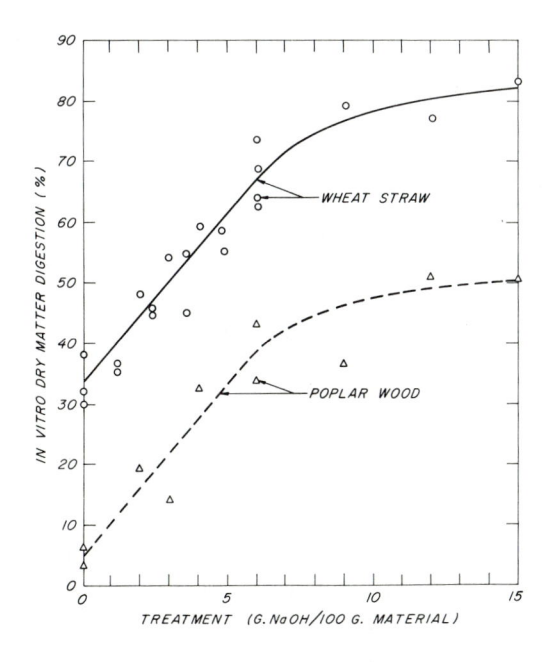

Figure 1. Effect of NaOH pretreatment on the percentage of dry matter digestion in vitro with rumen micro-organisms (49)

Recent work at the Forest Products Laboratory has shown that on treating sugar maple, a hardwood, with 1–2% NaOH and washing out the free NaOH, the rate of digestion of the wood by wood-rotting fungi is increased (Figure 2). The fungi used were a brown rot, *Lenzites trabea* (Madison, 617); a white rot, *Polyporus versicolor* (Madison, 697); and a soft rot, *Graphium sp.* (Madison, 11).

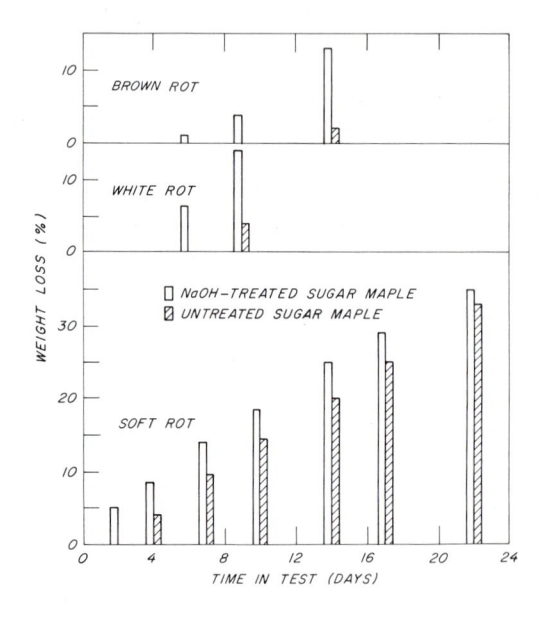

Figure 2. Comparison of rate of decay of dilute NaOH-treated (and washed) sugar maple with that of untreated wood by three types of fungi: White rot, Polyporus versicolor; brown rot, Lenzites trabea; and soft rot, Graphium sp.

Ghose and King (15) compared the digestibility of an average grade alfalfa hay with that of jute (a lignocellulose) and of jute modified by treatment with dilute NaOH. The organisms used were bacteria isolated from rumen fluid. Holocellulose made from the hay or jute had equally high digestibilities. This, of course, is consistent with the earlier observation that chemical pulps have high *in vivo* digestibilities (20, 24). Although the natural hay and jute have about the same lignin contents, the hay has an average *in vitro* digestibility of 24%, and the jute an average digestibility of 3.5%. Thus, although lignin is an important digestion-depressing factor (considering holocollulose), the location and perhaps the nature of the lignin bonding with the carbohydrate fraction are also important factors. The mild treatment of the jute with dilute NaOH

(1%), however, raised the average *in vitro* digestibility to 27%, about the same as that of the alfalfa hay.

Thus, evidence is considerable that mild NaOH treatment of selected lignocellulosic materials increases the digestibility of their carbohydrate fractions. To know the maximum benefits possible from this treatment, it is helpful to know the mechanism for obtaining the improvement.

Results

Chemical Changes that Occur when Hardwoods Are Treated with Dilute NaOH. Cellulosic materials swell in aqueous NaOH. Under certain conditions the swelling is intramicellar (as well as intermicellar) and leads to a transformation of Cellulose I to Cellulose II. Sisson and Saner (*30*) developed a phase diagram describing the combinations of NaOH concentration and temperature that produce this transformation in cotton cellulose. At room temperature the concentration must exceed about 15%. Because of poorer lateral order of crystalline regions in wood cellulose, the minimum concentration for wood is probably lower, perhaps 8%. With lesser concentrations the phase transformation does not occur although some intermicellar swelling does occur.

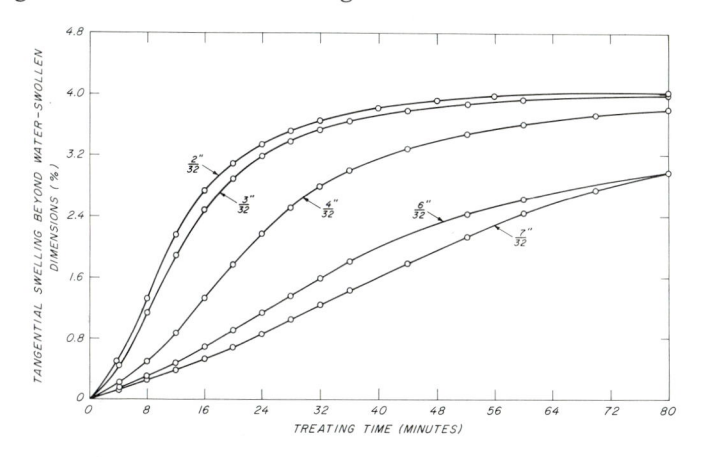

Figure 3. External tangential swelling beyond waterswollen dimensions (excess swelling) for five thicknesses of sugar maple cross sections in 1% NaOH at 40°C.

Figure 3 describes the external tangential swelling of green cross sections of sugar maple at 40°C. in 1% NaOH as a percent of the original water-swollen dimension. The rate at which this excess swelling occurs is influenced by the longitudinal thickness. The excess external swelling has a small reversible component and a pronounced irreversible component as shown by the retention of a considerable portion of the excess swelling after washing out the NaOH with water; this suggests a chemical change. It will be shown that the external swelling still does not describe the total irreversible swelling that occurs on treatment of hardwoods with dilute NaOH.

A significant clue about the nature of the chemical change is obtained by measuring the free carboxyl content by calcium ion exchange (Figure 4). The carboxyl content increases to a maximum value of 24 milliequivalents per 100 grams, although the rate of attainment of the maximum is much higher with 0.5% NaOH than with 0.1% NaOH. The increase in exchange capacity results from the saponification of ester groups. Sjöström *et al.* (*32*) found similar results. The carboxyl component of the esters in hardwoods is derived from 4-O-methyl-D-glucuronic acid present as pendant groups every 6–9 anhydro xylan residues along the xylan chains (*42*). The location of the alcoholic component is unknown at present; it may be on another xylan chain or on the lignin. The xylan chains also contain acetyl groups which saponify as readily as the 4-O-methyl-D-glucuronic acid esters (*42*).

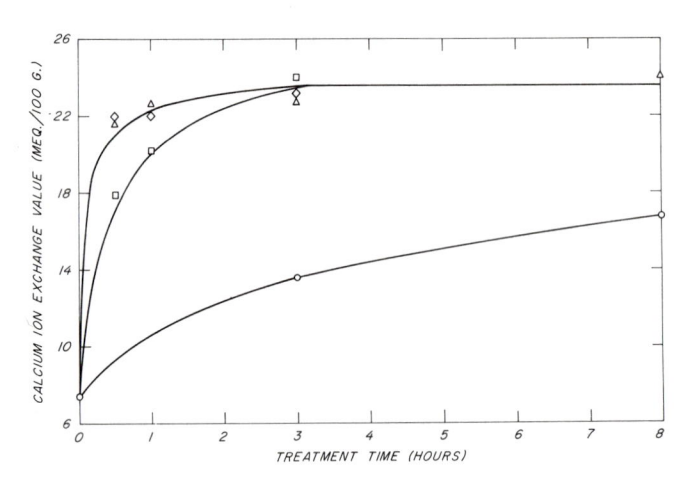

Figure 4. *Effect of treating sugar maple with aqueous NaOH at room temperature on values for calcium ion exchange (free carboxyl content)*

○ 0.1% NaOH
□ 0.5% NaOH
△ 1.0% NaOH
◇ 2.0% NaOH

Thus, NaOH at these low concentrations is involved in at least two reactions with hardwoods: Saponification of uronic esters and saponification of acetates. If these are the only reactions occurring, then the NaOH has the following fates:

(1) A portion is present as the sodium salt of the liberated carboxyl groups on the glucuronic acid residues attached to the xylan chains; the amount is readily computed from the free carboxyl content after removal of sodium by acid washing.

(2) A portion is present as sodium acetate; the amount left in the wood will depend on the extent to which the sodium acetate diffused out of the wood during treatment.

(3) A portion will be present within the free water in the lumens and at the same concentration as that of the treating solution if equilibrium was established.

Green maple cross sections (0.03 inch) were treated with aqueous NaOH at two concentrations for five minutes. The specimens were then lightly blotted, dried (total time out of solution was about one hour), ashed, and assayed for equivalent NaOH (Table I). The amount held by the xylan carboxyl groups (computed from calcium ion exchange value), of course, is independent of the external concentration. The amount held as sodium acetate should also be independent of the external concentration. The amount held in the lumen should be proportional to the external concentration provided equilibrium was established in five minutes of immersion in the aqueous NaOH. The sum of these component amounts should be the total equivalent NaOH in the treated wood provided none of the sodium acetate diffused out of the cross section during the five minutes.

Table I. Distribution of NaOH in NaOH-Treated Sugar Maple

NaOH Concentration, %	In Lumen, mols/100 g.	Held By Carboxyl on Xylan Chain, mols/100 g.	Held as Acetate, mols/100 g.	Total Including Only 15% Held as an Acetate, mols/100 g.	Total Observed NaOH Content, mols/100 g.
1	0.033	0.022	0.110	0.071	0.077
4	.133	.022	.110	.171	.191

In a supplementary experiment it was learned that under the conditions of the experiment the acetate content within the wood decreased by 85% (over the theoretical value based on the initial acetyl content). Thus, within the five-minute period in the treating solution, saponification of the acetate occurred, and 85% of the resulting sodium acetate had diffused out of the wood. This rapid diffusion is consistent with earlier work (37), which showed that the diffusion rate of water-soluble material through the cell walls is increased several times by NaOH treatment. The amount of sodium held as sodium acetate should thus be only 15% of that computed on the basis of the initial acetyl content. With this correction, the computed total equivalent NaOH content in the treated wood is given in column 5 of Table I. The observed value is given in column 6. The agreement is reasonably good. Thus, the chemistry by which NaOH in dilute solution, reacts with hardwoods is explained.

The treating times for these 0.03-inch cross sections was five minutes. If continued longer, the same reactions would occur, but some lignin or hemicellulose or both would be extracted from the wood. This would result in a decreased yield (Table II). The highest yield was 91%. However, the treating time was six hours. On the basis of results shown in Figure 3, 30 minutes would have been adequate to bring about the two saponifications in 1/32-inch-thick cross sections. The somewhat lower yields reflect solubilizations of some wood substance by the alkaline solution.

Table II. Effect of NaOH Treating Conditions on Yield (1/32-in. Sugar Maple Cross Sections)

NaOH Concentration, %	Temperature, °C.	Time, Hrs.	Yield, %	Theoretical Yield,[a] %
1	24	6	91	95
1	60	3	90	95
4	60	3	83	95

[a] Based on weight loss by deacetylation only.

After treatment with dilute NaOH, the free NaOH in the lumens and the remaining sodium acetate can be readily washed out with water. If deionized water is used, a retention of the sodium by the carboxyl groups on the xylan chains would be expected. Alkali-treated (1.5% NaOH, three hours) sugar maple cross sections were washed with deionized water. At various stages of washing, the residual sodium content was measured. The sodium content decreased very rapidly to a constant value of 26 milliequivalents per 100 grams. This, of course, represents the sodium combined with free carboxyl groups in the maple. It is in fairly good agreement with the value of 24 milliequivalents per 100 grams found by calcium ion exchange. The agreement confirms the findings of Sjöström et al. (31). It means that the carboxyl groups on the xylan chains are so disposed that a calcium ion can bridge two carboxyl groups. This is a significant finding and deserves additional attention.

An intriguing possibility is treating air-dry hardwood chips or other suitable lignocellulosic material with 2 to 4% NaOH for only the time that the material imbibes enough NaOH to bring about the two saponifications (49).

If the material is to be used as a feed supplement for ruminants, washing would probably not be necessary. On the basis of results in

Table I, using 4% NaOH and under the same conditions of treatment, the total NaOH requirement, including that equivalent to all the original acetyl content, is less than 10% of the weight of the wood.

Chemical Changes that Occur when Hardwoods Are Treated with Ammonia. Because the action of liquid NH_3 on wood could be of interest for increasing the digestibility of lignocellulosic material, the work of Wang et al. (48) becomes pertinent. They showed that ammonolysis does occur, and determined the relative amount of nitrogen present as ammonium salts and as amide groups in ammonia-treated birch.

In related work at the Forest Products Laboratory it was shown that the reaction of maple with aqueous NH_3 up to 20% concentration (25°C.) was very slow. The results of Wang (48) using liquid NH_3 were confirmed. Some of the kinetic factors will be presented later. In one experiment, sugar maple was treated with liquid NH_3 at +25°C. for eight hours. The sample was air-dried and washed with water. The nitrogen content as ammonium and amide was found to be 0.36 gram per gram. This is equivalent to about 25 milliequivalents nitrogen per 100 grams treated wood, or 25 milliequivalents of total carboxyl. This is in good agreement with the carboxyl values found by calcium or sodium ion exchange for NaOH-treated wood. Thus, the chemical action of liquid NH_3 on the hardwood is entirely consistent with that predicted from the action of very dilute NaOH on wood.

Physical Changes that Occur when Hardwoods Are Treated with Dilute NaOH. ADSORPTION ISOTHERMS. Because of the inability of very dilute NaOH to bring about phase transformations within cellulose, it would be expected that the moisture adsorption isotherm of a hardwood treated as described would be similar to that of the untreated wood. This was found to be essentially true up to at least 90% relative humidity (RH). Using a previously described technique, the equilibrium moisture content (EMC) of wood substance at essentially 100% RH was measured (11, 38); a very marked effect of the treatment was found.

The procedure used is analogous to that of gel permeation chromatography. Using a series of well characterized polyethylene glycols (PEG) of increasing molecular weight (10), the amount of internal moisture in the water-swollen wood substance to which the PEG is not accessible is measured (11, 38) (Figure 5). With increasing molecular weight of PEG, the nonaccessible water content (nonsolvent water (1)) increases to a value that then remains independent of further increases in molecular weight. This value represents the maximum internal moisture incapable of acting as a solvent for PEG with a molecular weight exceeding the critical, or cutoff, molecular weight. At these higher molecular weights, the PEG is completely excluded from the wood substance, or cell walls. The maximum nonsolvent moisture content is identical with the fiber saturation point (FSP). The significance of the critical molecular weight will be discussed later.

From Figure 5, it is noted that the FSP of holocellulose, a type of pulp, is considerably higher than the usually assumed value of 30 to 35% (33). The FSP of the NaOH-treated hardwood (designated "superswollen" wood) is approximately doubled by the treatment.

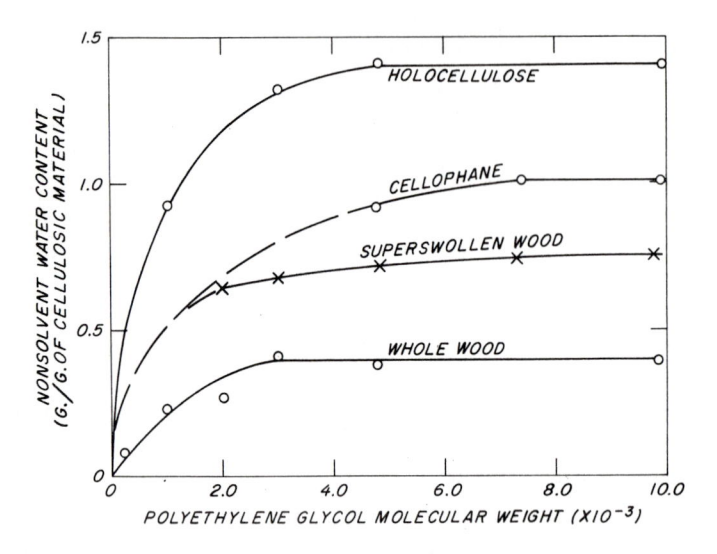

Figure 5. Relation of polyethylene glycol molecular weight to nonsolvent water content for waterlogged whole wood; for superswollen wood (NaOH-treated, acid-washed sugar maple); for cellophane (water washed, air-dried, and rewetted); and for holocellulose (Sitka spruce)

The excess tangential swelling of maple above the normal swelling shown in Figure 3 is a consequence of the increase in FSP of wood shown in Figure 5; yet, because of the considerable additional swelling that occurs internally into the lumens, the excess swelling revealed in Figure 3 does not describe the total additional swelling caused by the NaOH treatment. The true additional swelling, however, is measured by the increase in FSP. Specifically, for wood substance, the normal volumetric swelling between 0 and 100% RH is about 50%; following treatment of the hardwood with dilute NaOH and washing out the NaOH, it is about 100%.

Some aspects of the kinetics of the transformation to a modified wood substance with excess swelling capacity are shown in Table III. Following treatment with NaOH, the specimens were given two alternative wash treatments: Water followed by dilute acetic acid and water or water alone. The FSP of specimens given the water wash treatment is somewhat higher than that of the specimens given the acetic acid and water wash treatment.

Table III. Effect of NaOH Pretreatment and Washing on the FSP of Cross Sections of Sugar Maple

NaOH Concentration, %	Treating Time, Hrs.	Fiber Saturation Point After Washing In 5% Acetic Acid and Water, %	Water, %
0	—	41	42
0.5	0.5	—	58
1	1	—	78
1	3	73	87
1	7	69	89
2	1	—	82
2	3	—	86
4	6	68	—

Table IV. Change in Free Carboxyl Content and FSP (Three Species) Following NaOH Treatment (1.5% NaOH, 3 Hours, Room Temperature)

Species	Uronic Acid, %	Total Carboxyl Content,[a] meq./100 g.	Free Carboxyl Content,[b] meq./100 g.	Relative Free Carboxyl Content, %	FSP, %
Sugar Maple					
Original	4.40	25.0	7.4	30	42
NaOH Treated	4.10	23.2	22.4	97	87
Aspen					
Original	4.20	23.9	7.2	30	52
NaOH Treated	3.64	20.7	18.3	89	100
Beech					
Original	4.84	27.5	7.8	28	41
NaOH Treated	4.01	23.0	22.9	99	84

[a] Computed from uronic acid assuming molecular weight of 176 for uronic acid anhydride.
[b] Measured by calcium ion exchange (23).

A possible explanation for the increase in FSP following NaOH treatment is shown by the results in Table IV, which give the relative amounts of free carboxyl groups as computed from calcium ion exchange values. Figure 6 shows the schematic representation of free and esterified carboxyl groups. The fraction of the total carboxyl content that is free (unesterified) in the original wood is about 30%; following NaOH treatment, it is 90 to 99%. The chemical effect of the dilute NaOH treatment is one of breaking uronic ester bonds by saponification; in effect, cross-links between xylan chains and other polymeric units are broken. From polymer theory (*12, 13*), it would be expected that the interaction of water with polymer segments, with increased mobility and length between remaining resistant cross-links, would increase. In other words, the FSP should increase as is borne out by the values in Table IV. Recent work by Liu and Burlant (*22*) using NMR spectroscopy describes this increased mobility of polymer segments between cross-links of decreasing frequency.

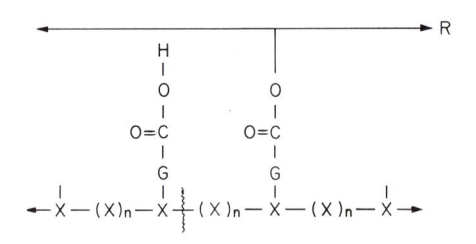

```
                                          ────────► R
                    H
                    |
                    O                 O
                    |                 |
                   O=C               O=C
                    |                 |
                    G                 G
        |           |                 |           |
    ◄── X ──(X)n── X ──(X)n── X ──(X)n── X ──►
```

Figure 6. Schematic structure of a hardwood xylan chain with a free carboxyl group and an esterfied carboxyl group functioning as a cross-link between the xylan chain and some other polymeric unit

Table III also shows that an additional factor is present in water-washed material that is responsible for the uptake of a slightly greater amount of moisture than is taken up by the acid-washed material. This factor probably derives from the polyelectrolytic character of the uronic acid formed on saponification. Washing with water following saponification merely removes free NaOH; it does not remove sodium ions associated with the carboxyl groups. Consequently, the polyuronic acid exists as a polyion. Flory has presented arguments (*12, 13*) for expecting a somewhat higher swollen condition with a polyion than when a polymer is in the undissociated condition (acid washed).

The somewhat smaller internal swelling (FSP) of the wood substance after acid washing shown in Table III is accompanied by a corresponding decrease in excess external swelling of the whole wood. Figure 7 shows the tangential swelling of a 1/16-inch-thick sugar maple cross section as a function of pH. Swelling starts around pH 9 and increases rapidly with no additional external swelling above pH 13. As the pH is then lowered below 13, some shrinkage occurs, but the original dimension is not recovered. When the pH is raised again to 13, the swelling increases again to its maximum value. Thus, the external dimension changes with pH are consistent with those observed in the FSP (total swelling) measurements. This modified wood behaves like a polyelectrolyte.

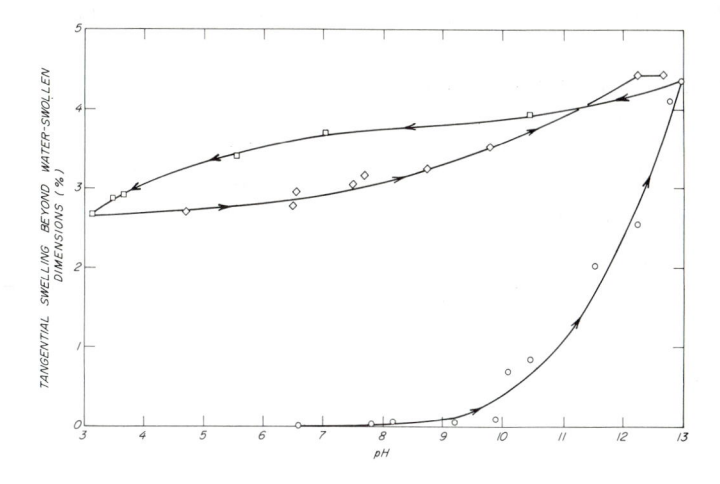

Figure 7. Effect of pH on the tangential swelling beyond waterswollen dimensions for 1/16-inch cross section of sugar maple

○ 1st pH increase
□ 1st pH decrease
◇ 2nd pH increase

Increases in FSP similar to those found for NaOH-treated maple were found with liquid NH_3-treated maple. Reaction times (20-mil-thick cross sections) at room temperature, however, were longer than those with aqueous NaOH. Maple cross sections treated with liquid NH_3 at 30°C. for 1 hour and 18 hours followed by thorough washing had FSP values of 51% and 70% respectively. The increase in FSP here also is caused by the breaking of uronic ester cross-links. Cleavage, or ammonolysis, of these groups does not lead to the formation of ionized carboxyl groups, but rather to amide groups (48). Thus, the maximum FSP obtainable by treatment with liquid NH_3 is similar to that found for NaOH-treated, acid-washed maple. As with the relationship of FSP with increased free carboxyl content following NaOH treatment, it would be expected that a relationship would exist between FSP and bound nitrogen content following treatment with liquid NH_3. The relationship is clearly shown in Figure 8.

Thus, the swelling behavior in water of liquid NH_3-treated hardwood is consistent with that of dilute NaOH-treated hardwood. In both, the marked increase in swelling characteristics in water is caused by the breaking of uronic esters functioning as cross-links. Other "cross-links" are undoubtedly present that limit the additional swelling of wood substance. These other cross-links are stable toward treatments described here; some, however, are broken during conventional pulping. This explains the considerably higher FSP of chemical pulps and holocellulose as shown in Figure 5. Breaking these additional cross-links, however, would add considerably to costs.

Sarkhar *et al.* (*29*) have shown that many of the reactions discussed also occur when jute is treated with dilute NaOH. Thus, it is predicted that the marked increase in FSP found with hardwoods would also result if jute and straw were treated with dilute NaOH or liquid NH$_3$.

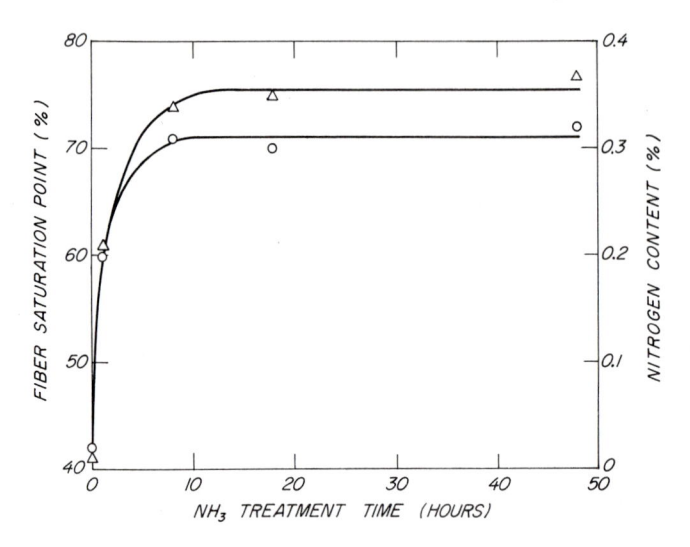

Figure 8. Effect of liquid NH$_3$ treatment (followed by thorough washing) on nitrogen content and FSP of sugar maple

○ *Fiber saturation point*
△ *Nitrogen content*

VISUAL APPEARANCE OF HARDWOOD WITH INCREASED FSP. Two yellow birch tongue depressors (1/16-inch thick) were waterlogged. One was immersed in 1.8% NaOH for several hours, then thoroughly washed in water; the other, the control, was not treated. Compared with the control, the treated specimen was very flexible and it could easily be wrapped around a glass tube with a 1-centimeter diameter. Furthermore, it was remarkably translucent indicating an increased adsorbed water-to-wood substance ratio, or higher FSP, in the treated specimen.

MECHANICAL PROPERTIES. Klauditz (*21*) found that the wet strength of dilute NaOH-treated birch was 50% that of untreated wet wood. The wet strength of the treated wood was about 35% that of the dry untreated wood.

PERMEABILITY. Thin, air-dry specimens of NaOH- or NH$_3$-treated hardwood when placed in water sank in seconds or minutes, whereas the corresponding untreated wood required overnight contact with water before sinking. This was observed repeatedly with cross sections as well as flat-grain 1/16-inch-thick veneers. The effect is probably related to the high FSP of the treated materials. The swelling resulting from this adsorption of moisture is, to a high degree, internal into the lumens; buoyancy is thereby reduced.

Discussion

It has been shown that a marked increase occurs in the swelling capacity of certain lignocellulosic materials in water after these materials have been subjected to certain mild treatments. Furthermore, the change in swelling capacity occurs with little loss in weight. The increase in swelling capacity is limited and complete when ester groups in the xylan chains acting as cross-links have been broken. We now consider how these transformations can be used to explain the increase in rate of degradation by various kinds of cellulolytic micro-organisms.

Table V was prepared from Figure 5. The critical PEG molecular weight is that molecular weight below which some penetration can occur into the substance. For untreated wood, it is about 3000. However, the meaning of this value must be interpreted with caution. According to the theory of gel permeation chromatography, the important single solute parameter is effective molar volume rather than molecular weight (9). The molar volume of a polymer depends on polymer–polymer and polymer–solvent interactions; this is illustrated schematically in Figure 9 (36). The total chain lengths are the same for all the polymer systems, yet the configurations and bulkiness vary considerably.

Table V. Critical Molecular Weight for Penetration and Maximum Swelling Capacity of Wood and Derived Materials in Water (FSP)

Material	PEG Critical Molecular Weight	Estimated Protein Critical Molecular Weight	FSP, %
Wood (Untreated)	3000	30,000	40
Dilute NaOH-Treated Hardwood[a]	4-5000	40-50,000	75-80
Holocellulose	5000	50,000	140
Cellophane	7000	70,000	100

[a] Acid-washed.

PEG is a flexible molecule with a configuration in solution corresponding to that of a random coil (39). Assuming the coil is impermeable to the solvent, a radius of gyration of 18–20 A. was computed from viscosity measurements (38). With the further assumption, this is the actual radius of the dissolved molecule, a "cloud" density of about 0.2 grams per cubic centimeter was computed for the PEG-3000 molecule in solution. The PEG-3000 molecule is obviously highly expanded. It can be assumed that the volume of this expanded molecule is also the limiting volume for penetration of hydrated proteins into the cell wall. Globular protein molecules generally have highly compact configurations; these may be helical or tightly folded structures. Their "cloud" densities should be considerably higher than those of PEG-3000. It would be of great interest to know the molecular weight of these proteins having the limiting molecular volume. This can be determined by gel permeation chromatography.

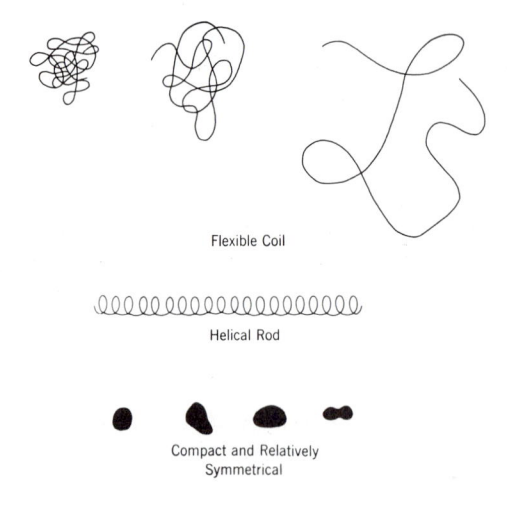

Flexible Coil

Helical Rod

Compact and Relatively Symmetrical

Figure 9. Schematic diagrams (courtesy of Charles Tanford) of the possible simple configuration of macromolecules in solution (36). Relative sizes give indication of relative space occupied in the solution for molecules of equal chain length

Andrews (3) determined the elution volumes of a number of globular proteins from Sephadex G-100. He found an inverse linear relationship between the elution volume and the logarithm of the molecular weight of the protein.

Although retardation in such chromatography can also be influenced by interactions between polymer and gel, the observed linear relationship indicates the retardation was caused mainly by a molecular sieve action. From these data, it is conservatively estimated that the cutoff molecular weight of the type of proteins used by Andrews on Sephadex G-100 is about 150,000.

Recently, Feist et al. (*10*) studied the gel permeation chromatography of various PEGs on Sephadex G-100. From these data, a PEG molecular weight cutoff of about 15,000 is computed. Thus, in aqueous solution, PEG-15,000 has approximately the same effective molar volume as a hydrated protein of molecular weight 150,000. Here is a dramatic illustration of the relatively highly expanded configuration of PEG (*28*). As a crude approximation, for equal molar volumes in solution, proteins have 10 times the molecular weight of PEG. Sephadex G-100 is a very open gel adsorbing 10 grams moisture per gram of dry gel. Suppose a much "tighter" gel is considered such as untreated wood substance adsorbing about 0.35 gram moisture per gram of wood. Because the PEG cutoff molecular weight of untreated wood is 3000, as a very crude approximation, the globular protein cutoff molecular weight is about 30,000 (Table V). Proteins or enzymes in solution with lower molecular weights should penetrate wood substance to a degree. This is not very helpful, however, for little is known about the molecular weight of cellulases. Micro-organisms probably produce a coordinated system of enzymes with differing molecular weights. At least three enzymes have been isolated from *Polyporous versicolor* (*25*). Their molecular weights range from 11,000 to almost 70,000. Much less is known about the enzyme systems produced by rumen-inhabiting bacteria.

On the basis of the data in Table V, there is the indication that the PEG cutoff molecular weight increases when wood substance is modified. Using the conversion factor of 10 (with some license), the cutoff molecular weight of proteins in NaOH-modified hardwood is increased to 40,000–50,000, and in holocellulose, to about 50,000. Some increase in digestibility can be expected if penetration into the untreated substance was marginal. Thus, the increase in cutoff molecular weight may be partly responsible for the substantial increase in digestibility of chemical pulps and for the modest increase in digestibility of dilute NaOH-treated lignocellulosic materials. Yet, other consequences of the treatments must be sought.

Also in Table V, some measured fiber saturation points are listed. The additional moisture at 100% RH in the modified materials is present in submicroscopic water pockets within the substance. It would now be expected that there would be a marked increase in the rate of diffusion of water-soluble materials through the swollen substance. This was described in a previous publication (*37*).

Proteins within enzyme systems, however, must not only be able to penetrate or diffuse through the substance, but in order to promote cleavage of carbohydrate bonds, the very few active centers on the enzyme molecules must come into proper stereo-relationship with the vulnerable bonds in the carbohydrate substrate. Only then can the enzyme-substrate complex be formed. To accommodate themselves to the substrate, the enzyme molecules must be able to rotate freely; this occurs through Brownian movement at a rate determined by the rotatory diffusion constant (2, 17). A high water-to-substance ratio, or high FSP, would provide conditions for the more rapid attainment of the normal rotatory diffusion constant and optimization of successful interactions between enzyme and substrate. These conditions very likely do not prevail in the unmodified wood. In addition, the increased internal surface area resulting from the creation of additional submicroscopic water-filled pockets would increase the area of carbohydrate surface accessible to the enzyme molecules. Recent work by Caulfield at the U. S. Forest Products Laboratory (7) has shown that the increase in *in vitro* digestibility of ball-milled cotton by a cellulase is caused mainly by the increase in specific surface area rather than by a decrease in x-ray-measured crystallinity.

A useful characterization of these modified lignocellulosic materials would be that of the size or of the size distribution of the water-filled submicroscopic pockets and of the additional internal surface area in the water-swollen condition created by the various modifications. Characterization such as this based on adsorption measurements has been made by Stone and Scallon (34). Additional characterizations based on x-ray-scattering measurements are being carried out at the U. S. Forest Products Laboratory.

Conclusions

Hardwoods and certain other lignocellulosic materials derived from annual plants respond to dilute alkali (below mercerization concentrations) in a manner that increases their digestibility by certain fungi and rumen-inhabiting bacteria. The important chemical reaction is a saponification of esters of uronic acid associated with the xylan chains. Because the esters bridge polymeric units, the effect of saponification is a breaking of cross-links. A similar breaking of cross-links is obtained on ammonolysis; that is, on treating the lignocellulosic materials with liquid ammonia.

The physical effect, which is measured quantitatively, is a marked increase in fiber saturation point, or moisture content of the substance at 100% relative humidity. With hardwoods treated with 1% NaOH, followed by washing, the swelling capacity of the cell walls is doubled. This increase in FSP not only provides for improved diffusion conditions of water-soluble materials, it also provides for improved enzyme-substrate interactions. In addition, there is an indication of a small increase in the maximum molecular weight of globular proteins capable of diffusing into the substance.

These maximum effects are obtained with a NaOH consumption of less than 10% based on the weight of the lignocellulosic material. The amount of NH_3 necessary to get the same effect will depend on the efficiency of the recovery system. Incidentally, the chemically combined nitrogen (as amide or ammonium) should contribute to the usefulness of this material for feeding ruminants.

Perhaps more important, however, is the *in vivo* performance of these modified materials as feed supplements. Work on this phase is under way with different ruminants.

An important question arises on using modified lignocellulosic materials as a feed-supplement for ruminants. How do micro-organisms have access to the modified tissues? Xylem tissues of annual plants and hardwoods have longitudinal vessels that are prominent conducting pathways with diameters of 50–100 microns. Probably these vessels provide an important pathway for the micro-organisms to penetrate the gross structure and approach the modified substance. In annual plants, these vessels are generally unobstructed, although with increasing maturity, there are some indications they become plugged with tyloses (19). In hardwoods, they may be open as in aspen or plugged with tyloses as in white oak. In white oak, penetration by micro-organisms would be impeded and probably no benefit would derive from modification of the xylem tissue as has been described. Thus, only certain species of hardwoods, may be useful for modification to a "foragelike" material. The quantity of tyloses-free hardwoods that accumulate yearly is conservatively placed at tens of millions of tons (46). Similar quantities of cereal straws and other annual xylem tissues (jute, hemp) are available for modification.

The development of an economic process for increasing the digestibility of these lignocellulosic materials should provide for disposal—in a useful manner—of large quantities of coarse fibrous materials. This type of process would provide the medium through which these coarse unused plant tissues would be transformed into protein-rich food supplements for human consumption.

Experimental

Values for Calcium Ion Exchange. These values were measured on 40-mesh (Wiley mill ground) ovendry samples using the technique of Millett *et al.* (23).

NaOH Treatment. Cross sections of waterlogged 20-mil microtomed wood (or other dimension, if specified) were treated with dilute NaOH and washed for several days in distilled water or washed in 5% acetic acid followed by washing in distilled water (37). These samples were used for chemical analyses as well as FSP measurements.

FSP Measurement. The FSP of treated and untreated samples was measured using the nonsolvent water content technique described by Feist *et al.* (*12, 37*).

Liquid NH$_3$ Treatment. Transverse cross sections of air-dry, 20-mil microtomed sugar maple were immersed in liquid NH$_3$ at 30°C. (*ca.* 150 p.s.i.). The samples were washed with fresh liquid NH$_3$, air-dried, and water-washed for several days.

Fungal Decay of Woods. Blocks of NaOH-treated and untreated sugar maple were inoculated with three test fungi—a brown, a white, and a soft rot—by the standard soil-block method (*8*). At various intervals, the test blocks were removed, the fungal growth washed off, the block sterilized, and ovendried to obtain weight losses.

Acknowledgment

The authors express appreciation to the late C. G. Duncan, who directed the tests to determine the digestion rates of wood-rotting fungi, and to R. C. Weatherwax, who performed the NaOH-distribution experiments. Both are of the U. S. Forest Products Laboratory, Madison, Wis. The authors wish to acknowledge the cooperation of the Tennessee Valley Authority, Division of Forestry Development, Norris, Tenn., in this work.

Literature Cited

(1) Aggebrandt, L. G., Samuelson, O., *J. Appl. Poly. Sci.* **8**, 2801 (1964).
(2) Alberty, R. A., Hammes, G. G., *J. Phys. Chem.* **62**, 154 (1958).
(3) Andrews, P., *Biochem. J.* **91**, 222 (1964).
(4) Baumgardt, B. R., Scott, R. W., Millett, M. A., Hajny, G. J., *J. Animal Sci.* (to be published).
(5) Bechmann, E., *Sitz. Preuss. Akad.* 275 (1919).
(6) Brown, L. R., "Conference on Alternatives for Balancing World Food Production and Needs," Ames, Iowa, 1966, p. 3-18, Iowa State Univ. Press, Ames (1967).
(7) Caulfield, D. F., Moore, W. E., *U. S. Forest Products Lab.*, Madison, Wis. (to be published).
(8) Cowling, E. B., *U. S. Dept. Agr. Tech. Bull.* 1258 (1961).
(9) Determann, H., "Gel Chromatography, Gel Filtration, Gel Permeation, Molecular Sieves; A Laboratory Handbook," p. 73, Springer-Verlag, New York, 1968.
(10) Feist, W. C., Southerland, C. F., Tarkow, H., *J. Appl. Poly. Sci.* **11(1)**, 149 (1967).
(11) Feist, W. C., Tarkow, H., *Forest Prod. J.* **17(10)**, 65 (1967).
(12) Flory, P. J., "Principles of Polymer Chemistry," p. 519, Cornell Univ. Press, Ithaca, N. Y., 1953.
(13) *Ibid.*, p. 585 (1953).
(14) Food and Agr. Org. U.N., "State of Food and Agriculture 1966," Rome, 1966.
(15) Ghose, S. N., King, K. W., *Textile Res. J.* **33**, 392 (1963).
(16) Hall, J. A., Saeman, J. F., Harris, J. F., *Unasylva* **10**, 7 (1956).
(17) Hammes, G. G., Alberty, R. A., *J. Phys. Chem.* **63**, 274 (1959).
(18) Hayward, H. E., "Structure of Economic Plants," p. 30, MacMillan Co., New York, 1938.

(19) *Ibid.*, p. 610 (1938).
(20) Hvidtsen, H., Homb, T., *Proc. 11th Int. Congr. of Pure and Applied Chem., London* **3**, 113 (1951).
(21) Klauditz, W., *Holzforschung* **11**(2), 47 (1957).
(22) Liu, K. J., Burlant, W., *J. Poly. Sci., Part A-1: Poly. Chem.* **5**, 1407 (1967).
(23) Millett, M. A., Schultz, J. S., Saeman, J. F., *Tappi* **41**(10), 560 (1958).
(24) Nordfeldt, S., *Proc. 11th Int. Congr. of Pure and Applied Chem. London* **3**, 391 (1951).
(25) Pettersson, G., Porath, J., Cowling, E. B., *Biochem. Biophys. Acta* **67**, 1 (1963).
(26) Pew, J. C., Weyna, P., *Tappi* **45**(3), 247 (1962).
(27) President's Science Advisory Committee, "The World Food Problem," 3 vol., U. S. Government Printing Office, 1967.
(28) Ryle, A. P., *Nature* **206**, 1256 (1965).
(29) Sarkhar, P. B., Chatterjee, H., Mazumdar, A. K., *J. Text. Inst.* **38**, T318 (1947).
(30) Sisson, W. A., Saner, W. R., *J. Phys. Chem.* **45**, 717 (1941).
(31) Sjöström, E., Enström, B., Haglund, P., *Svensk. Papperstid.* **68**, 186 (1965).
(32) Sjöström, E., Janson, J., Haglund, P., Enström, B., *J. Poly. Sci.* **11C**, 221 (1965).
(33) Stamm, A. J., "Wood and Cellulose Science," p. 157, Ronald Press, New York, 1964.
(34) Stone, J. E., Scallon, A. M., *Tappi* **50**(10), 496 (1967).
(35) Stranks, D. W., *Forest Prod. J.* **9**(7), 228 (1959).
(36) Tanford, C., "Physical Chemistry of Macromolecules," p. 127, John Wiley and Sons, New York, 1961.
(37) Tarkow, H., Feist, W. C., *Tappi* **51**(2), 80 (1968).
(38) Tarkow, H., Feist, W. C., Southerland, C. F., *Forest Prod. J.* **16**(10), 61 (1966).
(39) Thomas, D. K., Charlesby, A., *J. Poly. Sci.* **42**, 195 (1960).
(40) Timell, T. E., *Tappi* **40**(7), 568 (1957).
(41) Timell, T. E., *Tappi* **40**(9), 749 (1957).
(42) Timell, T. E., *Wood Sci. and Tech.* **1**, 45 (1967).
(43) Turner, H. D., *Forest Prod. J.* **14**(7), 282 (1964).
(44) *Unasylva* **20**, 1 (1966).
(45) U. S. Dept. Agr., Econ. Res. Serv., Foreign Regional Analysis Div., Rept. No. 19, "The World Food Budget 1970," Washington, D. C., 1964.
(46) U. S. Dept. Agr., Forest Serv. Rept. No. 17, "Timber Trends in the U. S.," Govt. Printing Office, Washington, D. C., 1965.
(47) Virtanen, A. I., *Science* **153**, 1603 (1966).
(48) Wang, P. Y., Bolker, H. I., Purves, C. B., *Can. J. Chem.* **42**, 2434 (1964).
(49) Wilson, R. K., Pigden, W. J., *Can. J. Animal Sci.* **44**, 122 (1964).

RECEIVED October 14, 1968. The Forest Products Laboratory is maintained at Madison, Wis., in cooperation with the University of Wisconsin.

Discussion

A. Beelik: "Were pectins and fatty acid resins—*i.e.*, other COOH-containing materials removed from maple wood before NaOH treatment?"

H. Tarkow: "No. Most of our work has been done with hard maple, where such extraneous carbonyl-containing materials are negligible. With other hardwoods, such as the oaks with their hydrolyzable tannin contents, such preextraction should be employed."

13

Digestibility as a Simple Function of a Molecule of Similar Size to a Cellulase Enzyme

J. E. STONE and A. M. SCALLAN
Pulp and Paper Research Institute of Canada, Pointe Claire, P. Q., Canada

E. DONEFER and E. AHLGREN
Department of Animal Science, Macdonald College, McGill University, Montreal, Canada

Cotton linters were swollen to an increasing extent by means of phosphoric acid of increasing concentration. Using a number of simple sugars plus a series of dextran molecules ranging in M.W. from 180 to 2.4×10^7 and diameter from 8 to 1600 A. it was possible to measure the pore volume and calculate the surface area within the swollen fibres accessible to all the molecules within this range. The substrates reacted with a commercial cellulase enzyme preparation, and the initial rate of reaction was compared with the accessibility of the substrate to molecules of various sizes. There was found to be a linear relationship between the initial reaction rate and the surface area within the cellulose gel which was accessible to a molecule of 40 A. diameter.

The ability of cellulolytic micro-organisms and of cell free cellulolytic enzymes to degrade cellulose varies very greatly with the nature of the substrate. Thus, the cellulose in untreated wood is virtually indigestible to ruminants whereas after delignification the cellulose is rapidly and completely digested. Similarly, forages of different species or of differing degree of maturity can have widely differing digestibilities, even though their cellulose contents may not be too dissimilar, and this fact has been a constant challenge to animal nutritionists to seek the basic cause and an economical means of increasing the digestibility of low grade forages.

For an enzymatic reaction to take place, direct physical contact between the enzyme and its substrate must occur, producing an enzyme-substrate complex which then breaks down into the products of the reaction. It is to be expected, therefore, that the rate of reaction should be a function of the surface area of the cellulose which is accessible to the enzyme. This has been recognized for many years and the influence of changes in the structure and composition of a variety of cellulosic substrates upon their susceptibility to enzymatic attack has been reviewed by Cowling (1). Such changes are (i) particle size reduction, (ii) delignification, (iii) change in crystallinity of the cellulose, (iv) hydrolysis and reduction of cellulose chain length, (v) swelling, and (vi) disruption of the lignocellulose complex by radiation (8) or ball-milling. In many cases a good correlation has been obtained between a particular parameter and the reactivity of the substrate, but there is also a good deal of ambiguity and lack of agreement between the various sets of data. For example, sprucewood ground to pass an 80-mesh screen or cut into 1 mm. cross sections and then crushed in a press was found to be no more susceptible to enzymatic attack than the original wood, whereas 10 minutes treatment in a vibratory mill produced considerable reactivity (11). Dehority and Johnson (3) ball-milled mature timothy for 0, 6, 24, and 72 hours and found that the digestibility increased markedly. Pure cellulose, however, wet milled for 6, 24, and 96 hours to cause a large reduction in particle size, showed no change in the rate of digestion (4). As far as crystallinity is concerned, it has been shown that when cotton is progressively decrystallized it becomes increasingly reactive (12), but Lee (9) has found with regenerated cellulose that the most reactive of his materials was also the most crystalline. A further demonstration that crystallinity of the cellulose is not of itself a controlling factor in reactivity is the common observation that a never-dried cellulosic material is much more reactive than one which has been dried and reswollen in water, even though the crystallinity is increased only slightly if at all by such a drying treatment. Lee concluded that the degree of swelling of the substrate is the most important factor because a higher degree of swelling increases the pore size above a certain critical value which is necessary to allow the enzyme molecule to penetrate the substrate. As will emerge from the present study, we agree in principle with this conclusion, but before continuing with a description of the experiments and the results obtained, a brief discussion of the surfaces and surface area development in cellulose will be given.

In a porous body, the surface area available to a reactant will depend upon the relative sizes of the pores and the penetrating substance. Thus, in the case of wood and its attack by microorganisms, the latter are so large in comparison to pit membrane pores that the initial attack must be restricted to the external surface of the wood. Grinding or cutting the wood to open all the lumens will expose considerably more surface, the amount reaching about 1.0 sq. meter/gram. Once the lumens are accessible, however, the large increase in surface which might be expected from further grinding will be found upon calculation to be illusory. This is because cell wall thickness is very small compared with its length, and exposing cut ends increases the total surface only by the cross-sectional area of the fibre. For example, typical sprucewood fibres cut into 100 pieces, thereby producing a very fine meal, will have increased in surface by only about 20%. If an enzyme molecule can reach a cell lumen *via* the pits, the grinding of even comparatively large pieces of wood will not appreciably increase the accessible surface and no increase in reactivity should be expected. If vibratory ball milling or other treatment does increase reactivity, it must be owing to changes within the cell wall itself and not to the size of the cell wall fragment.

Another example of the confusion which can arise is in equating the accessibility of the cellulose to its amorphous fraction. The amorphous fraction may determine the accessibility of the hydroxyl groups to water or the glucosidic bonds to dilute mineral acid but is not necessarily related to the accessibility of the cellulose to a large molecule such as an enzyme. For example, the amorphous material in wood is accessible to water and hydrogen ions but is completely inaccessible to an enzyme. If, therefore, a correlation is found between the amount of amorphous cellulose in a sample and its susceptibility to attack by cellulase, this is coincidental, and the true reason for the increase in reactivity must be caused by some other change which has occurred simultaneously with the change in crystallinity.

To further emphasize this point concerning accessibility and reactivity, an example may be taken from our own experience. One of the methods which has been used for measuring the surface area of a water-swollen fiber is the removal of the water by solvents and then drying in an inert atmosphere followed by a determination by nitrogen adsorption of the surface area of the aerogel so formed. Sprucewood gives a value of about 1 sq. meter/gram and grinding to an 80-mesh meal in a Wiley mill gives very little increase, as expected. Wet grinding in a disk refiner, however, as in the production of mechanical pulps for newsprint, leads to surface areas of over 40 sq. meters/gram after solvent exchange drying. This demonstrates the creation of a great deal of surface within the cell wall and it was expected that such pulp would have an appreciably higher digestibility to a ruminant than did sawdust or woodmeal. It was found, however, that the digestibility remained at zero. The large surface which had been created was accessible to a nitrogen molecule, therefore, but not to the enzyme produced by the rumen microorganisms, and as far as the enzyme was concerned, no surface had been created.

If it is hypothesized that the rate of reaction between an enzyme and its substrate is dependent upon the surface area which is accessible to the enzyme, it is necessary, in order to support this hypothesis, to measure accessibility with a solute molecule of the same size as the enzyme and under conditions which approximate those which exist during the enzymatic reaction. In recent years, we have been developing a technique which appears suitable for this purpose. Termed the "solute exclusion technique," it employs a series of solute molecules which range in size upwards from glucose with a hydro-dynamic diameter of 8 A. to a dextran of M.W. 24×10^6 and diameter of 1600 A. These act as a set of molecular probes of the gel structure and give a complete pore size distribution. The method has been applied to a study of the fiber saturation point of wood and wood pulps (13), to the change in cell wall porosity during pulping (14) and beating (15), to the formulation of a structural model for the cell wall (16) and to a study of the accessibility of regenerated cellulose gels (17).

The original motivation for the present investigation was the need to obtain a deeper understanding of the reason for the difference in digestibility of various forages by ruminants, and since it was known that the degree of lignification was an important factor, the first phase of the study used samples of sprucewood progressively pulped to lower yields as the substrate and a rumen inoculum as the source of enzyme. The reason for using sprucewood and pulps derived from the wood instead of a series of conventional forages was that it enabled the complete range of digestibilities from 0 to 100% to be achieved with fibers of the same morphology. The fibers were, however, changing in composition and a rumen inoculum is not a simple enzyme system, so for the second phase of the investigation a simpler, more easily interpretable system was used consisting of cotton, swollen to various extents with phosphoric acid, as the substrate and a commercial cellulase from *Trichoderma viride* as the enzyme.

Experimental

Wood Pulps. Black spruce (*Picea mariana*) shavings were extracted with alcohol-benzene and a small portion ground in a Wiley mill to pass through a 40-mesh screen. This represented the 100% yield sample. The remaining shavings were cooked in one liter bombs by the acid sulfite process (sodium base, 1% combined and 5% total SO_2) at a liquor-to-wood ratio of 5:1. The time at the cooking temperature of 135°C. was varied from 0 to 300 minutes, the contents removed, fiberized briefly in a Waring Blendor and washed with distilled water. Half of each sample was stored in water at 5°C. and is hereafter referred to as the never-dried sample. The other half was dried overnight at 105°C. and then ground in a Wiley mill to pass a 40-mesh screen. This is termed the dried sample. The yield of each pulp was obtained by enclosing a small weighed sample of wood within a wire mesh basket in each bomb and then drying and weighing the sample after pulping. The yield and lignin content of each pulp is listed in Table I.

Table I. Composition of Wood Pulps

Yield	% Cellulose (18, 19)	% of Total Lignin[a] Removed
100.0	54.1	0
96.4	58.2	2
90.1	62.0	11
81.3	66.2	16
70.0	73.5	36
62.6	79.9	54
57.0	88.0	71
52.5	93.4	81

[a] Klason plus U.V. lignin.

In Vitro **Rumen Fermentation.** The different pulps were digested by means of rumen organisms obtained from a fistulated steer, the procedure, preparation of inoculum and preparation of the basal medium being essentially that of Donefer, Crampton, and Lloyd (5). (A fistulated steer is an animal with a surgically prepared opening in its side through which material can be added to or removed from its rumen. The rumen is the large (40–60 gallons) forward compartment of the stomach in which the microbial digestion of forage occurs. A rumen inoculum is a solution of cellulolytic microorganisms obtained by filtration of the rumen contents to remove solid particles.) In each fermentation tube were placed a sample containing 200 mg. of cellulose, 20 ml. of the phosphate buffer extract of the rumen ingesta plus basal medium to give a total volume of 50 ml. The digestions were performed at 40°C. and pH 7. Cellulose analysis on the samples before and after digestion were carried out essentially by the procedure of Crampton and Maynard (2).

Cellulose Substrates. The source of cellulose was a chemical cotton supplied by Hercules Inc., Virginia. It was ground in a Wiley mill to pass a 40-mesh screen before use in order to prevent flocculation and permit easy mixing with the swelling agent and enzyme solution.

The cotton was divided into several portions, one of which was swollen in water for several days and the others in phosphoric acid of several concentrations to give a series of cellulose samples of various degrees of swelling. The treatment with phosphoric acid is essentially that of Walseth (18) except that, whereas he used 85% acid at 2°C. for different lengths of time, we used different concentrations of acid at 2°C. for fixed lengths of time. The reason for this change was to avoid the necessity of having to use very short times for the small degrees of swelling and the possibility of non-uniform swelling. It was found that the samples swelled to equilibrium in 20 to 30 minutes and thereafter remained constant. The range of acid concentration used was from 72.0 to 78.8% by weight and the degree of swelling or fiber saturation point was determined by the method described below. The relationship between acid concentration and swelling is depicted in Figure 1.

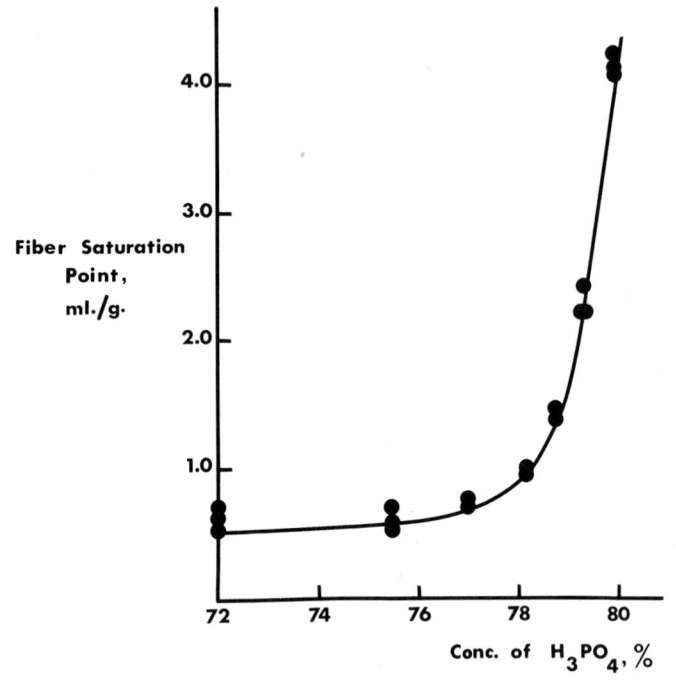

Figure 1. Swelling of cotton in phosphoric acid

The experimental details of the swelling treatment are as follows: 1500 ml. of acid of the required concentration were made up by weighing. The acid was cooled to 2°C. and 60 grams of air-dried cotton stirred in. After thorough mixing the cotton and acid were allowed to stand for one hour in an ice-water bath and the mixture was then poured into 10 liters of water at 2°C. The cellulose was filtered off, washed repeatedly with water, then with 1% sodium carbonate solution and finally with water until washings had the same refractive index as distilled water. The cellulose was drained free of water and stored moist at 5°C. until required.

Enzymatic Digestion. An industrial grade enzyme preparation "Meicellase" kindly supplied by the Meiji Seika Kaisha Company Ltd., Japan, was used in the enzymatic digestion of the cottons. The preparation is derived from the fungus *Trichoderma viride* and contains among other enzymes a cellulase which is active on native cellulose. The molecular weight of the cellulolytic enzymes is reported by the manufacturer to be about 60,000. A stock enzyme solution of about 2% concentration was prepared by stirring a quantity of the enzyme in sodium acetate buffer solution (pH 4.5) for one hour at 0°C., and then centrifuging off undissolved solids. The concentration was accurately determined by evaporating to dryness and weighing an aliquot of this solution.

The conditions chosen for the digestions were: pH 4.5, temperature 40°C., concentration of cotton 3% w/v and concentration of enzyme 1% w/v. An amount of wet cotton corresponding to 1.2 grams oven-dry material was placed in a 125 ml. Erlenmeyer flask, the appropriate amount of acetate buffer was added and the mixture was stirred by hand to make a homogeneous suspension. This was heated to 40°C. on a water bath, the enzyme solution added and the flask placed on a shaker in a chamber maintained at 40°C. The amount of cotton digested after various periods of time was determined by measuring the amount of oligosaccharides and glucose produced by the enzymatic reaction. For this, the orcinol method was used. At various times .05 ml. of sample solution was withdrawn and added to 3.0 ml. of the orcinol reagent (0.2% orcinol in 70% sulfuric acid) in a test tube. The covered test tube was then heated in boiling water for 20 minutes, cooled in cold water for 15 minutes and the absorbency of the solution measured at 550 mμ. From a standard curve prepared using pure glucose, the amount of glucose produced and hence the percentage cellulose solubilized was calculated. It will be appreciated that the sulfuric acid in the orcinol reagent hydrolyzes any oligosaccharides present to glucose, so that the glucose actually measured is the sum of that produced enzymatically and that produced from the oligosaccharides by acid hydrolysis. Very good agreement was obtained by the orcinol method and by the method of weighing the cellulose before and after the reaction.

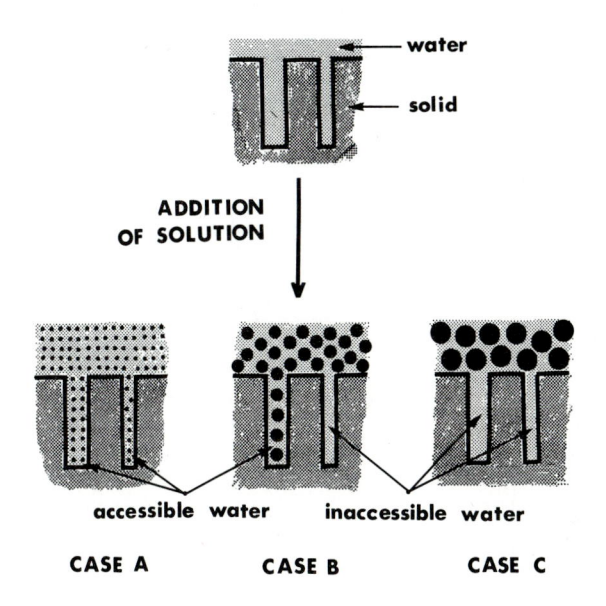

Figure 2. Sketch of the principle of solute exclusion

Porosity Measurements. The solute exclusion technique is best described in connection with a diagram such as that in Figure 2. A porous body, swollen in water and immersed in an excess of water, has added to it a known weight of a certain solution. The system is then thoroughly mixed. If all the water originally associated with the porous body is accessible to the solute molecules, it will all contribute to the dilution of the solution (Case A). If a solution of larger molecules is used (Case B) so that the smaller pores in the porous body are inaccessible to the solute molecules, then the water in these smaller pores is unavailable for dilution and the solution after mixing will be somewhat less diluted than in the first case. This difference in concentration is the basis of a simple calculation to give the amount of water inaccessible to the solute. When the solute molecule is so large that it cannot enter the porous body at all (Case C) the inaccessible water equals the total water of swelling, a useful quantity which is often very difficult to measure by other means.

A typical curve relating inaccessible water to solute diameter is shown in Figure 3. The fiber saturation point corresponds to the plateau formed with very large diameter solute molecules while the median pore size is that value above and below which one-half the total volume of

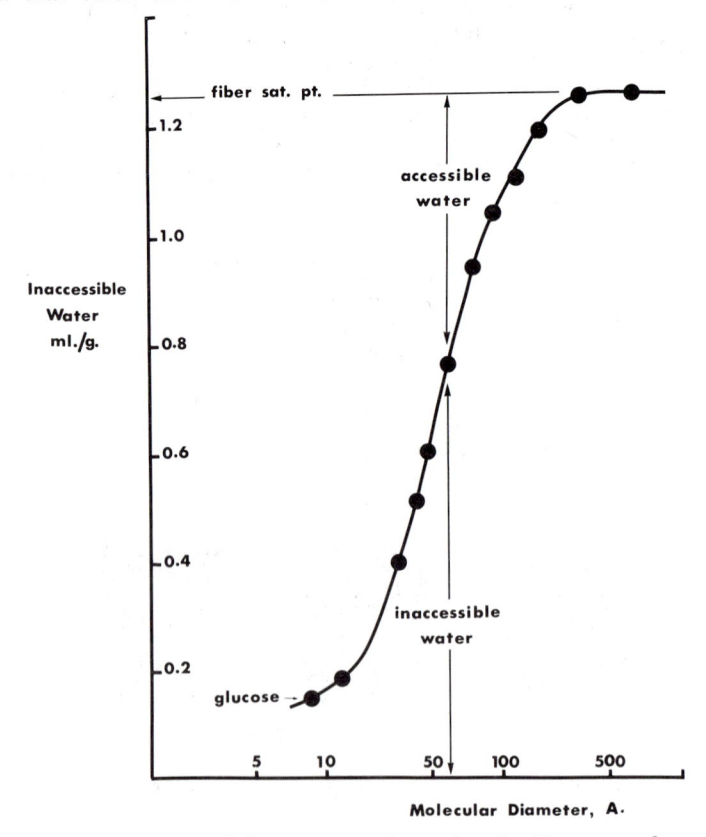

Figure 3. Accessibility of typical wood pulp fibers to solute molecules of various diameters

pores lies. Although the technique actually measures inaccessible water it will be appreciated that it is also a measure of accessible water. Thus:

Accessible water = Fiber saturation point − Inaccessible water.

In order to define this accessibility curve clearly it is necessary to use a number of solute molecules which range in size over the whole range of pore sizes anticipated in the swollen structure. We have found the most suitable solutes to be the dextrans marketed by Pharmacia (Uppsala) Ltd. supplemented by a few low molecular weight sugars. Grotte (7) has reviewed evidence to show that these dextrans behave in solution as hydro-dynamic spheres, and that the diameters of these molecules in solution may be calculated from their diffusion coefficients according to the Einstein-Stokes formula:

$$\text{Diameter} = \frac{RT}{3\pi \eta DN} \qquad (1)$$

where R is the gas constant, T is the absolute temperature, η is the viscosity of water, N is Avogadro's Number and D is the diffusion coefficient. The diameters of the dextrans used by us were found by interpolation of the data given by Granath (6) for a series of similar dextran fractions (B512Ph), and the diameters of the sugars were calculated from the diffusion coefficients of Longsworth (10). The molecules used, together with their diameters, are listed in Table II.

Table II. Properties of Macromolecules

Macromolecule	Molecular Weight M_w	M_w/M_n	Molecular Diameter in Solution A.
Glucose	180	1.0	8
Maltose	342	1.0	10
Raffinose	504	1.0	12
Stachyose	666	1.0	14
Dextran 2.6	2600	1.3	26
Dextran 4.3	4300	1.3	33
Dextran 6.1	6100	1.3	39
Dextran 10	11,200	2.0	51
Dextran 20	21,800	1.5	68
Dextran 40	39,800	1.5	90
Dextran 80	72,000	1.5	120
Dextran 150	150,000	1.6	165
Dextran 500	420,000	2.7	270
Dextran 2,000	2×10^6	—	560
Dextran 24,000	24×10^6	—	1600

The detailed procedure for making the porosity measurements has been described elsewhere (15, 16, 17) and will not be repeated here. The method of using the data to calculate surface areas has also been described (16); however, for the sake of convenience, a brief description will be given here.

The accessibility curve given in Figure 3 is effectively a cumulative pore distribution relating pore volume to pore size. With an assumption of pore shape it supplies all the information required to calculate the surface area of the pores. In previous publications we have postulated that the most basic particle shape within the cell wall is that of lamellae or sheets of microfibrils. The pores would then be the slit like spaces between these sheets as depicted in the simple model shown in Figure 4. In the model two sheets are separated by a distance w (pore width) and contain a pore volume Δv. If ΔA is the area of the pore bounded by the two sheets, then obviously $\Delta A/2$ is the area of one face of the pore and the three quantities, ΔA, Δv, and w are related by the equation:

$$A = \frac{2\Delta v}{w} \qquad (2)$$

Thus, if the complete accessibility curve is divided into a series of increments the surface area of each increment may be calculated in the same way, where w is now the average pore width within each increment.

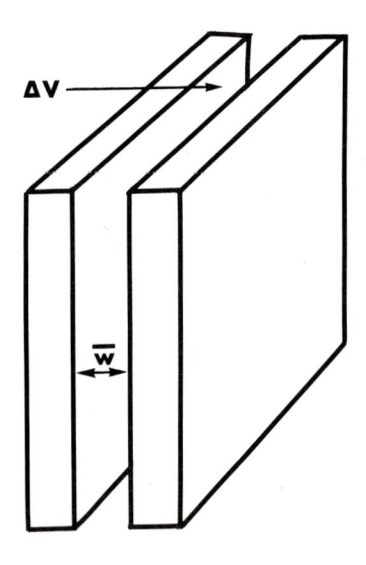

Figure 4. Model for calculating surface area

Starting with the contributions of the largest pores and working down towards the smallest, the increments of area may be added up to give the total surface area. The surface area accessible to a given sized molecule will be found by stopping the calculation at a pore width corresponding to the size of the molecule. The smaller the molecule the greater the surface area accessible to it in a given sample. The very large increase in accessible surface area with decreasing molecular size is shown in Figure 5.

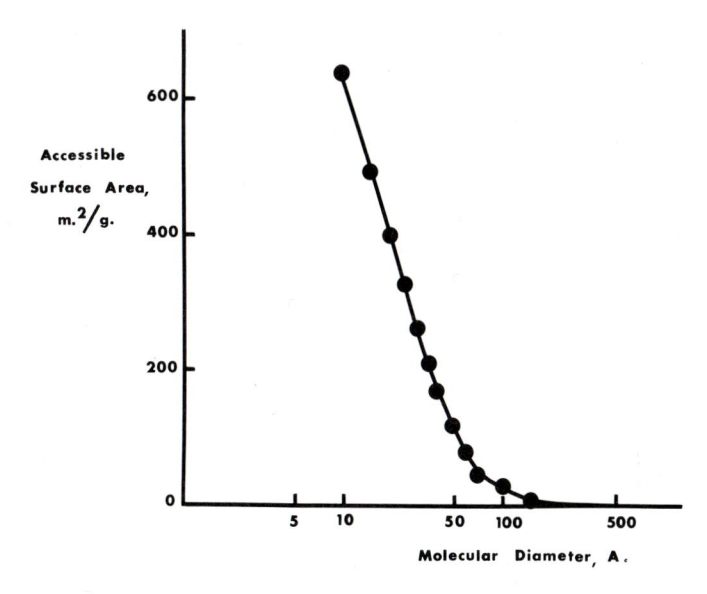

Figure 5. Accessible surface area calculated from the curve in Figure 3

Results and Discussion

Wood Pulps–Rumen Inoculum. In Figure 6 the percentage of cellulose solubilized after 24 hours treatment is shown as a function of the yield of pulp. It is seen that the original wood was virtually indigestible and that pulps produced from this wood and dried and reswollen in water remained indigestible down to a yield of about 75%, at which point about a third of the lignin had been removed. At 70% yield the digestibility of the cellulose in the dried samples became measurable and rose rapidly with further pulping to reach a value of over 90% at 57% yield and lower. The never-dried samples behaved differently in that the digestibility rose steadily as a result of pulping, reaching 50% digestibility at 72% yield (at which point the dried and reswollen sample retained the same indigestible character as the original wood) and reached about 100% digestibility between 60 and 50% yield. One conclusion to be drawn from this is that an economical forage for ruminants is unlikely to be obtained by a chemical treatment of sprucewood if the material is allowed to dry before use because it would be necessary to reduce the yield to about 60% before the cellulose digestibility approached that of conventional forages and pulps of this yield are relatively expensive to produce. Never-dried pulps are clearly more promising from this point of view. (Unpublished work in these laboratories has shown that by contrast with spruce, hardwoods such as birch increase in digestibility very markedly as a result of comparatively mild chemical treatments.)

Dried and never-dried pulps have the same lignin, hemicellulose, and cellulose content at the same yield, but very different digestibilities. The composition of a pulp is therefore a poor guide to its digestibility, some other factor obviously being of considerable importance, and, if this is true of the present series of samples, it is likely to be true also of forages of different species or maturity.

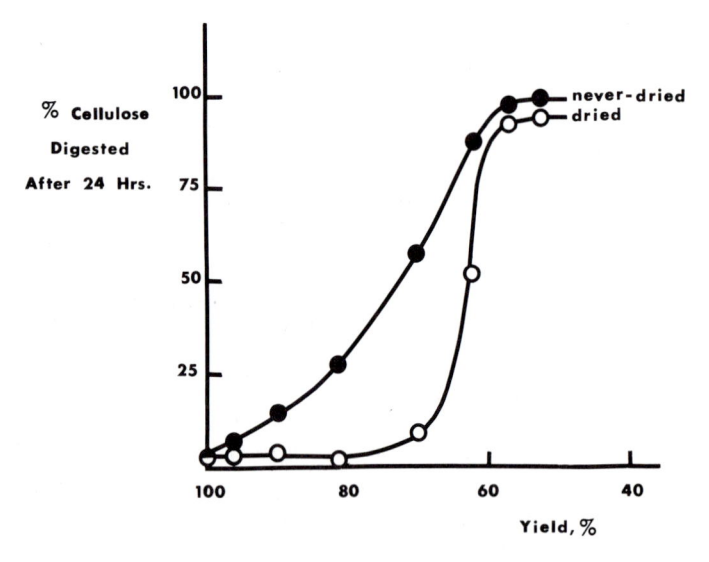

Figure 6. Digestibility of spruce sulfite pulps by a rumen inoculum

The crystallinity of the cellulose is often quoted as a factor of considerable importance in digestibility but it obviously has little bearing on the susceptibility to attack of the present samples because the crystallinity of cellulose certainly does not decrease as a result of pulping; if anything, it increases. Similarly, never-dried and dried and reswollen pulps have very similar crystallinities but very different digestibilities.

The extent of swelling of the samples will now be considered. As described in the *Experimental* section, the solute exclusion technique provides a simple and unambiguous measurement of the amount of water contained within the cell wall and in Figure 7 this quantity is plotted against the yield of pulp. Two features of these curves are worthy of note; one is that the never-dried pulps are more swollen than the pulps which have been dried and reswollen, and this is in accord with the different digestibilities of the two sets of samples, substantiating the

13. STONE ET AL. *Digestibility as a Simple Function* 231

belief that the degree of swelling is an important factor in digestibility. Secondly, it will be noted that the degree of swelling or fiber saturation point does not continually increase with decreasing yield in the same way that the digestibility increases. At lower yields the cell walls of the fibers shrink and are less swollen than at intermediate yields. No correlation is possible, therefore, between swelling and digestibility insofar as the present samples are concerned.

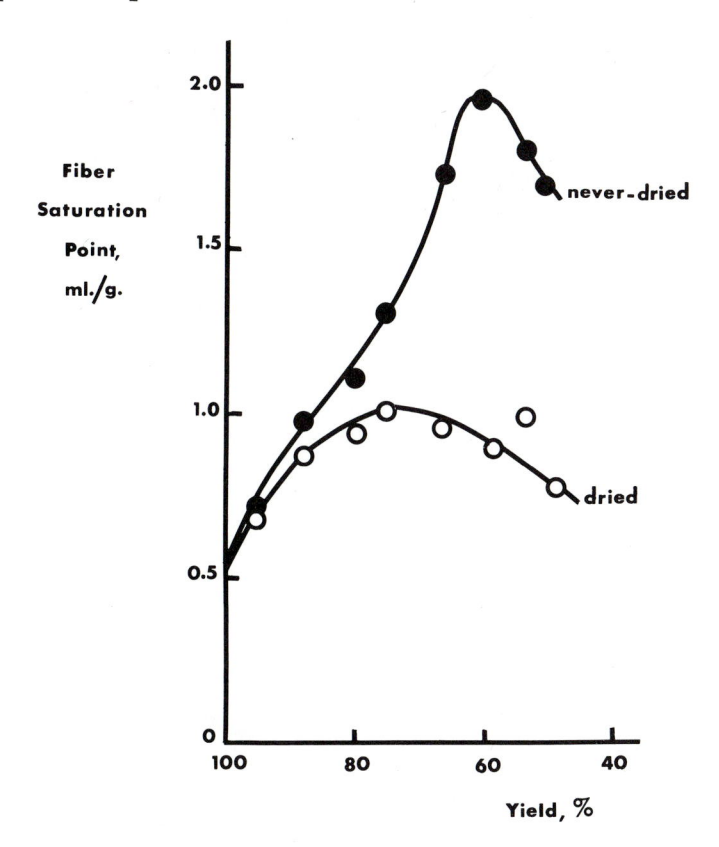

Figure 7. Fiber saturation points of the pulps in the dried and never-dried states

We now come to the question of cell wall porosity. Using the solute exclusion technique described in the experimental section, a complete pore size distribution curve was obtained for the wood and for all the pulps of different yield. These curves have been reported elsewhere (*14*) so will not be repeated here, but the median pore size—*i.e.*, the size of pore above and below which lies half the total pore volume, is plotted against pulp yield in Figure 8. Qualitatively at least, there is now seen to be a resemblance between the change in digestibility with yield (Figure 6) and the change in porosity (Figure 8). The porosity continues to increase with decreasing yield as does the digestibility, and the dried samples have a lower porosity than the never-dried samples, again in agreement with the digestibility data. In Figure 9 a correlation is attempted between median pore size and digestibility and it is seen to be very good in the case of the never-dried samples; no digestion occurs until the median pore size reaches a critical value of about 15 A. and then there is a proportional increase in digestibility for a given increase in pore size until almost complete accessibility to the cell wall and complete digestibility is reached. Although the same trend exists for the dried samples, the two curves are not, unfortunately, superimposable. There may be several reasons for this, one being that the porosity measurements are made on the substrate before digestion has commenced and if a correlation between porosity and reactivity is sought, the initial rate of reaction and not the extent of reaction after 24 hours should be used. The realization of this, coupled with the obvious complexity of a system involving samples of different composition reacting with a rumen inoculum, suggested the second phase of the present investigation. It is worth pointing out that Figures 7 and 8 demonstrate a very important point with regard to gel structure, that is, that it is possible for a gel to shrink—*i.e.*, contain less water per gram of solid substance, while at the same time the median pore size increases.

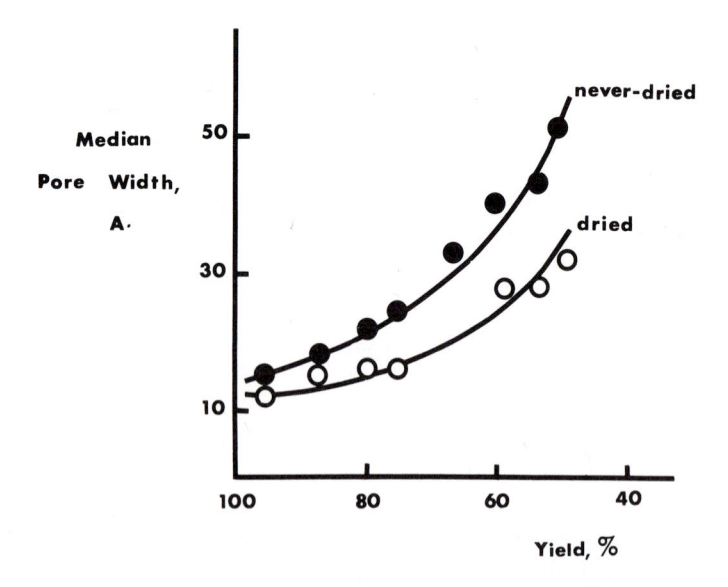

Figure 8. Median pore sizes of the pulps in the dried and never-dried states

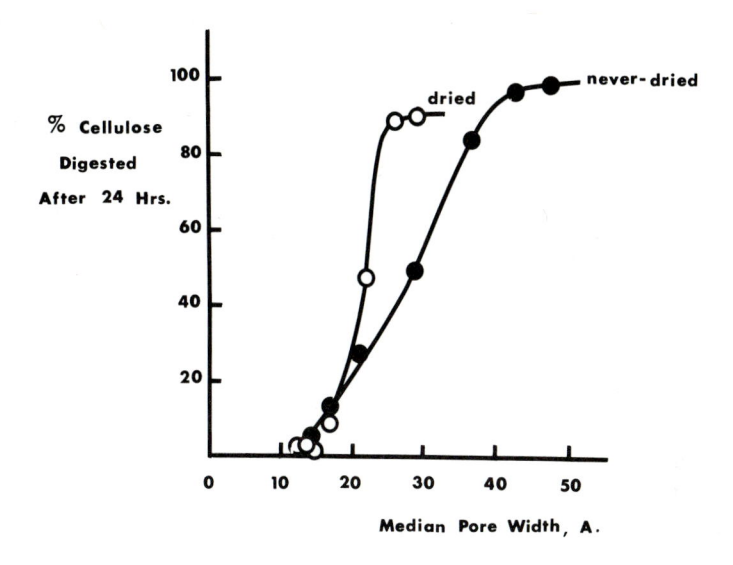

Figure 9. Wood pulps of various yields. Relationship between median pore size and the amount of cellulose dissolved in 24 hours by a rumen inoculum

Cellulose–Cellulase System. In Figure 10 are shown the accessibility curves of the cotton which had been swollen to different extents by phosphoric acid. The fiber saturation point, given by the amount of water inaccessible to a very large molecule, is seen to increase from 0.48 ml./gram for cotton treated with water alone to 1.40 ml./gram for cotton swollen in 78.8% acid. The median pore size increased from 23 A. to 52 A.

In Figure 11 are given the curves relating the amount of cellulose solubilized by the enzyme to the time of the reaction. The initial rates of reaction were obtained by drawing tangents to the curves at zero time. The initial rates were then plotted against various parameters of structure derived from the accessibility curves.

The first correlation attempted was against the total water in the cell wall—*i.e.*, the fiber saturation point (Figure 12). There is seen to be an excellent straight-line relationship but the line does not pass through the origin, an initial reaction rate of zero corresponding to a cellulose containing 0.25 ml./gram of water. Failure to pass through the origin is due to the fact that the fiber saturation point represents the volume within the cellulose accessible to the water molecule and a considerable amount of this water is in regions inaccessible to a molecule the size of a cellulase enzyme.

A more logical correlation is between the initial reaction rate and the volume of water within the cell wall accessible to molecules of larger

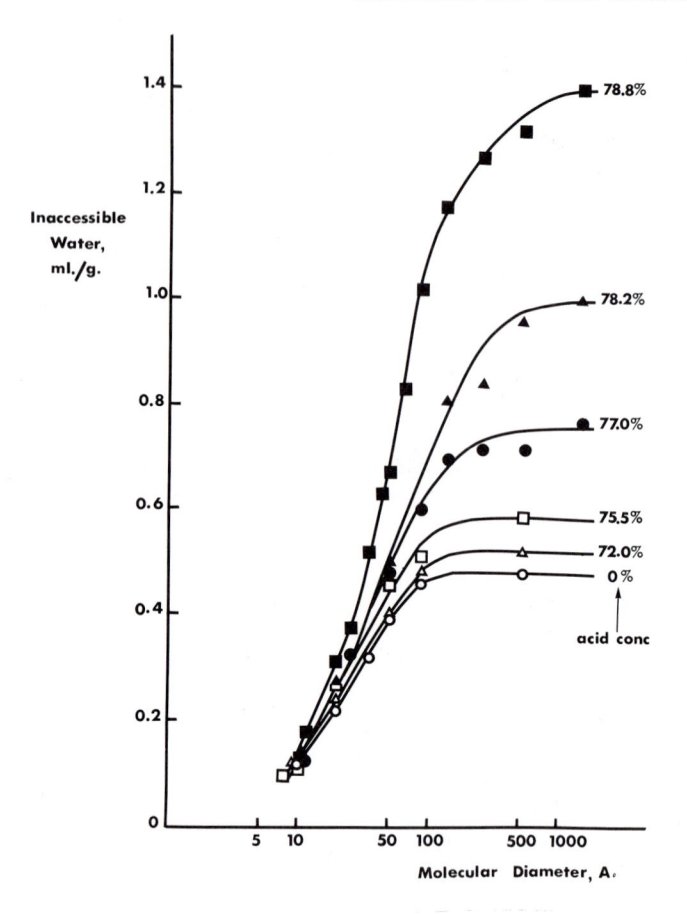

Figure 10. Accessibilities of swollen cottons to solute molecules of various diameters

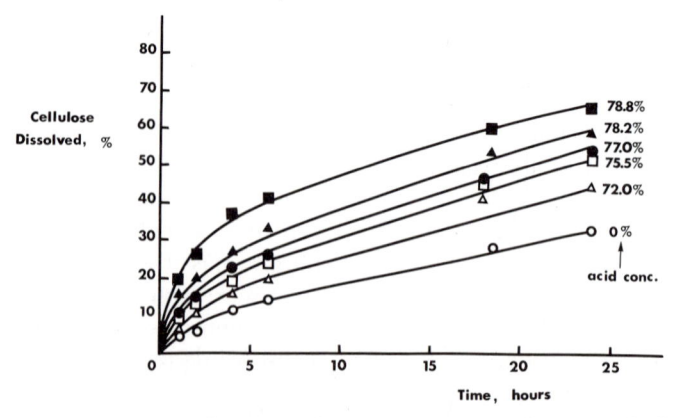

Figure 11. Rate of reaction between enzyme and cotton which has been swollen in phosphoric acid to different extents

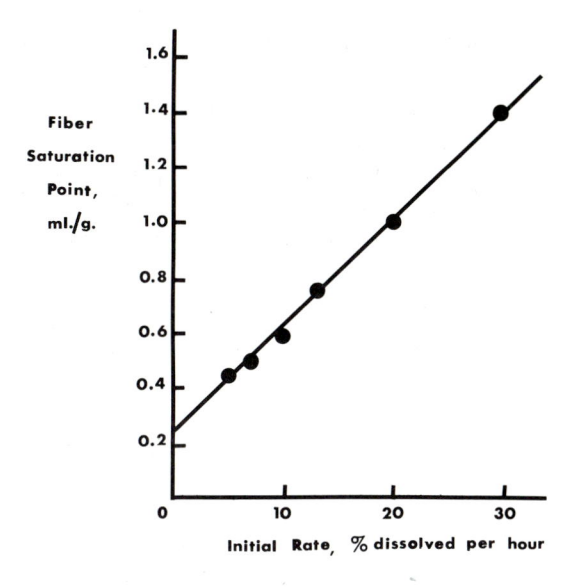

Figure 12. Relationship between the initial rate of reaction and the total water of swelling

sizes. This information is derivable from Figure 10 and the data are plotted in Figure 13. The best correlation is seen to be with a dextran molecule of 30 A. diameter, there being a straight-line relationship between the initial reaction rate and the volume of water accessible to this molecule, and the line passes through the origin. This suggests that the enzyme molecule has a diameter of 30 A., and when the pore volume accessible to it is zero, the reaction rate is zero, but for each increase in accessible pore volume there is a corresponding increase in reaction rate. Molecules smaller than 30 A. give a positive intercept on the y–axis because more water is accessible to them than to the enzyme molecule. Molecules larger than 30 A. give a negative intercept on the y–axis, indicating that less water is accessible to these than to the enzyme molecule.

Although a good correlation was found between the accessibility of the celluloses to a 30 A. molecule and their susceptibility to enzymatic attack, a further relationship was sought between reactivity and the surface area accessible to a molecule of a particular size. This would seem to be a more theoretically valid correlation to establish since the rate of a heterogeneous chemical reaction between molecules in solution and a solid is usually proportional to the available surface. The surface areas were calculated by incremental analysis of the accessibility curves in Figure 10, using the method discussed in the *Experimental* section and assuming that the surfaces in question are those associated with sheets of microfibrils or lamellae within the cell wall. (The assumption of a lamellar structure is not essential to the argument because the alternative possibility of rod-like units such as micro-fibrils or molecules would merely lead to multiplying these surface areas by two). The surface area–reactivity data are plotted in Figure 14 for molecules of 20, 30, 40, 60, and 100 A. diameter, and it is seen that there is a straight line relationship passing through the origin for a solute molecule of 40 A. diameter.

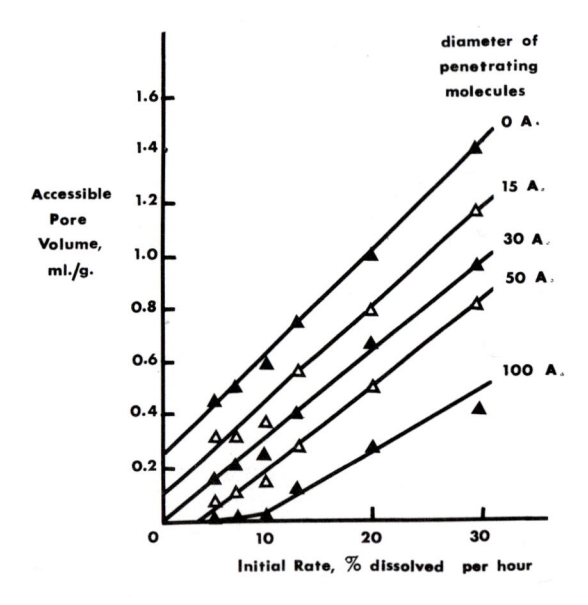

Figure 13. Relationship between the initial rate of reaction and the volume of water accessible to molecules of different sizes

Thus, it appears from the two methods of analysis that the digestibility of a series of celluloses of different degrees of swelling is directly proportional to the accessibility of a molecule of 30 to 40 A. diameter. This suggests that a cellulase enzyme molecule is likely to have a diameter in solution within this range. Little work appears to have been reported on the size of cellulase molecules but Whitaker (20), who made extensive studies on the cellulase from *Myrothecium verrucaria,,* concluded that it had a molecular weight of 60,000 (the same value assigned by the manufacturer to the enzyme preparation used in the present study), was ellipsoidal in solution with a length of 200 A. and diameter at its center of 33 A. This diameter is in good agreement with the present work and suggests an endwise penetration of the enzyme into the cellulose system. [Although this may seem inherently unlikely, the following points should be considered. (i) We visualize the pores in swollen cellulose as the

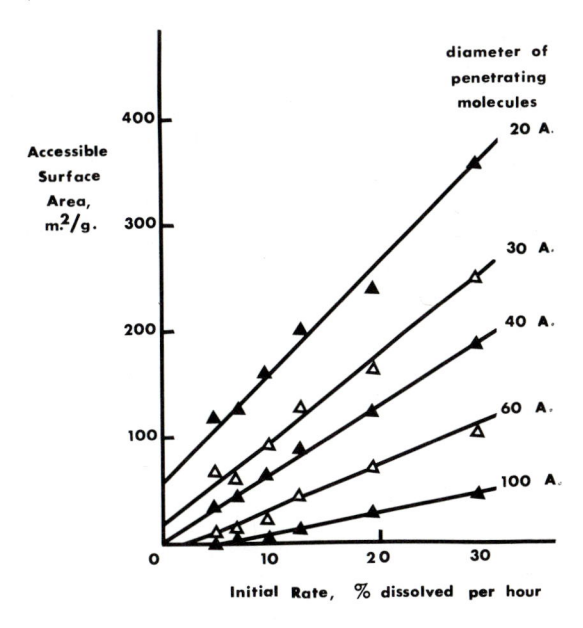

Figure 14. Relationship between the initial rate of reaction and the surface area of cellulose accessible to molecules of different sizes

slit-like spaces between lamellae, and not as being circular in cross-section. Sideways penetration of the enzyme therefore would be possible. (ii) The cellulose itself is in Brownian motion. (iii) Molecular sieves frequently separate ellipsoidal molecules according to their minor axes.]

The reaction between cellulose and cellulase is a heterogeneous one and it is to be expected that the rate of reaction should be proportional to the amount of cellulose surface which is accessible to the enzyme molecule. Using a recently developed method for measuring the amount of surface in a water-swollen gel which is accessible to a solute molecule of any particular hydrodynamic diameter, it has been shown, with a series of pure cellulose samples of differing porosity, that there is indeed a linear relationship between the initial rate of reaction and the initial surface accessible to a molecule of the same size as the enzyme molecule. The cellulosic substrates were, however, a homologous series, and before the relationship between reactivity and accessibility can be established as a general rule it will be necessary to show that it holds with a wide range of diverse substrates. For example, if one of the substrates used in the present study were dried and reswollen in water it would become less reactive and have a lower porosity, but would the new reactivity correspond to the new amount of surface accessible to a 40 A. molecule? Again, the rate of dissolution of cellulose decreases as the reaction proceeds, and it would be interesting to know if this decrease in rate was due to a decrease in the amount of surface available to a 40 A. molecule. Such questions are under investigation at the present time. If a general relationship can be established between reactivity and accessible surface for pure cellulose substrates, regardless of their history, an interesting extension of the study would be to an investigation of substrates containing lignin and hemicellulose. The solute exclusion technique measures surface regardless of chemical composition, so a lower reactivity for a given amount of surface in the case of substrates containing components in addition to cellulose might permit a calculation of the percentage of cellulose at the surfaces within the gel. Such information would be useful in following the changes which take place within the cell wall as a result of, for example, the chemical pulping of wood.

Literature Cited

(1) Cowling, E. B., "Advances in Enzymatic Hydrolysis of Cellulose and Related Materials," E. T. Reese, Ed., Pergamon Press, N. Y., 1963.
(2) Crampton, E. W., Maynard, L. A., *J. Nutr.* **15**, 383 (1938).
(3) Dehority, B. A., Johnson, R. R., *J. Dairy Sci.* **44**, 2242 (1961).
(4) Dehority, B. A., *J. Dairy Sci.* **44**, 687 (1961).
(5) Donefer, E., Crampton, E. W., Lloyd, L. E., *J. Animal Sci.* **19**, 545 (1960).
(6) Granath, K. A., Kvist, E., *J. Chromatog.* **28**, 69 (1967).
(7) Grotte, G., *Acta. Chir. Scand. Suppl.* **211**, 1 (1956).
(8) Lawton, E. J., Bellamy, W. D., Hungate, R. E., Bryant, M. P., Hall, E., *Tappi* **34**, No. 12, 113A (1951).
(9) Lee, G. F., M.Sc. Thesis, Syracuse Univ., N. Y. (1966).
(10) Longsworth, L. G., *J. Am. Chem. Soc.* **74**, 4155 (1952).
(11) Pew, J. C., Weyna, P., *Tappi* **45**, No. 3, 247 (1962).
(12) Reese, E. T., Segal, L., Tripp, V. W., *Textile Res. J.* **27**, 626 (1957).
(13) Stone, J. E., Scallan, A. M., *Tappi* **50**, 496 (1967).
(14) Stone, J. E., Scallan, A. M., *Pulp and Paper Mag. Can.* **69**, 288 (1968).
(15) Stone, J. E., Scallan, A. M., Abrahamson, B., *Svensk Papperstidn.* **71**, 687 (1968).
(16) Stone, J. E., Scallan, A. M., *Cellulose Chem. Tech.* **3**, 343 (1968).
(17) Stone, J. E., Treiber, E., Abrahamson, B., *Tappi* **52**, 108 (1969).
(18) Walseth, C. S., *Tappi* **35**, 228 (1952).
(19) *Ibid.*, **35**, 233 (1952).
(20) Whitaker, D. R., Colvin, J. R., Cook, W. H., *Arch. Biochem. Biophys.* **49**, 257 (1954).

RECEIVED October 14, 1968.

Discussion

H. Swenson (IPC): "Do you feel that the difference in porosity (to your model molecules) between dried and never-dried cellulose might be in differences in pore size distribution in the two cases? My point is that the median porosity in the dried case may be different from the one you measure in the undried sample."

J. Stone: "I think there is no difference in the pore size distribution. For a given pore size the dried material is more digestible than the never-dried material, but it is also a lower yield pulp—*i.e.*, it has less lignin. A possibility, therefore, is that in drying the sample the pores which remain open are more cellulosic. One has to bear in mind not only the size of the pores, but the nature of the surface of the pores. This matter is also involved in the question as to why the enzymatic reaction with cotton slows up after an initial rapid rate reaction. Again, I think we are getting into the question of the nature of the surface of the pores."

J. Marton (Westvaco Corp.): "It occurs to me that you and Dr. Tarkow are using a similar approach, the main difference that polyethylene glycol is an extended molecule, while crosslinked dextran is more compact. Would you comment on what are the basic differences between your conclusions and his?"

J. Stone: "I think there are no basic differences at all. Actually our dextrans are not crosslinked but similar to the P.E.G.'s in that they are long chain molecules which randomly coil in solution to form spherical molecules. One reason we prefer the sugars and dextrans is that they are optically active and thus we can use a precision polarimeter for measurement of concentration. We also have a very wide range of molecular sizes. The dextrans have been well characterized by the physical chemistry group at Uppsala and we use the data which we get from them. As far as our basic conclusions are concerned, I don't think there is any dispute at all. We both agree that wood isn't digestible because the pores in wood are too small for the enzyme to get in."

K.-E. Eriksson: "At the Gordon Conference held recently on chemistry of paper, I was talking about the enzymatic attack on fiber surfaces, and I had some microphotographs and some slides showing enzymic attack on never-dried and dried fibers and these fibers were damaged. The photomicrographs distinctly showed how the enzymes penetrated the damaged part of the fibers first and the never-dried fiber was penetrated and dissolved very quickly, while the dried fibers were hardly not at all attacked by the enzymes. I think there are some differences between your observations and our photomicrographs."

J. Stone: "I have seen those photomicrographs, and agree that the enzyme attacks at points of damage. The reason probably is that the points of damage are points of greater swelling and hence greater accessibility."

H. Tarkow (U.S. FPL): "Perhaps there was a difference in the methods of drying. How were your fibers dried?"

J. Stone: "Yes, this is possibly the reason for the difference between the partial digestibility of our dried samples and the indigestibility of Eriksson's. Our samples were dried overnight at 105°C. and reswollen in water for 24 hours before adding the enzyme, and possibly Eriksson's were not. Thus, his samples may not have had time to reswell and react before the micrographs were taken."

H. Tarkow: "In answer to a previous question regarding your work and ours, I would agree that our conclusions are quite similar. Your work is more quantitative. Both stress the difficulty encountered by enzymes in penetrating cell walls of wood substance because of limited submicroscopic pore sizes. We considered dextrans initially as our probe material. We found, however, that commercially available dextran preparations, such as dextran 10 and dextran 20, were highly dispersed as revealed by gel chromatography down a Sephadex column. Since one of the requirements for the solute exclusion procedure is a uniform molecular size, we sought another polymer system and found the polyethylene glycols to be suitable."

J. Stone: "Well, the first four molecules we use are glucose, maltose, raffinose and stachyose. They are certainly monodisperse. I agree it would be desirable if all the solutes used were monodisperse and this condition could be approached with the dextrans by appropriate gel fractionation."

E. T. Reese: "I don't think we should let the opening statement you made regarding Lee's work at Syracuse go unchallenged. You said that the more crystalline material was more rapidly hydrolyzed by the enzyme and this is true. But you didn't point out that this material was also the most porous of the materials that Lee used, and this was the reason for that increase in susceptibility."

J. Stone: "What I was pointing out was that digestibility is not solely a function of crystallinity, but that some other factor has to be involved. As you say, and as I have attempted to show in this paper, it is porosity."

P. J. Van Soest (Cornell Univ.): "Concerning the dried and undried pulps, were the lignin contents not different? If so, would not lignin account for the differences in digestibiilty of dried and undried pulp at the same pore size?"

J. Stone: "Yes, at the same pore size the dried and reswollen pulp had a lower lignin content than the never-dried pulp. It also had a higher digestibility. We feel that this is due to the difference in composition of the material lining the pores. Thus, digestibility is likely to be a simple function of pore size only for substrates of the same composition."

Query: "Do you associate lignin content with the difference in susceptibility?"

J. Stone: "Yes, the lignin content is important, but as I think I pointed out, if you take a never-dried pulp and dry it then it has exactly the same lignin content, but very different digestibility. So there are a number of factors bearing on the system."

T. K. Ghose: "It is not quite clear what is meant by drying in the specific context of increased digestibility."

J. Stone: "Well, we dry the material in the oven overnight, and then soak it in water for 24 hours before the digestion with the rumen inoculum. Our "dried" material is material which has been dried and re-swollen in water."

T. K. Ghose: "Contrary to your data, our experiments show that lignin-free wood pulp loses susceptibility to a considerable extent when heated to temperatures from 50°C. to beyond 200°C. for a short time, measured against unheated pulp as a control. Will you please comment?"

J. Stone: "This is not contrary to our data. It agrees with it. If you dry cellulose you reduce its pore size and you make it less digestible. The longer the time and the higher the temperature of drying the less digestible the material and the smaller the pore size. We have found that regardless of the sample's drying history the digestibility is related to its pore size."

J. F. Saeman: "One small question. Your data show rather thoroughly that physical accessibility is a controlling factor in the attack by cellulases. Do you have the feeling that physical or chemical modification of lignified tissues without the kind of delignification that you get in pulping might bring about a useful improvement in the digestibility of lignocellulose?"

J. Stone: "First, I should say that hardwoods behave very differently from softwoods. With softwoods it is necessary to pulp to low yields and remove the lignin to a very great extent to make the material digestible. But with hardwoods—birch is the wood that we have studied chiefly—comparatively mild chemical treatments can increase digestibility very markedly. The degree of swelling also increases very markedly. I would therefore say that coniferous woods are unlikely to be convertible into animal feeds at an economic cost. With hardwoods on the other hand, certain mild chemical or physical treatments without appreciable lignin removal may very well be economically feasible."

14

Utilization of Cellulose by Ruminants

B. R. BAUMGARDT

Department of Animal Science, Pennsylvania State University,
University Park, Pa.

Expressed in an oversimplified manner, this part of the Symposium on Cellulases and Their Applications will deal with the utilization of lignocelluloses in the production of human food using ruminant animals as intermediates.

Many arguments have been advanced for the relative merits of animal *vs.* plant agriculture in the production of food for man. Thorough consideration indicates that when properly used and integrated, these two forms of food production can be complementary rather than competitive. Livestock offer a way of utilizing land resources that cannot be used directly to produce human food; they can utilize waste products of the food industry, and many other industries, and they can, in the case of ruminants, synthesize proteins and vitamins, and release energy from low grade raw materials including cellulose and urea.

Ruminants include those animals that we normally think of as cud-chewing, or ruminating animals; common examples include cattle, sheep, and goats. The most unusual feature of these animals involves their digestive system, and in particular their multi-compartmented stomach. They are commonly referred to as having four stomachs. The true stomach, that part comparable to the stomach of man, is preceded by 3 other compartments. These are, in order, the rumen, reticulum, omasum, and abomasum.

When food is swallowed by a ruminant it first enters the Rumen and reticulum which are not well separated. This rumeno-reticulum is by far the largest part of the ruminant stomach, having 80–85% of the total stomach capacity. No digestive enzymes are secreted by the animal into the rumeno-reticulum, but this large organ, retaining a high moisture content, maintains a large bacterial and protozoal population under near anaerobic conditions, and functions as a fermentation vat. A large and strong musculature keeps the contents agitated and the animal secretes large quantities of alkaline saliva, perhaps 10 liter/day in a sheep and 85–100 liter/day in a cow, which helps neutralize the short-chain organic acids produced from the microbial fermentation. pH in rumen will range

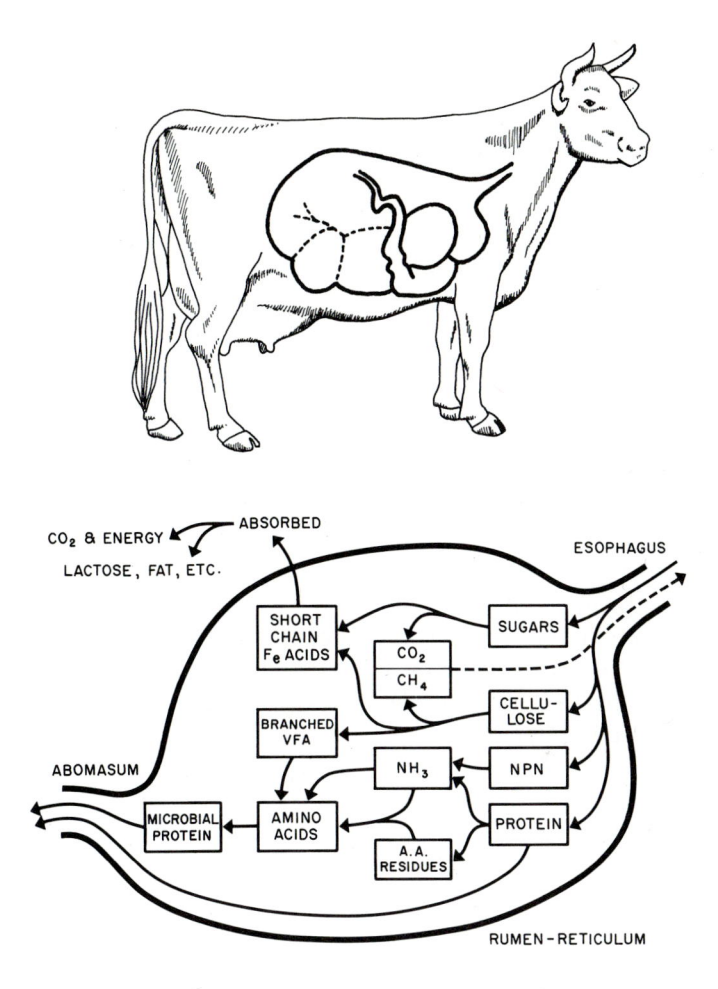

Figure 1. Digestive system of the cow

from 5 to 7 (depending upon diet and time after feeding) and the temperature is about 39°C.

The main features of the rumen phase of digestion include:

(1) Fermentation of dietary sugars, starches and cellulose to produce volatile fatty acids, principally acetic, propionic, and butyric acids, which are absorbed and used by the host animal for energy and various synthetic functions.

(2) Synthesis of microbial protein from protein and non-protein nitrogen in the diet. The microbial protein is later digested and the amino acids used by the host.

(3) Synthesis of the B-vitamins and vitamin K.

Two points should be mentioned in relation to this part of the symposium.

(1) To maintain normal rumen function and physiological activity some tactile or scratch stimuli must be provided in the rumen. This is normally provided by the "fibrousness" of roughage and fodder crops.

(2) Although cellulase activity of rumen microflora is high, lignocellulose feed stuffs are only partially degraded; that is, only part of the potential energy is made available for animal production.

Thus, we will be dealing with what might appear to be two divergent uses of lignocellulose in ruminant feeding.

One concerns the use of lignocellulose as a roughage substitute to provide the tactile stimulation necessary for rumen function, without particular regard to any nutritive contribution from the lignocellulose *per se*. Such possibilities find use in cattle feed-lot operations where grain feeding is practiced to produce high quality beef rapidly. The other aspect concerns modifications of lignocellulosic materials to provide new energy sources for ruminants, so that grains might be conserved for direct consumption by man.

These two needs, or aspects of the problem, are now separated largely on the basis of geography and agricultural productivity, but time may well dictate that obtaining new energy sources is the more critical and fundamental problem.

15

Lignocellulose in Ruminant Nutrition

W. J. PIGDEN and D. P. HEANEY

Research Branch, Canada Department of Agriculture, Ottawa, Ontario

Lignocellulose varies markedly in lignin, cellulose, and hemicellulose content depending on its plant source and stage of maturity. Nutritionally, cellulose and hemicellulose are the important components because they can be utilized by the rumen microflora and provide the host animal with a source of energy. As a plant matures, varying amounts of the cellulose and hemicellulose present are made unavailable to the rumen microflora by the chemical and/or physical properties of the lignocellulose complex. Within practical limits, reduction in particle size has little influence on the digestion ceiling of lignocellulose but greatly speeds up rate of processing in the rumen resulting in a corresponding increase in animal production. Chemical processing or gamma radiation disrupts the lignocellulose complex, allowing the rumen bacteria access to energy previously unavailable.

Lignocellulose is the complex of lignin, cellulose, and hemicellulose which exists in close physical and chemical association and accounts for most of the cell wall constituents of plants. As a proportion of the total dry matter it varies widely in forages, depending on species and stage of maturity. In wood products, where stage of maturity has little meaning, the lignocellulose content is relatively constant within a species but may vary somewhat between species (Table I).

Lignin is an aromatic compound which varies in content from about 2% in immature forages up to 15% in mature forages; in wood the percentage is somewhat higher (Table I). Its main function is to supply strength and rigidity to plant materials, but its nutritional significance lies in its indigestible nature. It not only acts as an inert diluent in feedstuffs but because of its close physical and/or chemical association with the cell wall polysaccharides it frequently acts as a physical barrier and impedes the microbiological breakdown of these compounds.

Table I. Content of Cell Wall Constituents in Lignocellulose of Various Plant Materials (as % of DM)

Plant Material	Hemicellulose	Cellulose	Lignin	Total Lignocellulose	Ref.
S23 rye grass (early leaf)	15.8	21.3	2.7	39.8	*41*
S23 rye grass (seed setting)	25.7	26.7	7.3	59.7	*41*
Western alfalfa (medium maturity)	6	25	7.2	38.2	*40*
Orchard grass (medium maturity)	40	32	4.7	76.7	*40*
Rye straw	27.2	34.0	14.2	75.4	*34*
Birch wood (*Betula verrucosa*)	25.7	40.0	15.7	81.4	*34*
Spruce wood (*Picea excelsa*)	20.9	46.0	24.1	91.0	*34*

Hemicelluloses are amorphous polysaccharides which include short chain glucans, polymers of xylose, arabinose, mannose, and galactose plus mixed sugar and uronic acid polymers (*4, 33*). Xylan is usually the main hemicellulose in forages. Hemicelluloses exist in close association with the cellulose and lignin and are generally separated from the cellulose by extraction with dilute alkali and acid (*22, 41*). They vary widely in content from one type of plant material to another with a range of about 6 to 40% (Table I). The availability of hemicellulose to the rumen microflora varies greatly but is generally in the range of 45–90%.

Cellulose is usually the most abundant polysaccharide of the cell wall constituents and the most insoluble. It is a polymer of glucose units and the degree of polymerization varies within and between sources of cellulose (*38*). In most isolation procedures employed for nutrition studies some polymers of xylose are included with the cellulose. Cellulose is available to the rumen microflora to a variable degree ranging from about 25 to 90%.

When born, the young ruminant is incapable of utilizing lignocellulose because it lacks cellulolytic enzymes. Over a period of time young ruminants normally develop, in the reticulo-rumen compartment of their stomach, a microflora with powerful cellulolytic activity so that by the end of several months the young animal can digest lignocellulose to almost the same degree as the mature ruminant. The end products of lignocellulose digestion are mainly acetic, propionic, and butyric acids, and bacterial proteins, which are utilized in the metabolic processes of the host animal.

Thus, because of the rumen microbial digestion, the ruminant animal can readily and efficiently extract most of the energy from the lignocellulose of immature plant materials but is poorly equipped to remove it from mature forage plants and from wood. The problem resolves itself into three components. Firstly, breaking down the cell wall material into particle sizes which facilitate bacterial attack and allow the feed to move through the alimentary tract efficiently; secondly, improving the availability of the energy in the lignocellulose fraction; and thirdly, providing adequate nutrients for the rumen microflora to utilize the available energy.

The lignocellulose complex accounts for most of the organic matter and, hence, gross energy of common forages and wood. Nutritionally, it consists of three distinct categories: (a) an unavailable fraction including compounds such as lignin which, for practical purposes, is not degraded by the rumen microflora; (b) the digestible energy (DE) fraction representing the carbohydrates which are normally available for bacterial degradation; and (c) the potentially digestible energy (PDE) fraction which includes the carbohydrates not normally available to rumen microflora because of chemical and/or physical associations within the lignocellulose complex, but which can be made available by appropriate treatment or by supplementation.

In young forages the PDE fraction is relatively insignificant because most of the carbohydrate material is available to the rumen microflora. As the plant matures and lignification proceeds, however, more of the carbohydrate is "tied up" and in unprocessed mature forages the PDE fraction may be as large as the DE fraction. The DE level is dependent upon several key factors, any one of which may be limiting. The major elements of this "limiting factor concept" include lignification, particle size effects, the effects of nitrogen and minerals on activity of the rumen microflora and effects of chemical processing.

This review is not exhaustive, rather it is presented to highlight the most important aspects of utilization of the mature cell wall constituents of forages and wood by ruminants and the related and limiting factors.

Digestion Ceiling Concept

This concept is based on the hypothesis that each lignocellulose material has its own characteristic "digestion ceiling" which is determined by the interaction of several factors (including lignification, level, and availability of the polysaccharides, particle size, nitrogen, and minerals) with the rumen microflora. Each digestion ceiling then represents the total amount of energy available from a particular forage under a given set of conditions.

The *in vitro* rumen system of Tilley *et al.* (*37*), as modified by Pritchard and MacIntosh (*31*), provides a convenient method for demonstrating this concept using finely divided (0.7 mm. screen) lignocellulosic materials. Minerals are added in the buffer mixture and the highly cellulolytic inoculum (strained rumen liquor from hay-fed animals) contains adequate nitrogen for the *in vitro* fermentation, so the assumption can be made that under these conditions lignification is the primary determinant of the digestion ceiling.

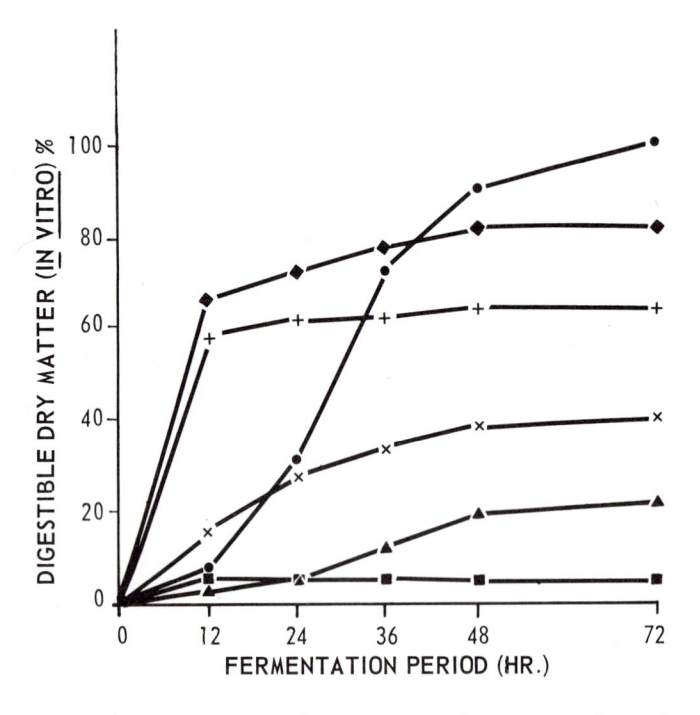

Figure 1. Digestion ceilings of several forages and wood products

● Bleached spruce sulfite pulp
▲ Ground aspen wood
■ Ground lodgepole pine wood
× Wheat straw
◆ Timothy
+ Alfalfa

Figure 1 shows the rate and extent of the breakdown of the lignocellulose as illustrated by the dry matter digestion *in vitro* of three forages and of three wood products. Four points emerge. First, each material has its own characteristic digestion ceiling where the extent of breakdown plateaus for all practical purposes. For timothy and alfalfa this plateau is much higher than for wheat straw and aspen wood. Second, the initial rate of digestion varies widely between the various materials. The timothy and alfalfa, representing high and medium quality forages, ferment quickly. Third, the curve for the lodgepole pine ground wood shows that some wood products are indigestible with essentially all of the energy "locked up" in the unavailable and PDE fractions. Lodgepole pine is not unique in this respect as black spruce, eastern hemlock, and balsam fir ground woods behave similarly. The level of DE in the ground aspen wood, slightly below 20%, is substantially less than that in wheat straw which is considered one of the lowest quality forages. Fourth, the curve for the bleached sulfite pulp demonstrates that with delignified material there is essentially no digestion ceiling because the material is completely available to the rumen microflora. It is, of course, true that pulping removes many materials in addition to the lignin. Nevertheless, this example clearly illustrates how the digestion ceiling can be raised by chemical treatment. Other workers (21, 23, 26) have shown that isolated celluloses are 90 to 100% available to rumen microorganisms. While it can be argued the isolation procedures could alter the physio-chemical nature of the cellulose, it has been shown that cotton fiber which had received only a mild de-waxing treatment was 97% available (17).

Comminution and Particle Size Effects

Forages are normally consumed by ruminants in the long or coarsely chopped form and by a combination of mastication and fermentation are reduced in the alimentary tract to a finely divided state. The extent of alimentary tract comminution is illustrated by the particle size distribution in fecal dry matter from a sheep consuming a high quality alfalfa hay, compared with two forages ground through a Wiley mill fitted with a 2.0 mm. screen:

Sieve		% of Total Weight Passing Sieve		
USBS No.	Mesh	Feces from Alfalfa	Alfalfa	Brome Straw
20	20	88.4	92.6	94.6
45	42	73.8	53.5	55.0
60	60	37.3	24.0	21.9

Cattle grind their feed less finely so particles passing the bovine digestive tract are somewhat larger.

Mechanical comminution of lignocellulosic materials beyond that achieved by animals tends to increase the digestion ceiling *in vitro*. This is illustrated by the percentage of the forage carbohydrate fermented *in vitro* at 48 hrs. using the method of Pigden (29) with two forages ground through 2.0 mm. and 0.2 mm. screens in a micro Wiley mill.

	Alfalfa		Brome Straw	
	2.0 mm.	0.2 mm.	2.0 mm.	0.2 mm.
% carbohydrate fermented	66.8	72.1	53.6	58.7

However, extremely fine grinding by ball milling for many hours degrades the lignocellulose complex and makes substantial amounts of the original PDE available to the rumen microflora, thereby raising the digestion ceiling—*i.e.*, part of the PDE is converted to DE. The following data for aspen wood and for *Panicum maximum*, a low digestibility pasture grass from W. Australia illustrate this effect.

		In Vitro DM Digestibility[a] (%)		
Material	Harvest Date	Normal Grinding	Ball Milled	Difference
Panicum maximum[b]	8/5/56	43.6	69.4	+25.8
Panicum maximum[b]	30/7/56	34.6	70.8	+36.2
Aspen wood[c]		23.4	42.8	+19.4

[a] Determined by technique of Tilley *et al.* (37).
[b] Ball milled 3-5 hrs. in a Spex 500 mill. Data courtesy of J. M. A. Tilley.
[c] Ball milled 24 hrs. in a Stoneware mill.

In some instances ball milling more than doubled the energy available to the rumen microflora. Although of considerable theoretical interest, this approach is of little practical significance in animal production because of the high processing cost as well as the fact that, *in vivo*, a high proportion of the extremely fine particles would not stay in the rumen long enough to be fermented.

So far we have considered only the digestion ceiling of materials confined in an *in vitro* system. In the ruminant animal the effect of lignification on the digestion ceiling of lignocellulose is modified substantially by physiological factors. Most of the fermentation of the lignocellulose occurs in the rumen. However, some feed particles pass out of the rumen and through the remainder of the gastro-intestinal tract before they have been exposed to the rumen microflora long enough to permit complete extraction of their DE fraction. This is because rumen retention time of a given feed particle depends chiefly on its size and specific gravity (3). For forages fed in the long form these two characteristics are largely governed by extent of mastication and microbiological attack but this is not true of mechanically masticated or ground forages. Grinding not only increases surface area and speeds up fermentation, but it reduces particle size to a point where many or all of the feed particles are small enough to pass the reticulo-omasal orifice immediately after entering the rumen. Consequently, they need only absorb moisture to attain the specific gravity which allows them to sink and reach the reticulo-omasal orifice which is the exit point located on the floor of the rumen. Balch (3) suggests the critical size is about 2 mm. Feeding high levels of ground material tends to force the feed through the rumen more quickly, thus reducing the time feed particles are exposed to the rumen microflora and lowering the digestion ceiling. The following data from Blaxter et al. (5) show these effects very clearly:

Physical Form	Daily Intake (grams)	DM Digestibility %	Average Retention Time (hr.)
long	600	80	103
	1500	79	68
ground	600	77	74
(1/4 inch screen)	1500	70	42
ground	600	76	53
(1/16 inch screen)	1500	65	34

The reduction in the digestion ceiling is chiefly caused by incomplete fermentation of the lignocellulose fraction, as indicated in the following table (19) by the marked drop in crude fiber digestibility by sheep of ground and pelleted timothy compared with the long form. Thus, the rapid rate of passage of ground materials from the rumen increases the fecal energy losses per unit of forage eaten. However, the increased rate of passage also markedly increases the level of voluntary consumption of the forage. The net result of these opposing effects can be a dramatic increase in the daily intake of digestible energy (kcal./$W_{kg}^{0.75}$), as shown by the following data:

	Physical Form of Timothy	Stage of Maturity Medium	Late
Digestibility of crude fiber %	Chopped	72	46
	Pelleted	54	31
	Difference	−18	−15
Intake of digestible energy (Kcal./$W_{kg}^{0.75}$)	Chopped	190	90
	Pelleted	250	175
	Increase %	+32	+94

Such increases in digestible energy intake can greatly increase the efficiency of animal production.

We conclude that while digestion ceilings established *in vitro* can provide a most valuable guide to the DE content of lignocellulosic materials, these ceilings are not constants and can be substantially modified *in vivo* by factors such as degree of mechanical comminution and level of feeding. Therefore, digestion ceilings of 85% digestible dry matter are about the maximum likely to be found *in vivo*, even with such materials as delignified wood pulp (Figure 1) which has a potential digestion ceiling of 100%, *in vitro*.

In practical feeding regimes the degree of mechanical comminution of forages may be varied to achieve a reasonable balance between the extent of fermentation and rate of passage. For materials which ferment very rapidly and where high animal production levels are wanted, it is an advantage to grind the material fairly finely to achieve higher rates of passage and higher intake levels. Materials which have high digestion ceilings but ferment very slowly are essentially unavailable under these conditions. For materials which ferment less rapidly and for low levels of production—*e.g.*, maintenance—the lignocellulose should be much more coarsely divided so that it can be retained in the rumen sufficiently long to extract most of the DE. Under some conditions, such as self-feeding, particle size could be used to control intake.

Most of the research concerned with the effect of particle size on availability of lignocellulose to ruminants has been carried out with forages hammermilled through various sizes of screens. However, grinding various forages through the same screen can result in tremendous differences in actual particle size distribution in the ground products (19). There is need for additional information on the effect of specific particle sizes of lignocellulosic materials on rumen retention times and digestion ceilings *in vivo* to achieve the most efficient combinations for specific ration purposes.

Factors Affecting Cellulolytic Activity of Rumen Microflora

Source of Inoculum. The diet of the donor animal can have a marked effect on the ability of the microflora to extract the DE from lignocellulose. Rumen inoculum from goats maintained on high quality hay diets was compared with inoculum from aspen silage fed animals, for the fermentation of aspen silage in the *in vitro* rumen system over a 72 hour fermentation period. The inoculum donor goats had been maintained on aspen silage for 3 weeks before the trials. The aspen silage was obtained from whole poplar trees (*Populus tremuloides*) which were put through a chipper, ground, and ensiled in a pit silo.

Figure 2 shows that much larger amounts of digestible dry matter of the aspen silage were fermented with the hay inoculum than with the aspen silage inoculum. When a small amount of hay (50 grams/animal/day) was added to the aspen silage diet the cellulolytic activity of the rumen microflora was stimulated (Figure 2) but was still much below the activity of inoculum from goats on the hay diet.

The inoculum donor goats could be maintained on aspen silage as the sole diet for only 3–4 weeks after which the animals began to go off feed and eventually refused to eat. With 50 grams hay/day supplement they continued to eat the aspen silage over extended periods.

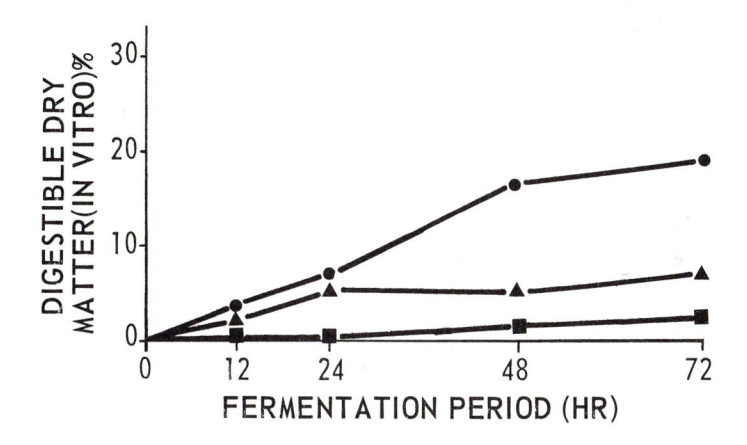

Figure 2. Effect of activity of rumen inoculum on digestion ceilings of aspen silage

● Inoculum—hay fed goats
▲ Inoculum—aspen silage and hay fed goats
■ Inoculum—aspen silage fed goats

The aspen silage was extremely low in nitrogen (0.1% DM basis) so that nitrogen deficiency in the inoculum from animals fed this material was probably a major factor in reducing the cellulolytic activity of the rumen microflora.

From a practical point of view the technique of mixing small amounts of high quality forage with poor quality lignocellulose materials permits more efficient utilization of the latter by ruminants.

Nitrogen. In this discussion nitrogen as a supplement will be discussed only in relation to its effect on rate and extent of breakdown of lignocellulose in the rumen and will take no account of nitrogen requirements of the ruminant *per se*.

In the ruminant animal the availability of nitrogen for breakdown of lignocellulose in the rumen is mediated by both feed and physiological factors. Nitrogen is available to the microflora immediately from the feed and in the form of urea which is recycled from the blood stream and enters the rumen in the parotid saliva, and also directly from the blood through the rumen wall. Thus, over short feeding periods of a few days the ruminant can compensate, to a degree, for lack of feed nitrogen and maintain nitrogen levels in the rumen adequate for lignocellulose breakdown. Over more prolonged feeding periods, if a shortage of nitrogen persists it can become a serious limiting factor in lignocellulose breakdown.

The minimum nitrogen requirement for efficient lignocellulose breakdown of roughages fed as the sole diet is claimed to be from about 0.6 to 0.8% nitrogen (7, 27, 45). Very small additions of urea amounting to 0.1–0.2% nitrogen added to a diet of oat straw containing 0.55% nitrogen resulted in a greatly accelerated rate of cellulose breakdown in the rumen of cows and a 26% increase in DM intake (10). Further increases of 0.3–0.5% nitrogen resulted in maximum increases in DM intake of 40% over the control and a 20% increase in the digestion ceiling of crude fiber (10). This is a clear demonstration of a situation where lignification was not the first limiting factor and nitrogen was clearly insufficient for efficient utilization of lignocellulose.

Conversely, there are many low nitrogen forages which do not respond to nitrogen supplementation. Under experimental conditions similar to those of Campling et al. (10) supplemental urea had no effect on the digestion ceilings or intake of wheat straw (0.64% N) or oat straw (0.52% N) fed to sheep and steers, respectively (28).

Others have reported similar negative results (18, 25, 44). In these cases it is doubtful if nitrogen was the first limiting factor.

Recent studies (13) showed urea supplementation of NaOH treated oat straw resulted in very large increases in intake by sheep and slight but non-significant increases in the digestion ceiling. Because this material was ground it may have moved through the rumen too quickly to reach its maximum digestion ceiling. In our experience digestion ceilings of this type of material are frequently substantially higher *in vitro* than *in vivo*. Supplementation of the nontreated straw with urea had no significant effect on digestibility or intake, thus raising the digestion ceiling by chemical treatment increased the nitrogen requirement.

From the available information we conclude that 1% of nitrogen is adequate for efficient breakdown of lignocellulose in the reticulo-rumen for materials containing no more than 50% DE. Where DE levels are higher as may be the case for chemically treated straw or wood the nitrogen requirement may increase to 1.5%. Where readily available carbohydrate (starch, sugar) is included in the ration at levels greater than 20% of total ration the nitrogen requirement for lignocellulose degradation will increase to about 2% (14).

Minerals. Most of our available information on the influence of minerals on lignocellulose utilization has been obtained with the *in vitro* rumen system. Molybdenum increased cellulose digestibility *in vitro* and *in vivo* (*15*). Sulfur, phosphorus, and iron have shown similar effects *in vitro* (*9*). In other *in vitro* studies sulfur, magnesium, and calcium stimulated cellulolytic activity whereas copper, cobalt, zinc, and boron depressed cellulose digestion (*20*).

Cobalt, while essential at low levels for vitamin B_{12} synthesis and efficient functioning of rumen microflora, will depress cellulolytic activity of rumen microflora *in vitro* at levels of 12 p.p.m. and above (*35*). Selenium inhibited cellulose digestion *in vitro* at the 0.3 p.p.m. level (*24*).

Most forages contain adequate mineral matter to support lignocellulose breakdown in the rumen. For specific problem materials such as processed wood, corn cobs, and some types of straw, their mineral contents may be inadequate to support efficient lignocellulose utilization. Swift *et al.* (*36*) found that alfalfa ash had a highly significant effect on digestibility of corn cobs, a material very high in hemicellulose. Davies *et al.* (*12*) showed calcium carbonate increased the availability of crude fiber in molina-fescue hay. Many mineral nutrients are recycled in the ruminant parotid saliva and inorganic phosphorus is maintained at relatively constant levels in the rumen even when sheep are maintained on a diet grossly deficient in phosphorus over prolonged periods (*11*).

It is highly probable that in the reticulo-rumen many of the mineral deficiencies and/or toxicities noted *in vitro* would not occur because of the recycling of minerals and the detoxification ability of the microflora. However, from evidence available, some minerals can have a pronounced effect on lignocellulose breakdown and are in some instances the first limiting factor in rate and/or extent of lignocellulose degradation.

Readily Digestible Carbohydrates. This category of carbohydrates refers mainly to starch and sugars which are fermented very rapidly and completely in the rumen.

There appears to be a requirement for a small amount of readily fermentable carbohydrate to stimulate fermentation of the lignocellulose complex. Burroughs *et al.* (*8*) suggests this is of the order of 5–10% of the ration. Larger amounts of readily fermentable material frequently have a very depressing effect on lignocellulose utilization. This is believed to be primarily caused by competition between cellulose digesting microorganisms and those which attack the more digestible material. Most forages and wood products do not contain enough readily available carbohydrate to depress lignocellulose utilization but there are some exceptions (*16*).

Fat. Fats tend to depress the cellulolytic activity of rumen microorganisms. For example 3% of corn oil was sufficient to depress the digestibility of crude fiber in rations high in corn cobs and cottonseed hulls, and 10% corn oil produced a pronounced depression (*42*). Addition of alfalfa ash offset this effect when corn oil did not exceed 5%

of the total ration. Similar cellulose depressing effects have been reported by numerous workers with other types of fat. Feeding additional calcium tends to offset this effect (43).

Chemical Composition of Lignocellulose and Availability to Rumen Microflora

Stage of maturity of the plant has a marked effect on availability of the cell wall constituents of most forage plants (Table II). The data in Table II for alfalfa indicate much less effect than normally expected, primarily because the availability of the cell wall constituents is lower than usual at the early stage of maturity. The general decrease in cell wall availability as most forages mature is believed largely caused by increasing lignification which encrusts and protects the cell walls from bacterial attack. Nevertheless, the availability of the cellulose in forages tends to remain relatively high as compared with hemicellulose, and may be influenced more by its xylan content than by lignification *per se*.

Table II. Digestibility of Cell Wall Constituents in Various Plant Materials

Material Plant	Hemicellulose	Cellulose	Crude Lignin	Reference
S23 rye grass (early leaf)	93	92	23	41
S23 rye grass (seed setting)	56	73	0	41
Alfalfa (early maturity)	67.2	66.7	20.0	22
(medium maturity)	62.8	67.2	15.3	22
White clover (medium maturity)	77.7	86.9	16.4	22
(late maturity)	66.0	75.0	14.9	22

The digestibility of the hemicellulose constituents is also markedly influenced by stage of maturity (Table III). At early growth stages digestibility of all the components is high but at later stages xylan and uronic acids are relatively indigestible. Evidence has been found for a chemical bond between wood carbohydrates and lignin; xylan may be the component involved (6). This could explain the depression in digestibility of xylans in grasses as they mature.

Purified celluloses from different plant sources differ in digestion rate, *in vitro* (2, 39). This is also true for cellulose in intact plant materials digested *in vitro* (23). Although a relationship between crystallinity and digestion rate was reported (2) it was not confirmed (39). Cellulose and hemicellulose were digested at a similar rate *in vivo* (1).

The availability of both the cellulose and hemicellulose in untreated wood sawdust is low and for conifers such as lodgepole pine, eastern hemlock, and balsam fir is essentially zero.

Table III. Digestibility of Hemicellulose Constituents of Forages by Sheep

Plant Material	Xylans	Arabans	Hexosans	Uronic Acids	Reference
Alfalfa					
(early maturity)	43.9	87.8	89.5	—	22
(medium maturity)	45.7	85.6	79.4	—	22
White clover					
(medium maturity)	70.0	84.5	79.0	—	22
(late maturity)	56.1	76.6	71.4	—	22
Orchard grass					
(early maturity)	61.1	82.9	88.9	—	22
(medium maturity)	52.1	68.1	83.6	—	22
S23 rye grass					
(early leaf)	91	95	96	92	41
(seed setting)	52	69	74	52	41

Chemical Processing and Irradiation

The very close physical and/or chemical association between lignin, cellulose, and hemicellulose acts as a barrier which impedes microbiological breakdown of the cell wall polysaccharides. When mature, low quality lignocellulosic materials are fed to ruminants, with or without prior mechanical reduction of particle size, the animals frequently excrete more PDE in the feces than is absorbed. Chemical treatment, or irradiation, of the lignocellulose complex can bring about dramatic increases in the digestion ceiling. Such treatments apparently increase the susceptibility of the lignocellulose complex to microbial breakdown by disrupting the chemical bonds, swelling the fibrils to "open" the ultrastructure, or both.

Figure 3 illustrates the effect of chemical processing by a method previously developed (46) on the *in vitro* digestibility of wheat straw. The lower curve shows the digestion level for the untreated wheat straw after a 48 hour fermentation period. The upper curve demonstrates how treatment of the wheat straw with 6% alkali has increased the DE (represented by 48 hrs. *in vitro* fermentation in our technique) by about 30%. In previous studies (*46*) wheat straw and aspen wood were treated with various amounts of alkali and treatment levels up to 9% caused progressive increases in the digestion ceiling *in vitro*, but above 9% alkali no further increases were obtained.

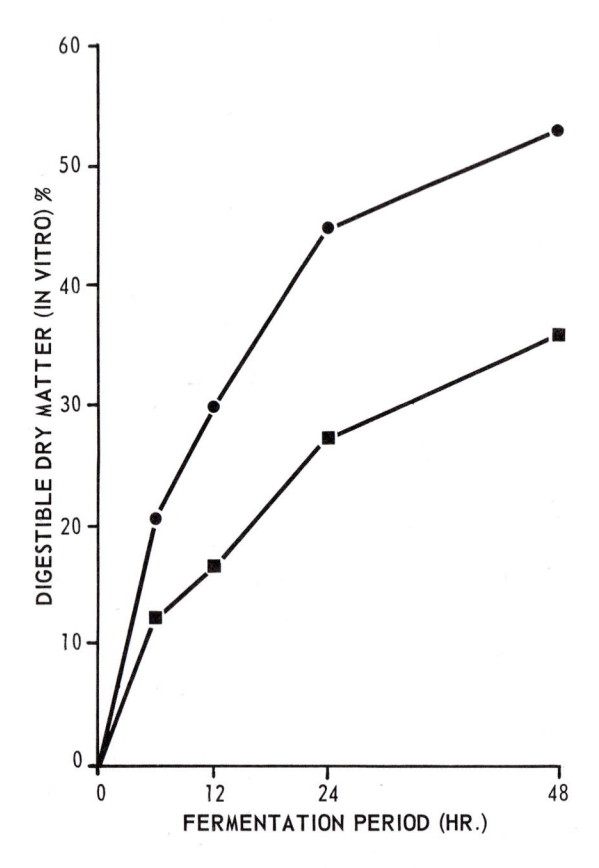

Figure 3. Effect of NaOH treatment on the digestion ceilings of wheat straw
● *NaOH treated straw*
■ *Untreated straw*

Gamma irradiation of wheat straw at levels of 1×10^8 and 2.5×10^8 rads increased the dry matter digestibility *in vitro* from 40% to approximately 70% (*32*). Increased volatile fatty acid (VFA) production during the *in vitro* fermentation indicated the breakdown products of the irradiation were largely available to the rumen microorganisms. In further work (*30*) fecal material from sheep fed low quality forages was subjected to gamma irradiation and the treated material digested in the *in vitro* system. The VFA production, shown below, demonstrated that substantial quantities of PDE were made available to rumen microorganisms by irradiation of the fecal material. In the case of timothy utilization of PDE from irradiated feces was about equal to the DE in the original feed.

Experimental	VFA Production m eq/g DM		
		Feces	
Material	Feed	Control	Irradiated[b]
Timothy[a]	3.3	0.0	3.0
Alfalfa[a]	3.6	0.0	2.0

[a] DM digestibility (sheep) timothy 45.5% and alfalfa 48.5%.
[b] 2.7×10^8 rads from a cobalt source.

With equipment presently available, the levels of gamma irradiation used are well above what are economically feasible for commercial feed processing.

The data from the irradiation and alkali treatment experiments amply support the concept that low quality forages contain substantial quantities of PDE which can be made available to rumen microorganisms.

Literature Cited

(1) Bailey, R. W., *Proc. N.Z. Soc. Animal Prod.* **25,** 85 (1965).
(2) Baker, T. I., Quicke, G. V., Bentley, O. G., Johnson, R. R., Moxon, A. L., *J. Animal Sci.* **18,** 655 (1959).
(3) Balch, C. C., "Digestive Physiology and Nutrition of the Ruminant," D. Lewis, Ed., p. 23, Butterworth, London, 1960.
(4) Bath, I. H., Ph.D. Thesis, University of Reading (1959).
(5) Blaxter, K. L., McC. Graham, N., Wainman, F. W., *Brit. J. Nutr.* **4,** 69 (1956).
(6) Bolker, H. I., *Nature, Lond.* **197,** 489 (1963).
(7) Burroughs, W., Frank, N., Gerlaugh, P., Bethke, R. M., *J. Nutr.* **40,** 9 (1950).
(8) Burroughs, W., Gall, L. S., Gerlaugh, P., Bethke, R. M., *J. Animal Sci.* **9,** 214 (1950).
(9) Burroughs, W., Latona, A., De Paul, P., Gerlaugh, P., Bethke, R. M., *J. Animal Sci.* **10,** 693 (1951).
(10) Campling, R. C., Freer, M., Balch, C. C., *Brit. J. Nutr.* **16,** 115 (1962).
(11) Clark, R., *Onderstepoort J. Vet. Sci. Animal Ind.* **26,** 137 (1953).
(12) Davies, R. O., Jones, D. I. H., Milton, W. E. J., *J. Agr. Sci. Camb.* **53,** 268 (1959).
(13) Donefer, E., Advan. Chem. Ser. **95,** 328 (1969).
(14) Ellis, W. C., Pfander, W. H., *J. Nutr.* **65,** 235 (1958).
(15) Ellis, W. C., Pfander, W. H., Muhrer, M. E., Pickett, E. E., *J. Animal Sci.* **17,** 180 (1958).
(16) Frens, A. M., *Soc. Chem. Ind. Monograph* No. 9, p. 202 (1960).
(17) Halliwell, G., Bryant, M. P., *J. Gen. Microbiol.* **32,** 441 (1963).
(18) Head, M. J., *J. Agr. Sci.* **43,** 281 (1953).

(19) Heaney, D. P., Pigden, W. J., Minson, D. J., Pritchard, G. I., *J. Animal Sci.* **22**, 752 (1963).
(20) Hubbert, F. Jr., Cheng, E., Burroughs, W., *J. Animal Sci.* **17**, 559 (1958).
(21) Hungate, R. E., *13th Symp. Soc. Gen. Microbiol.* p. 266 (1963).
(22) Jarrige, R., *Proc. Intern. Grassland Congr., 8th*, p. 628 (1960).
(23) Kamstra, L. D., Moxon, A. L., Bentley, O. G., *J. Animal Sci.* **17**, 199 (1958).
(24) Little, O., Cheng, E., Burroughs, W., *J. Animal Sci.* **17**, 1190 (1958).
(25) Louw, J. G., Van der Wath, J. C., Onderstepoort *J. Vet. Sci. Res.* **18**, 177 (1943).
(26) McAnally, R. A., *Biochem. J.* **36**, 392 (1942).
(27) McNaught, M. L., *Biochem. J.* **49**, 325 (1951).
(28) Minson, D. J., Pigden, W. J., *J. Animal Sci.* **20**, 962 (1961).
(29) Pigden, W. J., Ph.D. Thesis, University of Saskatchewan (1955).
(30) Pigden, W. J., Pritchard, G. I., Heaney, D. P., *Proc. Intern. Grassland Congr., 10th*, p. 397 (1966).
(31) Pritchard, G. I., *Animal Res. Inst., Mimeo Contrib.* **No. 195**, Canada Dept. Agr., Ottawa (1965).
(32) Pritchard, G. I., Pigden, W. J., Minson, D. J., *Can. J. Animal Sci.* **42**, 215 (1962).
(33) Roelofsen, P. A., "Handbuch der Pflanzenanatomie," **3 No. 4**, Gebruder Bonrtraeger, Berlin, 1959.
(34) Salo, M. L., *Mattalorest. Aikakawsh* **37**, 127 (1965).
(35) Salsbury, R. L., Smith, C. K., Huffman, C. E., *J. Animal Sci.* **15**, 863 (1956).
(36) Swift, R. W., Cowan, R. L., Barron, G. P., Maddy, K. H., Grasse, E. G., *J. Animal Sci.* **10**, 434 (1951).
(37) Tilley, J. M. A., Deriaz, R. E., *Proc. Intern. Grassland Congr., 8th*, **8**, 533 (1960).
(38) Timell, T. E., "Cellular Structure of Woody Plants," W. A. Cote, Ed., p. 127, Syracuse, New York, 1964.
(39) Tomlin, D. C., Davis, C. K., *J. Animal Sci.* **18**, 1516 (1959).
(40) Van Soest, J. P., *J. Animal Sci.* **23**, 838 (1964).
(41) Waite, R., Johnston, M. J., Armstrong, D. G., *J. Agr. Sci.* **62**, 39 (1964).
(42) Ward, J. K., Tefft, C. W., Sirny, R. J., Edwards, N. H., Tillman, A. D., *J. Animal Sci.* **16**, 633 (1957).
(43) White, T. W., Grainger, R. B., Baker, F. H., Stroud, J. W., *J. Animal Sci.* **17**, 797 (1958).
(44) Williams, N. M., Pearce, G. R., Delaney, M., Tribe, D. E., *Emp. J. Exp. Agr.* **27**, 107 (1959).
(45) Williams, V. J., Nottle, M. C., Moir, R. J., Underwood, E. J., *Australian J. Biol. Sci.* **6**, 142 (1953).
(46) Wilson, R. K., Pigden, W. J., *Can. J. Animal Sci.* **44**, 122 (1964).

RECEIVED October 14, 1968. This paper is Contribution No. 325, Animal Research Institute.

Discussion

T. K. Ghose: "Why do you prefer expressing the energy intake in terms of kcal./kg.$^{0.75}$ instead of normal expression of energy units like kcal. per kg. of body weight?"

W. J. Pigden: "Because the increase is more closely related to the 0.75 power of body weight than it is to body weight. It is a more valid way of comparing differences in intake between animals or between treatments than on a straight body weight basis."

16

Composition, Maturity, and the Nutritive Value for Forages

PETER J. VAN SOEST[1]

U. S. Department of Agriculture, Agricultural Research Service,
Animal Husbandry Research Division, Beltsville, Md. 20705

Nutritive value of forages depends in part upon the availability of nutrients in the plant for the animal. This availability is controlled by the chemical composition of the forage in respect to factors limiting the utilization of cellulose and hemicellulose. These include lignin, silica, and the total amount of plant cell wall substances. An important part of nutritive value is that of consumption or voluntary intake which is partly related to cell wall content and bulkiness of the forage. Consumption is much more variable due to the type and physiology of the animal and prediction from composition is much more difficult. The composition of the plant is controlled to a large extent by light, temperature, and maturity factors, but different plant species vary individually in this respect.

Forages are important in agriculture because they can be grown on lands where other crops are less easily grown and because they are the foundation of animal agriculture. A characteristic feature of forages is their fibrousness, which they have in common with by-products of the cereals, including straw, hulls, and bran. Paper and wood pulp also share these attributes of forages, in that they contain a potential supply of cellulosic carbohydrates for the ruminant animal.

Cellulose is important to herbivores possessing the intestinal microflora that can digest it (27, 28); it is of no value to animals that do not harbor these cellulytic organisms in their digestive tracts, for cellulases are not secreted by higher animals.

[1] Present address: New York College of Agriculture, Cornell University, Ithaca, New York, 14850.

The same biological facts hold true for the hemicelluloses, which cannot be digested by animals unless the requisite bacteria reside in their digestive tracts. Many grasses contain at least as much hemicellulose as cellulose. Thus, the large proportion of cellulose and hemicellulose in forages determines the place of these plants in the nutrition of animals, and a considerable amount of nutritional and physiological research has been aimed at understanding the utilization of cellulosic carbohydrates by ruminants and other herbivores.

Cellulose itself has sometimes been used as a negative measure of forage quality as an estimate of either fiber or of overall forage dry-matter digestibility in *in vitro* cellulose digestion. This usage has its weaknesses in that cellulose is seldom more than a third of the plant dry matter or roughly half of the total plant cell wall. The proportion of cellulose to lignin and hemicellulose varies widely, and thus cellulose may not be quantitatively representative of the fibrous constituents of the forage. In the discussion that follows, the term cellulosic carbohydrates will be taken to include both cellulose and hemicellulose.

Classification of Forage Components. In general, chemical constituents of plants may be divided into the structural components of the cell wall (lignin, cellulose, and hemicellulose) and the more soluble cellular contents (sugars, starch, fructosans, organic acids, lipids, and the nitrogenous fractions consisting of about two thirds protein and a third non-protein nitrogen compounds). Cellular contents often exceed 50% of the dry matter of forages and represent material of very high nutritive availability. The fibrous cell wall is much more insoluble and contains very nearly all of the matter that is truly indigestible (*see* discussion below).

Many tables of feed composition published during the past 100 years present proximate analyses involving ether extract, crude fiber, protein (total Kjeldahl nitrogen), ash, and nitrogen free extract (NFE). The proximate system does not properly separate the plant carbohydrates, most of the lignin and hemicellulose and some cellulose being extracted into the NFE, which is supposed to represent soluble and hence available carbohydrates. Crude fiber is composed largely of insoluble cellulose and does not recover fibrous fractions of the plant. Its use is thus limited by the same problems listed for cellulose above (*56, 61*). Recent efforts to find a replacement for crude fiber have originated in the ruminant field, where the amount and nutritional quality of fiber is of particular importance. The remainder of this discussion will explore aspects of plant composition and animal utilization that make the newer systems of forage analysis applicable.

Forage Nutritive Value

Analysis of the causal relationships between maturity, composition, and nutritive value is here approached through discussion of three types of problems—those involved in (a) defining and determining nutritive value; (b) relating chemical composition to various aspects of nutritive value; and (c) accounting for the effect of environment on plant composition and nutritive value. These problems are common to several fields, including nutrition, biochemistry, and plant physiology, and while it is impossible in a chapter to do justice to them all, it is important to catalogue them and to provide a perspective.

Problems in Measuring Nutritive Value. A basic problem in the evaluation of feeds, and particularly forages, is the integration of different and not always compatible factors into a single numerical scale. Everyone will agree that the best measure of nutritive value or feed quality is the animal productivity that can be obtained on a given feed. However, this may be broken down into digestibility, level of consumption, and efficiency—that is, an animal's production is limited by the digestible calories and nutrients that are consumed and the efficiency of use once absorbed.

The most commonly determined measure of nutritive value is digestibility, which may be expressed in various ways, often as a percentage of dry matter or organic matter. The traditional calculation of total digestible nutrients (TDN) which attempts to place fat, carbohydrate, and protein on an equivalent energy basis by means of factors, is gradually being replaced by a system in which energy digestibility is determined directly by bomb calorimetry of feed and feces.

Estimates of the various energy losses on the part of the animal lead to the net energy value, which is expressed as megacalories available per 100 lb. feed. Division of this number by a measure of digestible energy or TDN yields an estimate of energy efficiency. Values for TDN, net energy and efficiency are given in Table I for a number of feeds. As can be seen, TDN and net energy do not evaluate feeds identically (25, 43), a unit of TDN in forage having a lower net energy value than a TDN unit of a concentrate feed like corn or wheat. The reason for this difference is probably related to the energy lost in fermentation and rumination of cellulosic carbohydrates and in the fermentation of methane. Methane loss and decrease in efficiency are related to fiber content. The lower efficiency of cellulosic feeds relative to the concentrates is an economic factor limiting greater use of fibrous feedstuffs in the livestock industry in America at the present time. More research needs to be done on improving the efficiency of cellulose digestion by animals.

Table I. Relative Values of Feeds Expressed on a Total Digestible Nutrient (TDN) and Net Energy Basis (NE)[a]

Feeds	TDN lb./100 lb.	NE therms/100 lb.	NE as percent of TDN	Cell Wall Constituents
Corn	80	80	100	10
Barley	78	71	91	19
Oats	70	66	94	30
Alfalfa	51	41	80	45
Oat Straw	45	23	52	70
Wheat Straw	41	10	25	82

[a] Reference 42.

Despite the great value of efficiency and intake measurements, digestibility measurements predominate because of their ease and reproducibility under controlled feeding conditions. Net energy measurements (and therefore, efficiency estimates) are very expensive and time-consuming, and the much smaller body of net energy data has up to now been a primary factor limiting its wider use.

The remaining aspect of nutritive value, the amount of food eaten, is one of the most problematic areas in nutrition, because no satisfactory measure of consumption as an attribute of a feed or forage has been developed. The range in consumption among animals is dependent upon psychological and physiological factors that are not always related to the nature of the feed. Yet variation in consumption is more responsible for the range in animal productivity than is digestibility (3, 10, 11, 14).

There are two aspects of food consumption: palatability and voluntary intake (47, 48). Palatability connotes the appealing characteristics of the feed and is measured as the acceptability or relative amount eaten when a number of forages are offered in cafeteria style. Voluntary intake means the amount that is consumed when a single feed is offered to the animal. Often forages that are rejected in cafeteria feeding will be consumed to a great extent in a voluntary intake trial.

Ranking of forages by acceptability and voluntary intake will not be the same, but there is obvious mutual interdependence involving animals, since it is the total consumption of digestible energy that determines the limits of the animal for growth, fattening, or milk production.

Ranking of forages by acceptability and voluntary intake will not be the same, but there is obvious mutual interdependence involving animals, since it is the total consumption of digestible energy that determines the limits of the animal for growth, fattening or milk production.

The problem of intake measurement is that while reproducible values can be obtained for sheep (wethers) under specified conditions, the results will be correlated only slightly or not at all with those from beef cattle or dairy cattle (7, 29, 48). Dairy cattle present the greatest dissimilarity in that they consume much more feed in relation to body size than do other animals, including nonlactating dairy animals (11, 29). Intake measurements comparing dairy heifers and lactating cows rank forages differently (29). A further problem is that while all measurement trials consist of single-fed feeds or forages, practical feeding for production consists of carefully combined mixtures. Intake values of single feeds may not combine to give a reliable prediction of the intake of mixtures.

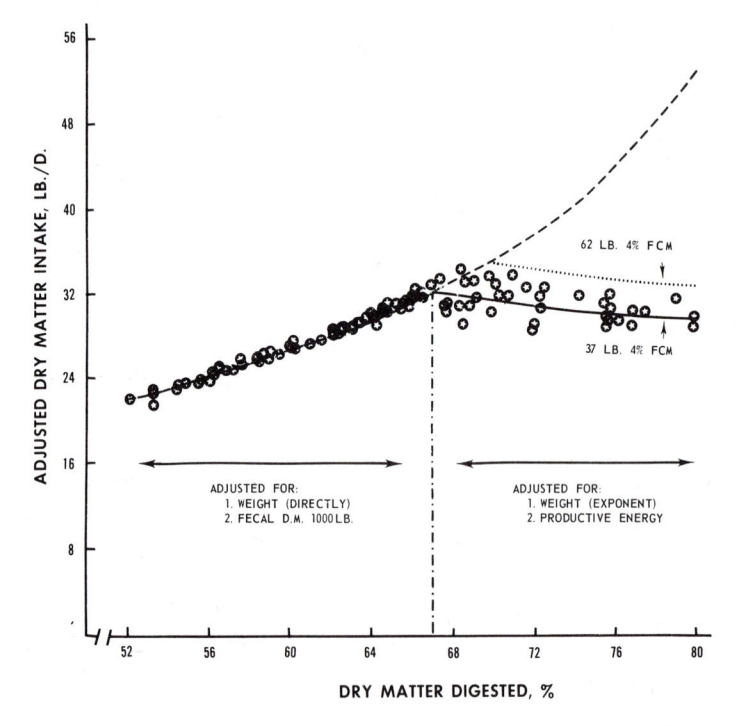

Figure 1. Showing the relationship of dry matter digestibility to adjusted feed intake (lb. per day) at two levels of milk production. Adjustments were proportional to weight below 66.7% dry matter digested and weight raised to the 0.51 power above 66.7% digested. Milk production expressed as pounds equivalent of 4% fat corrected milk (FCM). (Reference 13)

Much current research on the intake problem has characterized many forages as having low intakes relative to more concentrated foods. Consumption suggests a positive relationship with digestibility—that is, the more digestible forages will be consumed at a higher level. This tends to be an oversimplification. Conrad (12, 13) has shown that intake of highly digestible feeds is inversely related to digestibility, the dominating factor being the animal's requirement for metabolizable energy. Less of the more digestible feed is needed to meet this requirement. At lower levels of digestibility—*ca.* 65–70—there is an inflection in the curve (Figure 1) below which intake declines with digestibility.

Intake and Composition. A number of researchers (*14, 16, 17, 46*) have related the sloping part of the curve to the volume and rate of digestion of the cellulosic fraction of the forage. These parameters are not always highly correlated with overall dry matter digestibility. The volume of coarse forages is related to cell-wall content, which has been related negatively to voluntary intake (*57*); cell-wall content above 60% of dry matter is related to decline in consumption. Levels of cell wall below 60% are associated with diminished responses in intake, as shown in Figure 2. Some forages with very high cell-wall volumes (grasses) have low lignin contents, thus promoting relatively higher digestibility with low intake; legumes present an opposite picture of high lignin, low cell-wall content, and higher intake (*57*).

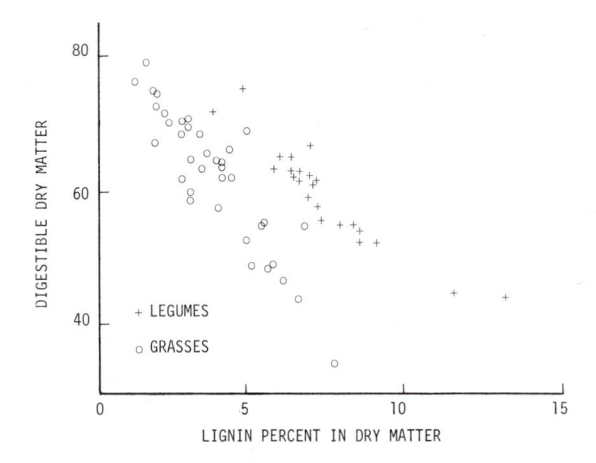

Figure 2. Relationship between voluntary intake and cell-wall constituents of 83 forages from West Virginia. Regression equation: $Y = 110.4 - 1716/(100-X)$. *Intake is expressed as grams per metabolic weight size, i.e. animal body weight to the .75 power. (Reference 57)*

A further instance in which volume affects voluntary intake occurs when feeds are ground and pelleted (*11, 41*). Pelleting of forages usually results in increased intake but the increase is greater with forages of high cell-wall content (*41*).

Grasses, with their high cell-wall and low lignin content, have a slower initial rate of cellulose fermentation than do legumes, which possess a lower cell-wall and a higher lignin content. Legume-cellulose fermentation plateaus sooner, while that of grass occurs over a longer time and produces a higher net digestibility. Studies on the *in vitro* digestion of cellulose have shown that a short digestion (12 hrs.) correlates with intake, whereas a longer time (48 hrs.) correlates well with digestibility and less well with voluntary intake (*15, 16, 17*). These observations are undoubted consequences of the cell-wall content and the lignin-to-cellulose ratio. Feeds fed at a restricted level are often more digestible than those consumed *ad libitum* (*6, 38, 39, 50, 64*). Since many standard digestion trials are conducted at restricted intake, the application of the results to productive situations at maximum intake can lead to errors. The decline in digestibility with increasing intake can be related to the lignin-cellulose ratio and the cell-wall content, as shown by the data of Riewe and Lippke (*50*) in Table II. Decline in digestibility with level of intake is greater in the grasses than in legumes.

Table II. Regression of Organic Matter Digestibility on Intake of Forages Fed to Sheep at Varying Levels of Intake[a]

	Number of Trials	r	Intercept	Slope[b]	Lignin/ Ligno- cellulose	Cell Wall %
Alfalfa	15	−.90	75	−2.7	22	50
White Clover	11	−.90	78	−3.3	16	41
Ryegrass	9	−.94	81	−3.5	11	56
Coastal Bermuda	13	−.97	67	−4.8	13	74
Sorghum	14	−.85	75	−7.6	9	57
Oats	10	−.96	86	−7.7	7	65

[a] Reference 50.
[b] Digestibility change per unit of intake as percent of body weight.

Composition and Digestibility. The relationship between composition and digestibility has been studied extensively because of the abundance of composition data deriving from digestion trials. Confusion has arisen over the factors which are responsible for lowered availability of nutrients as opposed to those that are secondarily associated with digestibility. The tendency for changes in forage composition to be highly associated with increasing maturity, during which digestibility declines, causes almost all components to be significantly correlated with digestibility. Many people have tended to use statistical significance as the main criterion for selecting regressions to predict digestibility. Many equations that are promising in the data from which they were derived perform poorly when applied to new populations of forages.

Hope for improvement lies in the identification of primary factors that limit availability of nutrients in plants. Ultimately relationships must be based on regression analyses, but primary factors should be more dependable predictors than those secondarily related in the cause-effect sequence. An instructive comparison is that between lignin (a primary factor) and protein (a secondary one). Within a plant species an increase in lignin content almost invariably results in a decrease in digestibility (37, 45, 52, 53). In the case of protein, nitrogen fertilization will increase the plant protein content without inducing an accompanying increase in digestibility. One of the reasons for this is that nitrogen fertilization also promotes lignin formation in the plant (21). Lignin is a primary factor limiting availability of certain fractions, but protein is associated positively with digestibility mainly because it decreases as lignin increases in the maturing plant.

The identification of primary factors is not a simple matter. Complex statistical designs (generally known as the Lucas tests) have been developed and applied in classifying chemical components of forages (35, 59). It is on the basis of such analyses that detergent-fiber methods were developed to provide a more ideal chemical fractionation of forages (61).

It must not be presumed that a primary factor like lignin will be consistently related to indigestibility of plant fractions, because other factors also operate. The relationship of lignin and digestibility varies with different plants (32, 52, 53, 56). The general picture is shown in Figure 3. Alfalfa has a higher lignin content than grasses of the same digestibility. Analysis by the Lucas test has shown that there is a variable fraction in forages which is free from lignification. This free fraction has a very high digestibility and is a smaller proportion of dry matter in grasses than in legumes. The lignified fraction corresponds to the plant cell wall, which contains quantitatively the indigestible fraction of the forage.

While the lignin fractions of grasses and legumes are quite different, grass lignin is more soluble and more easily removed than is legume lignin (61). The preponderance of evidence indicates that most of the difference shown in Figure 3 can be accounted for by a variation in the proportion of lignified plant tissue—that is, the lignin within the plant cell walls is distributed unevenly.

In the first place, plant cell walls of forages can be readily prepared by peptic and amylitic digestion of coarse-milled tissue (65) (20 mesh) or by neutral-detergent extraction of plant tissue (61), conditions which do not cause any removal of lignin. These observations also support the conclusion that a lignin-free fraction exists. Furthermore, ruminant feces appear to be composed of undigested plant cell walls in a fine state of division (50–70 mesh), plus much metabolic and bacterial matter (31,

58, 62). In studies of the true digestibility of forage protein and other soluble constituents, using classical nutritional methods (5, 35), most true digestibilities are over 90 and in many cases are not significantly less than 100 (24, 26, 55, 59). Some values are shown in Table III. True digestibility (availability) refers to the disappearance of the ingested food in the digestive tract; excluded from the digestion balance are metabolic fecal excretions and bacterial end-products containing soluble nitrogen and carbohydrate substances that cause apparent digestibilities of protein and soluble carbohydrates to be much less than complete (55, 59, 63).

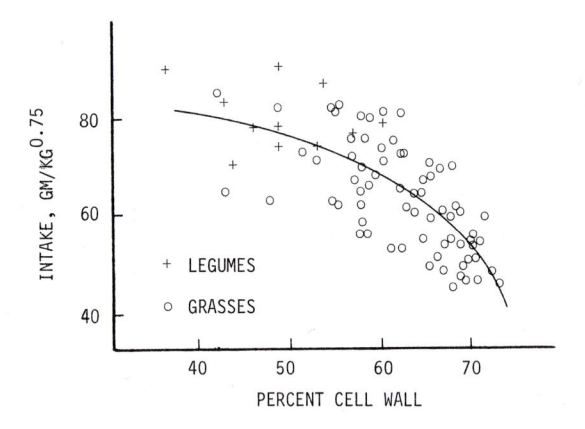

Figure 3. Relationship between lignin content of forages and digestible dry matter. Correlation between lignin and digestibility for grasses, r = −.82, P <.01; for legumes r = −.74, P <.01; overall r = −.40, P <.01. (Reference 56)

Table III. Estimates of Average True Digestibility and Correlations of Apparent Digestibility with Indices of Lignification for Different Forage Fractions[a]

Fraction	Estimated True Digestibility	Standard Deviation	Lignin/Cell Wall	Log. Lignin/ Lignocellulose
Crude Protein	93	±3.1	−.14	−.21
Cellular Contents	98	2.5	−.14	−.21
Cellulose	50	13.5	−.83[b]	−.93[b]
Hemicellulose	79	6.7	−.85[b]	−.90[b]
Holocellulose	73	11.7	−.87[b]	−.96[b]
Cell Wall	62	14.1	−.90[b]	−.98[b]

[a] Reference 59.
[b] Significant at the .01 level of probability.

It is apparent from the data in Table III that the effects of lignin are restricted to the cellulose and hemicellulose carbohydrates of the plant cell wall. One must reject the conclusion of Drapala (23) that availability of cellular contents is lowered by entrapment in lignified cells.

The encrusting of cellulose fibrils by lignin-carbohydrate complexes within the cell wall is still a possible mechanism for lowered cellulose availability. The nutritive availability of the different fractions of hemicellulose and cellulose is quite different as shown by the data of Gaillard (26) in Table IV. Xylan is generally less digestible than other fractions (26, 30), and Lyford et al. (36) have shown that the micellular pentosans have a low digestibility.

Table IV. Composition of Forages in Percentages of Dry Matter and Digestion Coefficients of Components[a]

	Fresh Grass		Fresh Lucerne		Fresh Red Clover	
	D.M.(%)	D.C.	D.M.(%)	D.C.	D.M.(%)	D.C.
Sugars (mono & disaccharides)	4.1	100	2.7	100	4.9	100
Fructosan	5.7	100	—	—	—	—
Starch	—	—	0.4	100	1.2	100
Polysaccharides Soluble in 0.5% am. oxal.:						
Anhydrous galacturonic acid	2.6	95	5.0	95	5.0	97
Anhydrous arabinose	1.8	99	0.6	97	0.5	97
Soluble in 5% and 24% KOH:						
Anhydrous galactose	1.3	85	0.5	82	0.6	78
Anhydrous glucose	1.9	88	0.9	91	0.8	96
Anhydrous arabinose	3.5	86	0.8	85	0.6	75
Anhydrous xylose	8.5	72	5.8	33	3.6	28
Cellulose residue:						
Anhydrous arabinose	0.3	38	0.4	79	0.4	60
Anhydrous xylose	0.6	60	0.3	29	0.1	<0
Pure cellulose	21.1	79	22.2	41	18.2	60
Aldobiuronic acid of xylan	7.3	79	7.3	52	7.1	64
Lignin	6.2	0.2	9.1	0.4	7.0	1.0
Organic matter	90.0	71.7	89.4	59	88.8	71

[a] Reference 26.
[b] D.M. = Dry matter.
[c] D.C. = Digestibility.

Most species of plants used as forages fall into the grass and legume families, although under range conditions a wider spectrum is often met. Legumes possess high ratios of lignin to cellulose even in the young plant (2, 37); grasses possess low ratios that increase markedly as the plant matures (44, 57). The ratio of hemicellulose to cellulose is much higher in the grasses than in the legumes and is relatively constant for a plant species at any stage of maturity (8, 54). Digestibility of hemicellulose is closely related to lignification (8).

The data in Table IV further illustrate the complete digestibility of the soluble carbohydrates. It should be noted that the pectin fraction of the cell wall also has a very high availability and is thus presumably free of the effects of lignification.

It follows that if the true digestibility of cellular contents (including protein and soluble carbohydrates) is complete, the fecal residue representing the true indigestible fraction consists solely of plant cell walls. The division into cellular contents and cell walls thus divides the dry matter of forages according to their availability. This situation precludes the use of single components to obtain precise prediction of digestibility because the cell-wall content does not vary consistently with lignification and other factors influencing availability of the cell wall itself (59).

Other Factors. Forage plants contain tannins (9, 51) and other substances which affect nutritive value. These are not ordinarily thought of in schemes for forage evaluation.

An important factor in many species of grasses is silica, which is metabolized and largely deposited in the cell wall with cellulose (34, 66). Silica depresses the digestibility of the cellulosic carbohydrates in a manner not unlike that of lignin (60). It is possible that one of the physiological functions of silica in the plant is to increase the resistance of the plant to pathogens and also to reduce the loss of water in transpiration (66). Silica content of grasses is related both to species and to the availability of silica in the soils in which the plants are grown (33). Legume species do not seem to metabolize significant amounts of silica. Variation in silica content of reed canary grass (*Phalaris arundinacea*) is shown in Table V. Correlation of digestibility with silica content of these 16 forages was highly significant ($r = -.86$). Correlation of lignin content with digestibility was only $-.58$.

Environmental Factors and Composition. The maturity of forages has often been related to their nutritive value through date of cutting, a decline in digestibility usually being associated with the growth of the plant. It is important to point out the context of these observations: the data are almost exclusively collected under the changing environment of spring and early summer, and environmental changes are generally increasing light intensity and temperature with time. Rainfall patterns and water supply undoubtedly contribute to the patterns also.

The associative changes with growth of the plant under these conditions make it difficult to discern causative relationships, for almost any factor that changes during the period of plant growth will probably be highly correlated with nutritive value. Relationships based on crude protein, crude fiber, lignin, and date of cutting tend to break down if one compares different forages, aftermath cuttings, or different forages cut on the same date at different places (46).

Table V. Average Composition and Digestibility of Reed Canary Grass According to Source and Composition[a]

Source	Number	Cell Wall	Lignin	In Vivo Digestibility	SiO_2
Penna.	3	62	3.8	62	0.6
Mich.	8	59	3.3	63	1.3
Mich.[b]	3	56	4.0	57	2.4
Iowa	2	60	3.8	45	5.4

[a] Reference 60.
[b] Siberian reed canary grass.

It has now become apparent that environmental changes are important in producing compositional changes, and therefore nutritive value. The principal factors influencing change in plant composition with time are light, temperature, and fertilization (1, 4, 18, 19). Changes in nutritive value are generally less marked than compositional changes because of compensatory effects that will be described below. There are, of course, considerable differences between plant species in the extent and nature of the responses to environmental factors. However, it is likely that the general effects of light, temperature, and fertilization within the grass family, at least, are in the same direction. Deinum has demonstrated consistent effects in both temperate and tropical grass species (18, 19, 20, 21).

Light intensity has a positive effect on the water-soluble carbohydrates and a negative one on crude protein, ash, and fiber components, as shown in Table VI. The result is a positive relationship between light intensity and nutritive value. Higher temperature decreases water-soluble carbohydrates and increases lignin and cell-wall content, resulting in a decline in nutritive quality. The negative effect of temperature far outweighs the positive effect of light (21), so that a general decline in nutritive value is associated with increasing light and temperature in the spring of the year.

Table VI. Effect of Light Intensity and Temperature on Composition of Rye Grasses Grown in a Greenhouse[a]

	Light Intensity[b]		Temperature[c]	
	Low 70 cal cm.$^{-2}$ day^{-1}	High 400 cal	Low 12.5°C.	High 22.5°C.
	Percent Dry Basis			
Crude Protein	23.7	15.0	21.0	18.3
Ash	11.1	6.0	8.2	8.9
Water Soluble Carbohydrates	8.5	24.7	18.5	14.5
Cell Walls	46.2	41.9	41.7	46.6
Cellulose	23.8	19.4	19.5	22.2
Lignin	2.3	1.7	1.8	2.3
Grass Dry Matter/POT	9.2	11.4	8.7	11.4
Est. Digestibility	79	81	80	79

[a] Reference 18.
[b] Mean temperature 17.5°C.
[c] Mean light intensity 235 Cal cm.$^{-2}$ day^{-1}.
[d] Wet basis.

In sudan grass grown in West Virginia (49) (see Table VII) nitrogen fertilization resulted in increased protein and lignin content. Nitrogen fertilization also tended to decrease cell-wall contents during intermediate stages of growth, so that digestibility was not consistently affected. However, at the latest stage of regrowth the increased lignification outweighed other effects and nitrogen fertilization resulted in a depression of digestibility. Deinum (18) has reported similar changes in composition with nitrogen fertilization of ryegrass, including associated decreases in digestibility. Depending on the balance of factors one might see digestibility changes in either direction. Further data (22) suggest changes within the cell wall, including increased lignin and cellulose but a decrease in hemicellulose with temperature, which might account for the depressed cell walls at the regrowth stages of sudan grass in Table VII.

Unpublished studies (22) suggest small improvements in composition during the fall while temperatures are decreasing. The declining temperature does cause the compositional change to retrace that of the spring because cell-wall components are relatively fixed and the slowed rate of growth in the fall does not rapidly dilute the already formed plant tissue. In spring it is otherwise, in that growth rate increases as the temperature increases.

The general variation of composition with environment and the concomitant compensatory effects between lignification, cell-wall, soluble carbohydrates and protein make apparent the difficulty of finding consistent effects in protein or fiber that will allow prediction of nutritive value. Accurate prediction on the basis of one or two chemical criteria or the date of cutting is not possible when mixtures of forage species are involved (46).

Table VII. Effect of Nitrogen Fertilization on Composition and Digestibility of Sudan Grass in West Virginia[a]

Date	Cutting	Protein LN	Protein HN	Cell Wall LN	Cell Wall HN	Lignin LN	Lignin HN	Digestibility LN	Digestibility HN
July 10	1st	15	20	53	49	2.1	2.5	77	79
Aug. 10	1st	7	8	66	68	5.3	5.8	62	65
Sept. 6	1st	4	6	72	68	6.2	6.9	64	63
Aug. 10	regrowth	10	12	61	61	2.7	3.3	71	72
Sept. 6	regrowth	7	9	67	66	2.4	4.5	69	61

[a] References 49 and 57.
[b] LN = 100 lb. nitrogen/acre.
[c] HN = 300 lb. nitrogen/acre.

The relationship between environmental temperature and plant composition may account for the generally low feeding value of tropical grasses (20). The problem of tropical forages is one of great importance to future world food supplies and is urgently in need of further research.

Many of the inferior characteristics of forages as feed arise from their high lignin and cell-wall contents. Various routes exist for their improvement. The simplest is management (harvesting when nutritive value is high), but it is apparent that the breeding of more efficient strains and the processing of low-quality forage into higher-quality feed will be more important in the future.

Summary

Composition of the plant cell wall with respect to proportions of cellulose, hemicellulose, and lignin is of considerable taxonomic significance. Lignification and the proportion of cell-wall to non-cell-wall constituents are influenced by environment, particularly temperature. Maturity and date of cutting are secondarily related to nutritive value as plant composition responds to changing environment. There is important variation between species of forage with respect to composition. In particular, contrasting behavior of legumes and grasses dominates the forage picture. Chemical composition will account for nutritive value, but causative relationships are complex. Lignin and silica reduce the digestibility of only the cell-wall fraction, which is itself an important variable influencing digestibility and the amount eaten.

Literature Cited

(1) Alberda, Th., *Neth. J. Agric. Sci.* **13**, 335 (1965).
(2) Archibald, J. G., Barnes, H. D., Fenner, H., Gersten, B., *J. Dairy Sci.* **45**, 858 (1962).
(3) Balch, C. C., Campling, R. C., *Nutr. Abst. and Rev.* **32**, 669 (1962).
(4) Blaser, R. E., *J. Animal Sci.* **23**, 246 (1964).
(5) Blaxter, K. L., Mitchell, H. H., *J. Animal Sci.* **7**, 351 (1948).
(6) Brown, L. D., *J. Dairy Sci.* **49**, 223 (1966).
(7) Buchman, D. T., Hemken, R. W., *J. Dairy Sci.* **47**, 861 (1964).
(8) Burdick, D., Sullivan, J. T., *J. Animal Sci.* **22**, 444 (1963).
(9) Burns, R. E., *Ga. Agr. Exp. Sta. Tech. Bull.* **N.S. 32** (1963).
(10) Campling, R. C., *Proc. Nutr. Soc.* **23**, 80 (1964).
(11) Campling, R. C., *Outlook on Agriculture* **5**(2), 74 (1966).
(12) Conrad, H. R., *J. Animal Sci.* **25**, 227 (1966).
(13) Conrad, H. R., Pratt, A. D., Hibbs, J. W., *J. Dairy Sci.* **47**, 54 (1964).
(14) Crampton, E. W., Donefer, E., Lloyd, L. E., *J. Animal Sci.* **19**, 538 (1960).
(15) Dehority, B. A., Johnson, R. R., *J. Animal Sci.* **23**, 203 (1964).
(16) Donefer, E., Crampton, E. W., Lloyd, L. E., *J. Animal Sci.* **19**, 545 (1960).
(17) Donefer, E., Niemann, P. J., Crampton, E. W., Lloyd, L. E., *J. Dairy Sci.* **46**, 965 (1963).
(18) Deinum, B., *Mededl. Landbouwhogeschool*, Wageningen, **66-11**, 1-91 (1966a).
(19) Deinum, B., *Proc. 10th Intern. Grassld. Congr.*, Helsinki (1966b).
(20) Deinum, B., Dirven, J. G. P., *Surin. Landb.* **15**, 5 (1967).
(21) Deinum, B., van Es, A. J. H., Van Soest, P. J., *Netherlands J. Agr. Sci.* (1968).
(22) Deinum, B., Van Soest, P. J. (unpublished data).
(23) Drapala, W. J., Raymond, L. C., Crampton, E. W., *Sci. Agr.* **27**, 36 (1947).
(24) Elliott, R. C., Topps, J. H., *Brit. J. Nutr.* **18**, 245 (1964).
(25) Flatt, W. P., Moe, P. W., Moore, L. A., Van Soest, P. J., *4th Symp. on Energy Metabolism, Warsaw, Poland*, in press (1967).
(26) Gaillard, B. D. E., *J. Agr. Sci.* **59**, 369 (1962).
(27) Halliwell, G., Bryant, M. P., *J. Gen. Microbiol.* **32**, 441 (1963).
(28) Hungate, R. E., *The Rumen and Its Microbes*, Academic Press, New York (1966).
(29) Ingalls, J. R., Thomas, J. W., Tesar, M. B., *J. Animal Sci.* **24**, 1165 (1965).
(30) Jarrige, R., *Proc. 8th Int. Grassld. Congr.* (1960).
(31) Jarrige, R., *Proc. 9th Int. Grassld. Congr.* (1965).
(32) Jarrige, R., Minson, D., *J., Ann. Zootech.* **13**, 117 (1964).
(33) Jones, L. H. P., Handreck, K. H., *Advan. Agronomy* **19**, 107 (1967).
(34) Jones, L. H. P., Milne, A. A., Wadham, S. M., *Plants and Soil* **18**, 358 (1963).
(35) Lucas, H. L., Jr., Smart, W. W. G., Jr., Cipolloni, M. A., Gross, H. D., **S-45 Report**, North Carolina State College (mimeo) (1961).
(36) Lyford, S. J., Jr., Smart, W. W. G., Jr., Matrone, G., *J. Nutr.* **79**, 105 (1963).
(37) Maymone, B., *Alimentazione Animale* **6**, 371 (1962).
(38) Meyer, J. H., Kromann, R., Garrett, W. N., "Physiology of Digestion in the Ruminant," Butterworths, Washington, p. 262, 1965.
(39) Moe, P. W., Reid, J. T., Tyrrell, H. F., *J. Dairy Sci.* **48**, 1053 (1965).
(40) Montgomery, M. J., Baumgardt, B. R., *J. Dairy Sci.* **48**, 569 (1965).
(41) Moore, L. A., *J. Animal Sci.* **23**, 230 (1964).

(42) Moore, L. A., *USDA-ARS 44-169* (1966).
(43) Moore, L. A., Irvin, H. M., Shaw, J. C., *J. Dairy Sci.* 36(2), 93 (1953).
(44) Mowat, D. N., Fulkerson, R. S., Tossel, W. E., Winch, J. E., *Can. J. Plant Sci.* 45, 321 (1965).
(45) Naumann, K., *Z. f. Tierernährung u. Futtermittelkunde* 3, 193 (1940).
(46) Oh, H. K., Baumgardt, B. R., Scholl, J. M., *J. Dairy Sci.* 49, 850 (1966).
(47) Reid, R. L., Jung, G. A., *J. Animal Sci.* 24(3), 615 (1965).
(48) Reid, R. L., Jung, G. A., Murray, S. J., *J. Animal Sci.* 25(3), 636 (1966).
(49) Reid, R. L., Clark, B., Welch, J. A., Doaza, L., Jung, G. A., *Final Report.* **USDA Contract 12-14-100-4524(24)** W. Va. Agr. Expt. Sta., Morgantown, W. Virginia (1962).
(50) Riewe, M. E., Lippke, H., *Proc. Assn. Southern Agric. Workers* (1968).
(51) Smart, W. W. G., Jr., Bell, I. A., Stanley, N. W., Cope, W. A., *J. Dairy Sci.* 44, 1945 (1961).
(52) Sullivan, J. T., *J. Animal Sci.* 18, 1292 (1959).
(53) Sullivan, J. T., *USDA Bull. ARS 34-62* (1964).
(54) Sullivan, J. T., *J. Animal Sci.* 25, 83 (1966).
(55) Turk, K. L., Morrison, F. B., Maynard, L. A., *J. Agr. Res.* 48, 555 (1934).
(56) Van Soest, P. J., *J. Animal Sci.* 23, 838 (1964).
(57) *Ibid.*, 24, 834 (1965).
(58) Van Soest, P. J., *Proc. Southern Past. Forage Improvement Conf.*, Blacksburg, Va., p. 24 (1966).
(59) Van Soest, P. J., *J. Animal Sci.* 26, 119 (1967).
(60) Van Soest, P. J., *J. Dairy Sci.* 50, 989 (1967).
(61) Van Soest, P. J., Wine, R. H., *J. Assoc. Off. Anal. Chem.* 50, 50 (1967).
(62) Van Soest, P. J., Wine, R. H., Moore, L. A., *Proc. X Inter. Grassld. Congr.*, Helsinki, Finland, **P-438** (1966).
(63) Virtanen, A. I., *Science* 153, 1603 (1966).
(64) Wagner, D. G., Loosli, J. K., *Memoir 400, Cornell University, Agr. Exp. Sta.*, Ithaca, N. Y. (1967).
(65) Weinstock, A., Benham, G. H., *Cereal Chem.* 28(6), 490 (1951).
(66) Yoshida, S., Ohnishi, Y., Kitagishi, K., *Sci. Plant Nutr. (Japan)* 8, 30 (1962).

RECEIVED October 14, 1968.

Discussion

M. Tarkow: "What is the mechanism whereby an organism gets into the cell wall in a forage?"

P. J. Van Soest: "I don't know. But we do know that in the rumen most of the cellulolytic bacteria are actually attached to the fibers."

M. Tarkow: "Yes, but how do they get into the interior? Is there any relationship between the porosity of the vessels within the plants and the ability of the organism to get into the cell wall? You have vessel elements, and these I understand are quite large in diameter."

P. J. Van Soest: "Yes, they might permit the organism to enter."

M. Tarkow: "And very effective. Now, might it be that the organism has an effect upon the formation of deposits within the vessels thereby impairing the movement of organisms to the cell walls?"

P. J. Van Soest: "This is possible. We have other observations that relate to this. We have been studying the particle size reduction that occurs naturally in the rumen. It is interesting that if you mill alfalfa, for example, with fresh feed you get a distribution of lignin which decreases with the particle size. However, if you look in the animal, the relationship between lignin and particle is just exactly the opposite to this. The finest particles of plant cell wall in animal feces are around 200–400 mesh, which is about 70 microns. If we look at these under a microscope there is a distinct suggestion that the laminal parts of the cell wall have been digested. We feel possibly that the bacteria themselves are helping us out, to a certain extent, by involving themselves in the particle size reduction in addition to the rumination process. We do know the bacteria are tightly attached to the particles, and they can be removed in part by putting rumen ingesta in a Waring Blendor and filtering through glass wool, but by no means do you get all of them. Some are rather firmly bound to the fiber."

B. Stone: "What is the silica content of wood cell walls?"

P. J. Van Soest: "I would like to know the answer to that question myself."

Comment

J. Saeman: "It can be very high. Some tropical species run up to 10%. It is so prominent that it is an extreme problem in sawing. In some species you can strike sparks from a saw as a result of sawing through the silica in the wood."

17

Use of Wood and Woody By-products as a Source of Energy in Beef Cattle Rations

W. D. KITTS, C. R. KRISHNAMURTI, J. A. SHELFORD, and J. G. HUFFMAN

Department of Animal Science, The University of British Columbia, Vancouver 168, B. C., Canada

> *The incorporation of sawdust in beef cattle rations has been investigated using three approaches.* In vivo *feeding trials with growing beef cattle indicate that there is no appreciable difference in the daily weight gain of steers when alder sawdust was substituted for hay in their rations. When hemlock sawdust, subjected to gamma irradiation up to a maximum of 1.46 $\times$ 10^8 rads, was used as a substrate for* in vitro *rumen fermentation tests it was observed that the per cent dry matter disappearance and cellulose digestion showed a steady increase with increasing irradiation levels. The amount of reducing sugars formed from irradiated sawdust by incubation with cell-free extracts of rumen microorganisms followed the same pattern as the* in vitro *fermentation tests.*

It has been recognized that the world population is increasing at an alarming rate. Cuthbertson (*14*) estimates the population to be about 4000 million by 1980 and about 6000 million by 2000. There are at least two aspects to consider in respect to this increase: (1) the problems associated with controlling reproduction and (2) the problems associated with providing adequate food and fiber to all concerned. Needless to say here that there is a great deal of research work under way on the problems of birth control while on the other hand much is being pursued to determine ways and means of increasing the world's food production. Both aspects are of paramount interest. However, in this article it is intended to discuss solely the subject of increased food production required to meet the requirements of the increasing population of the world.

It is evident that the domestic animal will eventually compete with the human for cereal grains and other foodstuffs that are being produced for use directly by man. It is possible, therefore, that the animal will have to use as its food supply, materials that cannot be utilized by the human. The ruminant is unique in that it fosters in its rumen a microbial symbiotic relationship required for its well being. The ruminant is adapted, physiologically and anatomically to roughage utilization; that is, to convert fibrous material containing large amounts of cellulose into food that is acceptable to man. Since wood and wood by-products contain mainly free cellulose, cellulose chemically associated with lignin and nitrogen, and pentosans often referred to as "hemicelluloses" it is reasonable to speculate that the rumen microflora and fauna could utilize these substances and in turn have the end-products of this fermentation transformed into food suitable to meet man's palate and appetite.

There is a large volume of work reported on several aspects of the utilization of wood by the ruminant and the relevant literature is reviewed in the following order for the sake of clarity: the use of wood and wood by-products as feeding materials for ruminants, factors influencing the availability of cellulose from these products, the physical and/or chemical procedures used by several workers to increase the availability of cellulose to rumen microorganisms, in vitro fermentation procedures designed to screen the efficiency of such treatments, and studies with cellulolytic enzymes from rumen microorganisms.

Interest was expressed at the turn of the century, in Germany, in the possibility of using wood as a feedstuff. Beckman (6) compared the food value of wood and straw through chemical analysis. In analyzing several different woods for nitrogen, fat, starch, and ash he found their value to be much lower than the values he obtained for straw. Haberlandt (23) discussed the possibility of using the starch, oil, and in some cases the protein and glucose stored in sap wood as a food source. He found this wood, being highly indigestible, had to be ground very fine before horses and cattle could utilize it to any degree.

That cellulose is a potential energy source for ruminants has been known for years. In 1965, Nehring et al. (42) reported on trials in which they compared purified cellulose and sucrose as energy sources. They found cellulose to be an equal if not better source of energy (net energy) for ruminants than sucrose. Bhattacharya and Fontenot (7) used peanut hull and wood shavings-poultry litter in a digestion trial with wethers. Energy yields were comparable with hay (244 kcal./kg. digestible energy and 22.7% digestible protein). The litter was incorporated into the ration at levels of 25% and 50%. Dry matter, nitrogen-free extract, and apparent digestible energy were lower for the rations containing the two levels of litter. Ammerman and Block (2) investigated the feeding value of rations containing sewage sludge and oakwood sawdust. In comparison with bermuda grass, the mixture of sewage sludge and sawdust did not appear nutritionally as good, but did appear to provide roughage to the diet. Bissell and Wier (8), using sheep and deer, tested the digestibility of chamise (*Adenostoma fasiculatum*) and interior live oak (*Quercus wislizenii*) as sources of protein and energy, using alfalfa hay as the control feed. They found chamise to be an inferior source of protein and energy when compared with alfalfa. The interior live oak appeared even less digestible.

Crampton and Maynard (12) reporting on the relationship between the cellulose and lignin content of feeds and their digestibility found that the digestibility of the cellulose varied inversely with the lignin content of the forage. Later, Ely et al. (16) studied the composition of lignin in orchard grass at various stages of maturity and found that the nitrogen content of lignin decreased as the plant matured. They also showed that the lignin in the orchard grass had digestion coefficients between 7.5 and 19.8. Sullivan (55) at a later date confirmed the findings of Ely et al. (16) and recorded from his work that depending upon the feedstuff the coefficients of digestibility for cellulose were between 56–89 and for lignin about 10.

Tomlin et al. (59) used the *in vitro* techniques to study the relationship between lignification and cellulose digestion. They found that lignin content was negatively correlated with *in vitro* cellulose digestion. Similarly Burdick and Sullivan (10) found a positive correlation between ease of acid hydrolysis of hemicellulose and its digestibility, indicating the interference of lignin in the hydrolysis of hemicellulose.

The inability of the herbivore to utilize fully the carbohydrate portion of wood and straw led investigators to study the effect of treating wood and straw to increase the digestibility. Lehman (39), treated straw with ammonium sulfate and sodium phospahte. It was then inoculated with desirable molds which after cultivation would fill the entire mass with desirable protein (biomass). As reported in *Papierztg.* (43) an attempt was made to increase the value of straw and bamboo shoots by cooking the roughage under pressure at temperatures up to 160°C. The organic acids released during cooking dissolved the intracellular substances, freeing the fiber. Small amounts of alkali were added, but not in sufficient amounts to neutralize the free acids. Honcamp and Hilgert (27) studied the decomposition of straw without chemicals. Steaming of the straw resulted in a loss in crude protein and an increased starch value. Hvidsten and Homb (29) conducted a survey on cellulose and Beckman-treated straw as potential livestock feeds. Using bleachable sulfite cellulose with a 1–2% lignin content and straw with a 1.2–1.5% sodium hydroxide solution they determined digestion coefficients with sheep. The values found were 87 and 66 for the "cellulose" and straw respectively.

The feeding of alkali-treated cereal straw was shown to be beneficial for the growth of young animals by Kehar (*31*). On the other hand, Sahai *et al.* (*46*) failed to observe any economic advantage of alkali treatment of straw in terms of augmentation of milk yield of cows. The treatment, however, helped the removal of deleterious substances from the straw and facilitated greater utilization of proteins. Gonzeleza (*22*) used the alkali method of treating straw employing modern techniques. After delignifying with alkali, washing with water and drying, the material was fed to lambs. It was apparent that the feeding value had improved. Similarly, Lampila (*36*) treated wheat straw with alkali and found from digestion trials with sheep, using urea as a protein source, the alkali straw to be as good as, if not better than, hay as an energy source for protein synthesis. Stone *et al.* (*53*) reported that cellulose in samples of bagasse washed with 1.5% NaOH was more available to cellulolytic enzymes than that in untreated samples. Later in 1965 they (*54*) investigated the effects of alternate treatments of bagasse with acid and alkali. The acid treatment did not have any beneficial effect, while the 4% alkali treatment yielded increased *in vitro* cellulose digestibility.

Stewart (*52*) studied the effect of various hydrolytic treatments on lignin, cellulose, and non-cellulosic fractions of wood. He showed that little lignin is attacked by dilute alkali at room temperature but a considerable portion of the non-cellulosic fractions is dissolved. He also found the reaction of mild alkali at these temperatures to be much more rapid than the reaction involving dilute acid at the same conditions. Saarinen *et al.* (*47*) compared the effects of different chemical treatments of birch wood and found that treatment with sulfate produced a pulp having a digestible carbohydrate yield of 51.5% of the oven dry wood. Using rams as the test animals, they later (*48*) reported that pulps obtained by alkali treatment were more digestible than those resulting from treatment with acid or chlorite and that the difference was brought about primarily by a difference in the digestibility of the cellulose-lignin fraction which varied from 34.7 to 84.7%, depending upon the method of preparation. Recently, Mellenberger *et al.* (*41*) studied the dry matter digestibility of wood and wood residue by introducing certain modifications in the *in vitro* method suggested by Tilley and Terry (*58*) for forage evaluation. They found that the digestion coefficients on sawdust, pulp, and ammoniated wood were 10–18, 55–70 and 25–33% respectively.

Other possible methods of breaking the lignin carbohydrate bond (*9*) and freeing the cellulose portion of wood might be attained by studying some of the physical and chemical properties of the various fractions of wood. Stamm (*50*) reported on the physical properties of lignin. At high temperatures lignin was found to behave much like a plastic, flowing at a temperature of 350°F. and pressure of 1500–2500 p.s.i. The possibility remains that with the added influence of steam this lignocarbohydrate bond at extreme temperature and pressure may be altered sufficiently to allow rumen microbial population to utilize, at least partially, the carbohydrate portion of wood. Krupnova and Sharkov (35) reported that milling of wood cellulose at high temperatures produced a substrate that was readily hydrolyzable with 10% H_2SO_4. Using this high temperature milling technique Katz and Reese (30) showed that the enzymatic hydrolysis of wood cellulose (Solka-Floc) thus treated can give concentrations of glucose (30%) comparable with the enzymatic hydrolysis of starch. Klopfenstein *et al.* (34) studied the effects of high pressure on the digestibility of roughages by subjecting them to a pressure of 28 kg./cm.^{2} in the presence of water, 0.5% HCl or 4% H_2O_2 and found increases in *in vitro* dry matter disappearance from 9 to 22%. Reducing the physical size of the wood or forage particles by grinding or ball-milling has been found by several workers (15, 62) to increase the digestibility of cellulose by rumen microorganisms. Sullivan and Hershberger (56) treated straw with dry chlorine dioxide and observed a decrease in the acid-insoluble lignin content and an increase in *in vitro* cellulose digestion.

Lawton, *et al.* (37) studied the effect of high velocity electrons on wood in relation to *in vitro* digestibility by rumen microorganisms. Maximum fermentability occurred at an irradiation of 10^8 rads. The volatile fatty acids formed constituted about 79% of the amount formed from an equal weight of filter paper. Possible reasons for the increase in fermentation were (1) that there is a definite bond between cellulose and lignin, which cannot be hydrolyzed by bacteria, and was disrupted by the high velocity electrons, (2) that there are present in wood natural bacteriostatic compounds which are destroyed by irradiation, and (3) that a disruption of the natural cellulose of wood which, unlike filter paper cellulose, is not susceptible to digestion by rumen bacteria. A comparison of the effects of cathode rays on sprucewood, isolated cell substance and cotton linters was made by Saeman *et al.* (45). They found that with the increase in the radiation dosage from 10^6 r. or onward there was an equal decrease in the degree of polymerization in all the three substances. At a dosage of 10^8 r., 14% of the cotton linters, 17% of the wood pulp, and 9% of the wood were degraded.

Mater (40) reviewed reports on high energy irradiation of cellulose, dextran, lignin, and sawdust. The studies related to cellulose were those of Lawton *et al.* (39). Dextrans under high energy irradiation were found to have a greatly reduced molecular weight. Lignin appeared to be little affected, the change being undetectable by bacteriological and chemical means. It was stated that up to 70% of the carbohydrate portion in Douglas fir sawdust could be made soluble by rumen microorganisms depending upon the treatment. This was compared with 1–3%

made soluble in untreated wood. The decomposition of wood constituents by different levels of gamma irradiation was studied by Seifert (*49*). Based on chromatographic and radiographic observations, he concluded that gamma irradiation first led to decomposition products of micelles that were still in their lattice-like order. This was found to be different from enzymic wood decay where the intermolecular bonds between the chains are attacked from the beginning. Using *in vitro* fermentation techniques, Pritchard *et al.* (*44*) investigated the effects of gamma radiation upon the feeding value of wheat straw. They found that 2.5×10^8 rads of gamma radiation was the optimum dosage for the release of nutrients from wheat straw and that the level was well above what is practical for commercial operations.

Since a proper understanding of the mechanism of breakdown of cellulose is essential to develop methods that would increase the availability of cellulose in low quality roughages efforts have been made to obtain cellulolytic enzymes from rumen microorganisms and study their mode of action. Kitts and Underkofler (*33*) first described the preparation of cellulolytic enzymes from mixed rumen microorganisms by grinding the bacterial cells from strained rumen fluid with alumina. They could not detect cellulolytic activity in the centrifuged and filtered rumen fluid suggesting that the cellulolytic enzymes were not present as such in the rumen fluid but were associated with the bacterial cells. Since glucose was the only major product of hydrolysis of carboxymethyl cellulose (CMC) they hypothesized that the cellulose degrading enzyme of the rumen microorganisms was a "celloglucosidase." Using the decrease in viscosity of CMC solution as the method of assay, Cason and Thomas (*11*) reported that the cellulolytic factor in rumen fluid was heat labile and influenced by changes in the ration. By viscosimetric methods Gill and King (*21*) demonstrated the presence of relatively thermostable cellulase found free in rumen fluid. Electrophoretically they were able to detect three components of which two had cellulolytic activity and the third B-glucosidase activity. Halliwell (*24, 25*) prepared cell-free extracts (CFE) from concentrated suspensions of rumen microorganisms by extraction with butyl alcohol or by grinding with alumina. The butyl alcohol extracts exhibited a higher activity on CMC than on insoluble cellulose powder whereas the alumina extracts were more active on insoluble substrates than on CMC. Based on these results he suggested a multienzymic theory of hydrolysis of cellulose. The study of Festenstein (*18, 19*) showed that two enzymes were involved in the breakdown of CMC by butyl alcohol extracts of rumen microflora, cellobiase, and carboxymethylcellulase, the former being selectively inhibited by D-glucono 1,4-lactone. Appreciable transferase activity was also detected in the extracts. Stanley and Kesler (*51*) obtained cell-free extracts by agitating rumen fluid in a Waring Blendor, followed by high speed centrifugation. The supernatant was found to be active on CMC and slightly on Whatman cellulose powder.

The increase in the carboxymethylcellulase activity of the supernatant after disintegrating rumen bacterial cells in French Pressure cell or treating with desoxycholate led King (32) to suggest that surface bound cellulases may occur in considerable proportions, in addition to the free diffusing extracellular ones. He also showed the presence of inhibitors of cellulases in crude rumen cell-free preparations and the capacity of iron and cobalt to stimulate cellulolytic action. Fernley (17) studied the action of CFE of sheep rumen microorganisms on cellulosic substrates prepared by coupling water-soluble reactive dyestuff with insoluble cellodextrins. He found that the substrates were hydrolyzed extensively and rapidly and that the enzyme was thermostable and not inhibited by gluconolactone.

The results of Abou Akkada *et al.* (1) indicated that extracts of *Polyplastron multivesiculatum* hydrolyzed cellulose slowly yielding glucose as the main product. The presence of cellodextrinases in the extracts of *Eremoplastron bovis* was shown by Bailey and Clark (5).

Only a few reports are available on the cellulolytic activity of culture filtrates of pure cultures of rumen organisms. Halliwell and Bryant (26) reported that whole cells of *Bacteroides succinogenes* strain S-85 solubilized up to 90% of insoluble cellulose substrates whereas CFE from the same organisms solubilized only to the extent of 3–9%. On the other hand *Ruminococcus flavefaciens* strain FD-1 yielded CFE which solubilized up to 46% of ground filter paper, though the whole cells were only about half as effective as *B. succinogenes*. Leatherwood (38) compared the cellulolytic activity of culture filtrates of *R. albus* with that of the enzyme extracted from mixed rumen organisms and found that the former had a K_m value lower than the latter with CMC as the substrate. Glucose was the main product of hydrolysis with the extracts of mixed rumen organisms whereas cellobiose was produced by the cellulase of *R. albus*.

The presence of cellobiose phosphorylase (EC 2.4.1.a) catalyzing the reversible phosphorolysis of cellobiose to α-D-glucose-1-phosphate and glucose was first demonstrated by Ayers (3, 4) in cell-free extracts of *R. flavefaciens*. Recently, Tyler and Leatherwood (60) reported on the presence of an epimerase catalyzing the conversion of cellobiose into glucosylmannose in the culture filtrate of *R. albus*.

The results of *in vivo* feeding trials and *in vitro* fermentation tests reported herein have indicated the possibility that a considerable amount of wood and wood by-products could be incorporated in the ration of beef cattle. However, because of the changes in the chemical composition of plants owing to age and species and the presence of natural inhibitors of cellulases, a preliminary physical and/or chemical treatment of these materials appears to be necessary. The prerequisites for the use of wood-products in practical feeding of ruminants are: (1) economic consideration; the finished product should be available at a competitive rate in comparison with the cost of other roughages, (2) treatment applied should allow more nutrients to become available to the rumen microorganisms without causing excessive depolymerization of polysaccharides or loss of micronutrients, and (3) treatment should effect the removal of natural inhibitors of polysaccharases, if any, and should not leave any residue which will be toxic to the microorganisms or the host.

Considering the fact that wood is a potential energy source for ruminants, research was initiated at the University of British Columbia to find ways and means of using wood sawdust, powdered wood, and wood by-products in the feeding of growing and fattening beef cattle. This investigation has been divided into three separate studies, (1) *in vivo* feeding trials, (2) *in vitro* fermentation tests, and (3) cell-free extract studies.

In vivo *Feeding Trials*

Experiment 1. In the first feeding trial alder sawdust replaced hay as a roughage source in the feeding of 32 Hereford steers during a feeding period of 63 days. In addition to the wood or hay supplement the animals were fed steamed rolled barley and a protein-mineral-vitamin concentrate (Table I). The nitrogen in the concentrate was primarily from soyabean, herring meal, or urea. The wood and hay were fed at a level of approximately 10% of the total daily feed intake. The animals consumed this amount of wood readily and showed no apparent digestive disturbances.

Table I. Composition of Protein-Vitamin-Mineral Supplement (PMV) Fed to Steers

Ingredient	PMV 1	PMV 2
	lbs./ton	
Soybean meal	1173	—
Alfalfa meal	280	—
Fish meal (75%)	190	—
Molasses	200	—
Barley	—	1223
Urea	—	300
Dehydrated grass meal	—	200
Bone meal	104	200
Cobalt-iodized salt	36	60
Zinc oxide	2	2
Vitamin A premix	10	10
Aureomycin	5	5

The lack of statistical difference between the daily gain of growing beef calves fed sawdust and those fed hay (Table II) indicates that alder sawdust replaced the hay successfully as a roughage and/or nutrient source. It is noticeable that the average daily gain of those animals consuming the sawdust along with barley and PMV #1 concentrate was only slightly less than their counterparts receiving hay (2.88 and 2.96 lbs. respectively). On the other hand, the animals fed sawdust, barley, and PMV #2 urea concentrate showed a lowered average daily gain and, therefore, less total gain for the 63 day feeding period.

Table II. Effect of Incorporation of Sawdust in Beef Cattle Ration

Group	Barley + PMV 1 + Sawdust (1)	Barley + PMV 1 + Hay (2)	Barley + PMV 2 + Sawdust (3)	Barley + PMV 2 + Hay (4)
Ration				
Barley (lb./day)	13.8	13.9	13.9	13.6
PMV (lb./day)	1.5	1.5	1.0	1.0
Alder sawdust (lb./day)	2.0	—	2.0	—
% sawdust	11.6	—	11.8	—
Mixed hay (lb./day)	—	2.0	—	2.0
Nutrient Composition				
Protein (%)	11.88	13.25	12.02	13.49
Calcium (%)	.259	.35	.25	.356
Phosphorus (%)	.454	.49	.447	.487
Digestible energy (kcal./lb.)	1358	1470	1341	1462
Performance				
No. of days on trial	63	63	63	63
No. of animals per group	8	8	8	8
Ave. initial weight (lb.)	720	740	709	699
Ave. final weight (lb.)	902	927	869	885
Ave. total gain (lb.)	182[a]	187[a]	160[a]	186[a]
Ave. daily gain (lb.)	2.88	2.96	2.54	2.94
Feed efficiency (lb. feed/lb. gain)	6.0	5.9	6.7	5.6

[a] Not significant.

Experiment 2. The second feeding trial was designed to study the extent to which non-processed, raw wood added to the ration at various levels would be utilized by the rumen microorganisms. Dry, ground alder wood was added to a basal diet (Table III) and was fed to groups of beef calves for a period of 182 days as outlined below.

Group A was fed a ration consisting of 35% dried ground alder wood in the basal ration (11.4% crude protein (C.P.)).

Group B was fed the same amount of basal ration as fed to Group A plus added wood (to 27.2% of ration—12.4% C.P.).

Table III.   Basal Ration

Steamed rolled barley	1,262 lbs.
Beet pulp	288
Soybean meal	66
Fishmeal	86
Cane sugar	100
Stabilized fat	100
Molasses	44
Urea	20
Limestone	15
Rock phosphate	5
Cobalt-iodized salt	15
Trace minerals	2
Vitamin A	6.8×10^6 I.U.
Aureomycin	25 gms.
Crude protein—calc.	15%
—actual	17.5%
Digestible energy—calc.	1,616 kcal./lb.
Ca:P	1:0.6

Group C was fed the same amount of basal ration as fed to Group A and B plus added wood (to 13.3% of total ration—14.8% C.P.).

Group D was fed basal ration to the level fed to Groups A, B, and C (17.5% C.P.). Acid-detergent fiber (ADF) and acid-detergent lignin (ADL) were determined according to the method of Van Soest (61). Statistical analyses were done by the analysis of variance.

The results of this trial, as shown in Table IV indicate that the ground wood was utilized to some degree by the animals. The group receiving the highest level of wood in the diet showed an average daily gain similar to those animals consuming only the basal ration. Animals in Groups B and C appeared to gain relatively faster—*i.e.*, average daily gains being 2.65 and 2.67 lbs. per day respectively) than those animals in Groups A and D (2.57 and 2.53 respectively). As the level of sawdust in the ration increased the feed conversion of the complete ration became less; on the other hand, calculations of the feed conversion of the basal ration alone excluding the sawdust showed that Groups B and C had a higher feed conversion than Groups A and D. This trend appears to indicate that the incorporation of sawdust at 13.3–27.2% had actually been beneficial to the growth of the animals.

The digestibility of the four rations described above was determined using mature wethers. The results of this study showed that as the percentage of the raw wood increased in the diet, the digestibility, as predicted, decreased as did the digestion of ADF and cellulose (Table V). It is of interest to note that the digestion of lignin in all the rations remained relatively constant. Since there is no statistical difference between the gain of animals receiving different levels of sawdust in their rations (Table IV), it may be inferred that the ground sawdust has actually been utilized by the animals. This is further substantiated by the per cent dry matter and cellulose digestibility of rations containing sawdust. The data, therefore, suggest that the beneficial effects of incorporation of sawdust in beef cattle rations are possibly because of the fiber stimulating rumination and supporting a more sustained fermentation in the rumen thereby causing a more efficient use of the other digestible fractions confirming earlier observations (*13, 28*). In addition, there appears to be a small degree of actual utilization of the alder sawdust by the ruminant as indicated by the per cent dry matter digestibility of rations containing 13.3% sawdust.

Table IV. Effects of Feeding Different Levels of Untreated Alder Sawdust[a] to Steers

	Groups			
	A	B	C	D
No. of animals	7	7	7	7
Per cent wood in diet	35	27.2	13.3	0
No. of days on trial	182	182	182	182
Average initial body weight	544	550	534	536
Average final body weight	1,012	1,033	1,020	996
Average total gain	468	483	486	460
Average daily gain	2.57[b]	2.65[b]	2.67[b]	2.53[b]
Feed conversion—complete ration (lb. feed/lb. gain)	9.13	8.03	6.71	6.19
Feed conversion—basal alone	6.08	5.90	5.86	6.19
Dressing percentage	55.1	56.5	57.1	59.2

[a] Proximate analysis of alder sawdust: Protein (N × 6.25) = 1.84%; Acid-detergent fiber = 69.19%; Lignin = 17.1%; Ash = 0.75% and ether extract = 1.64%.
[b] Not significant.

Table V. *In vivo* Digestibility of Nutrients by Wethers Fed Different Levels of Alder Sawdust

	Groups and Per cent Sawdust in Diet			
	A Sawdust 35%	B Sawdust 27.2%	C Sawdust 13.3%	D Sawdust 0%
	Per cent Nutrient Digestibility			
Dry matter	56.5	64.0	75.9	80.4
Acid detergent fiber	21.8	28.0	31.3	67.3
Lignin	22.5	24.0	23.6	21.6
Cellulose	21.7	29.0	34.0	75.0

Experiment 3. In this trial, the effect of feeding growing steers sawdust subjected to a temperature of approximately 350°F. and a pressure of 1500–2500 p.s.i. was investigated. The machine used for this purpose was designed by the Division of Engineering of the British Columbia Research Council originally for the production of fuel logs from wood wastes. When sawdust is fed into this machine it is subjected to a grinding motion and sufficient heat is built up mechanically to yield a product which will hereafter be referred to as "extruded wood." Using this machine approximately 200 lbs. of sawdust can be extruded in one hour.

The experiment was designed (1) to compare the utilization of untreated sawdust and extruded wood fed at different levels, and (2) to study the feasibility of substituting wood for hay as the roughage component in high concentrate rations. The experimental design and the rations fed are given in Table VI. All the rations were pelleted (11/64 inch) and fed *ad libitum*.

The performance of steers fed untreated sawdust, extruded wood, and hay at different levels is given in Table VII.

Table VI. Composition of Rations Fed to Fattening Steers

	15% Extruded Wood (1)	20% Extruded Wood (2)	15% Untreated Sawdust (3)	20% Untreated Sawdust (4)	15% Hay (5)	20% Hay (6)
Ingredients						
Barley	1427	1322	1427	1322	1440	1338
Urea	35	38.5	35	38.5	22	22
Molasses	100	100	100	100	100	100
Alfalfa meal	100	100	100	100	100	100
Salt	15	15	15	15	15	15
Premix (trace minerals)	5	5	5	5	5	5
Tricalcium phosphate	18	20	18	20	18	20
Limestone	0.5	—	0.5	—	—	—
Extruded wood (alder)	300	400	—	—	—	—
Untreated sawdust (alder)	—	—	300	400	—	—
Hay (12% crude protein)	—	—	—	—	300	400
Protein %	13	13	13	13	13	13
Energy kcal. D.E./lb.	1227	1145	1227	1145	1385	1357
Ca %	.5	.5	.5	.5	.5	.5
P %	.5	.5	.5	.5	.5	.5

Table VII. Effects of Feeding Untreated Sawdust, Extruded Wood, and Hay to Fattening Steers

	15% Extruded Wood (1)	20% Extruded Wood (2)	15% Untreated Sawdust (3)	20% Untreated Sawdust (4)	15% Hay (5)	20% Hay (6)
No. of days on trial	63	63	63	63	63	63
Average initial body weight (lbs.)	700	702	720	713	712	721
Average final body weight	870	881	877	886	965	961
Average total gain	170[b]	179[b]	157[b]	173[b]	253[a]	240[a]
Average daily gain	2.70	2.84	2.48	2.74	4.01	3.81
Average feed intake	1124	1165	1138	1201	1340	1342
Feed efficiency	6.61	6.64	6.67	6.94	5.29	5.58

[a, b] Values with differing superscripts in each line are significantly different ($P > .01$).

It is observed from the data the steers fed hay as the roughage component of the ration gained faster and had a better feed efficiency than those animals fed either the untreated sawdust or the extruded wood. Steers fed hay at the 15% level gained faster and were more efficient in feed conversion than those fed at the 20% level. There is an increasing trend in the total body weight gains and feed efficiency of animals fed extruded wood over those fed untreated wood. In all groups the beef cattle showed no palatability or digestive problems and appeared to eat with good appetite.

In vitro *Fermentation Tests*

In view of the encouraging results obtained by feeding alder sawdust to beef cattle, as was reported earlier, it seemed worthwhile to develop methods of treatment that would make the cellulose in low quality roughages more available to rumen microorganisms. In order to assess the merits of various physical and chemical treatments of wood and wood by-products, *in vitro* rumen fermentation tests were conducted and the extent of availability of nutrients to the microorganisms studied.

Hemlock sawdust was ground past a 0.8 mm. screen in a laboratory mill, and then subjected to five different levels of irradiation using a Gammacell 220 (Atomic Energy of Canada Limited, Ottawa) as the source of gamma irradiation. The irradiation dosages ranged from 4.85×10^6 to 1.46×10^8 rads.

The dry matter and cellulose digestibilities of all the samples were determined using a modification of the *in vitro* procedure of Tilley *et al.* (57). Rumen liquor collected from a rumen fistulated steer fed alfalfa hay served as the source of inoculum. Incubations were carried out in 125 ml. Erlenmeyer flasks fitted with a gas release valve for a period of 48 hours at 39°–40°C. At the end of the incubation period the residues were dried and dry matter disappearance determined. Residual cellulose was determined using the procedure outlined by Crampton and Maynard (12).

The results (Figure 1) show that the ADF and ADL fractions of hemlock sawdust decrease as the irradiation dosage increases from 0 to 1.46×10^8 rads. This is in agreement with the observations of Seifert (49). The *in vitro* per cent dry matter disappearance and cellulose digestion show a steady increase with increasing irradiation confirming the earlier report of Lawton *et al.* (37). Since even the physical effects of gamma irradiation on plant cell-wall constituents are not known fully, the primary objective in this study has been to fix the dosage that would be optimum in making cellulose more available to rumen microorganisms. Then only it may be appropriate to consider whether it is technologically and economically feasible to undertake such a treatment. Studies are underway to explore this possibility as a means of processing sawdust for feeding ruminants.

Cell-Free Extract (CFE) Studies

Ruminant nutritionists are primarily concerned with the microbial degradation and utilization of cellulose and other carbohydrates and the factors which will stimulate the same. This is in contrast to the interests of research workers in textile and wood industries whose main object is to prevent the decomposition of valuable products by cellulolytic microorganisms. Since a proper understanding of the mechanisms of breakdown of cellulose is essential to develop methods to prevent or stimulate the same, there is a convergence of interests at the enzymic level.

In the present investigation a procedure has been developed to obtain cell-free extracts of rumen microorganisms possessing considerable cellulolytic activity. The CFE were used for the initial screening of the efficiency of various treatments of wood and wood by-products in making cellulose more available to the rumen microbes. These initial screening procedures would facilitate a more efficient use of large animals for feeding and digestion trials.

Rumen contents were collected from a fistulated steer fed alfalfa hay and squeezed through eight layers of cheese cloth. CFE were obtained by ultrasonic vibration of the microbial cells released from the solid fraction of the rumen ingesta by gentle agitation with $0.1M$, pH 5.5 acetate buffer (enzyme fraction I). The liquor obtained after squeezing through cheese cloth was centrifuged and the pellet sonicated to yield enzyme fraction 2. The supernatant constituted enzyme fraction 3. Cellulase activity in the enzyme solutions was assayed by determining the amount of reducing sugars formed from cellulosic substrates using the dinitrosalicylic acid method (20). The incubation mixture for the cellulase assay consisted of 0.2 ml. of 3% $MgCl_2 \cdot 6H_2O$, 1.00 ml. of CFE, substrate and acetate buffer (0.1M, pH 5.5) to a total of 4.0 ml. Incubations were done at 39°C. under air without shaking. Cellulolytic activity in the three enzyme fractions obtained from rumen contents is given in Table VIII.

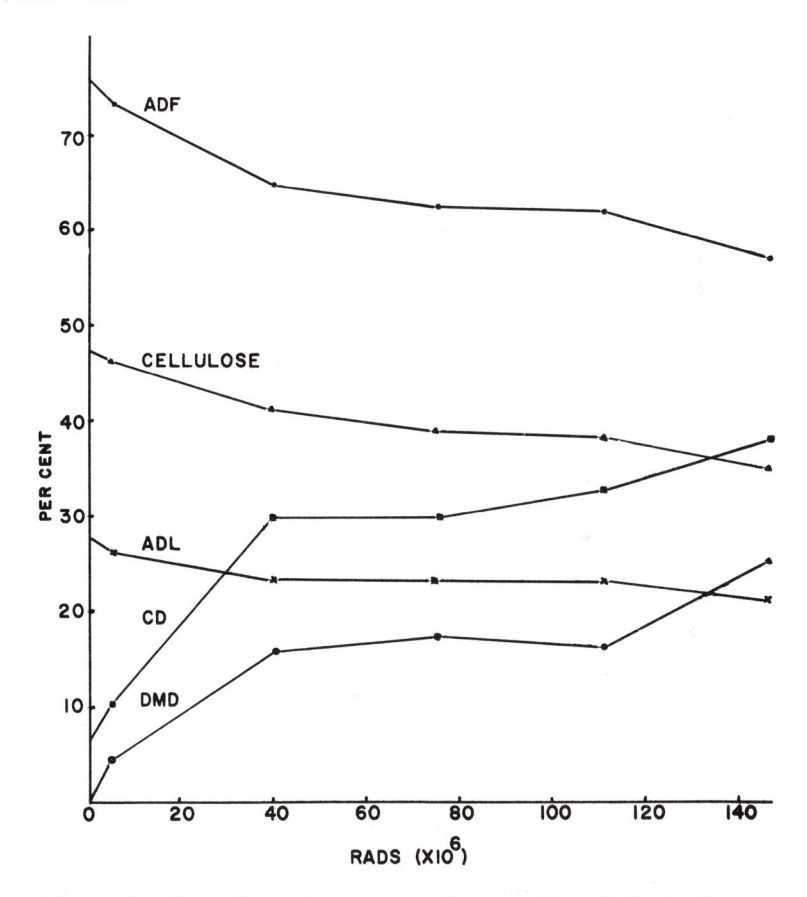

Figure 1. The effect of gamma irradiation of hemlock sawdust on certain of its chemical constituents and on its susceptibility to rumen microbial action in vitro. ADF = acid-detergent fiber; ADL = acid-detergent lignin; CD = cellulose digestion; DMD = dry matter disappearance

Under the conditions of the assay employed, cellulase activity could not be detected in the supernatant fluid (enzyme fraction 3) of rumen contents. Enzyme fraction 2 which was obtained by ultrasonic disruption of cells in the liquid portion of rumen contents exhibited appreciable activity on CMC but not on Avicel or Solka Floc. On the other hand, enzyme fraction I obtained by ultrasonic disruption of cells released from the solid portion of rumen contents by gentle agitation was very active on soluble and insoluble cellulosic substrates.

Table VIII. Distribution of Cellulase Activity in Rumen Contents

Specific Activity[a]

Substrate	Enzyme Fraction 1	Enzyme Fraction 2	Enzyme Fraction 3
Solka Floc	240	—	—
Avicel[b]	212	—	—
Carboxymethylcellulose (CMC-7HF)[c]	1211	813	—

[a] µg reducing sugar/mg. protein/hr./100 mg. substrate.
[b] American Viscose Division, FMC Corporation, Delaware.
[c] Hercules Incorporated, Delaware.

To overcome the variables associated with the use of rumen contents for the preparation of CFE, a cellulose enrichment culture of rumen microorganisms was prepared. The medium consisted of 15 ml. of mineral solution I (0.3% K_2HPO_4); 15 ml. of mineral solution II containing 0.3% KH_2PO_4; 0.6% $(NH_4)_2SO_4$, 0.6% NaCl, 0.06% $MgCl_2$ and 0.06% $CaCl_2$; 6 ml. of 6% yeast extract (DIFCO); 56 ml. of distilled water; 6 ml. of 6% aqueous solution of Na_2CO_3; 2 ml. of 3% sodium thioglycollate and strips of Whatman Number I filter paper. Rumen fluid collected from the fistulated steer was squeezed through cheesecloth and 15 ml. were inoculated into 300 ml. of sterile medium. After 4 days of incubation anaerobically at 39°C., microorganisms were harvested by centrifugation in the cold, washed, and disrupted ultrasonically. Cellulase activity in the sonicated fraction and in the supernatant of the culture medium was assayed as described above.

Two types of organisms, Gram-positive cocci and Garm-negative rods predominate in the cellulose enrichment cultures. Assay of cellulase activity after four days of incubation indicates that by ultrasonic vibration of organisms attached to cellulosic fibers as much as seven times the amount of activity in the supernatant could be obtained (Table IX).

The sonicated cell fraction from cellulose enrichment cultures was used to study enzymic hydrolysis of cellulose in alder sawdust subjected to gamma irradiation.

The amount of reducing sugars formed from alder sawdust by the activity of CFE of rumen microorganisms increase with increasing levels of gamma irradiation up to a dosage of 1.46×10^8 rads. (Table X). The results parallel very closely those obtained with *in vitro* fermentation tests on hemlock sawdust (Figure 1).

Table IX. Distribution of Cellulase Activity in Cellulose Enrichment Cultures

Substrate	Specific Activity[a] Sonicated Cell Fraction	Supernatant
Avicel	439	64
CMC	4659	647

[a] μg reducing sugar/mg. protein/hr./100 mg. substrate.

Table X. Effect of Gamma Irradiation on the Degradation of Alder Sawdust by CFE Prepared from Enrichment Cultures of Rumen Microorganisms

Irradiation Dosage (rads)	Reducing Sugars ug./ml.
0	79
4.85×10^6	86
4.0×10^7	199
7.52×10^7	190
1.11×10^8	175
1.46×10^8	256

In view of the rapidity with which assays and analyses can be done, CFE seem advantageous to *in vitro* fermentation tests to screen the efficiency of physical and chemical treatments of wood in making nutrients available to rumen microorganisms. The drawback to the use of CFE is that a preliminary washing of the irradiated material with water is necessary to remove the reducing substances liberated because of irradiation and to obtain lower blank values.

Acknowledgment

The authors wish to acknowledge the assistance of J. E. Breeze, Head, Engineering Department of the B. C. Research Council and of R. A. Sanders, Engineering Department, B. C. Research Council.

Literature Cited

(1) Abou Akkada, A. R., Eadie, J. M., Howard, B. H., *Biochem. J.* **89**, 268 (1963).
(2) Ammerman, C. B., Block, S. S., *J. Agr. Food Chem.* **12**, 539 (1964).
(3) Ayers, W. A., *J. Bacteriol.* **76**, 504 (1958).
(4) *Ibid.*, **76**, 515 (1958).
(5) Bailey, R. W., Clarke, R. T. J., *Nature* **199**, 1291 (1963).
(6) Beckman, E., *Sitzb. Kgl. Press Akad Wiss* 638 (1915); *Chem. Abstr.* **9**, 3309 (1915).
(7) Bhattacharya, A. N., Fontenot, J. P., *J. Anim. Sci.* **25**, 367 (1966).

(8) Bissell, H. D., Weir, W. C., *J. Anim. Sci.* **16**, 476 (1957).
(9) Bolker, H. I., *Nature* **197**, 489 (1963).
(10) Burdick, D., Sullivan, J. T., *J. Anim. Sci.* **22**, 444 (1963).
(11) Cason, J. L., Thomas, W. E., *J. Dairy Sci.* **38**, 608 (1955).
(12) Crampton, E. W., Maynard, L. A., *J. Nutr.* **15**, 383 (1938).
(13) Cullison, A. E., *J. Anim. Sci.* **20**, 478 (1961).
(14) Cuthbertson, D. P., "Progress in Nutrition and Allied Sciences," p. 434, Oliver and Boyd, London, 1963.
(15) Dehority, B. A., Johnson, R. R., *J. Dairy Sci.* **44**, 2242 (1961).
(16) Ely, R. E., Dane, E. A., Jacobson, W. C., Moore, L. A., *J. Dairy Sci.* **36**, 346 (1953).
(17) Fernley, H. N., *Biochem. J.* **87**, 90 (1963).
(18) Festenstein, G. N., *Biochem. J.* **69**, 562 (1968).
(19) *Ibid.*, **72**, 75 (1959).
(20) Gascoigne, J. A., Gascoigne, M. M., "Biological Degradation of Cellulose," p. 264, Butterworth, London, 1960.
(21) Gill, J. W., King, K. W., *J. Agr. Food Chem.* **5**, 363 (1957).
(22) Gonzaleza, B., *Anim. Inst. Invest. Vet. (Madrid)* **9**, 163 (1959).
(23) Haberlandt, G., *Sitzb Kgl. Press Akad.* 243 (1915); *Chem. Abstr.* **9**, 1516 (1915).
(24) Halliwell, G., *J. Gen. Microbiol.* **17**, 153 (1957).
(25) *Ibid.*, **17**, 166 (1957).
(26) Halliwell, G., Bryant, M. P., *J. Gen. Microbiol.* **32**, 441 (1963).
(27) Honcamp, F., Hilgert, H., *Landwirtsch Versuchs-Stat.* **113**, 201 (1931); *Nutr. Abstr.* **1**, 645 (1931).
(28) Hungate, R. E., "The Rumen and Its Microbes," p. 533, Academic Press, New York, 1966.
(29) Hvidsten, H., Homb, T., *Inst. Animal Nutr., Royal Agric. Coll., Norway* **62**, 14 (1947); *Nutr. Abstr.* **18**, 655 (1948-1949).
(30) Katz, M., Reese, E. T., *Appl. Microbiol.* **16**, 419 (1968).
(31) Kehar, N. D., *Ind. J. Vet. Sci.* **24**, 189 (1954).
(32) King, K. W., *J. Dairy Sci.* **42**, 1848 (1959).
(33) Kitts, W. D., Underkofler, L. A., *J. Agr. Food Chem.* **2**, 639 (1954).
(34) Klopfenstein, T. J., Bartling, R. R., Woods, W. R., *J. Anim. Sci.* **26**, 1492 (1967).
(35) Krupnova, A. V., Sharkov, V. J., *Gidrolizn. Lesokhim. Prom.* **16**, 8 (1963).
(36) Lampila, M., *J. Agr. Res. Centre, Helsinki* **2**, 105 (1963).
(37) Lawton, E. J., Bellamy, W. D., Hungate, R. E., Bryant, M. P., Hall, E., *Tappi* **34**, 113A (1951).
(38) Leatherwood, J. M., *Appl. Microbiol.* **13**, 771 (1965).
(39) Lehmann, F., *Deut. Zuckerind.* **40**, 317 (1915); *Chem. Abstr.* **9**, 3309 (1915).
(40) Mater, J., *Forest Prod. J.* **7**, 208 (1957).
(41) Mellenberger, R. W., Millet, M. A., Baker, A. J., Satter, L. D., Baumgardt, B. R., *J. Dairy Sci.* **51**, 975 (1968).
(42) Nehring, K., Schiemann, R., Hoffman, L., Klippet, W., Jentsch, W., "Energy Metabolism," K. L. Blaxter, Ed., Academic Press, New York, 1965.
(43) *Papierztg*, **33**, 444 (1908); *Chem. Abstr.* **2**, 1186 (1908).
(44) Pritchard, G. I., Pigden, W. J., Minson, D. J., *Can. J. Anim. Sci.* **42**, 215 (1962).
(45) Saeman, J. F., Millett, M. A., Lawton, E. J., *Ind. Eng. Chem.* **44**, 2848 (1952).
(46) Sahai, K., Johri, P. N., Kehar, N. D., *Ind. J. Vet Sci.* **25**, 201 (1955).
(47) Saarinen, P., Jensen, N., Alhojarvi, J., *Paperi ja PUU.* **40**, 495 (1958).

(48) Saarinen, P., Jensen, W., Alhojarvi, J., *Suomen Maataloustieteelisen Seuran Julkaisuja.* **94,** 41 (1959).
(49) Seifert, K., *Holz Roh-Werkstoff* **22,** 267 (1964).
(50) Stamm, A. J., "Wood and Cellulose Science," p. 355, The Ronald Press Company, New York, 1964.
(51) Stanley, R. N., Kesler, E. M., *J. Dairy Sci.* **42,** 127 (1959).
(52) Stewart, C. M., *Austral. Pulp and Paper Ind. Tech. Assoc. Proc.* **8,** 50 (1954).
(53) Stone, E. J., Girourd, R. E., Jr., Frye, J. B., Jr., *J. Dairy Sci.* **48,** 814 (1965).
(54) Stone, E. J., Morris, H. F., Jr., Frye, J. B., Jr., *J. Dairy Sci.* **48,** 815 (1965).
(55) Sullivan, J. T., *J. Anim. Sci.* **14,** 710 (1955).
(56) Sullivan, J. T., Hershberger, T. V., *Science* **130,** 1252 (1959).
(57) Tilley, J. M. A., Deriaz, R. E., Terry, R. A., *Proc. 8th Intern. Grassland Congr.* 533 (1960).
(58) Tilley, J. M. A., Terry, R. A., *J. Brit. Grassland Soc.* **18,** 104 (1963).
(59) Tomlin, D. C., Johnson, R. R., Dehority, B. A., *J. Anim. Sci.* **24,** 161 (1965).
(60) Tyler, T. R., Leatherwood, J. M., *Arch. Biochem. Biophys.* **119,** 363 (1967).
(61) Van Soest, P. J., *J. Assoc. Offic. Agr. Chemists* **46,** 829 (1963).
(62) Virtanen, A. I., *Nature* **158,** 795 (1946).

RECEIVED October 14, 1968. Supported by research grants from the National Research Council of Canada (A-132, A4913) and the Canada Department of Agriculture (OG 54).

18

The Value of Wood-Derived Products in Ruminant Nutrition

WILLIAM H. PFANDER, S. E. GREBING, GEORGE HAJNY, and WENDELL TYREE

Missouri Agricultural Experiment Station, Columbia, Mo. and Forest Products Laboratory, Madison, Wis.

Literature reviewed indicated that raw wood, as browse, and products such as leaves and beans can be used in ruminant rations. Raw woods are poorly utilized as energy sources. Cellulose prepared from wood is digestible, but particle size limits utility. Wood molasses and other hydrolytic products are well digested. Hemicellulose (HC) was obtained from mixed hard and soft woods. The distribution of the sugars was: galactose, 7.1; glucose, 14.4; mannose, 26.0; arabinose, 4.9; and xylose, 47.4. The HC was also neutralized and/or dried to 95% dry matter. All products tested at 30% of the ration were non-toxic to wethers for five years and to ewes through two reproductive cycles. The estimated digestible energy, in kcal./kg., for the products was: 65% DM from Laurel, 2900; 65% DM from Ukiah, 2420; 95% DM, 3100.

When an animal scientist looks at wood, he can elect to view it in several ways: (a) as an indigestible, highly abrasive material that increases fecal dry matter and nitrogen excretion, (b) as a mixture of cellulose and lignin from which, if, the lignin envelope could be removed, a ruminant could obtain a digestible cellulose, and (c) as a complex mixture consisting of (b) plus a rich supply of hemicellulose and a multitude of other compounds. In classical and current literature, there is support for each view. We will consider each in turn and then present in some detail the estimates of nutritional value and limitations of some derived products.

The bases for most ruminant rationing systems are rooted in the German schools of the 1800's. Variations of the Weende analytical scheme are included in AOAC (2) procedures and variations of the net energy scheme of Kellner (33) are used as the "ultimate" measure of energy yield of ruminant rations. Both systems have limitations and have survived numerous critiques.

Wood is a complex product of an unstable biological system. Its total chemical composition and the physical-chemical relationships remain to be elucidated. Parts of the tree are normally eaten at some seasons of the year by ruminants and other herbivores. Bark, browse, and leaves are the major items consumed (6, 19, 27, 28, 29). The reported composition of several components are shown in Table I.

Table I. The Composition and Nutritive Value of Wood Products

	DM %	Protein %	Fiber %	NFE %	DP %	Protein COD[b]	SE[c] TDN[d]	TDN[d]
Beech leaves	43	6.9	9.8	21.7	3.4	6	19.2	
Birch leaves	45	7.9	6.9	24.7	3.9		26.0	
Elm leaves	88.0	15.9	8.6	49.9	8.5	73	50.0	
Oak leaves, dried	94.0	9.3	29.9	45.3	0			17
Poplar leaves	84	10.8	17.4	39.6	3.4	56	26.7	
Brush wood	75	4.6	26.7	40.3	1.6		14.5	
Brush wood, beech	84.7	4.0			0.1		−12.9	
Acorns	85.0	5.7	11.6	61.6	3.8		69.0	
Locust beans	85.0	5.8	6.4	69.0	3.2		71.7	
Sawdust from pine	83.5	0.3	62.4	19.6			−3.3	
Molasses, wood	60.0	0.5	0	55.6	0			49
Molasses, cane[a]	74.0	3.0	0	62.0	0			54
Red clover hay[a]	83.5	15.3	22.2	35.8	7.0		35.6	54

[a] Common feeds for comparison.
[b] A reported value considerably different from that used to calculate the DP figure.
[c] Ref. 33.
[d] Adapted from Ref. 40.

Kellner's investigations which showed that fiber content limits the net energy value and that added sawdust depresses utilization of other components of the ration have discouraged research opportunities. Perhaps the most damaging data are those leading to the conclusion that the addition of 100 grams of sawdust had the same effect as removing 3.2 grams of starch (33).

A shortage of hays and concentrates in World Wars I and II stimulated research into the possibilities of preparing hydrolyzed sawdust or "fodder cellulose" from wood. Some reports showed good utilization of these products, but most of the work indicated that the value of the derived products was closer to that of sawdust than of hays (*3, 8, 22, 23, 25, 31, 38, 44, 47, 49, 50, 51*).

Generally, the rations used were not fortified with a full complement of required minerals, and some disturbances developed. More recently, sawdust has been used to limit feed intake or to serve as a source of "roughage" in the ration. Ten to 15% of sawdust can be fed without evidence of toxicity (*16, 45*). Nearly pure cellulose derived from wood has been used extensively in semi-purified diets (*20*). The digestibility of the product is limited in ruminants, but under *in vitro* conditions it is nearly 100% digestible if the reactions are continued for long periods (*36*). The availability of this product for microbial fermentation suggests that if an economical way of removing lignin were available, considerable wood would find its way into ruminant feeds. In fact, newspapers are now being used in experimental feeds (*34*).

Prior to and during World War II, research was initiated to develop a source of wood molasses for alcohol production. Later, these products of acid hydrolysis were fed to ruminants with good results (*5, 14, 15, 17, 24, 32, 35, 40, 41*).

Table II. Composition of Hemicellulose

	L 4%	L 43%	LHC 1	LHC 2
pH	3.80	4.04		4.08
Ash	0.15	1.52	2.17	2.17
Solids	4.00		64.00	64.00
Sugars (as is)	0.79	8.09		12.27
Sugars (after hydrolysis)	2.35	25.56	34.14	37.24
Methoxyl				
Lignin				
Acetyl			8.50	
Uronic Anhydride	0.35	2.00	3.75	
% of sugars				
As is				
Galactose	16.03	17.58		
Glucose	6.34	10.88		
Mannose	10.71	11.43		
Arabinose	45.09	31.95		
Xylose	21.83	28.16		
After hydrolysis				
Galactose	6.58	7.84	7.85	6.37
Glucose	15.54	14.19	14.34	14.57
Mannose	27.92	27.33	27.17	24.82
Arabinose	5.53	5.11	4.88	5.08
Xylose	44.43	45.53	45.76	49.16

[a] The following abbreviations are used: L—Laurel, Mississippi, Raw material about softwood; HC—Hemicellulose; N—Neutralized.

The main direction of research now appears to be toward the utilization of wood by-products. Wood generally contains from 50–60% cellulose, 20–30% lignin, 10–20% pentosans, 0–7% mannans and some acetyl linkages (usually 0.5–2% acetic acid) (48). This wide assortment of carbohydrates is of interest to ruminant nutritionists for at least three reasons. Ruminants need a source of readily available energy, usually obtained as hexoses derived from soluble carbohydrates. Most forages contain pentosans or hemicelluloses, but no reliable concentrated pure source of these materials is available and until recently it was not possible to obtain enough to evaluate the utility of these products in defined rations. Finally, the possibility of producing polymers from feed material is attractive.

Studies designed to determine the role of some of these by-products in ruminant rations are described in the sections which follow.

Methods and Experimental

The raw material was obtained as a by-product from the manufacture of hard board. In the usual process, logs are debarked and fed into a

Concentrates[a] **at Several Production Stages**

NLHC 1	NLHC 2	Dried LHC	NUHC	Dried UHC
7.00				
7.40		4.76	0.86	4.96
63.70	65.00	97.10	64.00	
9.78	11.64	18.72	24.50	18.08
31.03	34.98	58.40	42.48	56.26
1.92		3.06	1.14	2.41
5.88		5.35	0.92	7.44
7.09		7.36	4.26	9.13
4.15		6.86	2.32	5.76
9.66				
13.17				
12.55				
29.49				
35.13				
6.71	5.81	5.98	13.31	6.67
16.09	14.97	16.40	17.11	15.84
24.06	25.16	15.96	44.66	25.75
4.13	4.91	3.99	3.96	5.56
49.01	49.15	57.67	20.95	46.18

60% mixed hardwood and 40% pine; U—Ukiah, California, Raw material about 90%

chipper. The chips are subjected to live steam at 42 kg./cm.² pressure for one minute and washed in counter current washers. The washings contain about four percent dry matter. The washings are evaporated to 40% solids and then to 65%. The pH of the 65% solids product is about 3.7. It is referred to as hemicellulose (HC) (trade mark, Masonex, Masonite Corporation, Chicago, Illinois). If desired, the material may be neutralized (N) to pH 5.5 or spray dried. The typical raw material at the Laurel, Mississippi (L) plant contains 60% mixed hardwood and 40% pine. The Ukiah, California (U) plant processes about 90% softwood. Table II shows some typical compositions obtained by modifications of the methods of Saeman *et al.* (*46*) from products produced in various ways. Results of typical feed analyses are shown in Table III.

Palatability Trials. Two to five sheep were placed in a pen containing a feeder with four movable compartments. The compartments were numbered and their positions randomized each day. Two feeds were compared in each pen. Each day a known amount of feed was placed in each compartment and the position was determined at random. Twenty-four hours later the refusals were removed, weighed, and fresh feed added. A normal test period was seven days. The first trial was designed to determine the relative palatability of hemicellulose and molasses. The basal ration consisted of timothy hay, 68%; corn, 28.16%; 44% CP soybean oil meal, 2.31%; urea, 0.77%; dicalcuim phosphate, 0.26%; salt, iodized, 0.50%; and vitamin A and D concentrate to meet NRC requirements. Other rations were prepared by removing corn and increasing the soybean meal and adding 5% dry matter from cane molasses, LHC, dried molasses, or dried LHC.

Table III. Typical Results of HC Analyses According to Current Feed Control Methods

Component	HC	Dried HC
Protein	.5%	.5%
Fat	.5%	.5%
Fiber	1.0%	1.0%
Ash	6.0%	4.0%
Solids	65.0%	97.0%
Calcium	.50	.50
Phosphorus	.07	.07

The basal ration in Trials 2 and 3 was the same as in Trial 1 and the test materials were added at 5% of the ration dry matter with corresponding adjustments in corn and soybean meal to maintain isonitrogenous diets.

In Vitro **Studies.** A series of artificial rumen trials was run to test the effect of neutralization and hemicellulose source on normal cellulose digestion. The procedure used was that of Hargus (*21*).

Feeding Trials. TRIAL 1. Six lots of nine lambs each were used to compare hemicellulose and cane molasses. Each treatment was replicated three times. The rations used contained 66% ground ear corn, 16.6% mixed hay, 10.4% soybean oil meal, and 7% of either cane molasses or

HC. A mineral mixture was available. Gain, feed efficiency and carcass characteristics were determined.

TRIAL 2. Two lots of eight lambs each were used to determine the value of dried HC as a roughage. The basal ration contained 10% cottonseed hulls, corn, urea, minerals, Aureomycin and stilbestrol. The ration was calculated to supply 12.5% protein, 0.4% calcium, 0.3% phosphorus, 0.6% potassium, and 3.0% fat. In the test ration, 8% dried hemicellulose and 0.15% urea were substituted for 5% hulls and 3.15% corn.

About 100 pounds of long hay per lot was supplied during the first week the lambs were on test. From that time on, they were self fed their rations in sand covered lots that had automatic heated water cups.

They were killed after 66 days and hot carcass weights were obtained. Twenty-four hours later, cold carcass weights and grades were obtained.

TRIAL 3. Twelve percent of dried LHC replaced ten percent of cottonseed hulls in a fattening ration for lambs. The remaining ingredients were similar to those used in Trial 2.

Evaluation of Energy Availability. TRIAL 1. The rations shown in Table IV were designed to estimate the value of hemicellulose in rations for lambs and mature wethers. Each sheep received each ration for about 30 days. During the last 14 days of each period, animals were in digestion-metabolism stalls to determine nitrogen and energy use.

At the completion of each series the acceptability of rations was measured by the time required to change to the new treatment.

TRIAL 2. A 2 × 2 × 2 factorial design was used. The variables were hemicellulose from Laurel, neutralized or unneutralized and hemicellulose from Ukiah, neutralized or unneutralized, each fed at 15 and 30% of the ration. The rations are shown in Table V. There were four sheep per treatment. The standard procedure for digestibility trials as developed by the informal forage evaluation committee (30) was used. Each animal was fed individually on the ration for approximately three weeks during which time an estimate of the voluntary feed consumption was made and during days 21 to 30, a standard digestibility trial was completed.

Table IV. Composition of Rations Used to Evaluate Digestibility of HC, Trial 1

Ration	Basal %	Basal + Molasses %	Basal + HC %
Timothy hay	67.78	61.01	61.01
Shelled corn, ground	28.95	26.07	26.07
Soybean meal, 44%	1.81	1.61	1.61
Urea	.77	.69	.69
Dicalcium phosphate	.11	.10	.10
Ground limestone	.08	.07	.07
Salt	.50	.45	.45
Molasses	—	10.0	—
Hemicellulose	—	—	10.0
Vitamin A, 5000 IU/gm.	20 gms.	20 gms.	20 gms.
TOTAL	100.0	100.0	100.0

At the conclusion of the experiments, the animals were changed immediately to the next ration to be tested.

TRIAL 3. At the conclusion of the factorial design, four sheep received 50% Laurel hemicellulose and another group of four sheep received in sequence the high energy ration and then the high energy ration supplemented with 15 and 30% dried hemicellulose. The composition of the rations is shown in Table VI. Composition of the rations was determined by standard AOAC procedures and the energy content by methods previously reported by this station (21).

Table V. The Composition of Rations Used to Evaluate the Energy Value of Rations in Trial 2

Ingredient, %	Level of Hemicellulose			
	0	15	30	50
Cottonseed hulls	85.3	70.0	55.3	35.3
Hemicellulose	0	15.0	30.0	50.0

The rations all contained: soybean meal, 13%; dicalcium phosphate, 0.7%; salt, iodized, cobaltized, 0.5%; Vitamin A and D premix, 0.3%; sodium propionate (as a mold inhibitor), 0.2%.

Table VI. Composition of Rations Used to Evaluate Dried Hemicellulose

Ingredient, Kg	Level of Dried Hemicellulose		
	0	15	30
Corn, ground	955.00	809.00	655.00
Urea, 42% N	13.80	17.73	21.63
Hemicellulose, dried	—	150.00	300.00

Tests to Determine Long Term Effects. In 1963 wethers were placed on rations containing 30% HC. They have been maintained in confinement and have received no other foods.

Sixteen yearling ewes were allocated into two equal groups and fed the rations shown in Table VII. They were mated to Hampshire rams in September of each year and carried through two gestation–lactation periods in drylot. Observations were made on the general performance of the ewe and on the vigor and growth rate of the lambs produced. During the second year, lambs were killed and detailed tissue samples were obtained for further evaluation. Electrophoretic patterns on the ewe's blood and milk were obtained.

Results

Palatability. Results are shown in Table VIII. As the sheep increased in size, their feed intakes also increased. The various lots of sheep were not always consistent in their preference. Lot D seemed most sensitive and G least sensitive to ration variables. Some of each ration was always consumed and even the relatively unpalatable ration (4) was consumed in adequate amounts when the animals had no other choice. Sensitivity may have decreased as the experiment progressed. The sheep in these trials preferred hemicellulose and dry molasses. In two groups, D and G, there was a marked preference for molasses instead of the basal ration. No group expressed a strong preference for hemicellulose as compared to the preference for the basal ration. However, wet hemicellulose was well accepted by all lots in all comparisons.

The results of Trial 2 are shown in Table IX. In this test, the unneutralized product was more palatable than the neutralized and the basal ration was preferred to either product.

Results of Trial 3 are shown in Table X. If only overall averages are examined, the lambs appear to prefer the unneutralized products, but this is not a true indication of the values obtained. In the first place, the intake of these lambs declined as the weather became warm in July and August, and the unneutralized products were fed early in the trial. When the dry products were compared, the sheep did prefer the unneutralized product; but, of the wet products, they preferred the neutralized material.

In Vitro Trials. There were no significant effects on cellulose digestion where HC was added at levels of 1–15% to the standard assay system.

Feeding Trials. TRIAL 1. The data are summarized in Table XI. There were no differences between the ration containing HC and that containing molasses.

TRIAL 2. The results are shown in Table XII. The performance of the dried HC lot is equal or superior to the control.

Table VII. Composition of Rations Used for Reproduction Studies

Ingredients, Kg	Ration 1	Ration 2
Alfalfa, meal, reground pellets	100	100
Cottonseed hulls	500	500
Ground corn	208.5	206.5
Soybean meal	90	90
Cane molasses	75	—
LHC, 65% solids	—	75
Dicalcium phosphate	5	5
Iodized, cobaltized salt	5	5
Limestone	2	2
Sodium sulfate	0.5	2.5
Vitamin A & D—2250, 400 IU/gm.	1.0	1.0
Urea, 262 plus [a]	13.0	13.0
TOTAL	1000.0	1000.0

[a] Contains 45% N, product of E. I. du Pont.

Table VIII. The Palatability of Rations Containing Comparison

Groups	Rations 2 vs. 3 kg.	kg.	Rations 4 vs. 5 kg.	kg.	Rations 3 vs. 5 kg.	kg.	Rations 1 vs. 2 kg.	kg.
D	15.2	10.4	2.5	25.1	19.1	13.9	13.5	18.4
E	14.1	9.4	7.1	22.8	13.0	19.3	15.2	13.0
G	16.8	14.0	14.9	19.6	19.1	21.5	15.7	20.7
H	10.3	14.5	5.0	23.2	17.7	17.0	14.3	15.0
I	16.0	8.7	8.7	19.6	13.7	17.4	14.3	15.2
Sum	72.4	57.0	38.2	110.3	82.6	89.1	73.0	82.3
Mean/group	14.5	11.4	7.6	22.1	16.5	17.8	14.6	16.4
Mean daily/head	0.7	0.5	0.4	1.0[a]	0.8	0.8	0.7	0.8

Ration No. 1 = Basal
Ration No. 2 = Basal plus dried cane molasses
Ration No. 3 = Basal plus cane molasses
Ration No. 4 = Basal plus dried hemicellulose

Table IX. Intake of Feeds With and Without Hemicellulose

Lot	NL vs. UNL		Basal vs. NL		Basal vs. UNL	
D	19.7	19.5	21.1	15.6	25.2	11.0
E	18.7	21.4	20.6	9.4	19.2	19.0
G	15.6	15.2	23.7	9.3	19.4	21.2
H	17.3	20.5	23.5	15.2	23.8	14.1
I	9.0	10.4	27.0	8.5	15.6	10.4
Sum	80.3	87.0	115.9	58.0	103.2	75.7
Mean[a]	16.0	17.4	23.1	11.6	20.6	15.1

[a] In kg/lot/week.
NL = Neutralized Laurel
UNL = Unneutralized Laurel

TRIAL 3. The results are shown in Table XIII. The dried HC has apparently replaced all "roughage" in the lamb fattening ration.

In both direct comparisons, the lambs fed hemicellulose gained faster and more efficiently than those fed cottonseed hulls. Representative carcasses are shown in Figure 1. It is obvious that the hemicellulose fed lambs are equal or superior to the controls.

Evaluation of Energy Availability. TRIAL 1. The results are shown in Table XIV. No differences are significant ($P > .05$). The slightly

Five Percent Dry Matter from Molasses or Hemicellulose
Comparison

Rations 3 vs. 4		Rations 1 vs. 5		Rations 4 vs. 4		Rations 2 vs. 5		Rations 1 vs. 4	
kg.	kg.	kg.	kg.	kg.	kg.	kg.	kg.	kg.	kg.
30.6	3.7	22.0	17.3	16.2	17.1	21.3	23.8	20.3	12.9
13.5	17.1	15.8	17.7	15.8	16.4	22.0	14.3	22.7	20.4
15.0	21.5	23.2	22.2	18.5	20.5	24.5	18.8	12.8	18.6
17.8	15.7	19.1	16.4	16.6	17.2	23.0	14.8	18.6	18.9
14.4	13.6	16.7	16.8	14.8	16.6	19.1	15.0	27.8	10.0
91.3	71.6	96.8	90.4	81.9	87.8	108.9	86.7	102.2	80.8
18.3	14.3	19.4	18.1	16.4	17.6	22.0	17.3	20.4	16.2
0.9	0.7	0.9	0.9	0.8	0.8	1.0[b]	0.8	0.9	0.8

Ration No. 5 = Basal plus hemicellulose
[a] Significant difference between pairs, student's "t" < 0.01.
[b] Significant difference between pairs, student's "t" < 0.05.

lower protein digestibility of the molasses and HC rations is characteristic of such rations supplemented with carbohydrates.

TRIALS 2 AND 3. The results of Trials 2 and 3 are shown in Table XV, and the analysis of variance of the individual data are reported in Table XVI. There was a highly significant difference in level and source of hemicellulose, but no effects from neutralization. In Trial 3, there was a marked depression of digestibility at the 30% level of dried hemicellulose. It was calculated by the difference procedure that the digestible energy of the Laurel hemicellulose is 2900 kcal./kg., and of the Ukiah hemicellulose 2420 kcal./kg. An interaction between source and level complicates the interpretation of these data. Dried hemicellulose when fed at 15% of a high energy ration is estimated to provide 3100 kcal./kg., but the 30% level depressed digestibility and reduced the digestible energy content of the ration.

Long Term Effects of Feeding HC. The wethers have completed five years on their rations. No losses have occurred.

There were no differences in the reproductive performance of the two groups of ewes and no unusual disturbances were noted in any situation.

Discussion

The literature reviewed indicates that while wood is regarded as a possible component in emergency rations, it is not highly regarded for animal production. However, it has been stated that, "probably more

Table X. Intake of Lambs Fed Laurel

	July 2–7 Dry vs. Wet			July 27–Aug. 2 Wet vs. Wet		
Group	HC	HC	Total	HC	NHC	Total
G	1631	1483		1393	1568	
E	1403	1416		1226	1304	
I	1743	1430		1258	1912	
H	1735	1855		1438	1798	
Sum	6512	6184		5315	6582	
Mean[a]	1628	1546	3174	1329	1645	2974
Total, all dry		5460				
Total, all wet		6016				
Total, all neutralized		5492				

Table XI. Result of Feedlot Performance According to Kind of Ration

	Molasses	Hemicellulose
No. of lambs	27	27
Av. initial wt. (kg.)	35.49	35.48
Av. finished wt. (kg.)	47.01	47.01
Days on Feed	63	63
Av. daily gain (kg.)	.18	.18
Feed/kg. gain	4.09	4.11
Av. carcass grade	9.4 C+	9.2 C+
Av. carcass yield, %	52.88	52.83

Table XII. Lamb Performance when Dried Hemicellulose Replaces Part of the Cottonseed Hulls in a Lamb Finishing Ration

	Control	Dried LHC[a]
No. lambs	8	8
Starting weight, av.[b]	30.700 kg.	31.400 kg.
Finish weight, av.[b]	48.900 kg.	52.200 kg.
St. Louis weight, av.[c]	47.900 kg.	50.200 kg.
Carcass weight, av.[d]	25.600 kg.	26.900 kg.
Ave. daily gain	.275	.311
Percent shrink	1.740	3.650
Yield, percent	53.460	53.550
Grade	8	8
Condemned livers[e]	4	3
Feed efficiency	5.150	5.320

[a] Spray dried.
[b] Weighed after feeding or full weight.
[c] Weighed after 18 hours without feed. Used to calculate percent shrink to market and yield.
[d] Cold carcasses about 22 hours after killing. Used to calculate yield.
[e] "Saw dust" and/or tapeworms in controls. Tapeworms in HC lot.

Hemicellulose Processed in Different Ways

Aug. 16–21 Dry vs. Dry			Aug. 22–27 Dry vs. Wet		
HC	NHC	Total	NHC	NHC	Total
1468	1093		1212	1408	
1418	1193		963	1374	
1693	1206		1433	1408	
1343	1243		1062	1792	
5922	4735		4670	5982	
1481	1183	2664	1168	1496	2664

Total, all unneutralized 5984
[a] Each value is average of 3 lambs per lot.

Table XIII. The Value of Substituting Dried HC for Cottonseed Hulls in High Grain Rations

Lot #	Variable	ADG	Feed Efficiency
10	HC	.190	6.6
11	Control	.185	7.6
9	HC	.212	6.1
7	Control	.178	7.8

Figure 1. Representative carcasses from lambs fed hemicellulose (999 and 042) and cottonseed hulls (937 and 965)

Table XIV. Observations of Sheep on Digestion-Metabolism Trials to Determine the Value of HC and Molasses in a Maintenance Ration

	Basal	Basal + Molasses	Basal + HC
Av. daily gain, gm.	92.10	73.40	86.10
Feed effic., gm./gm. gain[a]	17.32	16.45	15.16
Time to change to ration, days	5.30	4.80	4.80
Digestible energy, % of gross	60.90	59.20	58.50
Digestible protein as %	64.90	61.30	58.50

[a] Not adjusted for DM content of rations.

animals feed on shrubs and trees, or on associations in which shrubs and trees play an important part, than on true grasses or grass, legume pastures, short and tall grass ranges, and steppes" (27, 28, 29). Perhaps we are either trying to use the wrong woods or have not properly treated or supplemented it. Limitations of previous rations are probably due in part to an inadequate mineral supply. With our current knowledge of the more detailed mineral requirements and mode of interactions, it is possible to provide the basic supplements which are needed for wood derived rations. Of particular interest are the relatively low levels of potassium, calcium, and some of the trace elements in wood products (39).

The second limitation in the use of wood is that microbial fermentation of wood products is slow. While it has been shown that Solka Floc (trade mark of the Brown Paper Co., Montreal, Quebec) and filter paper powder can be completely digested if exposed to rumen fluid for 72 hours, the particle size of these is such that they are likely to pass from the

Table XV. Composition of Rations and Coefficients

	DM		E.E.		Fiber	
Ration	%	COD[a]	%	COD[a]	%	COD[a]
07 Basal High Fiber	90.2	40.1 ± 3.2	0.9	76.0 ± 1.3	38.7	46.5 ± 3.1
39 + 15% LHC	90.1	47.6 ± 1.1	0.8	73.4 ± 0.5	33.5	42.8 ± 1.2
41 + 15% NLHC	89.8	46.1 ± 3.4	0.7	66.8 ± 4.8	31.6	37.1 ± 3.9
01 + 15% UHC	90.3	42.0 ± 2.0	0.6	60.9 ± 1.4	31.8	34.2 ± 2.3
03 + 15% NUHC	89.9	45.5 ± 2.3	0.6	48.9 ± 9.2	30.9	36.3 ± 2.2
40 + 30% LHC	90.3	52.6 ± 1.0	0.7	57.0 ± 3.2	27.7	40.1 ± 2.0
42 + 30% NLHC	89.4	54.7 ± 0.7	0.5	45.4 ± 0.8	26.7	40.5 ± 1.2
02 + 30% UHC	91.5	50.6 ± 1.3	0.7	71.3 ± 6.6	26.2	34.3 ± 1.2
04 + 30% NUHC	91.2	49.5 ± 1.0	0.6	62.0 ± 4.6	27.2	36.2 ± 2.1
29 + 50% LHC	70.9	51.8 ± 1.4	1.1	66.4 ± 6.6	14.5	21.3 ± 0.8
1 Basal Low Fiber	90.8	92.5 ± 0.4	3.8	90.9 ± 0.8	1.8	69.2 ± 4.7
2 + 15% Dried LHC	91.6	87.5 ± 1.9	4.8	88.5 ± 1.6	1.9	38.0 ± 13.2
3 + 30% Dried LHC	91.9	81.3 ± 2.1	3.5	84.3 ± 2.4	2.2	35.4 ± 14.7

[a] Mean ± standard deviation.

rumen before the complex hydrolytic processes are completed. Thus, most of these fine particles escape before being digested. The opportunity seems to present itself for developing a tailor-made roughage product from wood residues. Such products could either be inert or have relatively high nutritional value, depending upon the economics of production, distribution, and the comparative prices of competitive products.

The use of the hemicellulose fraction of wood offers great promise. Some of the results reported herein have been reported in preliminary reports (43) and confirmed in part in other species (1, 4, 8, 39, 42). Hemicellulose can be ammoniated (37). Other wood by-products have been tested with generally favorable results (18). Other reports of this symposium cover these aspects.

One should be alert to the possible toxicity of products fed at high levels. Such toxicity could be caused by other extractives from wood or to substrate inhibition. Repolymerization may occur if the products are heated (Table XV). However, our trials indicate no toxicity at generally acceptable levels of carbohydrate addition, and the dried product may serve as a roughage replacement.

A basic problem in ruminant nutrition which still needs additional work is the area of adaptation to rations and the possibility of providing specific microorganisms to handle the components in the mixture. While the addition of starch to ruminant rations has been shown to cause a marked shift in bacterial populations (26) similar information is missing on the build-up and adaptation to the addition of other carbohydrates. Highly soluble carbohydrates generally lead to a much more acid rumen and to the destruction of a number of protozoa. Fermentation patterns

of Digestibility of Rations Used in Trials 2 & 3

Crude Protein		Ash	NFE		TDN	GE
%	COD[a]	%	%	COD[a]	%	COD[a]
9.2	42.0 ± 1.9	4.4	37.0	31.2 ± 4.6	34.9 ± 2.8	
9.1	42.7 ± 1.2	4.0	42.7	52.4 ± 1.9	41.9 ± 1.1	47.0 ± 1.0
9.5	48.4 ± 3.3	4.7	43.3	51.5 ± 3.1	39.7 ± 2.8	45.6 ± 2.9
8.4	42.5 ± 0.8	3.8	45.7	47.6 ± 2.2	37.1 ± 1.8	40.7 ± 2.6
8.8	43.7 ± 1.3	4.2	45.4	52.1 ± 3.3	39.3 ± 2.0	41.8 ± 1.8
9.2	52.5 ± 0.9	4.0	48.7	60.0 ± 1.2	46.1 ± 0.9	51.5 ± 0.8
8.9	53.0 ± 2.4	5.4	47.9	62.3 ± 1.0	45.9 ± 0.6	53.2 ± 1.1
8.8	47.2 ± 1.1	3.9	51.9	59.6 ± 1.7	45.2 ± 1.3	49.4 ± 1.1
8.6	42.5 ± 5.8	4.8	50.0	57.0 ± 1.0	43.0 ± 0.9	48.8 ± 0.9
6.3	41.9 ± 3.0	5.3	43.7	62.1 ± 1.8	34.6 ± 1.1[b]	
14.3	87.4 ± 0.7	3.2	67.8	95.5 ± 0.3	86.9 ± 0.3	
12.8	78.7 ± 4.1	2.6	69.6	91.7 ± 1.3	84.7 ± 1.7	
14.7	75.2	3.6	67.9	85.4 ± 1.7	76.5 ± 1.8	

[b] Equivalent to 44% TDN on 90% dry matter basis.

Table XVI. Analysis of Variance of the TDN in Hemicellulose Rations

Source	DF	SS	MS	F
Sheep	3	2.13		
Area (Ar)	1	41.63	41.63	16.20[b]
Acid (AC)	1	3.19	3.19	
Level (L)	1	243.65	243.65	94.80[b]
Ar x L	1	30.01	30.01	11.68[b]
Error	23	59.18	2.57	
Total	30[a]	379.79		

[a] One degree of freedom used to calculate missing value.
[b] $p < 0.01$.

may also be changed. Thus, it appears that the organisms involved are constantly available in the population and it merely takes the presence of large amounts of starch to induce the changes. Cellulolytic organisms possibly require more negative oxidation-reduction potentials than those produced by the average ration. Some artificial means of reducing redox potentials might result in a better adaptation of rumen microorganisms.

Summary

The literature reviewed indicates that raw wood, as browse, and products such as leaves and beans can be used in ruminant rations, and may be the most important energy source for animals. Sawdust and other raw woods are poorly utilized as energy sources but may be used at the 10% level to replace hulls or roughage in high grain rations. Cellulose prepared from wood is digestible, but particle size limits utility. Wood molasses and other hydrolytic products are well digested.

Derived products containing hemicellulose were shown to be palatable and to provide utilizable energy for ruminants. Hemicellulose from various sources and treated by various methods was calculated to yield from 1800 to 3100 kcal. DE/kg. The drying of these products apparently formed some polymer which reduced the utility of the ration at high levels. The possibility of toxic effects when rations are supplied at the 15% level appears remote. Hemicellulose when supplying over 30% of ration dry matter and when dried or neutralized reduces palatability and lowers digestibility.

Acknowledgments

This paper is a contribution from the Missouri Agricultural Experiment Station, Columbia, Missouri and U.S.D.A. Forest Products Laboratory, Madison, Wisconsin.

The authors acknowledge the technical assistance of: Orval Lewis in care of management of experimental animals; Wayne Reichert in bomb calorimetry; C. W. Gehrke and staff for AOAC analysis; Harold Hedrick and staff for carcass data; R. L. Preston for computer programs used in data analysis; J. F. Saeman and staff for sugar analysis; Dale Galloway, Dale Turner, and T. R. Edgerton for many fruitful discussions leading to the designs of some of the trials reported.

The projects reported were supported in part by grants from Masonite Corporation, Chicago, Illinois. Supplies used were furnished by International Minerals, Chicago, Illinois; Calcium Carbonate Corporation, Quincy, Illinois; Hoffman La Roche, Nutley, New Jersey; Dawes Laboratory, Chicago, Illinois; and Thompson-Hayward, Kansas City, Missouri.

Literature Cited

(1) Algeo, J. W., Brannum, T. P., Hibbits, A. G., *Proc. Western Sec. Am. Soc. Animal Sci.*, 19th, Santa Ynez, California (1968).
(2) AOAC, "Official Methods of Analysis," Washington, D. C., 1965.
(3) Archibald, J. G., *Massachusetts Exp. Stat. Bull.* **230**, 160 (1926).
(4) Bartley, E. E., Farmer, E. L., Pfost, H. B., Dayton, A. D., *J. Dairy Sci.* **51(5)**, 706 (1968).
(5) Bergius, F., *J. Soc. Chem. Ind.* **52**, 1045 (1933).
(6) Bonsma, J. C., *Farming in S. Africa* **17**, 226 (1942).
(7) *Ibid.*, **17**, 259 (1942).
(8) Boren, F. W., Pfost, H. B., Smith, E. F., Richardson, D., *Kansas Agr. Exp. Sta. Bull.* **B 483** (1965).
(9) Breirem, K., *Papirjournalen* **28**, 101 (1940).
(10) *Ibid.*, **28**, 113 (1940).
(11) *Ibid.*, **28**, 119 (1940).
(12) *Ibid.*, **28**, 133 (1940).
(13) *Ibid.*, **28**, 152 (1940).
(14) Bunger, *Landwirtsch. Versuchs-Stat.* **126**, 1 (1936).
(15) Burkitt, Wm. H., Lewis, James K., Van Horn, J. L., Willson, Fred S., *Agr. Exp. Sta. Bull.* **498**, Bozeman, Montana (1954).
(16) Cody, R. E., Jr., Morrill, J., Hibbs, C. M., *J. Dairy Sci.* **51 (6)**, 952 (1968).
(17) Colovas, N. F., Keener, H. A., Prescott, J. R., Teeri, A. E., *J. Dairy Sci.* **32**, 907 (1949).
(18) Cullison, A. E., Ward, C. S., *J. Animal Sci.* **24**, 877 (1965).
(19) De Souza, A. J., *Bol. Minist. Agr. Brasil* **32**, 1 (1943).
(20) Ellis, W. C., M.S. Thesis, Univ. of Missouri, Columbia, Missouri (1955).
(21) Hargus, W. A., Ph.D. Thesis, Univ. of Missouri, Columbia, Missouri (1962).
(22) Herbst, Walter, *U. S. Patent* **2269665** (1936).
(23) Holtan, E. M., Holtan, M. D., *Papirjournalen* **28**, 157 (1940).
(24) Honcamp, F., Hilgert, H., Wohlbier, W., *Biochem. Ztschr.* **248**, 474 (1932).
(25) Hvidsten, H., Homb, T., *Proc. 11th Intern. Congr. Pure Appl. Chem.*, London, England **3**, 13 (1947).
(26) Hungate, "The Rumen and Its Microbes," Academic Press, New York, 1966.

(27) Imperial Agricultural Bureaux, *Imp. Agr. Bur. Joint Publ.* **No. 10**, viii (1947).
(28) *Ibid.*, **No. 10**, 231 (1947).
(29) *Ibid.*, **No. 10**, xxxiv (1947).
(30) Informal Committee on Forage Evaluation, unpublished data (1965).
(31) Jarl, F., *Kgl. Lantbruksakad. Tidskr.* **82**, 57 (1943).
(32) Jones, I. R., *Oregon Cir.* **181** (1949).
(33) Kellner, O., "The Scientific Feeding of Animals," The Macmillan Company, New York, N. Y., 1913.
(34) Kesler, E. M., Chandler, P. T., Branding, A. E., *J. Dairy Sci.* **239** (1967).
(35) Keyes, E. A., *Montana Cir.* **202** (1953).
(36) Lyons, D. T., M.S. Thesis, Univ. of Missouri, Columbia, Missouri (1958).
(37) Magruder, N. D., Knodt, C. B., *J. Dairy Sci.* **36**, 581 (1953).
(38) Mangold, E., Bruggemann, H., Theel, E., *Landwirtsch. Jahrb.* **78**, 649 (1933).
(39) Moore, J. D., M.S. Thesis, Univ. of Oklahoma, Norman, Oklahoma (1968).
(40) Morrison, F. B., "Feeds and Feeding," 22nd Ed., The Morrison Publishing Company, Ithaca, N. Y., 1956.
(41) Peet, H. S., Ragsdale, A. C., *Missouri Agr. Exp. Sta. Bull.* **605** (1953).
(42) Perry, T. W., Hiller, R. J., Shepard, J. P., Beeson, W. M., *Purdue Agr. Exp. Sta. No. 14, Re. Prog. Rep.* **153** (1964).
(43) Pfander, W. H., Ross, C. V., Preston, R. L., Vipperman, P. E., Tyree, W., *Univ. of Missouri, Suppl. Special Rep. Bull.* **38** (1964).
(44) Poijarvi, I., *Acta agral. fenn.* **57**, No. 1, 1 (1944).
(45) Preston, R. L., Kruse, C. G., unpublished data (1968).
(46) Saeman, J. F., Moore, W. E., Mitchell, R. L., Millett, M. A., *Tappi* **37** (8) (1954).
(47) Stahl, *Ztschr. Schweinezucht* **43**, 471 (1936).
(48) Stephenson, J. N., Ed., "Pulp and Paper Manufacture," Vol. 1, 3rd ed., McGraw-Hill Book Company, Inc., New York, N. Y., 1950.
(49) Virtanen, A. I., Koistinen, O. A., *Svensk kem. Tidskr.* **56**, 391 (1944).
(50) *Wisconsin Agr. Exp. Sta. Bull.* **323**, Annual Report (1920).
(51) Woodward, T. E., Converse, H. T., Hale, W. R., McNulty, J. B., *USDA Bulletin* **1272**, 9, 1924.

RECEIVED October 14, 1968. Journal Series No. 5479. Approved by Director (College of Agriculture) Univ. of Missouri.

19

Hardwood Sawdust as Feed for Ruminants

W. B. ANTHONY, JOHN P. CUNNINGHAM, JR. and R. R. HARRIS

Department of Animal Science, Alabama Agricultural Experiment Station, Auburn University, Auburn, Alabama 36830

> *Ruminant animals can utilize cellulosic materials as food because of a valuable symbiotic relationship with microorganisms present in the rumen section of their digestive tract. Cellulose pulps and wood waste have been fed as maintenance rations to cattle and horses during times of great national emergencies. Another potential use for waste wood is in rations for fattening cattle. Feeding trials were conducted with cattle and sheep using a basal, all-concentrate formula, with and without sawdust and oyster shell. Feeding treatments were: basal (B), B plus 2.5% of oyster shell, B plus 2.5% of oak sawdust, and B plus 10% of oak sawdust. Both lambs and steers made the best gain when fed B plus 2.5% sawdust. In another test with yearling steers, the basal mixture with 10% of oak sawdust produced equal gain with equal efficiency to the basal with 10% of ground coastal bermudagrass hay. These studies show value for sawdust in rations for fattening cattle and sheep.*

Cellulose is the most abundant organic substance occurring in the world. It forms the skeletal framework of all higher plants. In most plant materials, cellulose does not occur in a pure form, but is encrusted with other cell wall constituents, notably lignin (2, 35). The encrusting of cellulose by lignin and other cell wall constituents greatly affects the ease with which it can be degraded by microorganisms (7, 28). Because of the abundance of cellulosic materials and because the ruminant animal can use cellulose as a source of food through the action of rumen microorganisms (8, 16, 26), interest has long been directed toward improving cellulosic materials as feed for ruminants. In the late 1800's Kellner illustrated the value of cellulose as food for the ruminant (20, 21, 22). Kellner observed that the absorption products from one pound of digested cellulose were equal, for fattening of mature cattle, to those derived from one pound of digestible starch. Later studies on the digestion by ruminants of specially prepared cellulose material have confirmed Kellner's hypothesis (*18, 19, 35*). In 88 digestion experiments with cattle fed straw treated with 1.2 to 1.5% NaOH, Hvidsten (*19*) obtained an average digestion coefficient of 87 for the organic matter. Digestion coefficients for the cellulose obtained from spruce wood by the sulfite method averaged 95 (*18*). Much of the earlier work directed toward ascertaining the value of cellulosic materials for ruminants was reviewed by Anthony (*2*).

Wood fiber, although primarily cellulose, is encrusted with other cell wall constituents so that it is not degraded by rumen microorganisms unless it is modified by some chemical or physical process. In many European countries, notably Germany, researchers have chemically treated wood to render it more useful as an animal feed (*30*). Saarinen, *et al.* (*29*) made an intensive study of birch pulps as feed for sheep. Alkaline pulps were more digestible than were acid pulps or chlorite pulps. The objective of Saarinen's work was primarily to develop animal feed from small size wood and other low-grade wood materials. As reported by Saarinen, cellulose pulps were 80% or more digestible by ruminants if the lignin content was lowered to 10% or less.

This introduction reveals that much interest has been directed over a long time toward the utilization of various cellulosic materials as food for ruminants. Except in times of emergency (*9, 10, 18, 19, 24*), however, there does not appear to be a significant amount of wood waste used in feeding livestock.

In almost all instances where wood waste has been fed, its use was primarily to provide a maintenance ration for animals. As an example, during World War II the 1.5 million tons of sulfate and sulfite pulps fed to cows and horses in the Scandinavian countries was an emergency program to keep animals living until a more favorable time (*18, 24*). Since wood waste and pulps are relatively low in available food energy, it is obvious that their value for livestock would be greatest when they were used in maintenance rations. Indeed, an experiment reported by Ellis and Durham in 1968 (*15*) illustrated the use of ground mesquite wood in a wintering ration for pregnant beef cows.

Wood Waste-Roughage Substitute in Rations for Finishing Cattle

Another potential use for waste wood might be in rations for fattening cattle. For this purpose wood waste would not be expected to furnish energy, but rather aid in the feeding of all-concentrate rations.

Currently over 10 million cattle are finished for slaughter annually in the United States. Most of these are fed in lots having a capacity of 1000 head or more. These large concentrations of cattle place a strain on the available supply of roughage for ration formulation. Consequently, the trend has been to minimize the amount of roughage placed in the ration. It is not surprising, therefore, that finishing cattle on all-concentrate rations has been much publicized. An abundant supply of feed grains also has encouraged the use of grain instead of roughage in the finishing of cattle for slaughter, Durham et al. (13, 14).

Although cattle can be finished for slaughter on rations formulated without roughage (34), they usually perform better when fed high roughage rations (6, 32). Wise et al. (34) recently reviewed results of finishing cattle on all-concentrate rations. Incidence of bloat, founder, rumen parakeratosis, and abcessed livers caused significant problems where cattle are fed all-concentrate rations (1, 17, 27, 34). Inclusion of a small amount of ground hay in the ration largely overcomes these nutritional problems.

Because of the extreme difficulty of putting ground hay in cattle fattening rations, roughage substitutes have been sought. Oyster shell has been used without much success (25, 32). The availability of a large supply of wood waste suggested research to find a use for it in rations for fattening cattle. It has been estimated that in the several states drained by the Tennessee River and known as the Tennessee Valley Area there is annually available more than 400,000 tons of wood residue in the form of sawdust, shavings, and other products. Presently this large supply of wood residue has no economic use. It derives primarily from oak. In cooperation with the Forest Products Branch of the Tennessee Valley Authority, research was begun at Auburn University to ascertain if oak sawdust could be used as a roughage substitute in cattle and sheep fattening rations.

A series of feeding trials using sheep and cattle were conducted using oak sawdust. In addition, a measurement was made on the influence of the digestive process upon wood fiber. The feeding trials will be described and results reported in the order in which the tests were run.

Oak Sawdust vs. Oyster Shell. Feeding trials were conducted with lambs and yearling cattle to compare oak sawdust and "hen size" oyster shell in all-concentrate rations. The basal formula used in these trials is given in Table I. The formula provides a proper balance of protein, minerals and vitamins. The calcium and phosphorus ratio is near 1:1. Although calcium is lower in the mixture than might be expected, experience from many feeding tests reveal that more calcium would not improve the steer gain. Feeding treatments were (1) basal, (2) basal plus 2.5% oyster shell, (3) basal plus 2.5% oak sawdust, and (4) basal plus 10% oak sawdust. For the lamb feeding trial, a group of 12 lambs (mean weight 23.4 kg.) was allotted to four groups of 3 animals each and each group was fed *ad libitum*. For the cattle feeding trial, 48 yearling steers (mean weight 341.8 kg.) were allotted to 4 pens of 12 animals each. The feeding plan for the cattle was the same as described for the sheep. At the end of the feeding trials, the lambs and cattle were sold for slaughter and their rumens examined for parakeratosis. Also, digestibility data were obtained for feeds used in the four feeding treatments. Conventional digestibility data were obtained by use of yearling steers in two trials with two animals per feed in each trial. In addition, digestibility information was obtained by use of the small sample *in vivo* technique described by Lusk, *et al.* (23) and by the *in vitro* procedure described by Tilley, *et al.* (33). Cellulose was determined by use of the procedure developed by Crampton, *et al.* (12).

Table I. Basal Feed Formula

Ingredient	Per cent
Ground shelled corn	71.0
Salt, trace mineralized	1.0
Cottonseed meal (41%)	12.5
Defluorinated phosphate	0.5
Cane molasses	10.0
Condensed fermented corn extractives	5.0
Vitamin A	2203 IU/Kg. feed
Vitamin D	551 IU/Kg. feed
Aureomycin	70 mg./day/animal

Table II. Sheep Performance for Oak Sawdust Feeding[a]

Treatment	Average Initial Weight (kg.)	Average[b] Daily Gain (kg.)	Feed Dry Matter /kg. Gain (kg.)
Basal (B)	23.2	.147	6.48
(B) + Oyster Shell (2.5%)	22.7	.186	5.36
(B) + Oak Sawdust (2.5%)	24.1	.212	5.32
(B) + Oak Sawdust (10%)	23.2	.173	6.68

[a] Differences were not statistically significant.
[b] Days in test—136.

Results of the sheep trial are presented in Table II. The pen of lambs fed basal plus 2.5% oak sawdust made greater daily gain than the other pens and used the least feed per unit of gain. These differences, however, were not statistically significant. Including either oak sawdust or oyster shell appeared to improve animal performance over the basal (all-concentrate) ration. When the lambs were slaughtered, the rumens from all treatment groups were free of parakeratosis.

Rates of gain for treatments in the cattle feeding trial were more nearly equal than in the sheep trial, Table III. The highest average rate

Table III. Performance Data for Yearling Steers Fed Rations Containing Oak Sawdust

Treatment	Average Initial Wt. on Test (kg.)	Average[a] Daily Gain (kg.)	(kg.) Dry Matter /kg. Gain
Basal (B)	349.1	1.10	7.45
(B) + Oyster Shell (2.5%)	337.7	1.07	6.76
(B) + Oak Sawdust (2.5%)	340.5	1.12	7.74
(B) + Oak Sawdust (10%)	339.5	1.08	7.98

[a] Days on test—126.

of gain was produced by cattle fed the basal plus 2.5% oak sawdust. In contrast to the lamb trial, the lowest feed per unit of gain was for the pen of cattle fed oyster shell in the basal. Perhaps this was because this pen consumed a little less feed daily than other pens. Also, in contrast to data for lambs, rumen parakeratosis was present for a major portion of cattle in all treatment groups, Table IV. It is important to observe, however, that the most severe form of parakeratosis, hornified, did not occur in animals fed either level of oak sawdust, Table IV.

Table IV. Influence of Diet on Rumen Parakeratosis and Condemned Liver[a]

Number of Observations

Treatment	Papillae Clumped	Horni- fied[b]	Abcessed	Inflamed	Normal	Hair Balls	Livers Con- demned
Basal (B)	8	2	7	1	2	4	2
(B) + Oyster Shell (2.5%)	10	5	5	2	0	3	3
(B) + Oak Sawdust (2½%)	10	0	6	1	0	5	1
(B) + Oak Sawdust (10%)	7	0	6	0	4	0	2

[a] Cattle data—12 per treatment.
[b] Term used to identify extreme clumping of Papillae to a hard surface.

The relatively high dry matter digestion coefficients, Table V, emphasize the fact that tests rations were essentially concentrate except for the inclusions of oyster shell and sawdust. The conventional digestibility trial data and the *in vitro* data rank the rations in a similar order. No adequate explanation can be given for the low coefficient for the basal by the nylon bag technique. Although the small sample *in vivo* and the

Table V. Dry Matter Digestibility

Treatment	Conventional[a]	Nylon Bag[b]	Tilley-Terry[c]
Basal (B)	85.78	70.85	69.76
(B) + 2.5% Oyster Shell	83.56	74.95	69.91
(B) + 2.5% Oak Sawdust	81.28	76.10	67.02
(B) + 10% Oak Sawdust	71.46	75.25	61.93

[a] Mean of 2 trials; 2 animals per trial.
[b] Three bags in each of 2 fistulated steers; 24-hour incubation; values adjusted by use of standard forages.
[c] Values are not adjusted by use of standard samples.

Table VI. Conventional Cellulose Digestibility

Ration	Cellulose in Feed %	Cellulose Digested[a] %
Basal (B)	9.38	55.87
(B) + 2.5% Oak Sawdust	11.75	52.95
(B) + 10% Oak Sawdust	16.71	43.16

[a] Mean of two animals.

Table VII. Wood Waste Digestibility by Nylon Bag Technique

Kind of Wood Waste	Dry Matter Digestibility %
Dry mill waste through 3/16" screen	3.75
Dry mill waste through 1/8" screen	5.86
Wet chips through 1/2" screen	0.85
Wet chips through 5/16" screen	1.93
Wet chips through 1/4" screen	1.77
Wet chips through 3/16" screen	2.40

Table VIII. Feed Formula

Ingredient	%
Coastal hay or hardwood sawdust	10.0
Liquid supplement (30% protein)	10.0
Ground shelled corn	77.0
Soybean meal	2.5
Defluorinated phosphate	0.5
Stilbestrol	1.102 mg./kg.

in vitro procedures revealed important differences among the rations, the coefficients obtained by these procedures were much lower than coefficients obtained by animal balance trials. This emphasized the importance of using reference samples of feeds of known digestibility in small sample *in vivo* and *in vitro* tests. For the nylon bag data reported, forages of low and high digestibility were included in each run and data for these were used to adjust coefficients for the experimental feeds. Reference concentrate feeds were not available for use.

Cellulose digestibility declined with the additions of sawdust, Table VI. This was expected since extensive nylon bag digestibility data for various waste wood materials including oak sawdust revealed apparent dry matter digestion coefficients below 20, usually less than 5, Table VII.

The results of these feeding trials with lambs and yearling steers show oak sawdust to be a useful roughage substitute for use in rations constituted primarily of concentrates. Rumen parakeratosis in cattle was not prevented when the ration contained 2.5% of oak sawdust, but its incidence was materially reduced when 10% of sawdust was used. Adding 10% of sawdust to the ration for sheep and cattle did not adversely affect feed efficiency. Therefore, there was a net improvement in feed efficiency in terms of the concentrate portion of the ration.

Sawdust vs. Coastal Bermudagrass Hay. Coastal bermudagrass hay is a normal ingredient in cattle rations in the Southeast. The hay is a valuable ingredient in the feed mixtures, but major management problems are involved in grinding, storing, and conveying to mixing equipment. Sawdust could replace Coastal hay and eliminate some management problems if it supported equal animal performance. To compare sawdust and Coastal hay, tests with lambs and yearling cattle were conducted. In the lamb feeding trial the rations contained either 10% sawdust or 10% ground Coastal hay, Table VIII. One pen of eight lambs was fed each ration (mean initial weight of lambs was 33.1 kg.). For the cattle feeding trial, the test rations contained either 15% sawdust or 15% ground Coastal hay, Table IX. One pen of 10 yearling steers was fed each mixture (mean initial weight of cattle was 356.8 kg.). At the end of the trials the sheep and cattle were slaughtered and their rumens examined for parakeratosis.

Lambs fed the sawdust-containing ration had an average daily gain equal to lambs fed the coastal-containing ration and feed efficiency was similar, Table X. No palatability difference was observed and there was no sorting of the sawdust. The lamb trial was conducted during the summer and high ambient temperature adversely affected performance of both pens of lambs. Examination of the rumens showed no incidence of parakeratosis.

Table IX. Coastal Hay or Oak Sawdust in Steer Fattening Feed Formula[a]

Ingredient	%
Ground shelled corn	71.0
Cottonseed meal (41%)	12.5
Coastal hay or oak sawdust	15.0
Salt, trace mineralized	1.0
Defluorinated phosphate	0.5
Vitamin A	2203 IU/kg. feed
Vitamin D	551 IU/kg. feed
Aureomycin	70 mg./animal/day

[a] Animals received 36 mg. DES ear implant.

Table X. Oak Sawdust vs. Coastal Hay in Lamb Fattening Rations

Ration	Average Daily Gain[a] (kg.)	Feed/Unit Gain (kg.)
Coastal	.092	5.93
Oak sawdust	.095	6.20

[a] Mean of eight animals per treatment for 51 days.

The feeding trial with cattle using 15% sawdust was conducted specifically to determine if rumen parakeratosis could be prevented. Previous tests had tended to show some merit for including sawdust, but rumen parakeratosis was not entirely prevented when sawdust made up 10% of the diet. Results of the trial are shown in Table XI. Cattle fed sawdust did not grow as rapidly as those fed coastal hay. Also, the appearance of the sawdust-fed cattle was noticeably different from the coastal-fed animals. Rough hair coats of the sawdust-fed cattle distinguished them from the coastal-fed animals. The sawdust-fed animals consumed slightly less daily feed than the coastal-fed cattle. On examinations made at time of slaughter, the cattle fed coastal hay were free of rumen parakeratosis. In contrast, all the sawdust-fed animals showed some evidence of parakeratosis. However, there was no incidence of severe, hornified, parakeratosis.

In this section, therefore, the lamb feeding trial showed oak sawdust to be equal to ground coastal hay when each was used at 10% of the ration. This is a significant finding because oak sawdust can be put in a mixed ration with much less effort than can coastal bermudagrass hay. The hay requires field curing, storage, and grinding before it can be used. The sawdust received no special care in that it was brought directly from the sawmill and put in the feed.

For cattle in this test, 15% oak sawdust in the ration apparently was too much. Also, rumen parakeratosis was not prevented by this level of sawdust. In the trial reported above, a ration containing 10% of sawdust lowered the incidence of rumen parakeratosis and improved feed efficiency over an all-concentrate diet. Therefore, these data seem to indicate that sawdust can be added to steer rations with favorable results up to a level of 10% of the mixture. Perhaps fully to control rumen parakeratosis some other dietary factor will be required along with sawdust.

Table XI. Steers Fed Rations Containing 15% of Either Oak Sawdust or Coastal Hay

Item	Hay	Sawdust
Animals	10	10
Initial weight, kg.	339	339
Final weight, kg.	438	424
Weight gain, kg.	99	85
Days on feed	85	85
Av. daily gain, kg.	1.16	1.00
Feed/kg. gain, kg.	9.13	9.88
Daily feed, kg.	10.6	9.9
Ration as % body weight	2.72	2.59
Carcass grades:		
Choice, No.	2	1
Good, No.	8	9
Carcass yield, %	57.33	57.26

Sawdust-Protein-Mineral-Vitamin Pellet for Co-feeding with Whole Shelled Corn. Sawdust has a relatively low bulk density and poor flow characteristics in a conventional feed bin. Large cattle feedlots usually incorporate feed processing equipment designed for bulk conveying to mixing equipment. In its usual form, sawdust does not meet the requirements of a feed ingredient for cattle feed processing equipment. In order to make sawdust more useable for cattle feeders, the product was included as an ingredient in a protein, mineral, vitamin supplement. The total mixture was pelleted. The pellet formula was designed to be fed with whole shelled corn in the ratio of 85 parts of corn to 15 parts of the pellet. Fed at this level to yearling steers the pellet would provide the minimum daily requirements of protein, minerals, and Vitamin A. The formula of the pellet is given in Table XII. This pellet has been fed with corn to two groups of steers. The animals remained on a high level

Table XII. Formula of Wood Pellet Containing Vitamins, Minerals, and Protein for Co-feeding with Corn

Ingredient	Per cent
Urea	8.00
Cane molasses	10.00
Dicalcium phosphate	1.80
Ground limestone	.50
Salt (trace mineralized)	3.70
Soybean oil meal, 50%	12.00
Dry distillers solubles	4.00
Hardwood sawdust	60.00
Vitamin A to be added:	1,000,000 units/cwt.
Aureomycin:	350 mg./cwt.

Table XIII. Wood Waste—a Diluter for Corn Self-Fed[a]

Ingredient	%
Ground ear corn	68.0
Hardwood sawdust	30.0
Defluorinated phosphate	1.0
Salt	1.0

[a] Protein offered as liquid.

of feed intake without incidence of founder. The rate of gain and feed efficiency of the cattle compared favorably with comparable cattle fed conventional rations based on unreported data from this laboratory.

Further testing is in progress. But it is already clear that wood waste can be effectively used for feeding cattle by incorporating it in a pellet. The pellet provided the ration roughage factor and also the protein, minerals, and vitamins properly to balance a corn diet. By pelleting the sawdust its bulk handling property was greatly improved and it could be readily handled in standard feed mixing and conveying equipment.

Oak Sawdust as a Diluter for Corn Self-fed to Cattle. Frequently it is advantageous to limit intake of cattle on a concentrate mixture to prevent rapid growth or fattening. Wood waste appeared useful as a diluent in this type feeding program. Cody, et al. (11) used rations containing 25–45% wood fiber to control grain intake by dairy calves. A test was conducted using the feed formula listed in Table XIII. This feed mixture contains sufficient minerals for short term feeding. For extended feeding, it would be necessary to add trace minerals and Vitamin A. A group of "newly arrived" yearling steers was offered the feed mixture and a liquid protein supplement *ad libitum* for several weeks. The feed mixture was uniformly consumed and there was no selective refusal of oak sawdust. Physical condition of the cattle improved while on the feed and the feeding program was convenient and efficient for holding the new cattle until they were placed on scheduled finishing rations. This test, therefore, illustrates another value for wood waste in cattle feeding operations.

Table XIV. Sugar Extracted from Washed Feces Fiber of Wood Waste-fed Cattle

		% Total Sugar Extractables	
Treatment	Animal No.	Ethyl Alcohol	3% H_2SO_4
Basal (B)	1	.307	15.18
	2	.209	16.66
(B) + Oak Sawdust (2.5%)	3	4.97	22.44
	4	.734	23.38
(B) + Oak Sawdust (10%)	5	1.86	23.02
	6	1.19	10.65
Oak Sawdust only—		.781	5.68

Table XV. 3% H_2SO_4 Extractable Sugar[a]

	Individual Sugar[b]				
Treatment	Galactose	Glucose	Mannose	Arabinose	Xylose
Basal (B)	11.51	29.76	—[c]	13.24	45.49
(B) + Oak Sawdust (2.5%)		27.19[d]	10.11	20.91	38.58
(B) + Oak Sawdust (10%)		24.91[d]	5.42	8.52	63.86
Oak Sawdust	—[c]	11.08	15.51	17.10	64.87

[a] Mean of two animals per treatment.
[b] Individual sugars determined by paper chromatography.
[c] Sample lost.
[d] Galactose and glucose combined.

Change in Wood Fiber During the Digestive Process. Although wood waste used in steer finishing rations adds little or no energy (7), it might be made more bio-degradable during passage through the digestive tract. This could be important if the significant advantage of "wastelage" (3, 4, 5) is used. To test the changes made in wood fiber by the digestive process fecal fiber was washed and extracted (1) with ethanol and (2) with hot, 3% H_2SO_4. The results are summarized in Tables XIV and XV. Oak sawdust that had not been subjected to the digestive process yielded only 5.68% sugar by extraction with hot, 3% H_2SO_4. The high relative values for sugar extraction from fecal oak sawdust suggests that the fiber was changed during passage through the digestive system of the cow. A large portion of the sugar component was xylose, Table XV, which further suggests that the oak sawdust was partially hydrolyzed after the digestive process whereas it was only slightly solubilized by the same treatment before it passed through the digestive tract. This change in degradability of wood waste has important implications in organic waste disposal and in systems of waste re-use as livestock feed (31).

Summary

1. Oak sawdust proved more valuable than oyster shell as a "roughage factor" in all concentrate mixtures for slaughter cattle and sheep.
2. At a level of 10% of the total ration, oak sawdust proved equal in value to ground coastal bermudagrass hay.
3. Oak sawdust was effectively used in a pellet formulated for feeding with whole shelled corn.
4. Oak sawdust (30% mixture) effectively served as a diluter for *ad libitum* feeding a corn ration to cattle held for later finishing.
5. Oak sawdust is changed by the digestive process so that it is more susceptible to hydrolysis by 3% hot H_2SO_4.

Literature Cited

(1) Albin, R. C., Durham, R. M., *J. Animal Sci.* **26**, 85 (1967).
(2) Anthony, W. B., *Thesis, Cornell University, Ithaca, New York* (1953).
(3) Anthony, W. B., *J. Animal Sci.* **27**, 289 (1968).
(4) Anthony, W. B., "Proceedings National Symposium, May 5, 6, and 7, 1966," pp. 109-112, American Society of Agricultural Engineers, St. Joseph, Michigan, 1966.
(5) Anthony, W. B., Nix, Ronald, *J. Dairy Sci.* **45**, 1538 (1962).
(6) Anthony, W. B., Harris, R. R., Starling, J. G., *J. Animal Sci.* **20**, 941 (1961).
(7) Baker, Frank, *Sci. Progr.* **134**, 287 (1939).
(8) Barcroft, J., McAnally, R. A., Phillipson, A. T., *J. Exptl. Biol.* **20-21**, 120-129 (1943-45).
(9) Breirem, Knut, *Papir-J.* **28**, 101 (1940).
(10) *Ibid.*, **28**, 113 (1940).
(11) Cody, R. E., Jr., Morrill, J. L., Hibbs, C. M., *J. Dairy Sci.* **51**, 952 (1968).
(12) Crampton, E. W., Maynard, L. A., *J. Nutrition* **15**, 383 (1938).
(13) Durham, R. M., Zinn, Dale W., *J. Animal Sci.* **22**, 837 (1963).
(14) Durham, Ralph, Rasberry, J., Arnodt, B., *J. Animal Sci.* **27**, 1163 (1968).
(15) Ellis, Cotton, Durham, Ralph, *J. Animal Sci.* **27**, 1132 (1968).
(16) Elsden, S. R., *J. Exptl. Biol.* **22-23**, 51 (1945-47).
(17) Harbaugh, F. G., Ellis, G. F., Durham, R. M., Stovell, R., *J. Animal Sci.* **22**, 860 (1963).
(18) Hvidsten, H., *Norges Landbrukshogsk.* Foringsforsok., **60**, 172 (1950). Abstr. *Nutrition Abs. and Revs.*, **19**, 470 (1950).
(19) Hvidsten, H., Homb, T., *Royal Agr. Coll. Norway*, Repr. No. **62**, 14 (1947), Abstr. *Nutr. Abs. and Revs.*, **19**, 655 (1950).
(20) Kellner, O., *Chem. Ztg.* **23**, 828 (1899), Abstr. *Expt. Sta. Rec.* **11**, 770 (1899-1900).

(21) Kellner, O., "The Scientific Feeding of Animals," p. 80, Macmillan Co., Ltd., Toronto, 1914.
(22) Kellner, O., Kohler, A., *Landw. Vers. Stat.* **53,** 474 ((1900), Abstr. *Expt. Sta. Rec.* **12,** 1071 (1900-1901).
(23) Lusk, J. W., Browning, C. B., Miles, J. T., *J. Dairy Sci.* **45,** 69 (1962).
(24) Nordfeldt, S., *Pure and Appl. Chem.* **3,** 391 (1947).
(25) Perry, T. W., Troutt, H. F., Peterson, R. C., Beeson, W. M., *J. Animal Sci.* **27,** 185 (1968).
(26) Phillipson, A. T., *Nutrition Abs. and Revs.* **17,** 12 (1947).
(27) Powell, D., Durham, R. M., Gann, G., *J. Animal Sci.* **27,** 1174 (1968).
(28) Prjanischnikow, J. E., Tomme, M. F., *Biedermann's Zentr. B. Tierernahr.* **8,** 104 (1936), Abstr. *Nutrition Abs. and Revs.* **6,** 231 (1936-37).
(29) Saarinen, R., Jensen, W., Alhojarvi, J., *Paperi Ja Puu-Papper och Tra* **40,** 495 (1958).
(30) Schorgor, A. W., "The Chemistry of Cellulose and Wood," p. 496, McGraw-Hill, New York, 1926.
(31) Singh, Yugal Kishore, Anthony, W. B., *J. Animal Sci.* **27,** 1136 (1968).
(32) Starling, J. G., Anthony, W. B., *J. Animal Sci.* **26,** 208 (1967).
(33) Tilley, J. M. A., Terry, R. A., *J. British Grassland Soc.* **18 (2),** 104 (1963).
(34) Wise, M. B., Harvey, R. W., Haskins, B. R., Barrick, E. R., *J. Animal Sci.* **27,** 1449 (1968).
(35) Woodman, H. E., Stewart, J., *J. Agr. Sci.* **22,** 527 (1932).

RECEIVED October 14, 1968.

20

Effect of Urea Supplementation on the Nutritive Value of NaOH-treated Oat Straw

E. DONEFER, I. O. A. ADELEYE, and T. A. O. C. JONES[1]

Department of Animal Science, Macdonald College (McGill University), Prov. Quebec, Canada

> Ground oat straw was treated with a 13.3% NaOH solution (at a rate of 60 liters of solution per 100 kg. straw), neutralized with 16.7 liters of 50% acetic acid, and dried. Untreated or treated straw, each with or without urea (2.5% of ration) and/or sucrose (3.5% of ration) was fed ad libitum to sheep. Alkali treatment resulted in significant increases in energy digestibility, but had no consistent effect on voluntary intake unless urea was also fed. Urea supplementation of treated straw resulted in an average 160% increase in voluntary intake when compared with untreated, unsupplemented controls. Sucrose had no significant effect on either digestibility or voluntary intake. NaOH-treated, urea supplemented straw supplied 220% more digestible energy than the control ration.

The work of Woodman and Stewart (15) drew attention to the effect of lignification on limiting digestion of plant nutrients by the microorganisms of the rumen. Freudenberg (9) has aptly described lignin as being "comparable with the cement in reinforced concrete." The action of lignin in filling the intercellular spaces of the plant accounts for its ability to limit the availability of potential plant nutrients such as holocellulose and protein. Whereas the gross energy of less mature forage plants can be utilized (digested) to the extent of 50–70%, increased lignification with advancing plant maturity can result in only 20–50% of the potential energy being made available to the animal. Equally or of more importance than the effect of lignification in decreasing digestibility is the marked reduction in voluntary intake by the ruminant of plant

[1] Present address: Njala University College, Freetown, Sierra Leone.

material of advanced maturity. As a result of both these effects the ruminant animal is unable in many cases to obtain its maintenance energy requirements when fed highly lignified forage plants.

The utilization of low-quality forages takes on special importance because of present attempts and the clear necessity to increase world food supplies. A major factor limiting production of high-quality proteins through development of animal agriculture is the shortage of suitable feedstuffs for these animals. This problem is particularly severe in tropical areas as the climate encourages rapid plant growth and subsequent over-maturity from a nutritional point of view. In many areas, particularly in developing countries, highly lignified plant by-products such as straws, grain hulls, and sugar-cane bagasse, although available in large supply, are not efficiently utilized and in many cases are completely wasted.

The use of chemical procedures to effect delignification have been elaborated particularly through efforts of the pulp and paper industry. Efforts to increase the nutritive value of low-quality forages such as straw date back to the beginning of this century and were highlighted by the procedure reported by Beckmann in 1921 (3) in which treatment of straw with 1.5% NaOH solution effected a twofold increase in the amount of crude fiber utilized by ruminant animals. While the Beckmann process has proved popular in small-scale farm operations, wide-scale use has been limited owing to the large volume of dilute NaOH solutions required, the tedious washing operation to remove residual alkali, and the losses of soluble nutrients caused by washing. Recent work in Finland (13) and Canada (14) have been directed towards modification of the chemical delignification procedures with a view to their wider applicability in low-quality forage improvement schemes. The newer modifications which tend to overcome use of excess chemicals and water have been demonstrated to be as effective in increasing digestibility as the older procedures. In order to reduce chemical costs and to increase labor efficiency, delignification procedures might best be adapted to large-scale operations taking advantage of bulk chemical prices and utilization of grinding, mixing, and perhaps pelleting equipment.

The present experiments were undertaken to investigate modified NaOH-treatment procedures and to determine the effect of nitrogen supplementation on both the digestibility and voluntary intake of untreated and treated straw.

Experiment I. The Effect of NaOH Level and Dilution on the In Vitro Digestion of Oat Straw Cellulose

Methods. The effect of various NaOH levels and dilutions on cellulose digestion was determined by use of an *in vitro* rumen fermentation procedure (7) with bacterial inoculum (phosphate buffer extract) prepared from rumen contents obtained from a fistulated steer (11). The ground oat straw (*Avena sativa*) was treated in 50-gram batches in a 1000-ml. beaker. The appropriate NaOH solution was mixed with the straw and allowed to react for 24 hours, having observed in preliminary trials that a longer reaction period (five days) resulted in only slight increases in cellulose digested *in vitro*. Levels of NaOH used varied from 4 to 32 grams expressed per 100 grams of straw, with each NaOH level tested at three different dilutions, 30, 60, and 120 ml. of solution per 100 grams of straw. Actual solution concentrations thus varied from 3.3% (4 grams NaOH per 12 ml.) to 53.3% (32 grams NaOH per 60 ml.). After treatment the straws were dried and weighed into fermentation tubes with *in vitro* cellulose digestion determined after a 24-hour fermentation period. Prior to the introduction of the bacterial inoculum, acetic acid was added to each tube to adjust the pH of the NaOH-treated straw to neutrality.

Results and Discussion. The effect of the different NaOH levels and dilutions on *in vitro* cellulose digestion are illustrated in Figure 1.

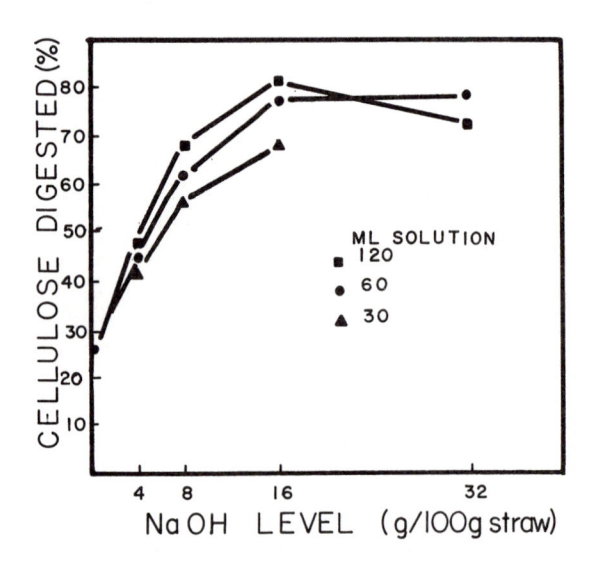

Figure 1. The effect of NaOH level and dilution on in vitro *cellulose digestion*

Ml. solution
■ 120
● 60
▲ 30

The effect of NaOH level was to increase cellulose digestion from the 25% observed for the untreated straw to a maximum of 81% as obtained with 16 grams NaOH in 120 ml. solution. At each treatment level, with the exception of 32 grams NaOH, the effect of the higher water dilution was to cause slight increases in cellulose digested, most possibly because of better mixing of the NaOH and straw and a more complete reaction at the higher moisture levels. At the 32 gram NaOH level, the levelling-off with 60 ml. of solution and the depression observed at the 120 ml. solution level is interpreted as due to an anti-microbial effect of the high salt levels contained in the substrate, as unreacted NaOH was not removed by washing although its effect on pH was neutralized by the acetic acid additions.

As the purpose of this *in vitro* experiment was to establish treatment guide-lines for subsequent animal feeding trials, the practicality of the various NaOH levels and dilutions was stressed. Ideally, both the amount of NaOH and water should be minimized in large-scale treatment procedures. Excess alkali either must be removed by washing as in the classical Beckmann procedure or alternatively, if neutralized by acid, the resultant sodium levels must not overload the animal's ability to maintain acid-base equilibrium (assuming acetate will be metabolized in the rumen). Use of large amounts of water should also be avoided as the final treated material would have to be dried prior to storage.

The combination of NaOH solutions and straw as used in this *in vitro* study resulted in mixtures varying in moisture content from 30 to 60% with no unabsorbed liquid present. In contrast, the Beckmann procedure (3) called for steepage of chopped straw in eight times its weight of 1.5% NaOH solution, with the treated material to be washed free of alkali and fed wet or dried. Although the strength of the NaOH solution used by Beckmann was low, the amount of NaOH used per 100 grams of straw was relatively high (12 grams) owing to the large amount of solution used. Lampila (13) proposed the use of 300 ml. of 2% NaOH solution per 100 grams straw, which is 6 grams NaOH per 100 grams straw or half the amount used in the Beckmann procedure. Lampila reported the *in vivo* digestibilities for crude fiber of his treated straw were similar to those obtained by Beckmann. Wilson and Pigden (14) reported the effect of NaOH levels up to 15 grams per 100 grams straw on the *in vitro* dry matter digestion of wheat straw. Their solutions which were all used at a level of 30 ml. per 100 grams straw, and thus quite concentrated, resulted in a maximum of 82% dry matter digested at their highest NaOH level.

As a result of the work cited and the *in vitro* trial herein reported, it was decided to prepare straw for *in vivo* trials by treating with 8 grams NaOH and 60 ml. of solution per 100 grams of straw. These levels although not resulting in the highest cellulose digestion in the *in vitro* experiment, minimized the quantity of alkali and water used with an anticipated marked improvement in the nutritive value of the straw (to approximately 60% cellulose digestibility).

Experiment II. The Effect of NaOH Treatment and Pelleting on the Nutritive Value of Oat Straw

Methods. TREATMENT OF STRAW. Baled oat straw harvested in 1965 was prepared for treatment by grinding in a hammer mill (Davis CR-2) (H. C. Davis and Sons Manufacturing Co., Bonner Springs, Kansas) to pass successively a 3 cm. and 0.35 cm. diameter mesh screen. The ground straw (100 kg.) was placed in a horizontal batch mixer (Davis S-20) and 60 liters of a 13.3 NaOH % solution (8 kg. NaOH per 100 kg. straw) was added to the straw while the mixer blades were turning. The NaOH solution was contained in two large polyethylene carboys placed on top of the mixer, with the solution slowly dispensed onto the straw from a spigot fitted on each container. The solution was administered over a 15-minute period with a total mixing time of one hour. The resultant mixture was left in the mixer for 24 hours, after which time 16.7 liters of 50% acetic acid was added and the contents mixed for one hour. The final mixture had a pH of 6 and a dry matter content of 60%. The treated straw was dried by spreading in a 5–10 cm. layer on a floor, using fans to increase air circulation.

Preparation of rations. In this experiment, straw was fed as the sole feedstuff with the animals having free access to water and salt (NaCl) containing cobalt and iodine. Treated and untreated (control) straw was fed in either the ground or pelleted form. Pellets were prepared by passing slightly moistened ground straw through the 7/16-inch (1.1 cm.) die of a pellet mill (Templewood Junior Provender Press, Distributed by Superior Processing Equipment, Hopkins, Minn.). The pellets, averaging 2 cm. in length, were air dried prior to feeding.

Animals. Eight female Cheviot lambs (initially 8–10 months old) were confined to individual digestion cages, designed to enable total collection of feces. Each feeding period was of 21-day duration, the first 14 days allowing for adjustment to the ration, with voluntary intake and digestibility measured over the last seven days. Voluntary intake was measured by feeding daily an amount of ration to insure an excess of at least 10% over expected intake, and determining actual intake by daily weighing of refused feed. A total collection of feces was made daily over the seven-day period with aliquot samples dried and retained for analysis. Four sheep were fed each of the rations.

Analysis and Calculation. Feed and fecal samples were analyzed for gross energy (Parr Oxygen Calorimeter), cellulose (7), dry matter, and crude protein (1). Apparent digestion of dry matter, cellulose, and energy was calculated as the difference between nutrient intake and fecal excretion, expressed as a percent of nutrient intake.

Relative Intake (%) was calculated as an expression of voluntary intake per unit animal metabolic size (gram dry matter/ sheep weight$_{kg}^{0.75}$) expressed relative to a "standard" high quality forage assumed to have a voluntary intake of 80 grams dry matter/weight$_{kg}^{0.75}$, according to the method of Crampton et al. (5). Nutritive value indices (NVI) were calculated as the product of relative intake and percent digestibility of energy (5) and represent a relative score of digestible energy (DE) intake which was also converted to absolute DE intake (DE_{kcal}/weight$_{kg}^{0.75}$) by multiplying each NVI by the feed gross energy content (kcal./grams) divided by 1.25 (6).

Statistical significance between treatment means was determined by Duncan's Multiple Range Test (8).

Results and Discussion. Experiment II results are summarized in Table I. NaOH treatment of the ground unpelleted straw resulted in significant increases ($P < 0.5$) in digestion in all cases except for cellulose, which approached significance. The depression in digestion because of pelleting largely counteracted the effect of the NaOH treatment. NaOH treatment had no effect on relative intake, which was significantly increased in each case because of pelleting. The decreased digestibility and increased relative intake because of pelleting resulted in almost identical nutritive value indices and digestible energy intakes for all ration combinations. Although digestion of the treated ground straw was significantly increased, a significant ($P < .05$) but unexplained depression in voluntary intake of this ration annulled any of the effect of the improved digestibility in terms of increasing energy intake. In this experiment the total ration was restricted to straw in order to study the effect of chemical and physical treatment on the microbial degradation of the forage uncomplicated by interactions with other nutrients. The results suggested nutrient deficiencies in the straw other than supply of energy were inhibiting rumen microbial action.

Table I. The Effect of NaOH Treatment and Pelleting on the Nutritive Value of Oat Straw (Experiment II)

Criteria:	Treatment: Form:	Untreated Ground	Pelleted	NaOH-treated Ground	Pelleted
Digestion Coefficients: (%)					
Cellulose		43.9[a,b]	38.2[a]	57.6[b]	34.6[a]
Dry Matter		36.4[a]	31.8[a]	54.4[b]	41.0[a]
Gross Energy		33.7[a]	28.4[a]	49.0[b]	34.3[a]
Relative Intake (%)		44.3[b]	57.7[c]	32.5[a]	49.4[b]
Nutritive Value Index		15.0[a]	16.4[a]	16.4[a]	16.9[a]
Digestible Energy Intake (kcal./Wt$_{kg}^{0.75}$)		53.1[a]	57.3[a]	53.7[a]	55.5[a]

[a,b,c] Means on the same line having different superscript letters differ significantly ($P < .05$).

Experiment III. The Effect of NaOH Treatment and Urea Supplementation on the Nutritive Value of Straw

Methods. An additional feeding trial was conducted utilizing straw remaining from Experiment II. Animals and all procedures were identical to the previous experiment. In order to test the possible effect of nitrogen deficiency as a factor limiting the energy utilization of NaOH-treated straw, a level of 2.5% feed grade urea (41.9% N) was added to both untreated and treated ground straws with three sheep fed each of the treatment combinations. Pelleting was dispensed with in this trial because the prior results did not appear to justify this method of preparation.

Results and Discussion. Composition and analysis of the rations used in Experiments II and III are presented in Table II. To be noted is the reduction of the gross energy content of the treated forages due to the diluting effect of the noncombustible NaOH. The ash content of the straw dry matter increased from an initial 5.7% to 13.8% after NaOH treatment and acid neutralization, with the decreased gross energy and cellulose content directly attributable to the increased ash content of the ration. It is of interest to note that the only abnormal symptoms observed in the case of sheep receiving the NaOH-treated straw was an approximate doubling of water intake and a concomitant increase in urine excretion. Urine pH rose from 7.9 to 8.6, and a slight rise in rumen pH from 6.8 to 7.2 was noted as an effect of the consumption of the NaOH-treated straw. There were no other observed clinical manifestations of the sodium stress imposed on the animal. The striking effect of the urea addition in increasing the crude protein content of the ration can also be noted in Table II.

Table II. Composition of Rations—Experiment II, III

Ingredients (%)	Untreated Straw	Treated Straw	Untreated Straw + Urea	Treated Straw + Urea
Straw—untreated	100	—	97.5	—
Straw—NaOH-treated	—	100	—	97.5
Urea (262% crude protein equivalent)	—	—	2.5	2.5
Analysis (Dry Matter Basis)				
Gross energy (kcal./g.)	4.36	4.04	4.38	4.02
Crude protein (%)	4.2	3.9	12.4	11.0
Cellulose (%)	42.8	40.1	42.1	38.9

Experiment III digestion and intake results are summarized in Table III. In all cases, NaOH treatment significantly increased cellulose, dry matter, and gross energy digestion coefficients. Digestibility coefficients obtained in this experiment were higher in all comparable cases than those observed in Experiment II, this difference being possibly caused by variations in the supply of straw. Also to be noted in the two *in vivo* experiments is the markedly higher cellulose digestion obtained with untreated ground straw (44.9 and 54.3%) as compared with 25% cellulose digestion observed *in vitro* (Experiment I). This observation might be attributed to differences in fermentation periods since although the *in vitro* periods were restricted to 24 hours the time could have been considerably prolonged *in vivo* owing to long rumen retention times characteristic of low quality forages. In contrast, the mean *in vivo* cellulose digestion of treated ground straw (62.0%) was identical to that obtained *in vitro* for the same NaOH level and dilution.

Table III. The Effect of NaOH Treatment and Urea Supplementation on the Nutritive Value of Oat Straw (Experiment III)

Treatment:	Untreated		NaOH-treated	
Criteria: Supplementation:	None	2.5% Urea	None	2.5% Urea
Digestion Coefficients: (%)				
Cellulose	54.3[a]	52.4[a]	66.4[b]	69.6[b]
Dry Matter	46.0[d]	44.2[d]	59.5[e]	62.5[e]
Gross Energy	44.2[a]	41.6[a]	53.8[b]	58.5[b]
Relative Intake (%)	54.0[b]	60.8[b]	30.4[a]	99.9[c]
Nutritive Value Index	24.3[d]	25.4[d]	16.4[d]	58.1[e]
Digestible Energy Intake (kcal. $Wt_{kg}^{0.75}$)	83.9[d]	89.5[d]	52.7[d]	188.0[e]

[a, b, c, d, e] Means on the same line having different superscript letters differ significantly: [a, b, c] $P < .05$; [d, e] $P < .01$.

In no case did the addition of urea have a significant effect on increasing digestibility although there was a trend in this direction in the case of the NaOH-treated straw. The addition of urea effected a slight increase in the relative intake of the untreated straw but the most striking result of this experiment was the significant increase in relative intake because of the combination of NaOH treatment and urea supplementation. The average relative intake of 99.9 for this treatment combination was similar to that which is observed for an excellent quality forage (5). This result indicated that although the delignification effect of NaOH resulted in increased energy digestibility, nitrogen was a limiting nutrient—*i.e.,* an adequate supply of nitrogen was necessary to increase the rate of microbial digestion and thus the voluntary consumption of forage by the animal. The rate at which the contents of the rumen

Table IV. The Effect of NaOH Treatment, and Sucrose and Urea

Criteria:	Treatment: Supplementation:	Untreated		
		None	Sucrose	Urea
Digestion Coefficents: (%)				
Cellulose		56.1[a]	57.0[a]	59.4[a]
Dry Matter		46.6[c]	47.7[c]	49.6[c,d]
Gross Energy		43.9[c]	44.6[c]	47.0[c]
Relative Intake (%)		35.6[c,d]	34.5[c,d]	40.8[c,d]
Nutritive Value Index		15.4[c]	15.5[c]	19.2[c]
Digestible Energy Intake kcal./Wt$_{kg}^{0.75}$		55.0[c]	54.8[c]	68.0[c]

[a, b, c, d, e] Means on the same line having different superscript letters differ signifi-

are reduced has been proposed as the major determinant of forage voluntary intake, with rate of digestion of fibrous components of forage largely contributing to reduction of rumen load (2, 5).

The apparent depresion in intake observed with the treated-unsupplemented straw, similar to that observed in Experiment II, may be an effect of reduced palatability owing to addition of alkali, although the results of this experiment lay question to the palatability concept. A classical reason for observed low voluntary intakes of poor quality forages (such as straw) has been attributed to their low "palatability" characteristics. The data in this experiment would indicate that the subjective measure of palatability probably plays a minor role in determining voluntary intake of such forages. The latter is demonstrated by the fact that the combination of three supposedly unpalatable substances (straw, NaOH, urea) has resulted in voluntary intakes similar to those observed for "highly palatable" forages such as alfalfa and clover hays. The present experiment substantiates the hypothesis that the rate of microbial degradation of the fibrous constituents of forage plants (as mediated through deficiencies in available energy, nitrogen, and possibly other required nutrients) is the main determinant of forage voluntary intake.

Since nutritive value index (NVI) is calculated as the product of energy digestibility and relative intake, the value of 58.1 observed for the NaOH-treated, urea-supplemented straw is a reflection of the marked improvement in these two criteria for forage nutritive value. Thus, the NVI and subsequent calculation of digestible energy intake indicates a more than twofold increase in digestible energy available to the animal because of increased digestibility and voluntary intake of the treated, urea-supplemented straw.

Supplementation on the Nutritive Value of Oat Straw (Experiment IV)

	Untreated		NaOH-treated		
	Sucrose + Urea	None	Sucrose	Urea	Sucrose + Urea
	58.3[a,b]	56.3[a,b]	61.9[a,b]	68.7[b]	67.9[b]
	50.5[c,d]	56.5[d,e]	61.7[e]	63.7[e]	63.6[e]
	46.9[c]	50.5[c,d]	55.3[d,e]	58.2[e]	58.5[e]
	47.0[d]	33.4[c,d]	29.6[c]	102.3[e]	109.8[e]
	22.1[c]	16.7[c]	16.2[c]	59.5[d]	64.2[d]
	77.2[c]	54.8[c]	52.7[c]	195.8[d]	211.2[d]

cantly: [a,b] $P < .05$; [c,d,e] $P < .01$.

Experiment IV. The Effect of NaOH Treatment and Sucrose and Urea Supplementation on the Nutritive Value of Oat Straw

Methods. Oat straw harvested in 1967 was treated with NaOH according to the method previously described. Eight female Cheviot lambs were used in this experiment which was designed to substantiate the results obtained in Experiments II and III, and to test in addition the effect of supplementation of straw with a soluble carbohydrate (sucrose) source. This experiment was designed as a 2 × 4 factorial with the effect of treatment (untreated vs. NaOH-treated) compared with four different supplementations (none, sucrose, urea, sucrose + urea). One sheep was fed each of the eight treatment combinations in each replicate, with six replicates constituting the experiment. As in the previous experiments, urea was added at a 2.5% level. Sucrose was added at a 3.5% level in three of the replicates, and an approximately equivalent amount of sucrose was supplied from a 7.0% level of molasses in the other three replicates. The percent of supplements added thus varied from 0–9.5% with the remainder of the ration (91.5–100%) consisting of treated or untreated straw.

Results and Discussion. The results for the eight treatment combinations are presented in Table IV. Since the source of sucrose (purified vs. molasses) did not contribute to any differences, the results for the two sucrose sources were combined. Neither sucrose nor urea supplementation resulted in any significant differences in the digestion coefficients observed for the untreated straw, whereas supplementation did significantly increase digestion of the treated straw, particularly in the case of added urea. The combination of sucrose and urea did not result in any greater increase in digestibility than urea alone, indicating the lack of an additive effect of these two nutrients.

Of the eight treatment combinations only two exhibited the large increase in relative intake observed in Experiment III and in both these cases NaOH-treated straw was supplemented with urea. The fact that sucrose supplementation in the absence of urea did not result in any significant increases in digestibility or voluntary intake was a clear indication that a source of readily available energy was not a factor limiting utilization of the treated straw. The amount of digestible energy intake, measured either directly (kcal. $DE/wt._{kg}{}^{0.75}$) or using nutritive value indices, indicated a three–fourfold increase owing to the combination of NaOH treatment and urea supplementation.

Variable results relating to urea supplementation of untreated straw have been reported with either increases (4, 10) or no effect (12) on voluntary intake and digestibility noted. As observed increases in energy intake were still too low to maintain body weight (10), this would suggest that the beneficial effects of nutrient supplementation of low-quality forages are limited unless accompanied by delignification procedures.

Since it is appreciated that the energy available to the animal for productive purposes (growth, milk, etc.) is not most accurately measured in terms of digestible energy, these trials indicate that further work involving more elaborate net energy measurements would be justified. Furthermore, it is recognized that straw is deficient in many nutrients in addition to protein—*i.e.*, essential minerals and vitamins. In these trials supplementation was limited to energy (sucrose) and/or nitrogen (urea) sources, although cobalt, iodine, and NaCl were supplied to all animals. Because of the short-term nature of these feeding trials it was assumed that deficiencies in nutrients such as calcium, phosphorus, vitamin A, etc., would not affect the performance of the animals. Any practical application of NaOH-treated straw as a major component of a production ration would have to take into consideration complete supplementation of all required nutrients.

Conclusions

This study has confirmed the significant increases in digestibility which result as an effect of NaOH treatment of straw. Although the amount of cellulose digested by the rumen microflora can be increased with progressive levels of NaOH, practical use of such a method to improve the feeding value of lignified forages would require minimal use of added alkali. A method has been described whereby the treatment of low-quality forages, such as straw, can be effected on a mechanical basis, utilizing a commercial feed mixer as the container for the delignification reaction.

Although NaOH treatment results in large increases in digestibility no effect on voluntary intake was achieved unless the treated straw was supplemented with a source of nitrogen. When a source of non-protein nitrogen (urea) was added to treated straw at a 2.5% level, the resultant increase in digestibility and voluntary intake by sheep contributed to a digestible energy intake equal to that expected for high quality forages such as legume hay. The addition of a soluble carbohydrate (sucrose) at a 3.5% level did not result in any additional increase in digestibility or intake over that observed for NaOH-treated, urea-supplemented straw. The results of this study indicate that NaOH-treated straw supplemented with nitrogen and other essential nutrients, might serve as a substantial portion of a ration fed to ruminant animals.

Literature Cited

(1) A. O. A. C., "Official Methods of Analysis," 10th ed., Association of Official Agricultural Chemists, Washington, D. C., 1965.
(2) Balch, C. C., Campling, R. C., *Nutr. Abstr. Rev.* **32,** 669 (1962).
(3) Beckmann, E., *Chem. Abstr.* **16,** 765 (1922).
(4) Campling, R. C., Freer, M., Balch, C. C., *Brit. J. Nut.* **16,** 115 (1962).
(5) Crampton, E. W., Donefer, E., Lloyd, L. E., *J. Animal Sci.* **19,** 538 (1960).
(6) *Ibid.,* **21,** 628 (1962).
(7) Donefer, E., Crampton, E. W., Lloyd, L. E., *J. Animal Sci.* **19,** 545 (1960).
(8) Duncan, D. B., *Biometrics* **11,** 1 (1955).
(9) Freudenberg, K., *Science* **148,** 595 (1965).
(10) Hemsley, J. A., *Proc. Australian Soc. Animal Prod.* **5,** 321 (1964).
(11) Johnson, R. R., Dehority, B. A., Bentley, O. G., *J. Animal Sci.* **17,** 841 (1958).
(12) Kay, M., Andrews, R. P., MacLeod, N. A., Walker, T., *Animal Prod.* **10,** 171 (1968).
(13) Lampila, M., *Ann. Agr. Fenn.* **2,** 105 (1963). (*Nutr. Abstr. Rev.* **34,** 575, 1964).
(14) Wilson, R. K., Pigden, W. J., *Can. J. Animal Sci.* **44,** 122 (1964).
(15) Woodman, H. E., Stewart, J., *J. Agr. Sci.* **22,** 527 (1932).

RECEIVED October 14, 1968.

Discussion

H. Tarkow: "Did you measure the lignin content following alkali treatment?"

E. Donefer: "There was only a slight decrease noted in lignin content when NaOH treated straw was compared with untreated straw. Since our procedure did not involve washing after treatment, the products of the alkali reaction were not removed, and thus the form which the lignin is in may be an important consideration when the different lignin contents are compared."

D. M. Updegraff (Univ. of Denver): "Do you know of any ruminant feeding experiment in which instead of the sodium hydroxide treatment of sawdust or wood products, waste paper was used, since this has approximately the same composition as the other products?"

E. Donefer: "I would refer this specific question to Dr. Baumgardt, as the work on waste paper was done at his institution. Before his answer, I would like to take this opportunity to make one point in reference to the use of wood products as animal feeds. In dicussing this question with Dr. Stone of the Pulp and Paper Institute, he mentioned that wood pulp is priced at over $100/ton. This type of material is thus immediately ruled out as a potential feedstuff since good quality hay can be purchased at $25–$30/ton and grain corn is available at approximately twice that price. Serious consideration must thus be made of the initial cost of potential feedstuffs and their chemical treatments, in relation to the nutritive value and costs of commonly available feeds. Corn grain is an ideal feed in the fattening ration of cattle and costs less than some of the potential wood pulp feed products. Certain waste wood products such as bark and sawdust might on the other hand justify improvement procedures because of their low initial cost."

B. A. Baumgardt: "This work was done by Dr. Kesler at Pennsylvania State. This involved a very minimum level of waste paper, in fact it was simply an attempt to find something as a carrier for liquid molasses. Paper was a very small part of the ration, but no problems were encountered. The research did not test the sort of thing we are looking at or talking about today. It was not intended to."

W. J. Pigden: "In connection with newsprint, we have run several sources through the rumen system and it usually gives us a digestibility of about 18–20%, which is very close to what we get with the ground aspen wood and newsprint is, of course, just ground wood."

Query: "Most newsprint is ground wood, but now they are putting quite a bit of sulfite and Kraft pulp in newspaper print. At times this may make a real difference in the digestibility of it, and I am wondering whether they did separate tests on ground wood *vs.* Kraft or sulfite pulp newsprint?"

W. J. Pigden: "We simply took samples of several Canadian newspapers."

J. Stone: "I would like to make a comment on newsprint. It is usually about 80% ground wood pulp and 20% chemical pulp. Ground wood pulp is indigestible by the animal, in fact, it may have a negative effect;

but the chemical pulp, the sulfite pulp, Kraft, is digestible. So, to that extent waste newspaper would be about 20% digestible."

Query: "May I ask a question of one of the other speakers who made a comment to the effect that as the cellulose in a ration increased the feed efficiency went down. I am wondering, have you ever tried a purified pulp to make this kind of a test, or has it always been roughage material?"

P. J. Van Soest: "Roughage material."

Comment: "It would be interesting to know what would happen with a purified pulp when other factors are not affecting digestibility."

Comment: "I am inclined to think that you would increase the digestible cellulose."

W. H. Pfander: "What is the nitrogen requirement that produces optimum digestibility of the carbohydrate?"

B. A. Baumgardt: "The figures that I have seen, eliminating the animal requirements for the moment, indicate the level required for cellulose digestion to be about 1% nitrogen."

Comment: "So, what really we are talking about here as far as feeding the animal is getting efficiency."

E. Donefer: "I would like to comment on the question of animal growth response and feed efficiency. We saw data presented today by Dr. Anthony in which beef cattle made weight gains of 2.5 to 4 lb./day. These gains were made very efficiently and reflect the full growth potential of such animals. On the other hand, cattle weight gains in the tropical climates characterizing most developing areas may only average 1/2 lb./day, and it would prove too expensive in terms of supplying an optimum ration to attempt to obtain the maximum growth rate and efficiency of these animals. In other words, the most efficient and nutritious ration may be far from the most economical one. To be economically feasible, livestock feeding systems must take into consideration cost of ration and value of animal product produced."

Comment: "Now just one second. This is very important. There are certain kinds of animals in our economy that can be carried on a maintenance ration. With brood cows, a cellulosic material not usable by steers becomes very usable by a mature non-producing beef cow."

Query: "Usable in what sense? I wonder if you might want to qualify that term. You said it was not usable by steers but was usable by a mature non-producing beef cow?"

Comment: "The rate of gain has to be at a certain level so feed efficiency is good. This has to do with efficiency of feed and cost per unit of performance. A brood cow is not performing, and she can go ahead and maintain herself on it. She is not producing for you at the particular time; but it doesn't adversely affect her performance at a later time."

W. D. Kitts: "I would like to comment on some of the feed trials that are taking place in the northern part of B. C. They have used ensiled wood chips, basically for maintenance of cows. I imagine there is an awful lot of heat that cannot be used for any other purpose but to warm the animal. This is where it comes in. The same thing reflects on tropical production as well. So, the consideration should be given more attention."

B. A. Baumgardt: "In summary, we are saying that there are more uses than meets the eye at first glance, and we must look at all possibilities, consider all geographic areas and all economic situations."

Query: "One of the speakers brought up the question of residence time of the ration in the rumen. Do any of you have any rough figures on what is a reasonable average resident time?"

B. A. Baumgardt: "Well, I will guess the figures while others are thinking up better answers. This does, of course, vary in relation to the particle size and the specific gravity which a particle must ultimately reach before it will pass on. Perhaps on a common type roughage ration we might say that the feed stays in the rumen about 50 hours with an additional 25 hours or so in the remainder of the tract."

Comment: "For the remainder of the digestive tract, I believe that the figure is 16 hours. Once it passes the rumen it needs up to 16 hours. I think that for high quality forage much less time is needed than that. Perhaps 80% of the forage is quickly digested with the more indigestible residues sticking around for a good many hours more."

21

Development of a Commercial Enzyme Process: Glucoamylase

L. A. UNDERKOFLER

Miles Laboratories, Inc., Elkhart, Ind. 46514

> *Development of commercial glucoamylase products has made possible universal use of this enzyme for manufacture of dextrose from starch. A mutant strain of* Aspergillus foetidus *was found which produced high yields of glucoamylase with negligible transglucosylase. Laboratory and pilot plant investigations established essential medium ingredients and fermentation conditions for high enzyme yields and the process for recovery as a commercial product. The plant process developed averaged six units/ml. in submerged fermentation at 32°C. with vigorous agitation and aeration on a medium containing ground corn and steep liquor. Subsequent medium improvements along with a newly selected strain of the organism doubled plant yields. Removal of traces of transglucosylase by treating with a suitable cation exchange resin has improved efficacy of the commercial glucoamylase.*

Microorganisms as sources for the practical production of enzymes have several advantages over other possible sources. A wide variety of enzymes from microbial strains, selected partially on the basis of their high yields of specific enzymes, can be obtained conveniently by fermentation in unlimited quantities and in a very reproducible manner. Nevertheless, the development of a process for production of a commercial microbial enzyme frequently is a long term endeavor and involves a number of steps. Glucoamylase, for example, was clearly established as a glucogenic enzyme in 1948, but it was not in commercial usage until 1959.

General Developmental Procedures

The first step in developing a commercial microbial enzyme production process is intensive culture screening to find appropriate organisms which produce the desired enzyme in good yield. A microbial source can usually be found for an enzyme which can accomplish any desired reaction within the limits of possible enzymatic catalysis. From the standpoint of industrial enzymology, the extent of such a search is determined by the potential economic importance of the reaction. Also it should be clearly understood that enzymes of similar type from different organisms may vary greatly in their properties. For example, the commercial α-amylases from various bacteria and fungi are quite different with respect to their temperature optima and stability, their pH optima, and the extent to which they degrade starch. Likewise, the cellulolytic enzymes from different organisms have great variability in temperature and pH relationships, and most importantly in their actions on native cellulose and cellulose derivatives. These differences in the cellulases are brought out in other chapters of this book.

Several sources of information and empirical guides may assist in screening microorganisms in the search for an enzyme to bring about a specific reaction. Frequently taxonomic descriptions in standard compendia of microorganisms or original literature references will give leads as to suitable organisms to investigate. Cultures for screening are obtained from standard culture collections, and are isolated from nature, particularly from sources rich in the substrate upon which the desired enzyme must act. Initial screening may be accomplished by plating the cultures on solid agar medium containing the potential substrate. Those colonies are selected for further study which show the largest zones of action on the substrate. They are then grown as pure cultures in liquid media and further selection is based on assays of the enzyme potencies produced. Alternatively, where the desired reaction cannot easily be detected on solid media in plate cultures, the initial screening may require growth in liquid media with enzyme assays on the culture liquid.

Having found organisms which have the ability to produce the desired enzyme, it is next necessary to conduct intensive fermentation studies in laboratory and pilot plant to develop the optimum medium and fermentation conditions resulting in maximum yields. Initial studies are usually made in the laboratory in shaken flask cultures and in stirred jar fermentors. These are followed by pilot plant fermentations in agitated and aerated closed fermentors of from 5 to 500 gallon capacity simulating the plant fermentors which may range between 1,000 and 30,000 gallon capacity. The final stage is scale-up to full plant operation. In spite of much research, it has not been possible to completely understand and control the complex, interdependent factors which affect the fermentation such as efficiency of aeration, agitator speed, impeller design, and power input in fermentors of different size, so that it is still not always possible to scale up directly to large systems from pilot data. It is necessary to determine rather empirically the optimum conditions of aeration, agitation, and power input in order to obtain maximum enzyme yields from individual plant fermentors. When once established for a particular fermentation and vessel, these conditions are thereafter rigidly adhered to in plant operations.

Parallel with the fermentation research must go experimentation on best procedures for recovery and stabilization of the enzyme product, as well as thorough characterization of the enzyme. Enzyme recovery on a large scale often presents major problems. Laboratory processing can be carried out quickly with readily available equipment for temperature control, centrifugation, filtration, vacuum evaporation, dialysis, precipitation, chromatography, and lyophilization. Appropriate equipment to carry out comparable operations on a large scale often is not available at all, or is difficult to engineer, or may be of inordinate size or cost. Furthermore, procedures which may be carried out in a matter of minutes in the laboratory with small volumes will often require several hours in the plant with the large volumes involved. This, of course, poses serious problems with regard to enzyme potencies and stabilities when converting from laboratory to plant scale operations. Through pilot plant trials the recovery and processing procedures to be employed in the plant must be adapted and developed to assure a workable plant operation.

Publications giving the details of the actual development work in establishing industrial microbial enzyme processes are non-existent. It is, therefore, of interest to describe the development of a commercial enzyme process for one of the most important industrial enzymes, glucoamylase. Although there are many patents and publications on glucoamylase, nowhere has there previously been an attempt to gather together the available published and unpublished non-confidential information into a coherent report.

Function and Use of Glucoamylase

The enzymatic production of sugars from starch is one of the oldest processes practiced by mankind. The saccharification of starchy materials by malt in occidental countries and by fungal koji in the Orient for beverage fermentations was conducted as early as there are written records. These saccharification processes were a basis for the earliest recognition during the nineteenth century of the biological catalysts we now know as enzymes.

Interest in the saccharifying amylase now known as glucoamylase began about 30 years ago (1938) at Iowa State College when intensive investigations of the use of fungal amylase to replace malt in grain alcohol fermentations were begun. Fungal koji had, of course, been used for centuries in the Orient, instead of malt, for saccharifying starchy mashes for alcohol production, and Takamine, in 1914, advocated the use of mold enzymes from *Aspergillus oryzae* in the distilling industry in this country (28). The work at Iowa State with numerous strains of molds from four genera showed that mold bran koji preparations obtained by semi-solid surface culture with strains of *A. oryzae*, *Rhizopus delemar* and *R. oryzae* gave excellent starch saccharification as evaluated by yields of ethanol produced by yeast fermentation (8). Based upon this work, *A. oryzae* mold bran was commercially produced and used during the last year of World War II in the government alcohol plant at Omaha as a partial replacement for malt (30).

A little later workers at the Northern Regional Research Laboratory examined a large number of molds of five genera for their potential for amylase production in submerged culture (21, 31). Only Aspergillus cultures gave satisfactory results; *A. niger* NRRL337 and *A. niger* NRRL330 were selected as most promising. Corman and Langlykke (4) studied the action of the amylolytic enzymes in the mold culture filtrates on the saccharification of starch, and found the action due mainly to an α-amylase and a glucogenic enzyme. This glucogenic enzyme system acted upon maltose, dextrins, and starch, producing glucose, and efficiency of saccharification was correlated closely with the glucogenic enzyme system rather than with the activity of α-amylase. These workers therefore were the first to clearly establish the importance of glucoamylase, although previously Kita (18) recognized that fungal koji must contain an enzyme which produces glucose directly from starch and Kerr and coworkers (16, 17) rather vaguely recognized the presence of enzymes of this type in commercial fungal amylase products used in syrup production.

Phillips and Caldwell (25, 26) starting with a koji preparation from *R. delemar* from Takamine Laboratory, and Kerr, Cleveland, and Katzbeck (15) with a submerged culture preparation from *A. niger* independently in 1951 reported on glucoamylase preparations free from α-amylase and characterized the general mode of action. These authors called the enzyme gluc amylase and amyloglucosidase, respectively. Although the latter name has been most commonly used, the official trivial name adopted in 1961 by the International Union of Biochemistry (11) is glucoamylase [α-1,4-glucan glucohydrolase, E.C.3.2.1.3].

Glucoamylase was characterized in 1951 as an enzyme which hydrolyzed α-$(1 \to 4)$ glucosidic linkages in starch so as to remove successive glucose units from the non-reducing ends of the chains. Subsequently, it was demonstrated that the enzyme also hydrolyzes α-$(1 \to 6)$ and α-$(1 \to 3)$ glucosidic bonds, although at much slower rates (23).

The work in the early 1950's aroused considerable interest in the possibility of producing dextrose from starch by means of glucoamylase, since the conventional acid hydrolysis process gives relatively poor yields of recoverable dextrose and an almost unusable hydrol residue containing high concentrations of salts and unpleasant tasting unfermentable reversion sugars.

Research in many laboratories, particularly in the United States and Japan, soon demonstrated that selected strains of *Aspergillus* and of *Rhizopus* had greatest potential for producing glucoamylase (7, 20). The results with the *Aspergillus* cultures available in the early 1950's were disappointing because of relatively low yields of glucoamylase, and, more importantly, their production of appreciable quantities of transglucosylase. This enzyme reduced the yields of dextrose from starch by causing the production of oligosaccharides containing α-(1 $\rightarrow$ 6) linkages.

In Japan efforts to find suitable *Aspergillus* strains were abandoned and work was concentrated on *Rhizopus* strains which produced glucoamylase essentially free from transglucosylases. Disadvantages of the *Rhizopus* cultures were that they would not give satisfactory glucoamylase yields in submerged culture (6) and the somewhat inferior properties of the *Rhizopus* glucoamylase as compared with the *Aspergillus* enzyme, notably lower temperature optimum, poorer stability, and narrower useful pH range (24, 29). Nevertheless, the Japanese succeeded in developing surface culture processes for the *Rhizopus* glucoamylase on a commercial basis, and these Japanese products have been extensively employed in Japan, and to some extent in Europe, for commercial dextrose production. Later Japanese investigators also developed processes for producing glucoamylase by submerged culture of an organism identified as an *Endomyces* species (9). This process is also employed commercially in Japan, but the *Endomyces* glucoamylase shows the same somewhat inferior properties as the *Rhizopus* glucoamylase.

In the United States, because of higher yields in submerged culture and the superior properties of *Aspergillus*-derived glucoamylase (higher temperature optimum and wider pH range for use), investigations with *Aspergillus* strains continued. Screening programs examined a large number of cultures, particularly mutants derived from promising strains, and resulted in the discovery in several different laboratories of strains that produced much higher yields of glucoamylase which were suitably low in transglucosylase (1, 3, 27). Within a short period all dextrose made in the United States was produced by the use of commercial glucoamylase derived from such cultures, and several new dextrose plants have subsequently been built to use the process.

Development of the Commercial Process

The balance of this paper is a report of work done in the Miles-Takamine research laboratories and plant in developing the commercial glucoamylase product marketed under the trade name Diazyme. The *R. delemar* koji product, from which Phillips and Caldwell originally isolated glucoamylase, was quickly abandoned because of its high cost. Numerous cultures of the *Aspergillus niger* group were screened, among which were found certain strains of *A. niger*, *A. phoenicis*, *A. awamori* and *A. foetidus* which produced glucoamylase. However, none of the original isolates tested gave sufficiently high yields to be attractive, and they also produced objectionably high levels of transglucosylase. Finally, early in 1961, through ultraviolet induced mutation and selection, a mutant culture of *A. foetidus* was obtained which had the desired characteristics. Table I shows the enzyme assays from an experiment in which this culture was compared with others. The cultures were incubated on a reciprocal shaker for 72 hours at 32°C. on 500 ml. of medium containing 15% whole ground corn and 2% corn steep liquor in 2800 ml. Fernbach flasks. Assay values of culture filtrates are expressed in Diazyme units (DU) per milliliter, one unit being the amount of enzyme which produces one gram of dextrose in one hour from a 4% soluble starch solution at 60°C. (*22*).

Table I. Glucoamylase Production by *Aspergillus* Cultures

Culture	Glucoamylase, DU/ml.
A. foetidus mutant	3.1
A. foetidus	1.9
A. niger NRRL330	0.9
A. niger NRRL337	0.7
A. phoenicis ATCC13157	1.0

Having found a promising culture, the further steps of laboratory, pilot plant and plant scale-up to develop an efficient fermentation production process were undertaken. Simultaneous with this were investigations leading to plant procedures for recovery and stabilization of the enzyme product, and studies on its characterization and applications. Within three months plant production was initiated. Changes and improvements continued to be made over a period of about a year based upon continuing research in pilot plant and plant until a "standard" process was firmly established.

Culture investigations regarding inoculum seed production were first made in the laboratory. In these studies spore cultures were employed for inoculation of shaken flask cultures, 5 ml. of which in turn were used to inoculate 500 ml. of sterile medium containing 15% whole ground corn and 2% corn steep liquor in 2800-ml. Fernbach flasks, which were then incubated on the reciprocal shaker for 72 hours at 32°C. and assayed for glucoamylase. The results in Table II show that Czapek-Dox agar medium was satisfactory for slant culture maintenance, and either mass spore transfer or single conidial head transfer could be employed.

The effect of initial pH of seed medium and the age of seed was studied in the same manner. The results are presented in Table III. It was demonstrated that it is not necessary to adjust the natural pH of the seed medium of about 4.0 before sterilization, and 24 hours is a suitable incubation time using 1% inoculum.

Table II. Effect of Stock Culture Medium and Transfer Technique on Glucoamylase Yields

Medium	Replicates	Transfer Technique	Glucoamylase, DU/ml.
Brer Rabbit molasses agar	3	mass	3.33
Wheat bran agar	3	mass	3.45
Czapek-Dox agar	3	mass	3.56
Wheat bran agar	10	mass	3.57
Czapek-Dox agar	10	mass	3.55
Wheat bran agar	7	single conidial head	3.21
Czapek-Dox agar	7	single conidial head	3.40

Table III. Effect of Seed Medium pH and of Seed Age on Glucoamylase Production

Adjustment, Before Sterilization, to pH	Time of Incubation, Hours First Flask	Time of Incubation, Hours Second Flask	Glucoamylase, DU/ml.
4.0	24	—	3.41
5.0	24	—	3.50
6.0	24	—	3.48
6.0	24	—	3.90
6.0	48	—	3.30
6.0	24	24	3.90
6.0	48	24	3.10

Experiments in shaken flask cultures showed optimum level of corn steep liquor was 2% in media containing 15% ground whole corn. With this level of corn steep liquor, enzyme yields suffered when the ground corn was replaced wholly or in part by starch or by corn bran plus starch. Likewise, glucoamylase yields were lower when corn steep liquor was replaced by Amber BYF yeast extract, Sheffield Peptone T, Promine D, Promine R, soybean meal, Pharmamedia cottonseed nutrient, or Distillers'

Dried Solubles. Supplementation with ferrous sulfate, magnesium sulfate, copper sulfate, or dipotassium phosphate caused no improvement in yield. It was, therefore, concluded that media containing only ground whole corn and corn steep liquor were appropriate for scale-up investigations.

Fermentations were conducted in laboratory 2.5-liter New Brunswick glass-jar, stirred fermentors. In these experiments, unless otherwise indicated, the medium contained 15% ground whole corn and 2% corn steep liquor, sterilized with vigorous agitation for 30 minutes at 121°C., and then transferred to the sterile jars. The media could not be sterilized by direct autoclaving in the jars because the corn meal settled and caked in the bottom of the jars. The fermentation temperature was controlled at 32°C. Typical New Brunswick fermentor results are shown in Table IV. Although somewhat variable, the New Brunswick fermentations indicated that higher mash concentrations gave higher glucoamylase yields, and relatively high agitation and aeration rates were necessary.

Development work was then concentrated on fermentations in 40-liter stainless steel pilot plant fermentors. Unless otherwise indicated, each fermentation employed 30-liters of medium which contained 15% ground whole corn and 2% corn steep liquor, sterilized under vigorous agitation for 30 minutes at 121°C. Inoculation was with 1 to 5% by volume of 24-hour seed culture, temperature was controlled at 32°C., 15 p.s.i.g. pressure was maintained, agitation was 400 r.p.m. and aeration was at one volume of air per volume of medium per minute. Three or four 30-liter fermentations were usually run simultaneously in order to determine the effect of different variables. In this way it was determined that the best fermentation yields were obtained when the head pressure was maintained at 15 p.s.i.g., and the temperature at 32°C. For example, yield was 4.29 DU/ml. at 98°F. (36.6°C.), and 5.55 DU/ml. at 90°F. (32°C.). In the glucoamylase fermentation, the head pressure (15 p.s.i.g.) has a function other than contamination control. Enzyme yield is significantly improved, and it is believed, that this is owing to better oxygenation in the very thick, viscous mash.

Since laboratory experiments had shown better glucoamylase yields with higher corn meal concentrations in the medium, it was desirable to try higher concentrations in the pilot plant. However, even 15% corn meal suspensions are quite thick and become thicker upon gelatinization during heat sterilization. This produces a heavy load on the agitator and also renders cooling of the mash slow and difficult. Thinning of the mash was successfully accomplished by adding Takamine HT440 thermostable bacterial amylase to the mash when the ingredients were mixed. During the heating process the starch is simultaneously gelatinized and liquefied producing a mash thin enough for efficient agitation. Initially 0.1%

(based on weight of corn) of HT440 was employed, but subsequent experiments showed a level of 0.05% was adequate. This enzymatic liquefaction did not adversely affect glucoamylase yields. A comparison of enzyme yields with media containing 20 and 15% ground whole corn is shown in Table V. The higher concentration of corn gave higher glucoamylase yields.

The effect of increased agitation when all other variables were held constant is shown in Table VI. The data indicate slightly better glucoamylase production with greater agitation, but the higher agitation rate caused a more rapid drop in pH.

Early in the pilot plant investigation stage, it was decided to attempt scale-up in the plant in a small fermentor. A medium containing 15% ground whole corn and 2% corn steep liquor in 1,000 gallons of water, and adjusted to pH 6.0 with caustic, was sterilized in the fermentor by heating to 250°F. (121°C.) with continuous agitation, holding for 30 minutes, and cooling to 90°F. (32°C.). It was inoculated with 40 gallons of 26-hour seed culture and incubated at 90°F. (32°C.) with agitation at 190 r.p.m., aeration at 115 c.f.m. and 15 p.s.i.g. head pressure. After 64 hours the final pH was 3.70 and glucoamylase assay 4.96 DU/ml. Another similar fermentor was run except that 20% ground whole corn was used with 0.05% HT440 added prior to heating to effect liquefaction, and aeration was at 50 c.f.m. After 117 hours the final pH was 3.50 and glucoamylase assay 5.87 DU/ml.

Subsequent scale-up to 2,000 gallon fermentations gave similar results. During a series of 2,000 gallon fermentations, corn concentrations of 15, 17.5 and 20% were employed with aeration varied between 60 and 75 c.f.m. The highest yield obtained in this series was 6.38 DU/ml. with 20% corn medium and aeration at 60 c.f.m. The average of 19 fermentors in this series was 5.0 DU/ml. and those containing 20% corn averaged 5.2 DU/ml.

The final fermentation process scaled up to 6,600 gallon fermentations averaged yields of about 6 DU/ml. in four to five days.

Along with the pilot plant fermentation work, recovery investigations were conducted. In the laboratory the fungal mycelium could be easily removed by filtration, and this was accomplished readily on the pilot plant vacuum precoat filter also. Washing the mycelium cake with a fan spray to give about 20% wash water volume dilution gave better than 95% recovery of the enzyme present in the beer in the pilot plant experiments. Again this was confirmed in the series of 2,000 gallon plant fermentations, where in about half the runs filtration recoveries were 95% or better; however, inadequate washing or other operational losses reduced the recovery in other cases so that the overall average recovery for the series was 93%.

Table IV. Glucoamylase Production in New Brunswick Fermentors

Corn, %	Agitation, r.p.m.	Aeration, V/V per min.	Glucoamylase, DU/ml. 72 hr.	96 hr.
7.5	400	1.0	2.6	—
7.5	400	1.2	3.2	—
7.5	400	1.4	3.2	—
15	400	1.0	3.8	—
15	400	1.2	4.0	—
15	400	1.4	3.7	—
15	350	0.6	1.9	2.2
15	350	0.8	2.8	3.7
15	400	0.6	2.9	3.7
15	400	0.8	2.0	2.8
15	400	1.2	3.3	5.0
15	400	1.2	3.7	4.3
15	400	1.2	3.0	5.0

Table V. Effect of Corn Meal Concentration on Glucoamylase Production

Time, Hours	15% Corn	20% Corn	20% Corn
20	1.49	1.31	0.95
36	3.59	3.62	3.23
48	4.69	4.99	4.84
60	5.49	5.79	5.71
72	5.63	6.48	6.90
80	5.83	6.65	—
97	6.42	6.90	—

Glucoamylase, DU/ml.

Table VI. Effect of Agitation on Glucoamylase Production

Time, Hours	Glucoamylase, DU/ml. 400 r.p.m.	480 r.p.m.	pH 400 r.p.m.	480 r.p.m.
20	1.28	1.54	3.81	3.40
32	3.16	3.55	3.55	3.15
44	4.43	4.89	3.40	3.05
52	5.13	5.57	3.40	2.90[a]
68	5.75	5.77	3.30	3.25

[a] pH adjusted back at this time to 3.40 with sodium hydroxide.

The relatively dilute glucoamylase filtrates made concentration advisable prior to further processing. The feasibility of vacuum evaporation was demonstrated in the laboratory by evaporation of filtrate from a New Brunswick fermentation at 30–40°C. to one-eighth original volume, with retention of 90% of the activity. On evaporation of 10 liters of pilot plant filtrate at 45–46°C. vapor temperature, 95% activity was retained. This stability was confirmed in plant evaporations of the series of 2,000 gallon plant fermentations, with an average recovery of 97% of the original glucoamylase activity.

Depending upon use and stability, commercial enzymes are marketed as liquid or solid preparations. It was found that the enzyme could be recovered as a solid product by the conventional method of alcohol precipitation employed for other solid commercial enzyme products. The most efficient process was found to be evaporation of the clear filtrate to one-third original volume, adjustment of pH to 3.2 by addition of hydrochloric acid, and addition of 3.5 volumes of cold special denatured alcohol, formula 35A. The precipitate was allowed to settle, clear supernatant decanted, the precipitate washed with an equal volume of alcohol, and filtered in a filter press. The solid was washed in the press with anhydrous alcohol, blown with dry air, and finally dried in a vacuum oven at 36°C.

However, for lower cost and greater convenience of use a liquid product has advantages. It was, therefore, necessary to determine the stability of liquid concentrates and their preservative requirement. A pilot plant filtrate assaying 4.39 DU/ml. was concentrated by vacuum evaporation. Samples of the filtrate along with concentrates containing 12, 24 and 36 DU/ml. were stored at 4°C., room temperature, and 36°C., and assayed at weekly intervals. The retention of activity after 21 days is shown in Table VII, indicating good enzyme stability, even when heavy yeast growth had occurred.

Table VII. Retention of Glucoamylase Activity in Liquid Solutions after 21 Days

	Retained Activity, %		
Sample	4°C.	Room Temperature	36°C.
Filtrate, 4.4 DU/ml.	98.2	94.2[a]	74.5[a]
Concentrate, 12 DU/ml.	—	90.8[a]	—
Concentrate, 24 DU/ml.	—	95.5	—
Concentrate, 36 DU/ml.	99.0	90.8	94.5

[a] Heavy yeast growth.

In order to prevent yeast growth the following potential preservatives were added individually to samples of glucoamylase liquid concentrate: 0.01% calcium propylhydroxybenzoate, 0.01% potassium sorbate, 0.01% sodium benzoate, 0.1% sodium benzoate, 30% glycerol, 20% propylene glycol. The samples were then inoculated with yeast from a previously contaminated sample and held for 21 days at room temperature. No yeast proliferation occurred in the sample containing 0.1% sodium benzoate, and full enzyme activity had been retained.

In the final operating procedure adopted for preparation of liquid Diazyme L30, the mycelium was removed on the vacuum precoat filter, the calculated amount of sodium benzoate to give 0.1% in the final product was added to the filtrate, it was evaporated under vacuum, standardized to 30 DU/ml. potency, and drummed for shipment. Stability assay data for ten different commercial batches are shown in Table VIII.

Recent Improvements

Following commercialization and the routine manufacture and sale of the glucoamylase product, in due course it became apparent that improvements were desirable and necessary. For example, lower costs were needed by both manufacturer and customers, and these could be achieved only by yield improvements. Two avenues were explored, (a) medium composition and (b) further mutation and strain selection.

Table VIII. Stability of Commercial Diazyme L30 Batches

Age, Months	Activity of Control Samples, % of Release Assay	
	Refrigerated	Room Temperature
1	—	100
1	—	100
2	—	98
2	—	99
3	—	95
3	—	100
4	—	100
4	—	97
5	—	95
5	—	90
9	101	90
9	100	91
10	99	94
10	99	93
11	99	88
11	97	86
12	99	87
12	102	91
13	98	94
13	93	86

In the original development work only limited investigations of medium composition were made since a simple medium of ground whole corn and corn steep liquor proved satisfactory, and both of these ingredients were indispensable. Careful examination of other possible ingredients now seemed justified. Several series of laboratory shaken flask fermentations were conducted to investigate a large variety of sources nesium sulfate, and monopotassium phosphate to the basal medium containing ground whole corn and corn steep liquor, the experimental ingredients were added alone or in combination with some variation in the proportions of the various ingredients. The addition of ground corn bran, soy peptone, Pearl Plus starch, Proto Soy, sodium citrate, magnesium sulfate and monopotassium phosphate to the basal medium appeared to increase glucoamylase yields. Subsequent extensive experiments in laboratory and pilot plant indicated the soy peptone, Pearl Plus starch and Proto Soy were of no advantage if levels of ground corn and corn steep liquor were suitably adjusted. A modified medium was thus developed containing 20% ground whole corn, 3% ground corn bran, 3.5% corn steep liquor, 0.9% sodium citrate dihydrate, 0.13% magnesium sulfate heptahydrate, and 0.065% monopotassium phosphate. This was thinned during sterilization with 0.02% HT440 bacterial amylase. With this medium, glucoamylase yields of 13 DU/ml. were obtained in 30-liter pilot plant fermentations with temperature maintained at 32°C., agitation at 400 r.p.m., aeration at 1.2 volumes per volume of medium and 15 p.s.i.g. head pressure. The first plant trial with this medium in a 6,600 gallon fermentation yielded 11 DU/ml.

Regular plant use of the modified medium along with a newly selected strain of *A. foetidus* resulted in doubling the routine plant yields of the commercial enzyme product. The new strain had little influence on maximum yields, but gave greater uniformity from batch to batch.

A second desirable improvement was a method for complete elimination of transglucosylase, since even the small trace of this enzyme present in the commercial enzyme product seemed responsible for producing small amounts of oligosaccharides, particularly isomaltose, which were troublesome in dextrose production. This had long been recognized and numerous methods have been found and patented for removing the traces of transglucosylase, some of which are listed in the references (*2, 5, 10, 12, 13, 14, 19*). A simple method had become available for determining the success of such treatments for removal of transglucosylase. This involves incubation of 100 ml. of a solution containing 10 grams of maltose and 5 DU of glucoamylase at 60°C. for 72 hours, then determining the final specific rotation at 25°C. If no oligosaccharides are present, the specific rotation should be that for pure glucose, 53.4°C. However, completely pure glucoamylase causes some polymerization of dextrose, the extent of this being larger with increasing dextrose concentration and higher enzyme levels (24, 29). At the level of sugar and enzyme used in this test, back polymerization is negligible, and it is considered that the enzyme is free of transglucosylase if the final specific rotation is less than 54°C.

In our own laboratory an efficient method for removal of traces of transglucosylase was developed involving use of a column of cation exchange resin (5). Typical laboratory results are shown in Table IX. The data show essentially no loss of glucoamylase activity with removal of the transglucosylase by the ion exchange treatment. Plant scale-up of the process gave no particular difficulties.

In this paper the development of a process for production of commercial glucoamylase has been detailed. The original process was developed through laboratory, pilot plant, and plant scale-up investigations. Later, distinct improvements were made, and plant yields were approximately doubled by means of medium improvements and further strain selection. Product quality was improved by complete removal of transglucosylase.

Table IX. Removal of Transglucosylase by Treating with Cation Exchange Resins

Resin	Specific Rotation Untreated	Specific Rotation Treated	Glucoamylase Recovery, %
Amberlite 200	55.4	53.7	97.2
Amberlite IRC50	55.2	53.8	100
Dowex 50	55.0	53.7	93.5
Duolite C-65	55.4	53.8	95.0

Acknowledgment

The experimental and developmental work upon which this report is based was conducted by E. G. Bassett, R. L. Charles, P. R. Casey, R. R. Barton, and G. B. Borglum of the Enzymology Research Laboratory, and by W. W. Windish and L. A. Zawodniak of the Takamine Plant, with the assistance of numerous other individuals.

Literature Cited

(1) Armbruster, F. C., *U.S. Patent* **3,012,944** (1961).
(2) Armbruster, F. C., Bruner, R. L., *U.S. Patent* **3,303,102** (1967).
(3) Bode, H. E., *U.S. Patent* **3,249,514** (1966).
(4) Corman, J., Langlykke, A. F., *Cereal Chem.* **25,** 190 (1948).

(5) Croxall, W. J., *U.S. Patent* **3,254,003** (1966).
(6) Fukumoto, J., private communication (1967).
(7) Fukumoto, J., "The Properties and Commercial Application of Gluc-Amylase," Osaka City University, Faculty of Science, Osaka, Japan, 1962.
(8) Hao, L. C., Fulmer, E. I., Underkofler, L. A., *Ind. Eng. Chem.* **35**, 814 (1943).
(9) Hattori, Y., *Die Stärke* **17**, 82 (1965).
(10) Hurst, T. L., Turner, A. W., *U.S. Patent* **3,117,063** (1964).
(11) International Union of Biochemistry, "Enzyme Nomenclature—Recommendations 1964 of International Union of Biochemistry," Elsevier, Amsterdam, 1964.
(12) Kathrein, H. R., *U.S. Patent* **3,108,928** (1963).
(13) Kerr, R. W., *U.S. Patent* **2,970,086** (1961).
(14) *Ibid.*, *U.S. Patent* **2,967,804** (1961).
(15) Kerr, R. W., Cleveland, F. C., Katzbeck, W. J., *J. Am. Chem. Soc.* **73**, 3916 (1951).
(16) Kerr, R. W., Meisl, H., Schink, N. F., *Ind. Eng. Chem.* **34**, 1232 (1942).
(17) Kerr, R. W., Schink, N. F., *Ind. Eng. Chem.* **33**, 1418 (1941).
(18) Kita, G., *Ind. Eng. Chem.* **5**, 220 (1913).
(19) Kooi, E. R., Harjes, C. F., Gilkison, J. S., *U.S. Patent* **3,042,584** (1962).
(20) Langlois, D. P., Turner, W., *U.S. Patent* **2,893,921** (1959).
(21) LeMense, E. H., Corman, J., Van Lanen, J. M., Langlykke, A. F., *J. Bact.* **54**, 149 (1947).
(22) Miles Laboratories, *Tech. Bull.* **9-245**, Elkhart, Ind. (1963).
(23) Pazur, J. H., Kleppe, K., *J. Biol. Chem.* **237**, 1002 (1962).
(24) Pazur, J. H., Okada, S., *Carbohydrate Res.* **4**, 371 (1967).
(25) Phillips, L. L., Caldwell, M. L., *J. Am. Chem. Soc.* **73**, 3559 (1951).
(26) Phillips, L. L., Caldwell, M. L., *J. Am. Chem. Soc.* **73**, 3563 (1951).
(27) Smiley, K. L., Cadmus, M. C., Hensley, D. E., Lagoda, A. A., *Applied Microbiol.* **12**, 455 (1964).
(28) Takamine, J., *Ind. Eng. Chem.* **6**, 824 (1914).
(29) Underkofler, L. A., Denault, L. J., Hou, E. F., *Die Stärke* **6**, 179 (1965).
(30) Underkofler, L. A., Severson, G. M., Goering, K. J., *Ind. Eng. Chem.* **38**, 980 (1946).
(31) Van Lanen, J. M., LeMense, E. H., *J. Bact.* **51**, 595 (1946).

RECEIVED October 14, 1968.

Discussion

E. T. Reese: "Few industrial researchers tell us exactly what they have been doing and how they have been doing it. I think Dr. Underkofler has been very frank here in covering the development of glucoamylase. Quite often we talk to industrial people and we don't get very much information in return. This process that he described, of course, is the thing that we have to compete with in our cellulase work. We are going to try to put him out of business by converting our cellulose to glucose . . . we hope."

Bruce Stone: "Did you imply that the purified glucoamylase and transferase activity were the same?"

L. A. Underkofler: "The glucoamylase does have polymerizing activity. The back polymerization depends upon the concentration of the enzyme and concentration of the glucose. With high levels of glucose and high levels of enzyme you get appreciable isomaltose. It's a reversible reaction."

K. E. Eriksson: "I would just like to know at what concentrations of glucose do these glucoamylases give transglucosylation?"

L. A. Underkofler: "Well, if you use 5–10% gelatinized product, you can get practically quantitative conversion to glucose. As you go higher, you begin to get a little isomaltose. If you use 30% starch paste you will probably get around 1/2%. If you use 50% glucose you will get several percent isomaltose."

L. E. Baribo (Wash.): "I would like to comment for those who think they are going to use this information to convert cellulose to glucose. First of all, the glucoamylase, when you put it in contact with granular starch, has absolutely no activity, which would be about the same as putting our cellulase in contact with pulp. Second, when you put enzyme in contact with soluble starch you still get such a low activity that for all practical purposes you couldn't use it in a commercial process. When you either acid hydrolyze or put an α-amylase in the starch to get a DE in the neighborhood of 10–12, only then do you get sufficient activity to make it worthwhile. So, we have a long way to go before we start hydrolyzing our cellulose to glucose."

L. A. Underkofler: "Yes, you do have to modify your raw starch, and you have to use a 30% or 35% starch concentration. You have to gelatinize and modify it. You are going to have to modify your cellulose."

22

Applications of Cellulases In Japan

NOBUO TOYAMA

Applied Microbiology Laboratory, Department of Agricultural Chemistry, Faculty of Agriculture, Miyazaki University, Miyazaki, Japan

> *Various applications of a cell separating enzyme from* Rhizopus *sp. and cellulases from* Trichoderma viride *and* Aspergillus niger *were developed by the author and other investigators in Japan. Vegetable tissues were degraded into individual cells with protopectinolytic activity of cell-separating enzyme preparation. A cellulolytic component capable of degrading filter paper contributed to the degradation of plant cell wall. Those enzyme preparations were useful for the following purposes: extraction of green tea components, soybean or coconut cake protein, sweet potato or corn starch and agar-agar; production of unicellular vegetables, vinegar from citrus pulp and seaweed jelly; for removing soybean seedcoat, modifying food tissue such as vegetables, rice, and glutinous rice; isolation of plant unicells and protoplast in an active state; decomposition of cooked soybean, enzymic digestion of excreta, and increasing tensile strength of paper.*

The existence of intercellular substances and cell walls influence markedly the digestibility, cooking quality, shape retention of vegetable foods, and the yield of nutrients such as starch and protein. In vegetable tissue, cells adhere to each other with protopectin. They are separable into unicells with cell separating enzyme (CSE) and the naked cell walls are then easily degraded with cellulase. Treating vegetable foods with CSE and cellulase improves them by loosening tissue, or by thinning and removing cell wall. A CSE from *Rhizopus* sp. and cellulases from *Trichoderma viride* and *Aspergillus niger* are being produced industrially in Japan.

Numerous attempts were made to utilize these three enzyme preparations for practical use. Cellulase preparations are being so far employed abundantly as an important member of combined digestive aids. These

Table I. Filter Paper Degradation

Time Required

Enzyme	15	30	45	60	75
Rhizopus Koji extract	−	−	±	+	++
Macerozyme	−	−	−	±	+
Cellulase onozuka P500	−	±	+++	++++	
Cellulase onozuka PE	+++	++++[a]			
Meicelase P	−	−	+++	++++	
Cellulosin AP	−	−	±	++	+++
Cellulosin AP with beta-glucanase	−	±	++	+++	++++

Filter paper degrading cellulase activity was measured by the author's (39) shaking method using 2 sheets of Toyo filter paper No. 2, 10 mm. × 10 mm. at 40°C., pH 5.0. Complete degradation was shown by ++++.

results were mostly published in the proceedings issued by Cellulase Association, a division of Japan Fermentation Technology Association. Already seven symposiums have been held in Osaka or Tokyo on "Cellulase and Related Enzymes." The 8th symposium was held in 1967 under the new title "Enzymatic Degradation of Cell Wall Substances."

Experimental

Cell Separating Enzyme and Assay Method. The edible tissue of higher plants consists of parenchymatous cells. These cells are combined with each other by cementing substances, particularly protopectin. When parenchyma was treated with CSE (protopectinase or macerating enzyme), the tissue was degraded into individual cells by the decomposition of insoluble protopectin to soluble pectinic acid. A CSE preparation from a *Rhizopus* sp. was developed by the author as a supplementary enzyme to cellulase and is being produced under the name of Macerozyme. The CSE activity can be determined as follows: To 0.5 grams of CSE preparation was added 50 ml. of 0.1M acetate buffer, pH 5.0. Impurities were removed by centrifuging. The central portion of succulent white potato was cut into a sheet, 10 × 10 × 2 mm. These two sheets were shaken with 5 ml. of CSE solution in Monod L type glass tube (inner diameter—18–19 mm., height of vertical tube—40–45 mm. and length of horizontal tube—143–148 mm.) in Monod shaking culture apparatus at 40°C., 60 r.p.m. Degree of potato degradation was shown as follows: residual diameter 9–10 mm. (−), 7–9 mm. (±), 5–7 mm. (+), 3–5 mm. (++) and 1–3 mm. (+++, complete degradation).

In the case of *Rhizopus* CSE, potato sheets were degraded from their surface into unicells containing starch granules, while *Aspergillus niger* preparation, having both CSE and cellulase activities, degraded them into isolated starch granules. Little degradation of potato occurred with *Tri-*

with Enzyme Preparations

for degradation (min.)

90	105	120	135	150	165	180
++	++	++	++	+++	+++	+++
++	++	+++	+++	+++	+++	+++ [b]

+++	+++	+++	++++

[a] After 20 min. incubation.
[b] Recent preparation has little cellulase activity.

Table II. Cell Separating Activity of Enzyme Preparations

Time Required for Degradation (hr.)

Enzyme	1	2	3	4	5	6	Product
Rhizopus Koji extract	—	±	+	+	++	+++	Unicell
Macerozyme	—	±	+	++	+++		Unicell
Cellulase onozuka P500	—	—	—	—	—	±	
Cellulase onozuka PE	—	—	—	±	±	±	Starch
Meicelase P	—	—	—	—	—	—	
Cellulosin AP	—	—	—	±	±	±	Starch
Cellulosin AP with beta-glucanase	—	±	++	++	+++		Starch

Cell separating enzyme activity was measured by the author's shaking method using 2 sheets of potato, 10 mm. × 10 mm. × 2 mm. at 40°C., pH 5.0. Complete degradation was shown by +++.
Macerozyme: *Rhizopus* CSE.
Cellulase onozuka: *Trichoderma viride* cellulase.
Meicelase: *Trichoderma viride* cellulase.
Cellulosin: *Aspergillus niger* cellulase.

choderma preparation. When 1% CSE solution was used, the time required for complete degradation of potato was usually 5 hr.

$$100,000/X \times 50 \text{ mg.} = 2,000/X$$

where X is time in hours and 50 mg. is the weight of enzyme preparation in 5 ml. of sample solution. Usual CSE activity of Macerozyme is 400 unit/mg. (33). The optimum pH of CSE in Macerozyme was found to be 4.5 to 5.0 and adequate pH values ranged from 4.0 to 6.5. The optimum temperature for CSE was exhibited at 45°C., and complete inac-

Table III. Production of Unicellular Vegetables and Fruits with CSE

Raw Material	Weight (g.)	1% Macerozyme Solution (ml.)	Incubation Time (hr.)	Weight of Unicells Wet (g.)	Weight of Residue Wet (g.)	Yield of Unicells Wet (%)	Weight of Residue in Control Wet (g.)
Carrot	50	60	5	14.4	3.3	28.8	55.0
Tomato	50	60	2	12.8	1.1	25.6	10.7
Apple	30	60	10	10.9	0.5	36.4	24.4
Cabbage	15	60	5	9.6	0.3	64.0	15.3
Spinach	10	60	5	5.3	1.1	53.0	7.9
New summer orange peel	30	60	5	27.3	0.9	91.1	46.0
Shaddock peel	30	60	5	18.1	5.4	60.4	44.3
Papaya	40	60	8	19.4	2.9	48.5	44.0
Onion	42	60	10	16.7	3.4	39.8	24.4

Table IV. Production of Unicellular Sweet

Raw Material	Enzyme Solution
Raw sweet potato Norin No. 2, cut 10 mm. × 10 mm. × 10 mm. 800 g.	1% Macerozyme pH 5.0, 800 ml. 0.1% Na-dehydroacetic acid
Raw sweet potato Norin No. 2, cut 10 mm. × 10 mm. × 10 mm. 800 g.	1% Macerozyme pH 5.0, 800 ml.
Air-dried sweet potato sliced into 3–4 mm. thickness and crushed by hand 50 g.	1% Macerozyme pH 4.5, 150 ml.
	1% Cellulase onozuka, pH 5.0, 150 ml.
	1% Macerozyme 75 ml. and 1% Cellulase onozuka 75 ml.
	Control 0.1M acetate buffer, pH 5.0, 150 ml.

Enzyme preparation: Cellulase onozuka P500 (*T. viride*) and Macerozyme (*Rhizopus*

tivation of CSE occurred at 60°C. for 10 min. The cellulase and CSE activities of enzyme preparations were compared (Table I and II).

Unicellular vegetables, fruits, or potato could be obtained by treatment with CSE from a variety of foods such as carrot, spinach, cabbage, potato, tomato, and green tea leaf. These unicell foods may be convenient for digestibility, preservation, transportation, combination, mixing with seasonings, flavors, and vitamins as well as cooking. In addition, in attempts to isolate various important materials such as starch, protein, and oil, CSE would exhibit marked improvement in cooperation with cellulase. The activities of CSE and cellulase are illustrated in Figure 1.

Production of Unicellular Vegetables with CSE. Substrates cut into 5 × 5 × 5 mm. (Table III) were incubated with 60 ml. of 1% Macerozyme (0.1M, pH 4.5 acetate buffer) in a 300 ml. Sakaguchi flask on a reciprocating shaker, amplitude 70 mm., 140 stroke/min. at 40°C. for indicated times. The residue was removed with gauze, and unicellular, but partly degraded, substrate in the filtrate was separated by centrifuging at 3,000 r.p.m. for 5 min. Little degradation of substrate was observed in control run with acetate buffer (33).

For example, almost all carrot pieces were degraded into viscous reddish orange paste, and the residue consisted of cuticle, crude fibers, and other insoluble matter. Production of unicellular sweet potato was exemplified (Table IV) by using Pancellase P500 (equivalent to Cellulase onozuka P500) with or without Macerozyme (33). Sweet potato pieces

Potato with CSE and the Effect of Cellulase

Incubation Method	Unicells	Residue	Remark
100 g. and 100 ml. in a 300 ml. Sakaguchi flask, at 40°C., 140 stroke/min. for 20 hr.	175 g. dried at 40°C. yield 21.9% as wet, 400 g.	25 g. dried at 40°C. as wet, 90 g.	Most of product unicellular
Glass cylinder 16.5 cm. × 21 cm. Propeller dia. 5 cm. and 11 cm. 800 r.p.m., 40°C. for 2 hr.	200 g. dried at 40°C. yield 25% as wet, 440 g.	90 g. dried at 40°C. as wet, 260 g.	Most of product unicellular
In a 300 ml. Sakaguchi flask, at 38°C., 120 stroke/min. for 3 hr., then filtered with Saran net and centrifuging at 4000 r.p.m. for 5 min.	25.9 g. freeze-dried	8.4 g. dried at 40°C.	Unicellular
	8.5 g. "	36.2 g. "	Starch granules
	27.6 g. "	8.7 g. "	Starch granules
	8.8 g. "	37.1 g. "	Starch granules

sp.) were employed.

were hardly degraded with Pancellase, however, remarkable degradation occurred with mixed enzyme solution giving rise to completely isolated starch granules.

Soybean seed coats were removed and unicellular soybean produced with CSE (Table V) (17). Both shaking and stirring were employed during degradation of the seed coat and cotyledon (at 40°C., pH 5.0). Stirring was particularly effective to shorten the incubation period. Removal of the external soybean seed coat gave good results, after cooking, in preventing the coloring of fermented soybean foods as Miso and soysauce. Freeze-dried unicellular beans are convenient for food processing.

Cellulase Capable of Degrading Plant Cell Wall. Cellulolytic activities in *T. viride* preparation could be separated into two fractions by the gauze column method, that is, non-adsorbed and adsorbed fractions. A typical procedure follows: Five grams of Cellulase onozuka P500 was dissolved in 15 ml. of McIlvaine citrate phosphate buffer, pH 2.8, after which its insoluble substances were separated by centrifuging for 10 min. at 4,000 r.p.m. The supernatant was placed on the column, after washing with 60 ml. of the same buffer, then 60 ml. of brownish eluate was collected by elution with the same buffer, pH 7.0 (= non-adsorbed fraction). The adsorbed fraction, colorless but somewhat turbid, was collected by elution with distilled water (up to 60 ml.). (Gauze column was prepared as follows: Pharmacopoeial gauze cut into 1×1 cm. sheets

Table V. Treatment of Soybean with CSE

Purpose	Condition of Enzymic Treatment	Remark
Removal of seed coat	In a 300 ml. Sakaguchi flask, 20 g. of air-dried soybean was shaken with 100 ml. of 1% Macerozyme solution, pH 5.0, on a reciprocating shaker at 140 stroke/min., amplitude 7 cm., 40°C. for 4 hr.	Complete removal of seed coat could be attained at 28°C., 200 r.p.m. with stirring for 4 hr. under similar conditions.
Production of unicells with shaking	40 g. air-dried soybean was soaked in water for 7 hr., after which it was shaken with 90 ml. of 1% Macerozyme solution, pH 5.0, in a 300 ml. Sakaguchi flask on a reciprocating shaker at 140 stroke/min., amplitude 7 cm., 40°C. for 43 hr. in the presence of 0.1% Na-dehydroacetic acid.	Freeze-dried unicells 12.3 g. (Control 0.3 g.) Yield of unicells 30.8% Residue dried at 40°C. 12.5 g. (Control 30 g.)
Production of unicells with stirring	250 g. air-dried soybean was soaked in water for 24 hr., after which it was incubated with 500 ml. of 1% Macerozyme solution, pH 5.0, in a glass cylinder, inner dia. 13 cm. × height 13 cm., radius of screw 3.7 cm., 3 blades, with stirring at 40°C. for 4 hr. in the presence of 0.1% Na-dehydroacetic acid, and 400 r.p.m.	Freeze-dried unicells including isolated protein 128 g. Yield of unicells 51.2% Residue dried at 40°C. 16.5 g.

Table VI. Enzymatic Activities of Adsorbed, and Non-adsorbed Fractions in *Trichoderma* Cellulase

Fraction	CMCase	Pectic Acidase	Amylase	Xylanase	Cellobiase
Non-adsorbed	10.89	2.72	23.65	3.52	0.45
Adsorbed	6.60	0.176	0.33	0	0

Fraction	CMCase	Filter Paperase	CMCase	Chitinase
Non-adsorbed	20.21	200 (unit/mg.)	9.7	38 (NAGA, ug./2 ml.)
Adsorbed	10.86	666	6.6	2

was loosened in Waring blender and then stuffed into a glass tube of adequate size.)

Various enzyme activities in each fraction were assayed using CMC (Cellogen WS, Daiichikogyo Seiyaku), pectic acid (Tokyo Kasei), soluble starch (Wako), cellobiose (Ishizu) and xylan prepared from rice straw in this laboratory. Relative activities (as glucose mg./10 ml.) are shown in Table VI. Most of the protease and β-1,3-glucanase were in the non-adsorbed fraction. Cell separating enzyme also was in the non-adsorbed fraction, while no maceration of potato was shown with the adsorbed fraction. The enzymes having little correlation with cellulose, such as pectic acidase (saccharifying activity to a commercial pectic acid), amylase, xylanase, cellobiase, chitinase, protease, β-1,3-glucanase and cell separating enzyme, could not be adsorbed by gauze column. Furthermore, it was shown that the addition of CSE to cellulase was necessary for the degradation of potato.

In the case of practical applications of cellulase in food processing, it is important to know which cellulolytic component, namely CMCase and filter paperase (filter paper degrading activity), is responsible for the degradation of parenchymatous cell walls. Correlation of filter paperase with potato cell wall degrading activity could be demonstrated by using the adsorbed fraction. Each 10 ml. of eluate from gauze column (with distilled water) was collected by fraction collector. Potato degrading activity was measured using 4 ml. of each fraction and 1 ml. of *Rhizopus* sp. wheat bran Koji extract by the author's shaking method. Selected CSE activity in the Koji extract was so weak that it could not degrade potato pieces after 4 hr. incubation. As can be seen in Table VII, in the presence of CSE, though its activity was very low, the most active degradation of potato was shown with the fraction having most filter paper degrading activity, although peak of this activity was somewhat different from that of potato degradation. No degradation of potato occurred with the fraction possessing only CMCase activity (20).

As described above, vegetable foods were completely decomposed with non-adsorbed fraction containing filter paperase. However, the non-adsorbed fraction including little or no filter paperase was also able to degrade vegetable foods. An examination under the microscope showed the existence of unicellular particles and no breakdown of cell wall was revealed in the latter case (35). Further examination was made to confirm the contribution of filter paperase to cell wall degradation by using potato, rice, sweet potato, carrot, and Shiitake *(Lentinus edodes)* mush-

Table VII. Peak of Filter Paper Degrading Enzyme Activity and Maximum of Potato Degradation in Adsorbed Fraction

Filter Paper Degrading Enzyme Activity

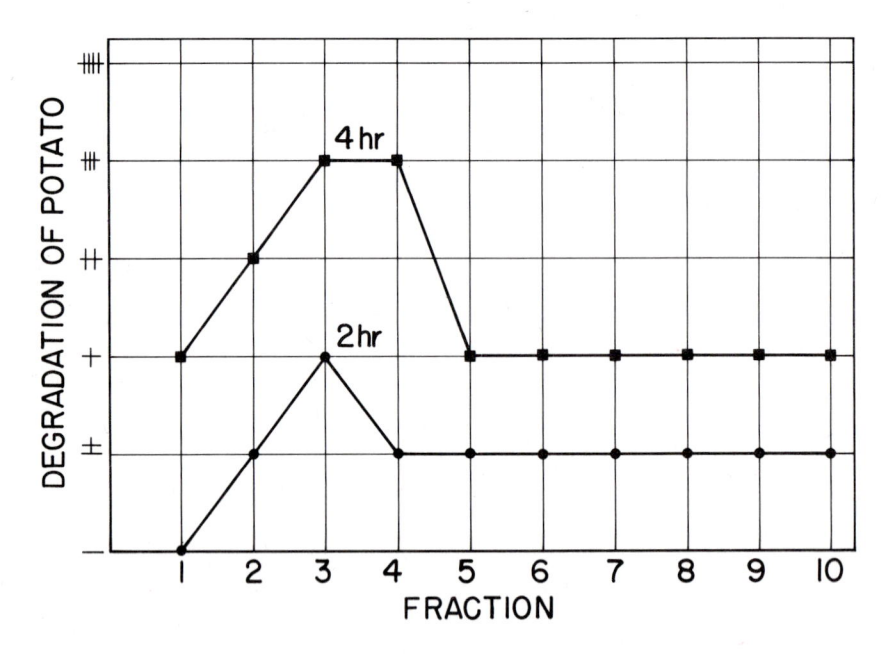

Incubation Time (min.)	1	2	3	4	5	6	7	8	9	10
15	−	−	−	−	−	−	−	−	−	−
30	−	++++	++++	+	−	−	−	−	−	−
45	±			++++	+++	±	±	−	−	−
60	++				++++	±	+	+	±	±
75	++					±	+	++	+	+
90	++					±	+	++	+	+
105	+++					+	++	+++	++	++
120	+++					+	++	+++	+++	+++
135	+++					+	++	+++	+++	+++
150	+++					++	++	+++	+++	+++

room. The employed enzyme solutions were made up in 0.1M, pH 5.0 acetate buffer solution. For the purpose of removing filter paperase, 100 ml. of enzyme solution was shaken with 1 gram of gauze in a 300 ml. Erlenmeyer flask on a reciprocating shaker at 40°C. for 24 hr., after which the gauze was removed. This enzyme solution was incubated at 70°C. in a water bath to inactivate residual filter paperase activity (38 min. for Cellulase onozuka PE, and 15 min. for Cellulosin AP). The degradation of vegetable foods was assayed as follows: 5 ml. of enzyme solution (however, in the case of adsorbed fraction with Macerozyme, 4 ml. of adsorbed fraction and 1 ml. of 1% Macerozyme) was incubated with 1 gram of polished rice grains, raw mushroom (1 × 2 cm.), or potato, sweet potato or carrot (1 × 2 × 0.2 cm.) in a test tube in the presence of toluene at 40°C. for 46 hr., followed by hand shaking (Table VIII).

As shown in this table, no degradation of foods occurred with heated enzyme solution, and CMCase did not contribute to the degradation. Filter paperase alone, even though it was active in adsorbed fraction, was incapable of degrading these foods. Shiitake mushroom was degraded with β-1,3-glucanase and chitinase in the non-adsorbed fraction. Similar activity of xylanase to that in non-adsorbed fraction or nontreated Cellulase onozuka PE was exhibited also in Macerozyme; however, these foods were degraded into only unicells with Macerozyme, while disappearance of those cell walls occurred with the adsorbed fraction containing little or no xylanase in the presence of Macerozyme (35).

Extraction of Green Tea Component. Various components in green tea leaves such as amino acids, carbohydrates, tannin, caffeine, saponin, and aroma are thought to be included within the cells. Manufactured green tea leaves were degraded readily, except for the stalks and veins, by the treatment with an enzyme solution composed of CSE and cellulase (1 : 1 by weight). After squeezing with gauze, a dark greenish and fragrant hydrolyzate and a residue consisting of stalks and veins were obtained.

The hydrolyzate could be separated by centrifuging into supernatant and unicells. The supernatant contained tea components 2 to 3 times as large in yield as by the usual infusion method. It was freeze-dried, spray-dried, or air-dried at 40°C. in the presence of supplements such as soluble starch. Unicellular green tea leaves (also containing tea components) were dried under the same conditions and made available as a manufactured green tea. It was also suitable for mixing with other foods. For example, 150 grams of Sencha, the tea most used in Japan, was stirred with 1,000 ml. of enzyme solution (0.3% Cellulase onozuka P500 and 0.3% Macerozyme, CSE 666 unit/mg.) at 40°C., 300 r.p.m. for 2 hr. in a Marubishi Jar-fermenter MBA-05 (5,000 ml.). The vein and stalk were removed as residue with two sheets of gauze, after which unicells were separated from soluble matter by centrifuging at 3,000 r.p.m. for 10 min. The soluble matter was dried with soluble starch at 40°C. under ventilation and the unicells were freeze dried. The yield of soluble matter was 32.9% and the yield of unicells was 8.4%. Consequently, 95.4 grams of the green tea powder containing 40 grams of soluble starch and 6 grams of enzyme preparations was obtained (Table IX) (36).

In these experiments, distilled water, pH 5.6, was used in place of acetate buffer to avoid an effect on tea flavor. Black tea leaf could also be degraded with wheat bran Koji extract of *Rhizopus* sp., but, little or no degradation occurred with Macerozyme alone.

Isolation of Soybean Protein. The cotyledon of soybean consists of numerous unicells containing protein and oil. Attempts were made to isolate soybean protein by using CSE and cellulase from air-dried soybeans and defatted soybean cakes. For instance, to 100 grams of air-dried defatted soybean cake (for soysauce brewing, crushed into 20

Table VIII. Influences of CMCase, Filter Paperase, and

	CMCase	Filter Paperase 60 min.	Filter Paperase 180 min.	Xylanase
Non-adsorbed[b] fraction (PE)	20.1	—	+	1.7
Adsorbed[b] fraction (PE)	3.6	++++ (90 min.)		0.2
Cellulase onozuka PE non-treated	21.2	++++		1.6
Cellulase onozuka PE heated	19.5	—	—	0
Cellulosin AP (beta-glucanase) non-treated	27.5	++++ (150 min.)		2.6
Cellulosin AP (beta-glucanase) heated	27.5	—	—	0.9
Adsorbed fraction (PE) and Macerozyme				
Macerozyme	6.1	±	+++	1.4
Control				

[a] Degraded into unicells.

Table IX. Extraction of Green Tea Component with CSE and Cellulase

	Gyokuro	Sencha
Raw material	150 g.	150 g.
Product	74 g.	95.4 g.
Soluble starch	40 g.	40 g.
Enzyme preparation	6 g.	6 g.
Soluble matter	28 g.	49.4 g.
Yield of soluble matter	18.7%	32.9%
Unicellular green tea	29.7 g.	12.6 g.
Yield of unicells	19.8%	8.4%
Yield of soluble matter and unicells	38.5%	41.3%

to 200 mesh) was added 360 ml. of $0.1M$ acetate buffer, pH 5.0, and 3 grams of Cellulase onozuka P500, or 3 grams of mixed preparation (P500 : Macerozyme = 1 : 1). The latter enzyme solution was inacti-

Xylanase on the Degradation of Vegetable Foods

Rice	Potato	Sweet Potato	Carrot	Mushroom
+++	+	+	+	+++
+	−	−	−	−
+++	+	+	+	+++
−	−	−	−	−
+++	+++	+	+	−
−	−	−	−	−
+++	+++	+++	+++	−
+++	+++ [a]	+++ [a]	+++ [a]	+
−	−	−	−	−

[b] Fraction from *T. viride* cellulase adsorbed—or non-adsorbed—on gauze column.

vated by boiling for 10 min. and employed as control. Three mixtures were incubated at 40°C., 250 r.p.m. with stirring for 5 hr. in each glass cylinder (9 × 22.5 cm.) held in a water bath, then the hydrolyzate was squeezed by nylon filter cloth (plain fabrics, 200 mesh/cm.^{2}) and the filtrate was separated into protein and supernatant after adjustment to pH 4.5 by centrifuging at 3,400 r.p.m. for 10 min. The residue was again stirred with 360 ml. of distilled water in the same vessel at 45°C., pH 6.2, 250 r.p.m. for 3 hr., following the same procedure (Table X). The amino acid in the supernatant was assayed by Pope and Stevens method (23). The yield of protein was increased 3.3 times with cellulase and 3.5 times with mixed enzyme as compared with the control.

The extent of denaturation of soybean by heating influences significantly the rate of protein isolation. The extent of denaturation can be shown by the NSI (nitrogen solubility index). It was shown to be adequate to use a mixed enzyme solution at concentrations of 0.2 to 1.0% (usually 1.0 to 3.0% of substrate). Although good results were obtained by enzymatic treatment of raw soybeans and defatted soybean cakes with NSI of 24 (for brewing of soysauce and fermented soybean

Table X. Protein Isolation from Crushed Control

	Weight (g.)	Total N (g.)	Total Sugar (g.)	Weight (g.)
Supernatant 1st	—	0.42	10.0	—
2nd	—	0.23	4.02	—
Protein	7.99	1.05	0.35	27.64
Residue	54.20	0.57	5.36	32.02

Protein and residue were dried at 40°C. under ventilation.

paste), the protein could be isolated from defatted soybean cakes with NSI of 80 readily by treatment with hot water alone (13).

Isolation of Sweet Potato Starch. After degradation of parenchyma into individual starch cells with CSE (Figure 1), the whole surface of the starch cell can easily be attacked with cellulase, resulting in complete isolation of the starch granules. The effect of mixed enzyme on the isolation of starch from raw sweet potato was examined. First, 800 grams of raw sweet potato (about 65% moisture) was disintegrated by a continuous juicer and separated into starch slurry and starch waste. Starch was recovered from the slurry and from the filtrate obtained by washing the waste 2 times with 1,000 ml. of water, and centrifuging at 4,000 r.p.m. for 5 min. In a glass cylinder (16 × 20 cm.), 395 grams of waste (gained by squeezing with gauze) was stirred with 400 ml. of *Rhizopus* CSE Koji extract (pH 5.0) and 400 ml. of 1% Meicelase P (pH 5.0) at 200 r.p.m, 40°C. After such a short treatment as 2 hr., the yield of starch amounted to 23.1% corresponding to that obtained by usual process (Table XI)

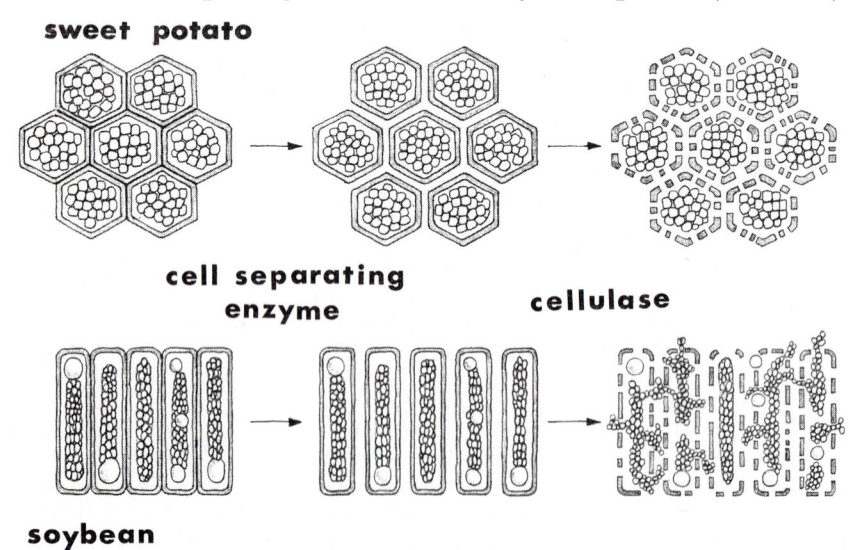

Figure 1. Enzymic degradation of sweet potato and soybean

Defatted Soybean Cake with Enzymes

Cellulase			Cellulase and CSE	
Total N (g.)	Total Sugar (g.)	Weight (g.)	Total N (g.)	Total Sugar (g.)
1.09	9.24	—	1.78	9.77
0.49	3.75	—	0.51	3.99
3.49	1.26	29.40	3.73	0.83
0.26	7.08	23.23	0.15	5.75

	Amino-nitrogen (mg.)		Protein Content in Isolated Protein (%)
	Supernatant 1st	Supernatant 2nd	
Control	22	2	82.4
Cellulase	34	6	78.8
Cellulase and CSE	52	8	79.3

(34). The important role of CSE in starch isolation may be clear from the following results (Table XII) (34). Air-dried sliced sweet potatoes were immersed in water for 2 hr., then cut into 1 × 1 × 0.3 to 0.4 cm. pieces. Each 1000 grams of substrate was shaken with 100 ml. of 1% Meicelase P solution, 50 ml. of 1% Meicelase P solution and 50 ml. of *Rhizopus* CSE Koji exeract (air-dried Koji : water = 1 : 5) or 100 ml. of water adjusted to pH 4.0 with sulfuric acid in a 300 ml. Sakaguchi flask on a reciprocating shaker at 40°C., 140 stroke/min. for 2 hr. After incubation, residue was removed by squeezing with gauze. Crude starch in filtrate was separated by centrifuging at 4,000 r.p.m. for 10 min.

Influence of combined enzyme preparations upon the recovery of starch from raw waste was also investigated (Table XIII, (34)). Each 1 gram of enzyme preparation was dissolved in 100 ml. in 0.1M acetate buffer, pH 4.5. Fifty grams of raw starch waste (moisture 55%, starch 15.1 g.) was shaken with 100 ml. of enzyme solution at 40°C. for 1 to 2 hr. in a 300 ml. Sakaguchi flask. It was noteworthy that no degradation of sweet potato or of starch waste occurred with Cellulosin AP (with β-glucanase), although this preparation was able to degrade effectively white potato, and all cellulase preparations were capable of degrading efficiently starch waste in cooperation with Macerozyme. The quantitative analysis of starch in sweet potato or potato would be feasible using Cellulase onozuka PE or Cellulosin AP (with β-glucanase) along with Macerozyme.

It was thought that starch might be isolated with mixed enzyme from air-dried sweet potato powder prepared by the dry milling process. Starch would be produced throughout the year from sliced air-dried sweet potato. Air-dried sliced sweet potato was disintegrated in a Waring Blendor and fractionated with pharmacopoeial seven sieves. Four representative fractions were selected and 40 grams of each was shaken with

Table XI. Enzymic Starch Isolation from Sweet Potato and Its Waste

Raw Material (g.)	Treatment	Isolating Method	Starch (g.)	Starch Waste (g.)	Yield of Starch (%)
Raw sweet potato 800	Mechanical disintegration[a]	Washing 2 times with each 1000 ml. of water and centrifuging at 4000 r.p.m. for 5 min.	130	—	16.3
Raw starch waste 395	400 ml. of CSE Koji extract and 400 ml. of 1% Meicelase P	Glass cylinder, inner dia. 16 cm., height 20 cm., propeller dia. 7 cm., at 200 r.p.m., pH 5.0, 40°C. for 2 hr.	55	25	13.9

[a] Sweet potato was disintegrated by a continuous juicer, Fuji Denki JC 281. Starch was recovered from starch slurry produced continuously by a juicer and from filtrate obtained by washing with water.

100 ml. of enzyme solution including 1% Cellulase onozuka P500 and Macerozyme in 0.1M acetate buffer pH 4.5, at 40°C., 150 stroke/min. in a 300 ml. Sakaguchi flask on a reciprocating shaker for 3 hr. After incubation, starch was separated from crude protein and residue by centrifuging at 3,000 r.p.m. for 5 min., followed by drying at 40°C. After drying, the starch was disintegrated by pestle and mortar, sifted with 200 mesh sieve and weighed (Table XIV) (37). A 200 gram portion of 200 mesh sweet potato powder was stirred with 400 ml. of 0.5% Cellulase onozuka P500 and Macerozyme solution, pH 4.5, at 40°C., 300 r.p.m. for 4 hr. in a glass cylinder (8 × 18.5 cm.). After being separated by centrifuging, isolated crude starch was treated with 300 ml. of 0.5% ammonium oxalate for 30 min. at room temperature (Table XV) (37). As a result of further experiments, the following method was proved to be practical. Sliced and air-dried sweet potatoes were disintegrated and fractionated by sifting into a 200 mesh powder and residue. By a ratio of 2 : 1, a 0.25% solution of mixed enzyme preparation was added to the powder and by a ratio of 3 : 1, a 0.33% enzyme solution to the residue. After 4 hr. incubation with stirring at 300 r.p.m., 40°C., pH 4.5, the starch recovery was 76% in the case of sweet potato powder. The recovery from residue was 32%, about four times the amount of starch as compared with the nonenzyme control. In this process, a small quantity of enzyme preparation, about 0.3% of the weight of raw sweet potato, was sufficient. The quantity of water used for starch isolation, purification and washing was half as much as in the usual process (37).

The enzymatic process was improved as follows. After soaking airdried sliced sweet potato in water for 3 hr., they were disintegrated by a disposer (Hush Waste King Universal 3000) for separating starch

Table XIII. Influence of Combined Enzyme Preparations upon the Sweet Potato with Cellulase

Enzyme	Crude Starch (wet)		Residue (wet)
Rhizopus CSE Koji extract	54.0 g.	(unicells)	32.0 g.
CSE extract and Meicelase P (1 : 1 by volume)	46.0	(starch)	40.0
Meicelase P	3.0	(starch)	101.5
Control	3.0	(starch)	109.0

Table XIII. Influence of Combined Enzyme Preparations upon the Recovery of Starch from Raw Waste[a]

Enzyme	Incubation Time (hr.)	Starch Recovered (g.)	Residue (g.)	Recovery %
Control	1	6.0	13.4	39.7
	2	5.8	13.1	38.4
Cellulosin AP (with β-glucanase) and Tetrase C	1	6.2	11.1	41.1
	2	6.3	11.0	41.7
Cellulosin AP (with β-glucanase) and Meicelase P	1	6.3	12.0	41.7
	2	6.1	11.8	40.4
Macerozyme and Cellulosin AP (with β-glucanase	1	15.3	1.0	101.3
	2	15.2	0.9	100.7
Macerozyme and Meicelase P	1	9.5	6.7	62.9
	2	11.1	5.0	73.5
Macerozyme and Cellulase onozuka PE	1	16.3	1.0	108.0
	2	15.3	0.7	101.3

[a] Raw starch waste was produced by a continuous juicer.

Table XIV. Influence of Particle Size on Enzymatic Isolation of Starch and Protein from Sweet Potato Powder

Fraction		Starch (g.)	Residue (g.)	Crude Protein (g.)
4 mesh (4.76 mm.)	Control	0.5	34.1	1.32
	Enzyme	10.8	15.2	2.80
10 mesh (2.00 mm.)	Control	1.2	34.8	0.80
	Enzyme	7.2	20.1	2.60
100 mesh (0.149 mm.)	Control	5.6	27.6	0.39
	Enzyme	15.0	5.2	2.20
200 mesh (0.074 mm.)	Control	5.6	27.9	0.90
	Enzyme	19.4	4.9	5.30

slurry from raw waste. Each 200 grams of raw waste was stirred with 400 ml. of 1% enzyme solution (0.5% Macerozyme and 0.5% Cellulase onozuka P500; 0.5% Macerozyme and 0.5% Cellulosin AP; or 1% Cellulosin AP) at pH 4.0, 40°C., 600 r.p.m. for 3 hr. in a glass cylinder (8 × 18.5 cm.) (Figure 2) (38). Good results were revealed by the use of mixed enzyme preparation and similar result was also obtained as shown in Table XVI (38).

Table XV. Isolation of Starch with Enzyme from Sweet Potato Powder

	Starch (g.)	Residue (g.)	Proteinous Residue (g.)	Yield of Starch (%)
Control	35.6	89.6	28.3	17.8
Enzyme	102.3	10.5	23.3	51.2

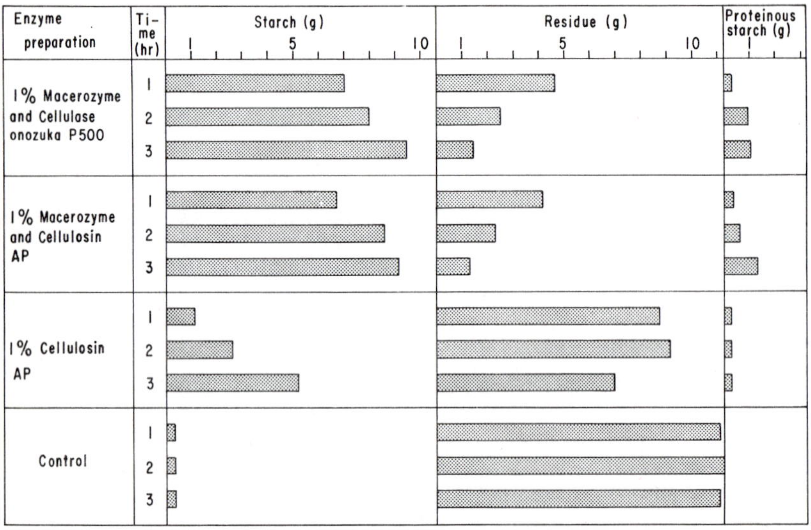

Figure 2. Effect of mixed enzyme preparation upon raw starch waste obtained from air-dried sliced sweet potato by disposer

Seaweed Jelly. Konbu (*Phaeophyceae, Laminaria*) is an important food for Japanese, because of its high content of calcium and iodine, as well as a source of flavoring. As the cells are bound to each other with insoluble alginic acid or its salt, the digestibility of seaweeds is low. An attempt was made to dissolve the insoluble intercellular substance with organic salts and to eliminate naked cell walls with cellulase to produce soluble and digestible seaweed jelly. Figure 3 shows the structure of tangle (kelp) and its gelatinization with organic salt and cellulase. Sufficient gelatinization was revealed using 1% sodium citrate solution after 30 min. boiling or 0.5% for 45 min.

Appreciable increases in quantities of L-glutamic acid, total sugar, calcium, total nitrogen, and iodine were obtained after treatment with

Table XVI. Enzymic Starch Isolation from Raw Waste Prepared by Disposer

Recovered Starch

	Yield (g.)	Yield (%)	Starch Value (%)	Rate of Recovery (%)
0.5% Macerozyme and 0.5% Cellulase onozuka P500	95.4	23.9	85.90	72.7
Control	25.5	6.4	83.44	18.9

Residue

	Yield (g.)	Yield (%)	Starch Value (%)	Rate of Starch Remaining (%)
0.5% Macerozyme and 0.5% Cellulase onozuka P500	20.4	5.1	74.0	13.1
Control	107.0	26.8	74.3	70.5

Raw waste 400 g., enzyme solution 800 ml., incubated with stirring at pH 4.0, 40°C., 600 r.p.m. for 3 hr.

Laminaria

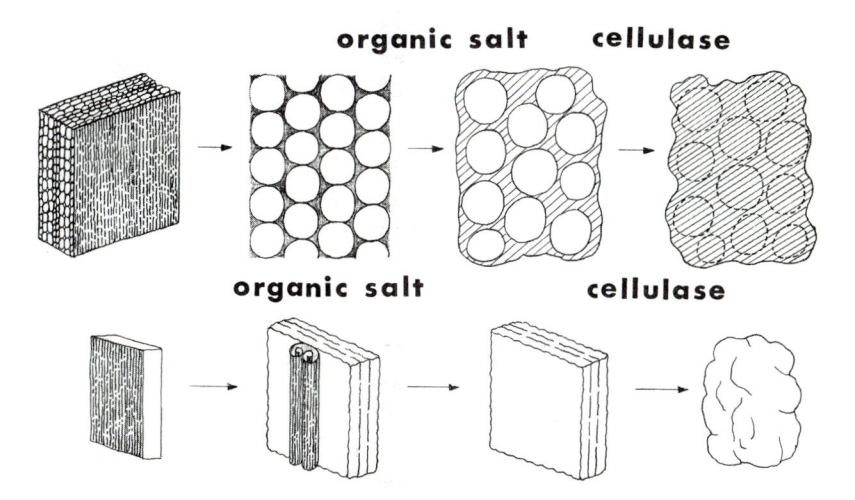

Figure 3. Gelatinization of seaweed with organic salt and cellulase

sodium citrate (Table XVII). In this experiment, 15 grams of air-dried konbu was gelatinized by treatment with 1,000 ml. of 1% sodium citrate solution, while 15 grams of konbu was immersed in 1,000 ml. of water for 1 hr. at room temperature, after which konbu extract was prepared by boiling for 5 min.; (this is the usual method for obtaining flavor from konbu in Japan). Enhancement of aroma and flavor was clearly attested (Table XVIII). After such treatments, however, numerous unicells were detected as precipitate. More homogeneous jelly could be prepared by dissolving the cell walls with cellulase. The efficiency of extracting seaweed components by sodium citrate with or without cellulase was evaluated (Table XIX) (7). In this experiment, 100 grams of air-dried konbu was jellied with 2,000 ml. of 1% sodium citrate. After diluting the jelly two times with water, 20 grams of Cellulosin AP were added. The reaction mixture was incubated at pH 6.0, 40°C. for 20 hr. on a rotary shaker, and insolubles were removed by centrifuging at 8,000 r.p.m. for 10 min. Control was run similarly but without enzyme.

Table XVII. Effect of Treatment with Sodium Citrate

Substance	Konbu Jelly (g./1000 ml.)[a]	Control (g./1000 ml.)[b]	a/b
L-Glutamic acid	0.47	0.28	1.7
Total-sugar	0.72	0.01	72.0
Calcium	0.095	0.02	4.8
Total-nitrogen	0.16	0.07	2.3
Iodine	0.048	0.02	2.4

[a] Air-dried kelp, 15 g., was jellied with 1000 ml. of 1% sodium citrate.
[b] Air-dried kelp, 15 g., was immersed in 1000 ml. of water for 1 hr., then boiled for 5 min.

Table XVIII. Organoleptic Tests on Aroma and Flavor

Dilution Rate of Konbu Jelly	Aroma Konbu Jelly	Control	Dilution Rate of Konbu Jelly	Flavor Konbu Jelly	Control
4	15[a]	0	1	14[a]	1
8	13[b]	2	2	12[b]	3
16	13[b]	2	4	12[b]	3
32	11	4	8	7	8

[a] Significant at $\alpha = 0.1\%$.
[b] At 1.0%.

The content of each component was calculated as weight in grams in the extract from 100 grams of air-dried konbu. The increase in the amount of amino-nitrogen may be ascribed to protease activity in the Cellulosin AP.

Modifying Tissue of Vegetable Foods. As a result of removal of cell walls and pectic substances, air-dried carrot treated by cellulase solution—e.g., Cellulase onozuka F400, T. viride, filter paperase 500 unit/mg.—regained their original form easily when warm water was poured over them, and the tissue softened moderately without cooking. The same was true of glutinous rice to which boiling water was added (5). Cell walls in rice tissue disappeared by the action of cellulase, giving rise to increased soaking efficiency and homogeneous water absorption. Air-dried glutinous rice treated with a T. viride cellulase preparation became rice cake after a few minutes' steaming. Instant rice may be produced by such a treatment.

Cell walls in rice tissue are so weak that they are degraded with weak cellulolytic activity, such as exists in a *Rhizopus* CSE preparation,

Table XIX. Efficiency of Extracting Seaweed Components by Treatment with Organic Salt Along with or without Cellulase

Substance	Konbu Jelly Treated with Cellulase (a) (g./100 g. dried konbu)	Konbu Jelly (b) (g./100 g. dried konbu)	a/b
Total-nitrogen	0.600	0.401	1.5
Amino-nitrogen	0.302	0.086	3.5
Total-sugar	8.177	7.078	1.2
Calcium	0.848	0.788	1.1

after a long incubation period. The *Rhizopus* preparation also contains a potent amylase capable of saccharifying raw rice. Rice cell walls could not be completely decomposed with Taka-diastase derived from *Aspergillus oryzae*. Rice grains could be effectively gelatinized after treating with cellulase or CSE preparation, whereas non-gelatinized starch grains were partly observed in non-treated rice tissue (6). In addition, puffed rice cake having soft tactile impression could be obtained by using rice grains treated with *T. viride* cellulase preparation.

Among other plant tissues analogous phenomena were observed. Especially interesting features were exhibited on freeze-dried foods after treatment with CSE with or without cellulase. For example, porous, voluminous and fragile dehydrated soybean could be produced using defatted soybean cake by treating with 1% Macerozyme, or 0.5% Macerozyme and 0.5% cellulase solution. After incubation of defatted soybean cake with an adequate quantity of enzyme solution at pH 5.0, 40°C. for 20 hr., the cake was freeze-dried. Little or no change was observed in the tissue of cake run as control, while cell walls in the tissue of enzyme-treated cake were thinned, degraded, or disappeared and the remaining cell walls were easily degraded by finger pressure releasing protein particles. When the treated or non-treated cake was boiled with N-hydrochloric acid for 15 min., noticeable difference could be seen, and such an enzyme-treated soybean might be digestible. Under similar conditions, readily reconstitutable dehydrated vegetables could be obtained. It is worthy of notice that the color of vegetable such as carrot and spinach does not fade, but rather heightens, during enzymic treatment. The activities of CSE and cellulase are shown in Figure 4.

Discussion

The parenchyma cells in the edible portion of most fruits, vegetables, and roots are held together by cementing substances, particularly protopectin. The walls of parenchyma cells in young plants are composed almost entirely of cellulose fibrils, however, the primary wall of many cells is also rich in pectic substances believed to be protopectin. It has been found that very large members of protoplasts and vacuoles may be released from the thin-walled, parenchymatous cells of tomato-fruit placental tissue by treatment with fungal polygalacturonase in 20% sucrose. These observations suggest that pectic substances are important for the structural integrity of the cell walls in tissues such as flavedo of citrus fruits and soft tissues of fruits (8). Recently it was found that the culture fluids of the soft-rotting bacterium *Erwinia aroideae* contain polygalacturonase activity and cause maceration of parenchymatous tissue. It was shown that close correlation exists between the pectate transeliminase and macerating activities of the different samples. It was also recognized that partially purified pectate transeliminase can degrade cell wall material, as well as extracted pectic substances. Those results suggest that maceration of potato tuber tissue by culture fluids of *E. aroideae* is caused primarily by pectate transeliminase (4). On the other hand, the lack of macerating activity of the cellulase was demonstrated by heating Seitz-filtered juice of sweet potatoes rotted by *Rhizopus stolonifer* for 10 min. at 55°C. and comparing the maceration activity of this juice with an unheated preparation. The heating destroyed both the polymethylgalacturonase and the macerating activity of the juice without inactivating the cellulase (26).

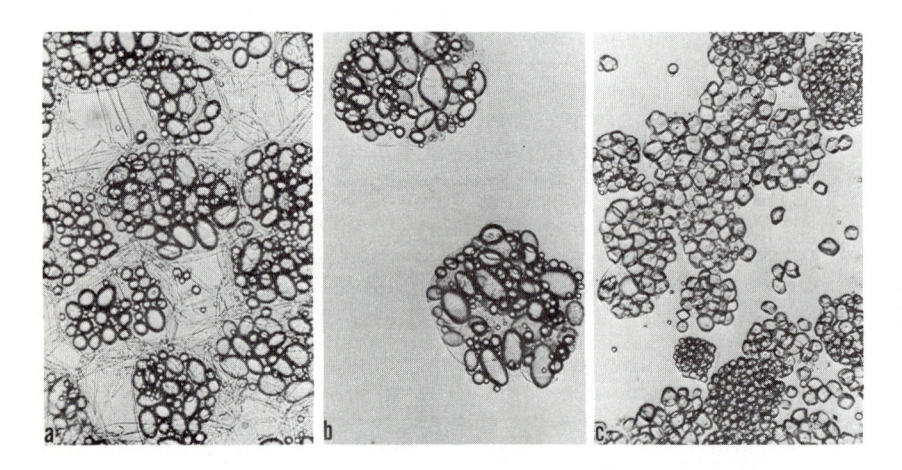

Figure 4. Effect of enzyme action on potato tissue. Photographs × 150
(a) Control in acetate buffer; no enzyme
(b) Unicells produced by Macerozyme (1%)
(c) Starch granules produced from sweet potato by Cellulase onozuka P500 (1%)

The term "maceration" is frequently used for describing the degradation of plant tissue owing to enzymic solubilization of middle lamella pectate and cell wall. The author exhibits a preference for the term "cell separating enzyme" to "macerating enzyme," because the former means an enzyme having only protopectinolytic activity.

The nature of a cell separating enzyme (CSE), Macerozyme, was studied by Suzuki *et al.* The maceration extent was measured by the decrease in weight of fresh tissue disks during enzyme treatment. Upon column chromatography of Macerozyme, a strong correlation was found between macerating enzyme and polygalacturonase activities. The products from radish root disks by the action of a macerating enzyme fraction were partially degraded pectic substances composed mainly of polygalacturonic acid. It was concluded that polygalacturonase of an endotype must principally contribute to the maceration of plant tissues (27).

Further chemical changes of plant tissues during enzymatic maceration have been studied using radish root and a purified endo-polygalacturonase preparation from Macerozyme. The major products of enzymatic maceration are water-soluble polygalacturonides with at least DP 50, which contain a considerable amount of neutral sugars. It was concluded that random splitting by an endo-polygalacturonase may cause release of water soluble fragments of pectic substances and consequent weakening of various secondary bonds among polysaccharide chains constituting intercellular cement, and that this process may be the most important reaction in maceration by the enzyme preparation, Macerozyme (28).

Since 1961, *Trichoderma viride* and *Aspergillus niger* cellulase preparations are being produced industrially. With purification (39), cell-separating enzyme was lost from *T. viride* cellulase preparations, and it was necessary to develop a new CSE preparation capable of dissolving the cementing substances of middle lamella. In 1963 to 1964, a CSE derived from a *Rhizopus* sp. was investigated by the author and produced industrially under the name of Macerozyme. It was considered that a *Rhizopus* sp. producing active CSE on wheat bran might be an ideal mold for making this enzyme, since the mold secreted very weak cellulase.

With regard to production of unicellular plants with CSE, Takebe *et al.* successfully isolated tobacco mesophyll cells in intact and active state. The maceration medium contained 0.5% Macerozyme, 0.8M D-mannitol and 0.3% potassium dextran sulfate, pH being adjusted to 5.8 with 2N HCl. Two grams (fresh weight) of stripped leaf pieces were placed in 20 ml. of the maceration medium in a 100 ml. Erlenmeyer flask and were infiltrated with medium by evacuating the flask for 3 min. The flask was then shaken in a water bath at 25°C. on a reciprocating shaker. After 15 min. of preincubation, during which broken cells and their contents were released from cut surfaces of leaf pieces, the reaction fluid was removed by decantation and was replaced by 20 ml. of fresh maceration medium. Thereafter, the maceration medium was renewed in the same way at 30 min. intervals and incubation was continued until maceration of mesophyll was practically complete (usually 2 hr.). Mesophyll cells isolated were washed several times with 0.8M mannitol containing 0.1mM CaCl$_2$ and were suspended in a 2% solution of Cellulase onozuka P1500 in 0.8M mannitol (pH adjusted to 5.4 with 2N HCl). The suspension was incubated at 36°C. for 2 to 3 hr. with occasional gentle swirling. Under the conditions described above, 50 to 90% of the released cells were morphologically intact and converted into spherical protoplasts by cellulase treatment. Cells isolated in this way from tobacco mosaic virus-inoculated leaves supported multiplication of the virus (*32*).

Takahashi (*30*) had developed the author's idea by using his strain of *Rhizopus* sp. as source of CSE. After drying *Rhizopus* wheat bran sawdust Koji, it was extracted with water and followed by treating with Asmit-173 N resin for the removal of color and odor. The extract was used as an enzyme solution resembling Macerozyme. The following cases are good illustrations of his work. The pulp of *Citrus unshiu* was boiled with 2 volumes of water for various periods to remove bitter substances, after which it was filtered, washed, and dehydrated. Dried pulp was treated with the equal weight of CSE solution at 30°C., pH 4.0 for 1 to 1.5 hr. Good yields of unicells were obtained.

He attempted to produce fruit vinegar using citrus pulp treated with CSE and achieved good results. He also succeeded in obtaining pectin from the pulp of summer orange *(Citrus natsudaidai)*. Excepting CSE, pectinase can be inactivated at low pH. Thus, to the pulp of summer orange adjusted to pH 2.0 to 2.4 was added 2 volumes of the mixture of CSE and *T. viride* cellulase solutions. Complete maceration was shown after incubation at 30°C. for about 5 hr. Seeds and residues were removed and markedly coagulable pectin could be prepared by ethanol precipitation, or by spray-drying the filtrate. As the action of enzymes proceeded at very low pH values, it was desirable to use boiled pulp.

In addition, he investigated enzymic production of unicellular vegetables with CSE. The enzyme solution was available two to three times repeatedly in the treatment of Japanese radish greens, Chinese cabbage, carrot, and onion. Undesirable taste caused by calcium and magnesium salts was removable by the addition of chelating agent in enzyme solution and by thoroughly washing unicells with physiological saline solution. The seed coat of soybean is separable mechanically in large scale factories, but this method is hardly applied in small factories, such as the brewing of fermented soybean paste (miso) and soy sauce, as the machine requires sizable investment. The removal of seedcoat is profitable for decreasing brownish color being caused by the reaction between amino acids and pentose derived from hemicellulose in seedcoat. The seedcoat (outer and inner) is eliminated when soybeans are subjected to machine, resulting in the separation of cotyledon.

The outer layer is composed mainly of hemicellulose and pectic substance and the inner layer is rich in cellulose, so that the outer layer can be easily eliminated with CSE, and both layers with the mixture of CSE and cellulase. In the course of enzymatic treatment, soluble matter such as sugars causing the reaction with amino acid are formed, and those can also be readily discarded. Nakayama and Takeda (*18*) developed the author's enzymatic method as follows: To the soaked and swelled soybean was added two volumes of 1% Macerozyme solution, pH 5.0. After incubating at 40°C. for about 3 hr. with stirring, the outer seedcoats were degraded and separated cotyledon. The inner seedcoats were degraded by further incubation. During incubation with Macerozyme, the seedcoats around hilum were separated accompanying hilum and such seedcoats were scarcely separable with water or by screening. Under similar conditions only hilums remained after incubating raw soybeans with the mixed enzyme solution containing 0.5% Macerozyme and 0.5% Cellulase onozuka P500. An apparatus was recommended in their report. During enzymatic treatment in the apparatus, the hilum and degraded outer seedcoat were removable through the screen provided at the bottom. The enzyme solution could be used twice, but the second treatment required a somewhat longer incubation period.

With the view of elucidating the cellulolytic component responsible for the degradation of plant cell wall, Imai and Kuroda (*16*) investigated the breakdown of cell walls of unicellular potato with *T. viride* cellulase. Potato unicells prepared with *Rhizopus* CSE were incubated with the fraction containing filter paper degrading cellulase and CMCase, which was obtained from *T. viride* cellulase preparation by using a cellulose column akin to the author's gauze column method.

Furthermore, *T. viride* cellulase was fractionated by Amberlite CG 50 column into filter paper degrading cellulase, CMCase, and xylanase fractions. It was conclusively shown that only filter paper degrading cellulase contributed to the degradation of cell walls of unicellular potato.

Regarding the extraction of green tea components with cellulase, Igi and Oshima (*15*) stated that the addition of *T. viride* cellulase preparation at 0.3, 0.6 and 1.2% concentrations to dried leaves facilitated the extraction of tea components at low temperature such as tannin, caffeine, soluble nitrogen, soluble matter, and tasting substance from freshly picked green tea leaves, manufactured green tea leaves, and green tea leaves dried by infrared ray lamps which was employed with the purpose of producing instant tea throughout the year.

In regard to isolation of soybean protein, the following result was obtained in a laboratory of a food company under the author's guidance. A 50 grams portion of defatted soybean cake was soaked in water for 2 hr., after which it was cooked with pressure for 30 min., followed by cooling (130 grams). The 130 grams portion of cooked cake was incubated with 400 ml. of acetate buffer, pH 5.0, in the presence of potassium sorbate (control) or with 400 ml. of 3% enzyme solution (containing Cellulase onozuka P500, Cellulosin AP and Macerozyme each at 1%)

for 4 hr. with stirring at 40°C., 500 r.p.m., and maintained at 40°C. for 72 hr. (Table XX).

Soysauce brewing requires about one year for its production mainly owing to slow digestion of mash under the high concentration of sodium chloride and the rate of utilization of raw materials is low. In addition, various equipment with large capacity such as the hydraulic press and numerous workmen are indispensable for the separation of much soysauce filter cake. An attempt was made to decompose cooked soybean, cooked defatted soybean, or cooked soybean inoculated with *Aspergillus oryzae* by the use of cellulase preparations from *T. viride* or *A. niger*. The most remarkable effect was obtained after mixing raw materials at 1,000 r.p.m., 40°C., and pH 4.0 for 10 to 15 hr. with enzyme solution containing cellulase at concentrations of 0.4 to 1.0% (*2*). Furthermore, attempts were made to enhance the rate of utilization of raw materials with *T. viride* cellulase, *A. niger* protease preparations, along with wheat bran Koji extract of *Aspergillus oryzae* under the same conditions as reported. Those materials were degraded almost completely within 6 to 8 hr. at 40°C., pH 5.0 with stirring at 1,500 r.p.m. (*1*).

Table XX. Decomposition of Cooked Defatted Soybean Cake with Mixed Enzyme Solution

	Time (hr.)	Total Nitrogen (g.)	Volume of Liquid (ml.)	Total Nitrogen (%)	Total Nitrogen in Raw Material (g.)	Solubility of Total Nitrogen (%)
Control	24	0.309	490	0.063	4.039	7.6
	48	0.370	480	0.077	4.031	9.2
	72	0.362	470	0.077	4.023	9.0
Enzyme	24	2.761	472	0.585	4.039	68.4
	48	2.994	462	0.648	3.974	75.3
	72	3.119	452	0.690	3.905	79.9

pH: Control, 5.57; enzyme treated 4.92 (initial pH 5.0).

An attempt to digest the associated fiber of coconut cake was made by using *T. viride* cellulase preparation, Meicelase. Two kinds of coconut cake from the Central Food Technological Research Institute, India were used as samples, namely brown and white materials which had been obtained in the form of a fine powder through an expeller-press and Krauss-Muffei processes, respectively. When the brown coconut cake was incubated with 0.5% Meicelase P solution at a concentration of 5% in a reaction mixture at 40°C., nearly 80% of the initial fiber was decomposed in 48 hr. Furthermore, it was found that more than 50% (more than 70% with stirring) of the initial protein was extracted after stationary incubating at 32° to 45°C. for 40 hr. A sample of the hydrolysis product was obtained as soluble fine brown powder by spray-drying the filtrate from the digest of the brown coconut cake. Thus, the results (Table XXI) suggest a possibility of the utilization of coconut cake for human food when it is treated by cellulase preparation (19).

Table XXI. Extraction of Protein from Coconut Cake by Meicelase

Percentage of Protein Extracted

Temp. (°C.)	White Cake			Brown Cake		
	Meicelase P	Meicelase F	None[a]	Meicelase P	Meicelase F	None
32	56.0	54.6	—	57.6	56.5	—
40	55.7	54.3	2.2	57.5	56.1	3.8
45	57.6	56.2	—	59.4	58.1	—

[a] Water was added in place of enzyme solution.

The enzymic isolation of starch was investigated by other workers. Hirao, Urashima, and Kuroda (14) treated crushed cereals with *T. viride* cellulase solution at 40°C. for 24 hr., followed by screening with 30 mesh sieve. In all cases, the amounts of isolated starch were increased by enzymic treatment (Table XXII).

Table XXII. Efficiency of *T. viride* Cellulase upon the Isolation of Starch

Cereal	Starch (mg./g.)		Residue (mg./g.)		Sugar (mg./g.)	
	Non-treat.	Treated	Non-treat.	Treated	Non-treat.	Treated
Milo	30	52	932	686	18.2	228.5
Rye	52	300	852	246	46.2	362.5
Barley	26	130	890	530	41.0	274.5

Takahashi, Ojima, and Yoshimura (31) published a report on enzymatic isolation of corn starch by using *T. viride* cellulase preparation (200 unit/mg.). In cereal starch production, wet milling process requires a long time steeping in SO_2 solution as a pretreatment. In corn starch manufacture, as long as 40 to 48 hr. of steeping is necessary to soften corn grains and to loosen the starch from the gluten matrix. In their paper, steeping conditions were studied and the possibility of reducing the steeping time was found by the addition of cellulase preparation to the steeping solution containing SO_2. Under the best conditions, 15 to 24 hr. of steeping was found to be enough (Table XXIII).

Table XXIII. Typical Results of Fractionation by Laboratory Wet Milling of Steeped Corn

Fractions	Steeping, Hours	12	15	24	48	No Added Enzyme, 50°C., 48 hr. Steeping
Recovery of starch (%)		77.8	79.5	77.4	82.3	77.8
TN in starch (%)		0.10	0.06	0.07	0.05	0.06
Reducing sugar in fiber (%)		1.8	1.9	0.8	1.8	1.8
TN in fiber (%)		1.5	1.4	0.8	1.0	1.7
Recovery of gluten TN (%)		38.4	44.4	43.2	36.7	34.0

TN: Total nitrogen measured by Kjeldahl method.
Condition: Steeped at 45°C., 0.2% sulfur dioxide and 0.5% (to substrate) enzyme added after 6 hr. from the start.

Since the author's investigation on the extraction of agar-agar from *Rhodophyceae* after treatment with *T. viride* cellulase (40), Tagawa and Kojima (29) tried increasing the yield of agar-agar. In order to improve the efficiency of extraction, the influence of cellulase treatment on *Gelidium amansii* was investigated using *T. viride* cellulase preparation (Cellulase onozuka F400). These results are summarized as follows: 1. Residue after extracting agar-agar was readily decomposed with cellulase and yielded considerable amounts of reducing sugar. Powdered agar-agar, however, yielded little reducing sugar with cellulase treatment. 2. After cellulase treatment of *G. amansii*, no appreciable morphological change was shown, but changes were observed after staining seaweed section by I_2-KI solution. They stated that there was no difference in appearance between treated seaweed and its control and no increase of reducing sugar was shown in enzyme solution. However, marked distinction was observed after boiling seaweed. The seaweed treated with cellulase was easily degraded and the residue was much softened.

The viscosity of agar obtained enzymatically was higher than that of control, so that it was difficult to separate agar from the residue by squeezing with cloth. If it would be possible to separate agar completely from the residue, the yield of agar might increase and the amount of residue might decrease. In their next experiment, *G. amansii* harvested in Portugal was employed. After heating at 60°C. for 3 hr. with 1.5% sodium hydroxide solution, the treated seaweed was cut in pieces and dried. Five grams of the seaweed and 200 ml. of Meicelase P solution, pH 4.6, at various concentrations of cellulase, were incubated at 40°C. for 24 hr. in the presence of toluene. The enzyme treated seaweed was washed and dried, followed successively by extracting agar-agar with 200 ml. of distilled water by direct heating for 3 hr. under reflux.

After squeezing with cloth, the residue was squeezed two times with boiling water, then all filtrates were gathered, and the agar-agar was recovered by freeze-drying. Control was run using acetate buffer only. No reducing sugar was shown when G. *amansii* was treated directly with cellulase, whereas appreciable amounts of reducing sugar were formed from the seaweed with cellulase after alkali treatment. The seaweed in the control was hardly degraded and the viscosity was not increased significantly by boiling, while cellulase treated seaweed was markedly degraded after boiling for 10 min.

Hachiga and Hayashi (9) studied the effects of Cellulase onozuka F500 *(T. viride)* on agar-agar extraction and in the case of *Gelidium amansii*, the following condition was found to be adequate: boiling for 2 hr. after treatment of seaweed with 1% cellulase solution at pH 4.0, 40°C. for 6 hr. Extraction with boiling water was compared with those done with boiling water after cellulase treatment and with boiling water slightly acidified with sulfuric acid. In each extraction, agar has been prepared by freezing and thawing and their yields and jelly strength were determined. Hachiga and Hayashi (10) reported enzymatic agar-agar extraction after alkali treatment using *Gracilaria convervoides*, important in industrial agar-agar production. The seaweed was treated with 3% NaOH solution at 23°C. for 3 hr., followed by washing and air-drying, after which to 10 grams of each sample was added 3% cellulase solution (pH 4.2, Cellulase F500, *A. niger*, Ueda Chem. Ind.) so as to give 4% substrate concentration. After incubating at 40°C. for 24 hr., treated seaweed was separated with cloth and washed, to which was added 500 ml. of water, followed by boiling for 2 hr. The agar-agar in filtrate was prepared by freezing and thawing (Table XXIV).

Table XXIV. Yield and Jelly Strength of Agar-Agar Produced from G. *convervoides* by Freezing and Thawing Method

Sample	Yield (%) A	B	C	Jelly Strength (g./cm.^{2}) A	B	C
Harvested in						
Argentina	40.4	26.0	28.6	250	180	110
Africa	34.9	20.0	21.8	450	220	300
Chili	14.0	11.2	10.3	60	10	40

Yield of agar-agar: The amount of freeze-dried agar-agar was shown as percentage of dried alkali-treated seaweed.
A: Boiled in water after treatment with cellulase.
B: Boiled in water.
C: Boiled in 0.003N H$_2$SO$_4$.

Furthermore, Hachiga and Hayashi (11) investigated the degradation of cellulose isolated from different seaweeds by *T. viride* cellulase preparation (Cellulase onozuka F1500). They stated that agar block, including cotton threads, 15 to 20 in parallel, cut into 7 × 4 cm., was incubated in 1% cellulase solution at 40°C. for 24 hr., after which remarkable decrease in tensile strength was shown, suggesting permeability of cellulase through agar-agar. Recently Hachiga and Hayashi (*12*) reported on microscopic observation of *Gracilaria verrucosa* treated with various reagents such as alkali, acid, and cellulase.

With regard to modifying tissue of vegetable foods, Saito (*24*) stated that after steam cooking polished rice, undegraded starch granules were revealed in the outside of rice grain, while gelatinized starch granules were shown predominantly inside the grain. On the other hand, ungelatinized cells were recognized within the rice grain and the starch granules in the outer layer were well gelatinized after cooking with water. It was assumed that the amount of absorbed water gave a significant influence to the physical properties and the quality or existence of cell walls might be directly responsible for water absorption. After incubation of rice grains with cellulase solution, the outer and inner layers were homogeneously cooked or gelatinized by cooking with steam or with water, so that rice-crackers with good feeling to teeth and tongue could be produced. Similar applications such as the improvement of qualities of foreign rice, preserved rice, non-irrigated rice, and other cereals may be possible by treatment with cellulase solution along with the manufacture of instant rice and glutinous rice. Saito, Baba, and Sato (*25*) moreover stated that the cooked rice treated with cellulase preparation was readily degraded during cooking. Therefore it could be supposed that the cell wall and intercellular substance in rice grain played an important role in the maintenance of its shape. The cellulase preparation, however, included other enzymes such as amylase, protease, and pectinase, and the influences of these enzymes on the cooking of rice were investigated.

In this experiment, the following enzyme preparations were employed: Cellulase 5000 (*T. viride,* Ueda Chem. Ind.), Pectinase (Takamine, USA), Pronase P (Kaken), liquefying amylase and saccharifying amylase (Daiwa Kasei). These preparations were used as 0.1% enzyme solution, except for liquefying amylase (0.02%). There were two distinct zones in cooked rice treated with cellulase, that is, nearly completely degraded outside zone and non-degraded hard central zone. Those rices treated with other enzymes were cooked homogeneously as well as control under the same conditions. The effect of cellulase treatment was particularly observed in the case of hard quality rice.

As present septic tank digestions require very long incubation periods, Omori and Kawanishi (*22*) attempted to improve efficiency of digestion in a septic tank by the addition of enzymes. An enzyme preparation including fungal cellulase, lipase, bacterial carbohydrase, and protease was developed by Toyo Brewing Co. Practical treatment of excreta by the addition of enzyme preparation was attempted in a plant having the planned capacity, 20 kl./day, and the practical one, 35 to 45 kl./day. The temperature in septic tank was 30 to 33°C. and the digestion required 30 days by stirring with gas in two tanks system. Secondary treatment was done by using distributing filter bed. The amount of enzyme preparation was 0.05% on an average to the quantity of excreta introduced per day, ranging from 0.02 to 0.1% according to excreta quantity. The results obtained by work from April to October in 1966 was compared with that done by work without enzyme preparation from April to October in 1965 as control. The removal of digested sludge would be decreased from 170 to 60–80 times in the case of enzyme treatment, considering dehydrated sludge per 1 ton of digested sludge was able to decrease about one half. The smell of raw excreta was lost at the third day after the addition of enzyme preparation and that of isolated solution from sludge was also remarkably decreased at around 15th day. It was concluded that time required for digestion could be shortened to 10–15 days by adding enzyme preparation, Cleanzyme (Table XXV).

Table XXV. Effect of Enzyme Preparation upon Digestion of Excreta in Septic Tank

	Dehydrated Sludge (kg.)/Original Excreta (m.^{3})	Dehydrated Sludge (kg.)/Digested Sludge (m.^{3})	Removal of Digested Sludge (times)
Control	9.8	97.7	229
Enzyme added	3.3	52.1	170

Finally, particular mention must be made of the application of cellulase in pulp industry such as done by Ogiwara and Arai (21). In their paper, the feature of the reaction and the phenomena of the hydrolysis of bleached sulfite pulp with Cellulase onozuka P500 were compared with those with hydrochloric acid. They stated that in the case of acid hydrolysis of the pulp pretreated with enzyme, two different results were obtained. With pulp slightly pretreated with enzyme, the rate of hydrolysis decreased, but it increased when the pulp was prehydrolyzed up to a certain extent, as compared with that of the original pulp. Nevertheless the shape of DP distribution curve and the average DP of the pulp changed remarkably with acid hydrolysis. Very small changes were observed when the pulp was treated with enzyme. Whereas pulp fiber was changed easily into fine powder with acid hydrolysis, the changes of the fiber in length and width with enzymatic hydrolysis were very small. Recently Arai and Ogiwara (3) studied changes in high alpha-sulfite pulp caused by treatment with Cellulase onozuka P500. Tensile strength increased initially with the time of cellulase treatment and reached a maximum value. The maximum became considerably higher with beating degree of the pulp. Tear strength decreased with the time of cellulase treatment. However, in the case of beaten pulps, a smaller decrease in tear strength was observed as compared with that of unbeaten pulp treated with cellulase. When the beaten pulp (with beating degree 37° SR) was treated with cellulase, the maximum breaking length was about 75% higher than that obtained with beating, whereas the tear factor was almost unchanged. It was concluded that these changes in tensile and tear strength might be interpreted by the fact that only a part of pulp fibers was hydrolyzed with cellulase treatment and a number of fibers were formed which, with regard to interfiber bonding, played an important role particularly in such materials as the high alpha-sulfite pulp which was less swellable.

Aforementioned results gained by the author and other investigators in Japan demonstrate in these later years that *T. viride* and *A. niger* cellulase preparations developed by the author have a wide variety of uses and these are giving us an effectual means especially in food processing along with *Rhizopus* cell separating enzyme preparation. The attempts to supplement them in feed are being carried out. Up to the present, however, the problem has not reached a satisfactory solution. CSE preparation should be employed as a feed supplement with cellulase preparation to be fully efficient. Recently these cellulase preparations are being exported to Europe from Japan at a rate of 500 kg./month. Industrial production of cellulase and CSE preparations in Japan gives ample opportunity for studying the enhancement of the value of foods and for making the best possible use of agricultural products. The use of these enzyme preparations is extending steadily.

Acknowledgment

The author's thanks are due to Kinki Yakult Co. Ltd., Meiji Seika Co. Ltd. and Ueda Chem. Ind. Co. Ltd. for kindly supplying Cellulase onozuka, Macerozyme, Meicelase, and Cellulosin for these studies. The author wishes also to acknowledge support in part by a Grant in Aid for Developmental Scientific Research of the Ministry of Education (1964 and 1965), a Grant in Aid for Fundamental Scientific Research of the Ministry of Education (1966, 1967 and 1968) and with the aid of funds supplied by the Waksman Foundation of Japan Inc. (1960) and by Miyazaki Prefecture (1966).

Literature Cited

(1) Akatsuka, S., Migita, R., Toyama, N., *J. Ferment. Technol.* **42**, 356 (1964).
(2) Akatsuka, S., Toyama, N., *J. Ferment. Technol.* **41**, 542 (1963).
(3) Arai, K., Ogiwara, Y., *J. Japan. Tappi.* **22**, 245 (1968).
(4) Dean, M., Wood, R. K. S., *Nature* **214**, 408 (1967).
(5) Fujii, N., Toyama, N., *J. Ferment. Technol.* **42**, 105 (1964).
(6) Fujii, N., Toyama, N., *J. Ferment. Technol.* **45**, 681 (1967).
(7) Fukinbara, I., Toyama, N., *J. Food Sci. Technol.* **15**, 79 (1968).
(8) Gregory, D. W., Cocking, E. C., *Biochem. J.* **88**, 40 (1963).
(9) Hachiga, M., Hayashi, K., *J. Ferment. Technol.* **42**, 207 (1964).
(10) Hachiga, M., Hayashi, K., *J. Ferment. Technol.* **43**, 440 (1965).
(11) Hachiga, M., Hayashi, K., *J. Ferment. Technol.* **44**, 753 (1966).
(12) Hachiga, M., Hayashi, K., *J. Ferment. Technol.* **46**, 375 (1968).
(13) Harada, Y., Toyama, N., *J. Ferment. Technol.* **44**, 835 (1966).
(14) Hirao, K., Urashima, Y., Kuroda, A., *J. Ferment. Technol.* **41**, 288 (1963).
(15) Igi, N., Oshima, J., *J. Ferment. Technol.* **41**, 292 (1963).
(16) Imai, T., Kuroda, A., *J. Ferment. Technol.* **44**, 854 (1966).
(17) Nakayama, S., Takeda, R., Toyama, N., *J. Ferment. Technol.* **43**, 648 (1965).
(18) Nakayama, S., Takeda, R., *Ann. Rep. Kagawa Pref. Fermented Food Exp. Station* No. 59, 11 (1966).
(19) Niwa, T., Zushi, S., Murase, M., *J. Ferment. Technol.* **43**, 750 (1965).
(20) Ogawa, K., Toyama, N., *J. Ferment. Technol.* **43**, 661 (1965).
(21) Ogiwara, Y., Arai, K., *J. Japan. Tappi.* **21**, 209 (1967).
(22) Omori, H., Kawanishi, S., *Proc. 24th Meeting Japan Pharmacol., Div. Hygienic Chem., Kyoto* (April, 1967).
(23) Pope, C. G., Stevens, M. F., *Biochem. J.* **33**, 1070 (1939).
(24) Saito, S., *Rep. The Niigata Food Res. Inst. Special Rep.* (Dec. 1962).
(25) Saito, S., Baba, M., Sato, Y., *Rep. The Niigata Food Res. Inst.* No. 8, 79 (1964).
(26) Spalding, D. H., *Phytopathology* **53**, 929 (1963).
(27) Suzuki, H., Abe, T., Urade, M., Nisizawa, K., Kuroda, A., *J. Ferment. Technol.* **45**, 73 (1967).
(28) Suzuki, H., Henmi, N., Nisizawa, K., Kuroda, A., *J. Ferment. Technol.* **45**, 1067 (1967).
(29) Tagawa, S., Kojima, Y., *Bull. Shimonoseki Univ. of Fisheries* **14**, 9 (1965).
(30) Takahashi, K., *Food Ind.* **9**, No. 14, 57 (1966).
(31) Takahashi, R., Ojima, T., Yoshimura, K., *J. Ferment. Technol.* **44**, 842 (1966).
(32) Takebe, I., Otsuki, Y., Aoki, S., *Plant & Cell Physiol.* **9**, 115 (1968).
(33) Toyama, N., *J. Ferment. Technol.* **43**, 683 (1965).
(34) Toyama, N., Fujii, N., Ogawa, K., *J. Ferment. Technol.* **43**, 756 (1965).
(35) Toyama, N., Ogawa, K., *J. Ferment. Technol.* **44**, 741 (1966).
(36) Toyama, N., Owatashi, H., *J. Ferment. Technol.* **44**, 830 (1966).
(37) Toyama, N., Fujii, N., Ogawa, K., *J. Ferment. Technol.* **44**, 731 (1966).
(38) Toyama, N., Fujii, N., Ogawa, K., *Rep. Appl. Microbiol. Lab. Facult. Agric. Miyazaki Univ.* (1967).
(39) Toyama, N., "Advances in Enzymic Hydrolysis of Cellulose and Related Materials," E. T. Reese, Ed., p. 235, Pergamon Press, New York, 1963.
(40) Toyama, N., *J. Ferment. Technol.* **40**, 199 (1962).

Received October 14, 1968.

Discussion

Comment—G. Haas (General Foods): "We have been following with great interest Dr. Toyama's, Dr. Reese's, and Dr. King's work in this area, and we have done a little work ourselves with *Trichoderma viride* and with *Pestalotiopsis westerdijikii*. We have grown both these organisms on various fiber rich substrates such as bran and coffee grounds, which are two prime candidates for cellulase application, and we isolated the enzymes. These were as active, but not more active than those reported in the literature. So on the whole we have only repeated some of the work of the above investigators and we really didn't get any further. We would be highly interested in getting an enzyme which would be more active and at an economical cost. We can see possibilities, particularly of upgrading feed stuffs, of salvaging wastes such as oat hulls and coffee grounds, and of making certain foods softer, such as baby foods or geriatric foods. We are interested in seeing Dr. Toyama's pictures and in the possibility that rehydration may be improved in that way too. There is also the possibility that protein separation at higher yields can be made in some extractions.

Ghose: "Dr. Toyama, I would be interested to know the total amount of cellulase produced in Japan from *Trichoderma*, in what form it is produced in your country, whether some of it is exported outside, and for what purposes most of it is being used?"

Toyama: "*Trichoderma* cellulase is being produced by two companies at the rate of 3 tons per month. And another cellulase is being produced at the rate of 2 tons per month. About 500 kg. per month are exported to Europe (West Germany). *Trichoderma* cellulase is also being exported to Australia as feed supplement (for cows) at the rate of 100 kg. per month. The major sources of these enzymes in Japan are:

Macerozyme	of *Rhizopus* sp. pectic enzymes	All Japan Biochem. Co. Nishinomiya, Japan
Cellulase onozuka	of *Trichoderma viride* cellulase	
Meicelase	" " "	Meiji Seika Co. Tokyo, Japan

23

The Production of Cellulases

MARY MANDELS and JAMES WEBER

Food Microbiology Division, US Army Natick Laboratories, Natick, Mass. 01760

> *Many fungi are cellulolytic, but only a few produce cell-free enzymes that will attack solid cellulose.* Trichoderma viride *grown on cellulose medium produces a stable cellulase complex including C_1. This enzyme is capable of extensive degradation of solid celluloses. Conditions for producing high yields of the enzyme in shake flasks and in a laboratory fermenter are described. Filtrates from these cultures readily hydrolyzed nine cellulose substrates of varying resistance. It is suggested that such culture filtrates could be used to hydrolyze waste cellulose. The hydrolyzate could be used to produce single cell protein or some other fermentation product.*

Uses for cellulase have been found (34, 35, 36), and more ambitious proposals wait only on a supply of enzyme of sufficient activity, stability, and economy. Companies like Wallerstein, Miles, and Rohm and Haas in the United States and Meiji Seika and Kinki Yakult in Japan are marketing cellulases derived from *Aspergillus niger, Trichoderma viride,* and other organisms. While these preparations are good, they are limited both in rate and in extent of action on solid cellulose.

Recently, we have been interested in the problem of finding new food sources for our expanding populations. It has been reported that cellulose is the most abundant organic substance on earth. Unfortunately man and most other animals cannot digest it. To use it as food it must first be hydrolyzed to glucose. Under wartime conditions wood was hydrolyzed by acid, and food yeasts produced on the hydrolyzate. The process requires strong mineral acids acting at elevated temperatures over extended periods of time. Special corrosion-proof equipment must be used. In addition, the digests must be neutralized and the sugar recovered and separated from excess salt and toxic by-products such as furfural. The process has not proved economical under normal conditions. Recently, however, some Russian workers have been reinvestigating the acid hydrolysis process (8). Their innovation is a pretreatment of the cellulose by heating and grinding to render it partially soluble and more susceptible to acid hydrolysis (15).

Enzymatic hydrolysis of cellulose, if it can be achieved efficiently, could give us glucose at moderate temperatures and pH conditions. Ultimately, the sugar could be recovered as is done in the enzymatic hydrolysis of corn starch (37). Indeed Katz and Reese (12) have shown that it is possible, using *Trichoderma viride* cellulase, to obtain cellulose hydrolyzates containing as much as 30% glucose. To do so, it was necessary to heat and grind the cellulose as done by the Russian workers, to employ very high enzyme concentrations, and to incubate for 15 days. Initially it may be more practical to aim for lower glucose concentrations, using the digest to grow yeast or other microbial cells for human or animal food. Microorganisms that grow directly on cellulose rather than on the hydrolyzate have not yet been found that are appealing as food. Another problem is separation of organisms from the cellulose residues.

For simplicity and economy it would be desirable to produce enzyme, glucose, and microbial cells in a unified and continuous process. The enzyme would be produced in the first step by growing the cellulolytic organism on the same cellulosic material to be used later as the substrate for the production of sugar. By this method, we would hope to produce the proper complex of enzymes, that is, one containing all necessary components for attack on the particular cellulose substrate. The mycelium and residual cellulose would be separated by centrifugation or filtration and the culture filtrate containing the enzyme transferred directly into the cellulose digester. In this way costly processing of the enzyme and losses in activity would be eliminated. In the third step sugar containing effluent from the cellulose digester would be fed into the fermenter in which microbial cells were produced.

To realize the above objectives two major problems must be solved: the activity of the enzyme must be increased, and the susceptibility of the cellulose substrate maximized. In this paper we discuss efforts to produce highly active cellulase in large amounts.

Materials and Methods

The fungus *Trichoderma viride* QM 6a was maintained on potato dextrose agar slants at 25°C. The basic medium for shake flask and fermenter culture contained in grams/liter KH_2PO_4 2.0, $(NH_4)_2SO_4$ 1.4, urea 0.3, $MgSO_4 \cdot 7 H_2O$ 0.3, $CaCl_2$ 0.3 and in mg./liter $FeSO_4 \cdot 7 H_2O$ 5.0, $MnSO_4 \cdot H_2O$ 1.56, $ZnSO_4 \cdot 7 H_2O$ 1.4, $CoCl_2$ 2.0. The pH was adjusted to 5.0. The cellulose source was usually 1.0% Solka Floc SW 40A (Brown Co., Berlin, N. H.) a purified spruce wood cellulose. Proteose

APRIL

S	M	T	W	T	F	S
1	2	3	4	5	6	7
8	9	10	11	12	13	14
15	16	17	18	19	20	21
22	23	24	25	26	27	28
29	30	.	.	.	.	.

Thursday
12
April

8.00
8.30
9.00
9.30
10.00
10.30
11.00
11.30
12.00
12.30
1.00
1.30
2.00
2.30
3.00
3.30
4.00
4.30
5.00

102　　　1973　　　263

Mandels, M & Weber,
The Production of
Cellulases
Adv in Chem Ser no 95
pp 391-414 (1969)

Peptone (Difco) [0.1%] and Tween 80 [0.1%] (Atlas Chemical Industries, polyoxy ethylene sorbitan mono-oleate) were sometimes added as noted. Shake cultures were inoculated with a spore suspension, and grown at 25°–29°C. on reciprocating or rotary shakers.

Fermenter studies were carried out in a New Brunswick continuous culture apparatus with 15-liter stirred fermenter jars for culture vessel and nutrient reservoirs and a 20-liter carboy for a harvest vessel. Temperature of the culture was maintained at 28°–29°C. Impeller speed was kept at 120 r.p.m. with a 3-inch diameter set of four impeller blades low in the culture, and a second impeller just above the liquid level to aid in foam control. A few drops of Dow Corning Antifoam AF at 25 mg./ml. (emulsion) were added manually as required to control foam. The rate of aeration was 1 to 3 liters per minute. Contamination through the impeller shaft port was prevented by a loose fitting sleeve through which live steam was continually passed (10). Samples taken from the fermenter were streaked on potato dextrose agar, and Bacto plate count agar, incubated for 7–10 days and inspected for contamination.

Cellulase activities were determined on filtrates of culture samples by production of reducing sugar measured as glucose by a dinitrosalicylic acid (DNS) method (32) as follows:

a. **$C_x\beta(1 \to 4)$ Glucanase.** Add 0.5 ml. of approximately diluted enzyme to 0.5 ml. of 1.0% carboxymethylcellulose (CMC) 50T (Hercules Powder Co.) in 0.05M Na citrate buffer pH 4.8. Incubate 30 minutes at 50°C., add DNS reagent and determine reducing sugar.

b. **C_1-Cotton.** To 50 mg. of absorbent cotton add 1.0 ml. 0.05M Na citrate buffer pH 4.8, and 1.0 ml. of appropriately diluted enzyme. Incubate 24 hours at 50°C., add DNS reagent, and determine reducing sugar.

c. **Filter Paper.** To 50 mg. (1 × 6 cm.) of Whatman No. 1 filter paper add 1.0 ml. 0.05M Na citrate buffer pH 4.8 and 1.0 ml. of appropriately diluted enzyme. Coil the paper strips in the assay by touching the tube to a vibratory mixer. Incubate 1 hour at 50°C., add DNS reagent, and determine reducing sugar.

The number of units per ml. in each case equals the inverse of the dilution required to give 0.50 mg. of reducing sugar as glucose. Filter paper (F.P.) and cotton activity is also expressed as the mg. of glucose produced by undiluted culture filtrate, or a solution of precipitated enzyme at 5 mg./ml., in the standard test. All values were corrected for appropriate enzyme and substrate blanks.

Enzyme preparations, CMC solutions, and buffers were preserved from contamination with 0.05% Merthiolate (Eli Lilly Co.).

Protein was measured as bovine plasma albumin by the method of Lowry et al. (19).

Cellulose substrates used in digestion studies were (ground) Solka Floc; alpha cellulose, a non-nutritive fiber used for formulating animal diets (Gen. Biochem. Co.); Avicel, microcrystalline cellulose (American Viscose); heated milled Solka Floc (7) and (not ground) Whatman No. 1 filter paper, cotton sliver (25) and absorbent cotton (White Cross Co.).

Selection of Organism

Cellulases are produced by insects, molluscs, protozoa, bacteria, antinomycetes, and fungi. Rearing insects or molluscs and then extracting enzymes from them is scarcely feasible except for some specialized use, so we must turn to the microbe and the techniques of the fermentation industry for a practical means of producing cellulase in quantity. Cellulolytic bacteria and protozoa are frequently anaerobic and difficult to grow, and the bacteria tend to produce large quantities of slime. The cellulase is associated with the bacterial cell and must be extracted. Cellulolytic fungi and actinomycetes usually grow readily on simple media and secrete their cellulases into the medium so that the enzymes can easily be separated from the cells by centrifugation or filtration. This gives a nice clean preparation free of cell contents that are released when grinding or other extracting procedures are required.

Many fungi are cellulolytic (1, 2, 6, 16, 39, 40), and most of these produce cellulase when grown on cellulose. However, only a few produce culture filtrates that extensively degrade solid cellulose. Over the years, we and many others have compared thousands of microorganisms for their ability to degrade cellulose and hundreds for their ability to produce cell free enzymes that can be used to hydrolyze cellulose. In general, the enzyme containing filtrates are rather inactive when one

Table I. Production of Cellulase by Various Fungi

QM No.	Preparation	C_1 Units	C_x Units	Hydrolysis of Cotton, % Sugar	Hydrolysis of Cotton, % Wt. Loss
	Buffer (0.05M acetate)	0	0	0	1
6a	Trichoderma viride	50.0	50.0	58	53
826	Chrysosporium pruinosum	30.0	70.0	11	19
137g	Penicillium pusillum	27.0	110.0	22	23
1224	Fusarium moniliforme	NT	3.5	39	39
72f	Aspergillus terreus	5.0	36.0	28	24
806	Basidiomycete	5.0	75.0	15	23
94d	Stachybotrys atra	1.0	8.0	5	6
B814	Streptomyces sp	0.7	40.0	9	12
38g	Fusarium roseum	0.7	10.0	9	10
381	Pestalotiopsis westerdijkii	0.7	60.0	4	8
460	Myrothecium verrucaria	0.4	28.0	2	2
459	Chaetomium globosum	0.2	0.5	NT	NT

Cultures grown on Solka Floc, except *Penicillium pusillum* 137g. grown on cotton duck.
C_1 units—action on cotton silver for 24 hours at 40°C. (25).
C_x units—*see* text.
NT—Not tested.
Hydrolysis of 1% cotton sliver 35 days at 29°C. (4 changes of enzyme).

considers the high rate of destruction by the growing organism. Some of the more active enzyme solutions are produced by the organisms which grow less rapidly on the cellulose (Table I). For hydrolysis of solid cellulose, the C_1 factor is required. In our experience and that of others, *Trichoderma viride* is the most reliable producer of a stable cellulase complex including this component (*10, 14, 25, 30, 34*). Although other organisms such as *Chrysosporium pruinosum* and *Penicillium pusillum* also produce filtrates having high C_1 unit values, the enzymes are less stable, and they are inferior to *Trichoderma viride* filtrates for the extensive hydrolysis of cotton (Table I).

A number of commercial cellulases, samples from other workers, and acetone precipitated powders were compared (Table II). All of these demonstrated good activity on carboxymethylcellulose. In fact a number of these, notably the enzyme from *Poria*, were more active on CMC than the *T. viride* enzyme. About half of the enzymes showed good activity on filter paper. But *T. viride* enzymes show the highest activity on cotton, the most resistant substrate. In case inactivation at 50°C. affected the activity on cotton and filter paper, these enzymes were also tested for hydrolysis at 25°C. (data not shown). The relative activities were the same as those found at 50°C. Selby (*31*) made a similar study of a number of cellulase preparations. He also found that the cellulase of *Trichoderma viride* had the highest activity on cotton.

We have therefore focussed our attention on *Trichoderma viride*. When someone discovers a better source or a superior cellulase, we will gladly transfer our allegiance. The results reported in this paper were obtained with the Army Quartermaster Strain QM 6a. This strain does not produce gliotoxin (*38*). We have tested over one hundred other strains of *Trichoderma viride* and closely related organisms. Most appear to produce a similar cellulase complex (Table III), but in lesser amounts.

Measurement of Enzyme Activity

Cellulase is a complex of enzymes showing various types of activities. Cellulose substrates include highly resistant crystalline forms such as cotton, various types of microcrystalline cellulose such as Avicel and hydrocellulose, sulfite pulps such as Solka Floc, as well as filter paper and cotton fabrics. More susceptible substrates include swollen or reprecipitated cellulose, cellophane, and ball-milled cellulose. Most susceptible are the soluble derivatives (of low D.S.) such as carboxymethylcellulose and cellulose sulfate. It is not surprising that there are many assay methods to detect or measure cellulase (*9*). These methods differ markedly in sensitivity, and in cellulase components detected, depending on the substrate used, the effect measured, and the duration and conditions of

Table II. Activities of

Source of Cellulase	Composition	
	Reducing Sugar	Protein
Trichoderma viride QM6a crude filtrate[a]	0 %	12.1%
Trichoderma viride QM6a acetone powder[a]	2.8	28.0
Trichoderma viride Meiji Seika (powder)	49.6	17.6
Trichoderma viride Kinki Yakult (powder) Onozuka P500	6.0	24.0
Aspergillus niger Wallerstein (powder) (T. Cayle)	53.2	12.0
Aspergillus niger Rohm & Haas (powder) 19AP	4.4	10.4
Aspergillus niger Miles 9X (powder)	5.2	25.2
Trametes sanguinea (powder) P. Horikoshi	0	32.0
Poria SEAB France, P. Laboureur (powder)	13.5	32.4
Chrysosporium pruinosum QM826 acetone powder[a]	2.0	22.0
Basidiomycete QM806 acetone powder[a]	5.2	20.8
Myrothecium verrucaria QM460 acetone powder[a]	4.0	12.0
Streptomyces sp QMB814 acetone powder[a]	0	15.2
Helix pomatia (Snail) digestive juice Ind. Biol. Francais	2.4	15.7

[a] Our preparation
Reducing sugar as glucose
Units (*See* text)

Protein as bovine plasma albumin
SEAB = Society of Biochemical Studies and Application, Jouy en Josas

Table III. Cellulase Activity of Some *T. Viride* Strains

QM No.	C_1 Units/ml.	C_x Units/ml.	Hydrolysis Solka Floc %
6a	33	44	57
317	31	30	80
3161	28	19	49
2098	25	18	65
7733	25	17	47
4093	19	7	23
2647	8	16	36

Cultures grown on 1.0% Solka Floc.
C_x units—*see* text.
C_1 units—action on cotton sliver for 24 hours at 40°C. (25).
Hydrolysis of 6% Solka Floc, 42 days at 29°C.

the assay. Principal methods have included:

a. Reduction in viscosity of a solution of a cellulose derivative. This is an extremely sensitive assay since a few random breaks in a chain cause a marked decrease in average chain length. This assay measures β-glucanase action independent of C_1 action.

b. Production of reducing sugar from a soluble cellulose derivative. This assay also measures β-glucanase independent of C_1. Considerably

Cellulase Preparations

Activity at 5 mg./ml.		Units per mg. Protein		
Filter Paper	Cotton	CMC	Filter Paper	Cotton
1.84 mg.	4.28 mg.	54.	8.0	40.5
2.82	2.80	74.	6.7	7.7
0.22	0.22	52.	0.5	0.5
1.38	1.30	18.	3.1	6.5
0	0	123.	0	0
0.21	0.17	33.	0.8	0.7
0.78	0.67	62.	1.5	3.2
1.07	0.67	25.	2.3	2.7
1.66	0.10	610.	3.6	0.1
1.20	0.46	71.	2.9	0.8
1.26	0.18	90.	3.5	0.3
0.45	0.12	60.	1.5	0.4
0.98	0.76	79.	4.1	1.5
0.22	0.22	4.8	0.6	0.5

Activity as mg. glucose produced in Standard Test at 50°C. (See text)

greater cellulase action is required to produce detectable reducing sugar than to cause a drop in viscosity.

  c. Production of reducing sugar or loss of residue weight from solid cellulose. These methods require much stronger cellulase preparations and measure the combined action of C_1 and β-glucanase. The role of C_1 becomes more important as the resistance of the substrate increases. Usually cotton is considered to be the most resistant substrate, and the best indicator of C_1 action (25). More extensive degradation is required if weight loss is measured than if production of reducing sugar is used.

  d. Reduction in optical density of a cellulose sol or suspension. With hydrocellulose as a substrate this method has been used to measure C_1 (14). It appears probable that β-glucanase is also required for this effect.

  e. Measurement of less clearly defined activities as reduction in breaking strength of thread, yarn, or fabric (29), swelling of cotton (28) or paper (4), maceration of filter paper (11, 34), and microfragmentation of cellulose micelles (10, 17).

  For quantitative work enzyme activity is measured in units based on (a) the extent of action by an enzyme in a unit time, (b) the time required for an enzyme preparation to accomplish a unit effect (34), or (c) the concentration of enzyme required to achieve a unit action in a unit time. All assay methods are based on relatively short hydrolysis periods and limited action by the enzyme on the most susceptible or accessible portion of the cellulose substrate. The enzyme may be very dilute. It is difficult to extrapolate the values obtained to predict more extensive degradation by concentrated enzymes acting over longer time periods when increasing resistance of the residual substrate, and possible instability of the enzyme become important factors.

In producing cellulase for a specific application, the ideal assay should be simple and closely approximate the condition under which the enzyme is to be used. Since our purpose is to produce sugar from waste cellulose, we measure enzyme activity as production of reducing sugar from filter paper in one hour at 50°C., pH 4.8 (Figures 1 and 2). The time is short, pH and temperature are optimum, and filter paper as a substrate is about as susceptible as the sulfite pulps or ground cellulose with which we will be working. Cotton is too resistant and CMC too susceptible for a reliable measure of the type of activity we require. The paper can be cut in strips of standard dimensions, eliminating the labor of weighing out a ground substrate. Contact of paper strip with enzyme in an unshaken tube is good. It is not necessary to remove the paper when the sugar is to be determined. Finally, this is a substrate that can be readily duplicated at other times or in other laboratories. Although

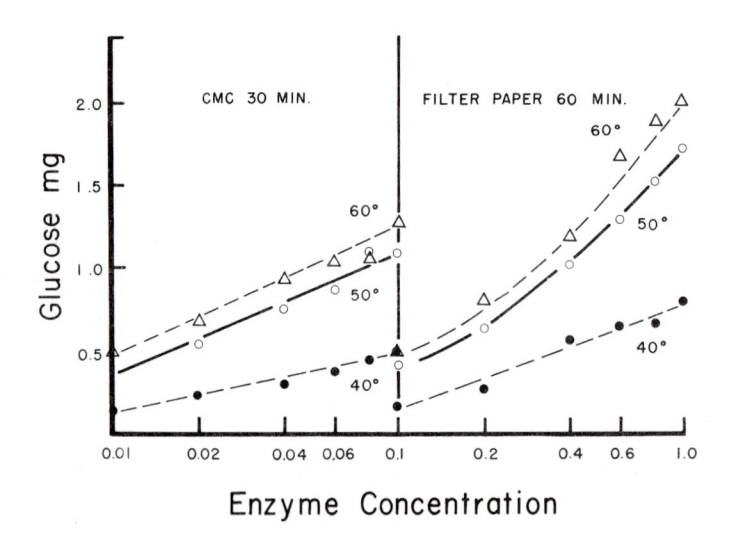

Figure 1. Effect of enzyme concentration and of temperature on hydrolysis of carboxymethylcellulose and filter paper by T. viride *cellulase*

Crude filtrate, 67 C_x units per ml., 7.1 filter paper units per ml., F. P. activity 1.74. CMC 1% 0.5 ml. + 0.5 ml. enzyme. 30 min. Filter paper 50 mg. + 1.0 ml. buffer + 1.0 ml. enzyme. 60 min. pH 4.8

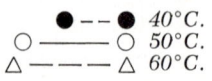

● -- ● 40°C.
○ —— ○ 50°C.
△ ---- △ 60°C.

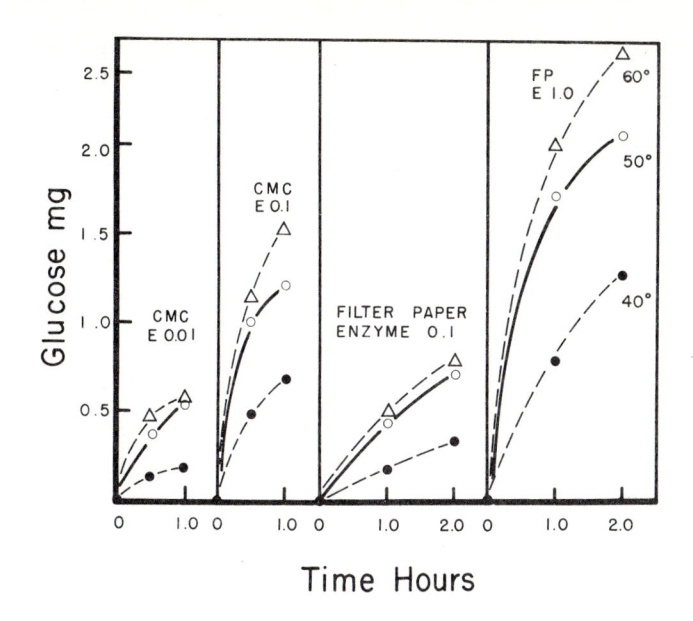

Figure 2. Effect of time and temperature on hydrolysis of carboxymethylcellulose and filter paper by T. viride cellulase

Crude filtrate, 67 C_x units per ml. 7.1 filter paper units per ml., F. P. activity 1.74. CMC 1% 0.5 ml. + 0.5 enzyme. Filter paper, 50 mg., + 1.0 ml. buffer + 1.0 ml. enzyme. pH 4.8

● - - - ● 40°C.
○ —— ○ 50°C.
△ — — △ 60°C.

filter paper is more susceptible than cotton, few enzyme preparations can produce significant levels of reducing sugar from it in one hour.

Induction of Cellulase

Cellulase in fungi is an adaptive enzyme. *Trichoderma viride* grows readily on most carbon sources, but cellulase is produced only when the fungus is grown on cellulose, on glucans of mixed linkage including the $\beta(1 \to 4)$ and on a few oligosaccharides (22, 23, 24). The true inducers of cellulase for a fungus growing on cellulose are the soluble hydrolysis products (Table IV) of the cellulose, especially cellobiose (23).

The role of cellobiose is complex (Figure 3). It is an inducer of cellulase, and can act also as an inhibitor of cellulase action (27). In addition, high concentrations (0.5–1.0%) of cellobiose or other rapidly metabolized carbon sources such as glucose or glycerol strongly repress cellulase formation (23, 24), so that cellulase does not normally appear in a cellobiose culture until after the sugar has been consumed. These same sugars added (at 0.5–1.0%) to a culture producing cellulase on cellulose (Figure 4) will inactivate the enzyme already formed (23). If the culture is young, the enzyme activity may reappear after the sugar has been consumed (Figure 4).

Table IV. Induction of Cellulase in *T. viride*

Inducer	C_1	C_x	Ratio C_1/C_x
Cotton Sliver	8.8	4.3	2.0
Cotton Duck	10.0	8.3	1.2
Solka Floc	8.8	10.0	0.9
Filter Paper	7.0	8.5	0.8
Lichenin	NT	7.8	
Cellobiose	2.5	3.8	0.7
Sophorose	1200.	2800.	0.4
Lactose	1.0	4.8	0.2
Glucose (purified on charcoal)	0	0	
Starch	0	0	
Glycerol	0	0	

(Units Induced per mg. Inducer)

C_x units—*see* text.
NT—no test.
C_1 units—action in cotton sliver for 24 hours at 40°C. (25).

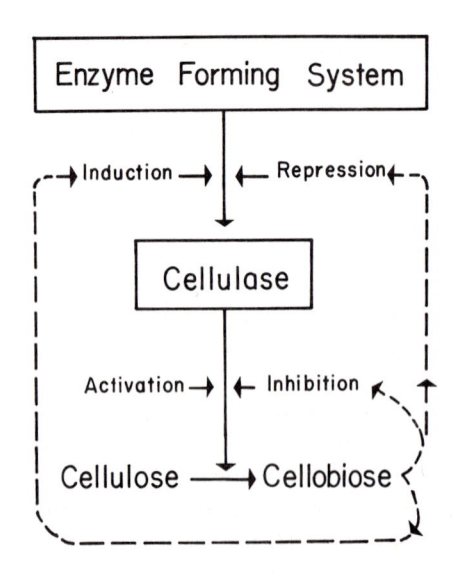

Figure 3. Enzyme forming system

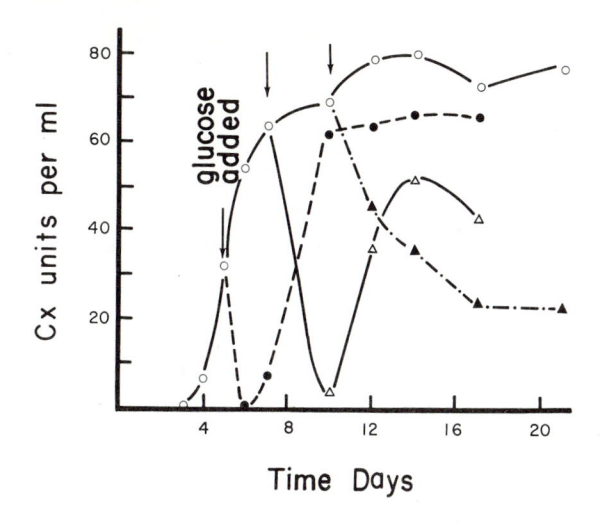

Figure 4. Inactivation of cellulase by adding glucose to a culture of T. viride *growing on cellulose*

○ ———— ○ *Control culture grown on 1.0% Solka Floc. Glucose (0.5%) added at 5 days* ● - - - - - ●, *7 days* △ ———— △, *10 days* ▲ - - - - - ▲

These effects are related to rapid metabolism of the sugar and can be prevented if the sugar is metabolized slowly. *Trichoderma viride* will produce as much cellulase on cellobiose as it does on cellulose if its metabolism is slowed by (1) suboptimum temperature and aeration, (2) deficiencies of mineral nutrients such as calcium, magnesium, or trace metals, or (3) excess of a mineral such as cobalt (23).

Finally, when *Trichoderma viride* is growing on insoluble cellulose or when cellulase is being induced on low levels of a highly active inducer such as sophorose, the addition of small amounts (0.1%) of readily metabolizable substrates such as glucose, glycerol, or peptone reduce the lag and increase the level of cellulose produced (24).

Lactose, $a\beta(1 \to 4)$ galactoside, induces cellulase in *T. viride* and a few other fungi. Sophorose, $a\beta(1 \to 2)$ glucoside, is a very powerful inducer of cellulase for *T. viride* only. These two sugars are the only known cellulase inducers that do not have $a\beta(1 \to 4)$ glucosidic linkage.

Soluble inducers are valuable in basic studies aimed at elucidating the nature of cellulase induction. As yet we have not found a good method to use them to obtain large quantities of cellulase. For producing cellulase on a large scale, the fungus is grown on cellulose.

Development of Cultural Conditions for the Production of High Enzyme Levels

Trichoderma viride grows rapidly on simple media and has no requirements for growth factors. In agitated culture it produces a suspension of short mycelial threads, rarely forming pellets. This trait is of great advantage if we are to attempt large scale growth and continuous cultivation. On carbohydrate media it is a strong acid producer and will continue growth until the pH reaches 2.5 or even lower. On peptone media the pH rises, and growth continues to about pH 7.5.

Cultures can be inoculated with a spore suspension, or with a small volume of cellulose culture containing mycelium. Enzyme yields are equal with either inoculum, but the mycelial inoculum usually gives more rapid growth and earlier enzyme development. Spore inoculum is usually more convenient for shake flask experiments.

Normally there is a lag in cellulase production after inoculation with a pH rise, particularly if peptone is present. As the cellulose is consumed the pH falls to about 3.0, and cellulase appears in the medium. Activity on filter paper appears early, and rises more rapidly than activity on CMC. Finally, the cellulose is consumed, pH rises, and enzyme production ceases.

T. viride will produce cellulase in a shake flask on nutrient salts plus cellulose with no additive, but higher and more stable cellulase yields, particularly on solid cellulose, are often obtained when a small quantity of a soluble carbon source is added to the medium (Table V, Figure 5). The optimum cellulose concentration is 0.5–1.0%, depending on the type of activity measured, and the level of peptone in the medium (Figure 5). Peptone or other protein derivatives are more satisfactory additives than glucose or glycerol (Table V). Optimum peptone concentration is 0.1 to 0.2%, depending on the cellulose concentration, with concentrations of 0.5% or higher strongly inhibitory to cellulase production (Figure 6). Proflo is stimulatory even at 1.0% (Figure 6). The maximum filter paper activity we have obtained in a shake flask was 4.04 on 1.0% Solka Floc plus 0.2% peptone. The maximum C_x activity was 154 units per ml. on 1.0% Solka Floc plus 1.0% Proflo. Tween 80 (26) also stimulates cellulase production, particularly when peptone is also present (Table VI).

When cellulase acts on cellulose, the most susceptible portions are rapidly digested, and the residue becomes increasingly resistant to enzyme action. Cellulases produced on different substrates vary in their ratios of C_x to C_1 action (Table IV). We hoped, therefore, that use of the resistant residues remaining after enzyme attack, as a substrate for enzyme production, might result in a preparation having a high activity on resistant cellulose. This would have additional advantages in a practical process. Cellulose residues from the hydrolyzer could be used for enzyme production, and the finely ground and digested residues from the hydrolyzer would be much easier to pump in a continuous process, than would the original pulp.

Table V. Effect of Additives on Production of Cellulase by *T. viride*

Additive (0.05%)	Maximum Cellulase Activity	
	Filter Paper Activity	C_x Units/ml.
None	1.02	46
Dow Corning Antifoam A. F.	1.10	38
Glucose	1.04	76
Glycerol	0.90	32
Proteose peptone	1.90	64
Proflo (cottonseed flour)	1.80	100
Yeast extract	1.64	84
Casein hydrolyzate	1.68	74
Soy extract	1.10	60
Phytone	1.86	78

Spore inoculum, shake flask cultures grown 18 days at 29°C., 1.0% Solka Floc. Filter paper activity = mg. glucose produced in standard test, *see* text.

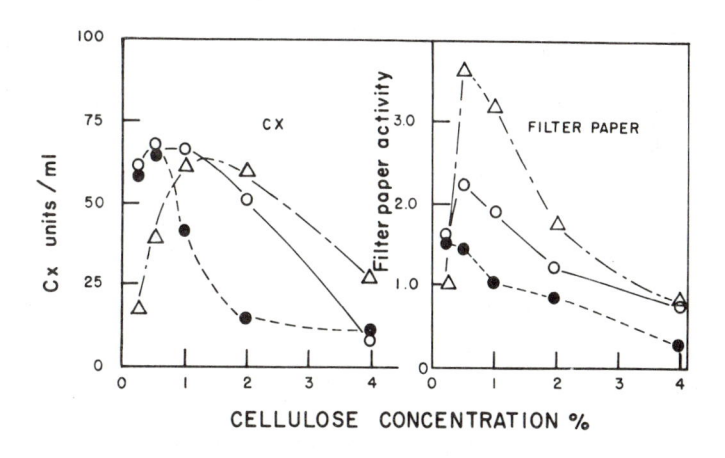

Figure 5. Effect of cellulose concentration on production of cellulase

Shake flasks, grow 18 days at 29°C., ● - - - ●, no peptone, ○———○, 0.05% peptone, △ — · — △, 0.20% peptone

As substrates for enzyme production we compared Solka Floc; heated and milled Solka Floc (25 minutes at 200°C., pot milled 24 hours); and the residues of each after treatment with *Trichoderma viride* cellulase (20% digestion of the Solka Floc, 16% of the heated milled Solka Floc). The residues were washed and dried at 70°C. Cellulase was then produced by growing the fungus on each of the four cellulose substrates at the 1.0% level. The milled or digested Solka Floc produced filtrates having equal or greater activities on CMC and on filter paper than the filtrates produced on the original Solka Floc (Table VI). Residues from the hydrolyzer would thus be suitable substrates for enzyme production.

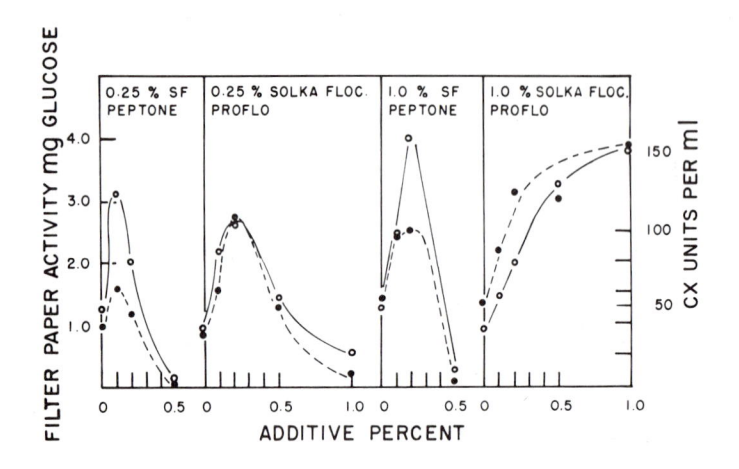

Figure 6. Effect of concentration of Solka Floc, proteose peptone, and Proflo (cottonseed flour) on cellulase production by T. viride

Spore inoculum, shake flask experiment, grown 18 days at 29°C.
○ ——— ○ Filter paper activity (mg. glucose produced in 1 hr. at 50°C.) by crude filtrate. Maximum value
● - - - ● C_x units per ml. crude filtrate. Maximum value

Conditions Affecting Long Term Hydrolysis

The enzymes produced on Solka Floc, Milled Solka Floc, and the residues after digestion were diluted to equal activities on filter paper and then tested against nine substrates of varying resistance. The cellulose was suspended in the enzyme filtrate at 5% and incubated unshaken at 30°, 40°, and 50°C. for 15 days. There were no significant differences in rate or extent of action by the enzymes produced on the four substrates. It appears that for one substrate (Solka Floc) increasing susceptibility by heating and milling, or removal of the most susceptible portions by partial enzyme hydrolysis, has little effect on the enzyme inducing properties.

Table VI. Effect of Solka Floc, Treated Solka Floc, Peptone and Tween 80 on Production of Cellulase by *T. viride*

Cellulose 1.0%	Peptone 0.1%	Tween 80 0.1%	Maximum Cellulase Activity Filter Paper	C_x Units per ml.
Solka Floc	−	−	1.22	64
	−	+	1.74	72
	+	−	1.90	88
	+	+	3.72	118
Residue of Digested (20%) Solka Floc	−	−	1.12	36
	−	+	1.26	64
	+	−	2.07	96
	+	+	3.80	148
Heated Milled Solka Floc	−	−	2.01	86
	+	−	1.60	76
	+	+	2.28	104
Residue of Digested (16%) Heated Milled Solka Floc	−	−	1.60	38
	+	−	2.34	50
	+	+	2.85	100

Spore inoculum, shake flask experiment, grown 18 days at 29°C. Heated milled Solka Floc (25 minutes 200°C. ball milled 24 hours). Digested samples treated *T. viride* cellulase (see text).

Table VII. Effect of Substrate on Extent of Hydrolysis by *T. viride* Cellulase at 50°C.

Substrate	1 Hour	1 Day	2 Days	4 Days	7 Days	15 Days
Solka Floc	2.6	15.9	27.8	37.0	54.3	68.7
δ-Cellulose	2.6	16.2	31.6	38.8	56.9	62.4
Avicel	0.9	10.8	18.0	26.9	36.1	51.8
HMSF	2.6	15.7	24.0	31.1	37.0	48.8
Filter Paper	3.8	12.9	19.2	28.4	32.9	36.4
D Solka Floc	0.4	6.8	11.7	21.4	29.4	34.7
DHMSF	0.6	7.2	14.2	22.0	29.6	32.8
Cotton Sliver	0.3	3.2	7.2	9.6	12.0	14.4
Absorbent Cotton	0.6	3.2	5.8	8.8	9.7	11.8

5% suspension in unshaken test tube (see Figure 7).

Data for the Solka Floc enzyme at 50°C. over the 15 days (Table VII) and for two of the enzymes at 30°, 40°, and 50°C. at seven days (Figure 7) are shown. In the early stages of hydrolysis filter paper, Solka Floc, and the alpha cellulose were quite susceptible, the digested Solka

Floc, Avicel, and cotton were more resistant. After longer hydrolysis, the Solka Floc and alpha cellulose were most susceptible (54–57% hydrolysis in seven days at 50°C.), digested Solka Floc, filter paper, and Avicel were intermediate (30–40% hydrolysis) and the cotton still very resistant (10–12% hydrolysis). For all substrates the initial rate of hydrolysis is rapid and then falls off (Table VII, Figure 8). The shape of the hydrolysis curve is not affected by resistance of the substrate or concentration of the enzyme (Figure 8). Reducing sugar production increases with increasing cellulase concentration at least to 5% (data not shown). Optimum pH for extensive hydrolysis at 50°C. is 4.8 (Figure 8a).

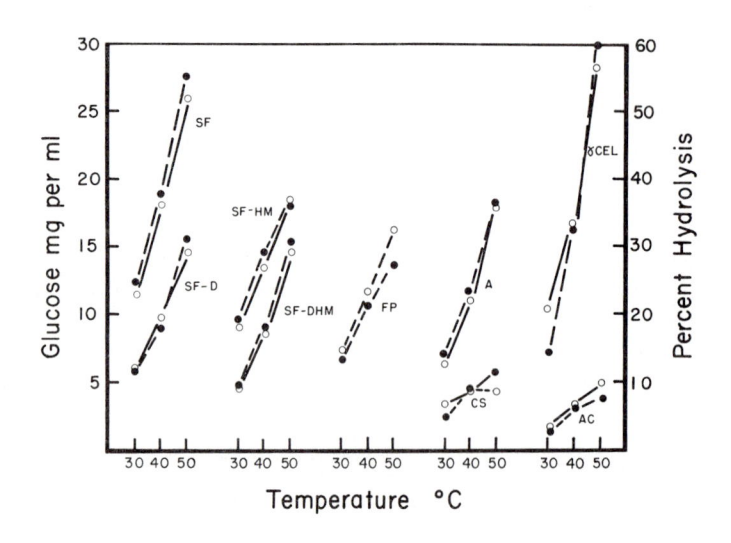

Figure 7. Effect of substrate on extent of hydrolysis by T. viride *cellulase at different temperatures*

500 mg. cellulose, 10 ml. enzyme in unshaken test tube, incubated at pH 4.8. 7 days

SF	Solka Floc
SF-D	Digested Solka Floc—i.e., residue after hydrolysis (20%) by T. viride *cellulase*
SF-HM	Heated milled Solka Floc (25 minutes at 200°C., ball milled 24 hours)
SF-DHM	Digested heated milled Solka Floc—i.e., residue after hydrolysis (16%) by T. viride *cellulase*
FP	Whatman No. 1 filter paper strips
A	Avicel—microcrystalline cellulose
CS	Cotton sliver, unground, dewaxed
δ Cel	Cellulose fiber, non-nutritive. Ground sulfite pulp
AC	Absorbent cotton, unground
○——○	Enzyme produced on Solka Floc, filter paper activity 1.88, 74 C_x units/ml.
●---●	Enzyme produced on digested Solka Floc, filter paper activity 1.90, 76 C_x units/ml.

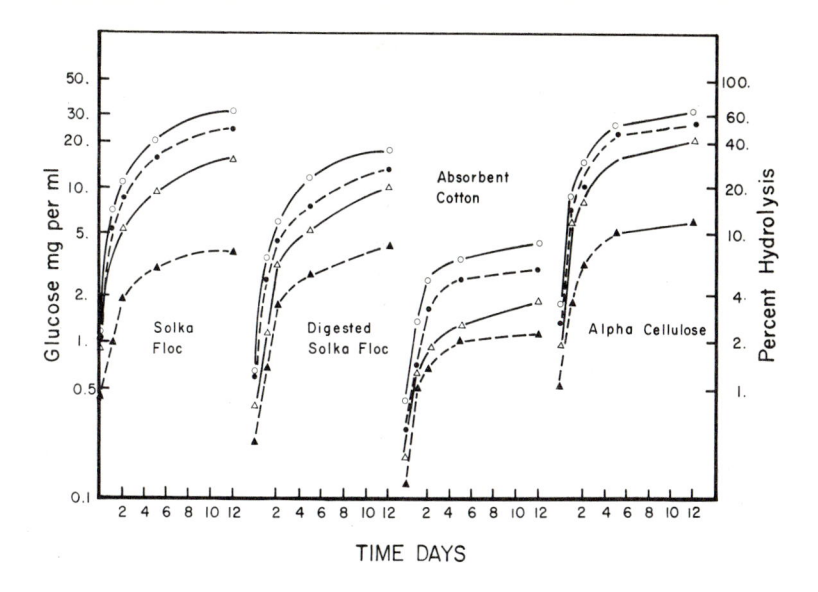

Figure 8. *Effect of enzyme concentration on extensive hydrolysis of cellulose by* T. viride *cellulase*

Crude filtrate from Solka Floc culture, F. P. activity 1.94 76 C_x units, 8.3 F. P. units, 10 C_1 units per ml. 500 mg. substrate + 10 ml. enzyme in unshaken test tube. Incubated at pH 4.8, 50°C. Substrates—see Figure 7. First point represents glucose at one hour

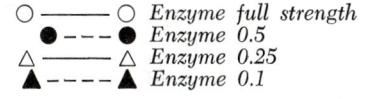

○────○ Enzyme full strength
●----● Enzyme 0.5
△────△ Enzyme 0.25
▲----▲ Enzyme 0.1

Production of Enzyme in Quantity

In Japan, a koji process is used for the production of *Trichoderma viride* cellulase. In this process, wheat bran is steamed, piled in trays, and inoculated with a spore suspension. The trays are sprayed during growth with a water mist, and forced air is circulated among them. The koji is turned to insure good aeration. When growth is completed, the koji is extracted, the extract filtered, and the enzyme precipitated with ammonium sulfate. If further purification is desired, the enzyme is redissolved, passed through an ion exchange resin, and precipitated with solvent (34). This process produces a good cellulase, but considerable handling of the enzyme is required. A submerged fermentation should be simpler.

Two patents have been issued in Japan for production of cellulase by submerged fermentation on nutrient salts plus 12.5% wheat bran (13) or 3% soybean cake plus 5% molasses (20). Other studies have investigated cellulase production by *Aspergillus* strains under submerged fermentation on nutrient salts plus filter paper, milled straw, or sawdust (*5, 18*). Pilot plant runs up to 100 liters were made (*5*).

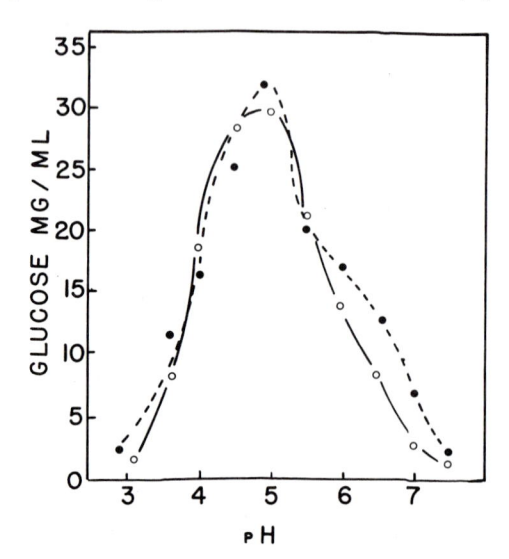

Figure 8a. Effect of pH on long-term hydrolysis by T. viride

5% suspension Incubated 10 days at 50°C. unshaken

○────○ Solka Floc
●------● Alpha cellulose

While we have not investigated fermentation technology in detail, we have used a laboratory fermenter to produce large quantities of *T. viride* cellulase in a semi-continuous system. The effluent from the fermenter has been used by Ghose in his digestion experiments (*7*) with no treatment other than filtration through glass wool and pH adjustment.

In one experiment, for example (Figure 9), 10 liters of medium (1.0% Solka Floc, 0.05% proteose peptone) in the 15-liter culture vessel was inoculated with a 100 ml. shake flask culture. The culture was allowed to grow for six days to reach a high cellulase value. Two liters of the culture was then harvested, and two liters of fresh nutrient pumped in from the nutrient reservoir to keep the volume 10 liters. Harvesting and replacement with fresh medium continued on a regular basis, with fresh nutrient reservoirs added, and the harvest vessel replaced as required. Solka Floc remained at 1.0% but peptone level was varied from 0.01% to 0.2%. Tween 80 at 0.1% was included after day 12. The maximum filter paper activity was 1.86, and maximum C_x 86 units/ml.

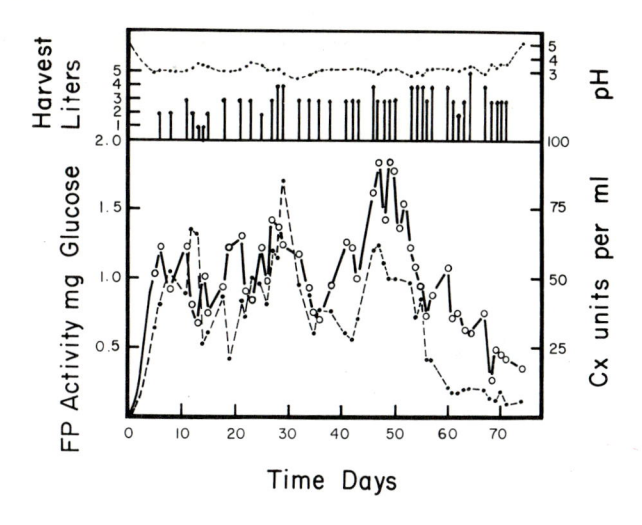

Figure 9. Production of T. viride *cellulase in a semi-continuous laboratory fermenter*

Culture volume 10L, T. 29°C., air 3 liter/min., impeller 120 r.p.m.
Nutrient (reservoir) Solka Floc 1.0% throughout; Tween 80- 0 day 1-12, 0.10% day 13-74; peptone 0.05% day 1-12, 0.10% day 13-28, 0.01% day 29-36, 0.10% day 37-48, 0.20% day 49-61, 0.10% day 62-74
Inoculum 100 ml. shake flask culture, F. P. Activity 1.38, C_x 19 units per ml.
○——○ *F. P. activity mg. glucose in one hour 50°C.*
●- - -● *C_x units per ml.*

Over 100 liters of active cellulase filtrate was harvested. After 60 days the culture became contaminated (with *Aspergillus niger*) and the yield declined. Other fermenters have maintained a good yield of cellulase for over 90 days without contamination.

This appears to be a practical means of producing cellulase on a large scale to supply a digester. *Trichoderma viride* grows vigorously, and the pH of 3.0 and cellulose substrate do not encourage contaminants. Most contaminations, in our experiments, occurred during the first few days or when peptone (or other soluble additive) was raised above 0.1%. Foam was a problem in some experiments and it did not respond well to the antifoam. When Tween 80 was added to the medium the foam became light, did not persist, and ceased to be a problem. To date, the best yields from a fermenter have been a filter paper activity of 2.26, and C_x 68 units/ml., only about one half the activity produced under our best conditions in shake flasks. In the shake flasks, however, 12–18 days growth was required for the highest activity. In the fermenter with a dilution rate of 0.1 to 0.3 per day, the culture remained young. The filtered effluent from the fermenter is clear and light straw color. It contains about 5 mg. per ml. of solids, about 0.5 to 1 mg. of this is protein. There is little or no reducing sugar present. If the filtrate is not used directly, the enzyme can be precipitated with ammonium sulfate or solvent (3).

Use of Cellulose Hydrolyzates

Trichoderma viride produces an extracellular enzyme capable of extensive degradation of solid cellulose. The activity of this enzyme may be sufficient for a practical conversion of waste cellulose to sugar. If the sugar is to be recovered, the rate and extent of hydrolysis, and the sugar concentration in the digest must be high. This may require that interfering substances such as lignin be removed, and that the cellulose be pretreated to increase its susceptibility. Using concentrated enzyme, hydrolyzates containing 30% glucose have been produced in very small scale (one ml.) experiment (12). In larger scale experiments, 8% glucose has been produced in a stirred vessel using untreated enzyme (7), and 10% concentrations appear feasible. In both cases, the cellulose was pretreated by heating and fine grinding.

If sugar recovery is not practical, the crude digest could be used directly to produce a fermentation product or single cell protein. We have grown yeast on a crude digest of Solka Floc containing 1.8% glucose. At 25°C. 45 of 47 cultures tested by us grew well, six of these consumed all the sugar. At 40°C. 14 of the cultures grew well, and three consumed all of the sugar (Table VIII). Apparently, there are no substances in the crude autoclaved enzyme toxic to the yeasts. We have not yet tested unautoclaved digests. The enzymes in these might attack the yeast cell walls (34).

An active cellulase will find many uses other than production of sugar (34, 36). These include removal of unwanted cellulose impurities (33) and breakdown of cell walls to improve extraction of a cell component, or to increase digestibility or nutrient availability of feeds (6, 35). Such uses require a preparation free of enzymes that might attack the residue being purified or the substance being extracted.

Many problems remain to be solved. We would like to have a more active enzyme. Should this be approached by improving the yield from *Trichoderma viride,* by modifying the growth conditions, or possibly through mutation? An obstacle to the latter approach has been that *Trichoderma viride* colonies spread so rapidly that techniques involving plating out have not been feasible. Recently, we have found that if conidia of *T. viride* are plated on desoxycholate agar (Baltimore Biological Laboratory), which contains 0.1% sodium desoxycholate, growth

Table VIII. Growth of Yeasts on Enzymatic Digests of Cellulose

Residual Sugar mg./ml.

QM No.	Name	at 25°C.	at 40°C.
8242	Torulopsis laurentii	0.6	0.8
8448	Candida tropicalis	0.2	0.9
8229	Saccharomyces fragilis	2.8	1.0
8209	Candida parapsilosis	4.3	4.2
8246	Trichosporon cutaneum	2.4	4.7
8240	Candida utilis	3.8	4.8
8233	Schizosaccharomyces pombe	5.4	4.8
8228	Saccharomyces elipsoideus	5.2	5.1
8226	Saccharomyces cerevisiae	5.4	5.4
8224	Rhodotorula sugami	3.7	6.7
8234	Schwanniomyces occidentalis	5.6	6.7
8250	Saccharomyces (unidentified)	5.9	6.7
8231	Saccharomyces lactis	1.9	8.8
8243	Torulopsis pulcherrima	0.3	9.5
5752	Pullularia pullulans	0.4	11.2
8450	Candida intermedia	1.0	15.6
8208	Candida guillermondii	1.2	17.7
8222	Rhodotorula mucilaginosa	2.7	10.0
8232	Saccharomyces lactis	4.0	10.4
8215	Rhodotorula bronchialis	4.1	11.0
8238	Sporobolomyces sp.	4.1	11.0
8216	Rhodotorula branchialis	4.3	11.0
8223	Rhodotorula sanniei	4.3	11.2
8236	Sporobolomyces sp.	4.4	10.8
8245	Trichosporon asteroides	4.5	10.2
8449	Candida lipolytica	4.5	16.5
8218	Rhodotorula glutinis	4.7	10.2
8235	Sporobolomyces salmonicolor	4.7	17.1
8219	Rhodotorula glutinis	4.8	9.6
8221	Rhodotorula mucilaginosa	4.8	9.8
8217	Rhodotorula glutinis	4.9	9.6
8237	Sporobolomyces sp.	4.9	10.0
8247	Trigonopsis variabilis	5.0	10.2
8248	Willia sp.	5.0	11.8
8212	Mycoderma cerevisiae	5.0	12.6
8239	Torula sphaerica	5.2	16.3
8249	Saccharomycete (unidentified)	5.2	17.1
8039	Ustilago maydis	5.3	11.4
8214	Rhodotorula aurea	5.5	16.5
8230	Saccharomyces lactis	5.7	15.3
8227	Saccharomyces cerevisiae	5.9	16.5
8241	Torulaspora fermentati	5.9	16.8
8252	Zygosaccharomyces acidifaciens	6.3	16.8
8213	Pichia membranefaciens	6.7	18.3
8220	Rhodotorula minuta	7.6	10.2

Cultures grown four days in shake flasks on *T. viride* cellulase digest of Solka Floc (filtered, pH adjusted to 5.0, sugar concentration 18 mg./ml., 0.03% urea and 0.01% yeast extract added).

is restricted so that colonies do not exceed 2 mm. in diameter even after ten days incubation. Although most viable conidia appear to form colonies, and sporulation is good, colonies do not appear until after five or six days of incubation. If desoxycholate lactose agar (BBL), which contains 0.05% sodium desoxycholate, is used for plating, the colonies appear earlier, but are still restricted to about 4 mm. in diameter. Therefore, we have begun a program attempting to induce mutations that may lead to a more productive strain. Should we look for some new and more potent organism? Many papers report strong cellulases (*16, 31*). Just how strong are they? And we still need a magic treatment of the substrate that will greatly increase susceptibility without too much cost. In this paper we are proposing a system. Final judgment will have to be made after further experimentation.

Acknowledgment

We are grateful to Elwyn T. Reese for his many helpful suggestions, and for obtaining many of the samples tested in this work; and to Hamed M. El-Bisi and Robert O. Matthern for their interest in and support of this project.

This paper reports research undertaken at the U.S. Army Natick (Mass.) Laboratories and has been assigned No. TP. 512 in the series of papers approved for publication. The findings in this report are not to be construed as an Official Department of the Army position.

Literature Cited

(1) Bilai, V. I., Pidoplichko, N. M., Taradii, G., Lizak, Yu, *Mol. Osn. Zhiznnenngkh Protessov. Dokl. Sess. Otd. Biokhim., Biofiz. Fiziol. Akad. Nauk Ukr. SSR Probl. Mol. Biol., Kiev, 1965*, p. 204 (1966).
(2) Bucht, B., Eriksson, K. E., *Arch. Biochem. Biophys.* **124**, 135 (1968).
(3) Cayle, T., *U. S. Patent* **3,075,886** (1963).
(4) Chang, W. S., Usami, S., *Hakko Kyokaishi* **25**:8, 349 (1967).
(5) Cholakov, G., Kolev, D., Banikova, St., Nikolov, T., Dimitrov, M., Petkov, P., *Kongr. Bulg. Mikrobiol., 1st, Sofia, 1965*, p. 151 (1967).
(6) Fugono, T., Nara, K., Yoshino, H., *Hakko Kogaku Zasshi.* **42**:7, 405 (1964).
(7) Ghose, T., Advan. Chem. Ser. **95**, 415 (1969).
(8) Gorokhu, G. I., Korsakav, I. V., The manufacture from wood of crystalline glucose. *Gidrolizh i Lesokhim. Prom.* **13**:5, 26 (1960).
(9) Halliwell, G., "Advances in Enzymic Hydrolysis of Cellulose and Related Materials," E. T. Reese, Ed., p. 71, Pergamon Press, London, 1963.
(10) Halliwell, G., *Biochem. J.* **100**, 315 (1966).
(11) Iwasaki, Y., Niihara, R., Matuda, Y., *Hakko Kogaku Zasshi* **45**, 1072 (1967).
(12) Katz, M., Reese, E. T., *Appl. Microbiol.* **16**, 419 (1968).
(13) Kawaji, S., Ishikawa, T., Saito, K., Inohara, T., Kubo, A., Tejima, E., *Japanese Patent* **2986** (1964).

(14) King, K. W., *J. Ferm. Tech. (Japan)* **43,** 79 (1965).
(15) Krupnova, A. V., Sharkov, V. I., *Gidrolizn i Lesokhim. Prom.* **17:3,** 3 (1964).
(16) Kuroda, H., *Hakko Kyokaishi* **25:7,** 283 (1967).
(17) Liu, T. H., King, K. W., *Arch. Biochem. Biophys.* **120,** 462 (1967).
(18) Loginova, L. G., Tashpulatov, Zh., Kalmykova, G. Ya., L. N. Sergeeva, *Ferment, Rasshcheplenie Tsellyul., Akad, Nauk SSSR, Inst. Biokhim.* 66 (1967).
(19) Lowry, O. H., Rosebrough, N. J., Farr, A. L., Randall, R. J., *J. Biol. Chem.* **193,** 265 (1951).
(20) Maejima, K., Yoshino, H., *Japanese Patent* **2985** (1964).
(21) Mandels, M., Matthern, R. O., El-Bisi, H., *Techn. Rept., Food Lab., US Army Natick Laboratories, Natick, Mass.* (1968).
(22) Mandels, M., Reese, E. T., *J. Bact.* **72,** 269 (1957).
(23) *Ibid.,* **79,** 816 (1960).
(24) *Ibid.,* **83,** 400 (1962).
(25) Mandels, M., Reese, E. T., *Developments Indus. Microbiol.* **5,** 5 (1964).
(26) Reese, E. T., *Can. J. Microbiol.* **14,** 377 (1968).
(27) Reese, E. T., Gilligan, W., Norkrans, B., *Physiol. Plantarum* **5,** 379 (1952).
(28) Reese, E. T., Gilligan, W., *Text. Res. J.* **24,** 663 (1954).
(29) Selby, K., "Advances in Enzymic Hydrolysis of Cellulose and Related Materials," E. T. Reese, Ed., p. 33, Pergamon Press, London, 1963.
(30) Selby, K., Maitland, C. C., *Biochem. J.* **104,** 716 (1967).
(31) Selby, K. Investigation of currently available cellulase preparations. *Rept. Comm. Organ. Econ. Coop. Develop.,* 5 pp. 3 fig. (1968).
(32) Sumner, J. B., Somers, G. F., "Laboratory Experiments in Biological Chemistry," Academic Press, New York, 1944.
(33) Tazaki, R., Ouye, K., *Hakko Kogaku Zasshi* **40,** 195 (1962).
(34) Toyama, N., "Advances in Enzymic Hydrolysis of Cellulose and Related Material," E. T. Reese, Ed., p. 235, Pergamon Press, London, 1963.
(35) Toyama, N., *Hakko Kyokaishi* **21:10,** 415 (1963).
(36) Underkofler, L. A., "Advances in Enzymic Hydrolysis of Cellulose and Related Materials," E. T. Reese, Ed., p. 255, Pergamon Press, London, 1963.
(37) Underkofler, L. A., ADVAN. CHEM. SER **95,** 343 (1969).
(38) Wiley, Bonnie, personal communication (1968).
(39) Yamane, K., Suzuki, H., Yamaguchi, K., Tsukada, M., Nisizawa, K., *J. Ferment. Technol. Japan* **43,** 721 (1965).
(40) Yerkes, W. D., *U. S. Patent* **3,310,476** (1967).

RECEIVED October 14, 1968.

Discussion

N. King: "Do you find any effect of the inoculum size on enzyme yield?"

M. Mandels: "Not a great effect. We have tried inoculation with mycelium. I should have mentioned that the *Trichoderma* is a nice fungus to grow in a fermenter. It forms a diffuse type of mycelium. So, we can inoculate with mycelium, or we can inoculate with spores. I do not find there is any great effect. We never use what one would consider a large inoculum. We normally inoculate with a milliliter of spore suspension for 50 ml. of medium, about 10^6 conidia."

Query: "Do you have any relative activity on wood or modified wood *vs.* pure cellulose?"

M. Mandels: "The activity on wood containing lignin is much lower, and so usually the wood has been ball-milled to increase its digestibility by enzymes."

Haas: "Apart from its influence on growth, how important is aeration to enzyme preparation?"

M. Mandels: "We have never really investigated that, but in our fermenter we have used an aeration rate of about 0.3 volume per volume per minute, which I think is low for a fungus. We had a lot of trouble with foam at first, until we began putting Tween 80 in the media. The Tween 80 made the foam collapse."

Comment: "In regard to the use of cellulase on wood, the literature reports that when you irradiate wood you degrade the cellulose more rapidly than the lignin because the irradiation seems to be absorbed by the phenol groups in the lignin. So, the possibility might exist that you could degrade your cellulose considerably in natural wood to a point where the cellulase would be more active."

M. Mandels: "I think the problem of lignin is probably accessibility, *i.e.*, it interferes with migration of enzyme to the cellulose."

Comment: "When you irradiate wood you increase the digestibility in ruminants quite markedly. You are actually opening up the structure by one means or another."

ize: 24

Enzymatic Saccharification of Cellulose in Semi-and Continously Agitated Systems

TARUN K. GHOSE[1] and JOHN A. KOSTICK

Microbiology Division, Food Laboratory, U. S. Army Natick Laboratories, Natick, Mass. 01760

> Enzymatic saccharification of cellulose pulp (Solka Floc) in semi- and continuously agitated systems has been studied using untreated enzymes obtained from submerged fermentation of Trichoderma viride (T. viride). Pretreatment of the cellulose substrate by heating, heating followed by grinding, and grinding followed by heating, leads to considerable difference in the rate of saccharification. It is possible to reactivate the residues from digestion to a high degree of susceptibility to enzymatic hydrolysis by heat treatment and milling. Saccharification of cellulose has been shown to be possible by milling the substrate in contact with Tv cellulase. Semi-continuous hydrolysis of heated and milled Solka Floc (10% w/v) by enzyme (with daily replacement of 40% of the reaction volume with fresh enzyme substrate suspensions) maintained the sugar level in the effluent over a period of several days at 5.1–5.6%.

A large number of rumen and soil microorganisms representing extreme ranges of aerobic and anaerobic species are found to possess the capacity of degrading native cellulose into a wide range of products, including cellobiose, glucose, methane, alcohols, and organic acids. Yet, not a single commercial undertaking exists today to produce any of these chemicals through this microbial pathway. The desirability of having cellulose converted into glucose by microbes has nevertheless been emphasized because of the numerous problems involved in the chemical means of doing it. Although there is much uncertainty concerning the nature and mechanism of the action of enzymes, there is considerable

[1] NRC Senior Visiting Scientist on leave from Jadavpur University, Calcutta, India.

knowledge of what they do. Generally, the slow rates, the diverse endproducts, and the problems of economic recovery of the products are some of the factors which stand in the way of biochemical utilization of cellulose. The success with which enzymatic conversion of starch into sugars has been attained (12, 18) motivates us to consider what might be similarly achieved in the case of cellulose.

Cellulolytic enzymes exist in nature, and many have been studied. Perhaps the best performing ones are not yet known to us, but the ones already produced from *Trichoderma viride* are good enough to saccharify cellulose in a manner comparable with that by which starch is converted into sugar. The success of such a program would lead to the creation of a tremendous new source of fermentable sugars from agricultural wastes. Besides discovering a more active source of cellulolytic enzyme, one of the most formidable goals to be accomplished lies in the modification of the native cellulose into a form more susceptible to hydrolysis. In addition to the complete conversion of cellulose into glucose, cellulase has been used to break down the cell walls of cereals, legumes, a large number of vegetables, and fruits, to make their edible contents more available to human digestive system. Other food uses of cellulase enzyme include clarification of citrus juices, removal of fiber from edible oil press cakes, higher yields in starch recovery from sweet potato, and cassava roots, extraction of proteins from leaves and grasses, tenderizing fruits and vegetables prior to cooking, extraction of essential oils and flavoring materials from roots, seeds, and barks containing large contents of crude fibers, etc. These applications are more fully described elsewhere (20).

A description of the difference between the mechanism of acid and enzymatic saccharification of cellulose has been reported (22). It has been shown that enzymatic hydrolysis of cellulose causes a slower loss in degree of polymerization (DP) than the acid hydrolysis. The most probable reason for such a difference is the relative sizes of the two catalysts (enzyme and mineral acid) and their ability to permeate the fine structure of cellulose. Large cellulase molecules (25,000–67,000 mol. wt.) are able to penetrate only into the larger intercrystalline space and on account of their high catalytic activity, all chain linkages (β-1,4 glucosidic) which can be contacted are readily hydrolyzed to form soluble sugars. On the other hand, the relatively small acid molecules are capable of diffusing into the smaller intercrystalline spaces and then attack glucosidic bonds which are otherwise not available to the enzymes. As a corollary to this, a finely ground cellulose would be expected to result in a higher conversion into glucose units by cellulose because of the reduction in crystallinity on one hand, and proportional increase of the surface on the other (3, 4, 7, 8).

A comparison (15) of activities of acid and enzyme catalysts on three cellulosic substrates at 50°C. shows that 100,000 times as much acid is required to bring about the same degree of hydrolysis. At the molecular level the difference is further increased because of the disparity in mol. wt. [(HCl–36, cellulase 63,000 (23)], so that approximately 10^8 HCl molecules are required to perform the work of a single enzyme under certain conditions. This is a point strongly in favor of enzyme hydrolysis of cellulose.

Microorganisms have no difficulty digesting cellulose (Figure 1). They accomplish it rapidly and effectively. Why is it then that we cannot utilize their systems to develop a practical conversion of cellulose to sugar? The answer is rather simple; we can—if we pour into this problem the effort it rightly deserves.

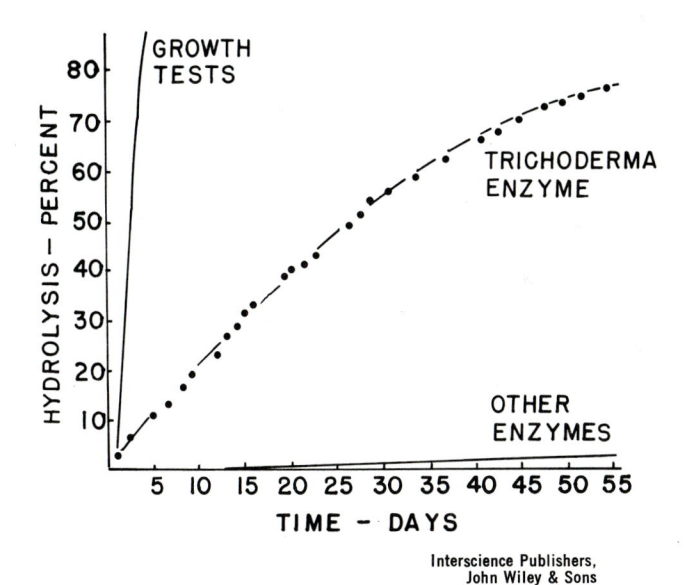

Figure 1. Hydrolysis of cotton by organisms and by cell-free enzymes (17). Growth tests—Consumption of cellulose by active cellulolytic fungi. Trichoderma enzyme— (●-●-●-● - weight loss of cellulose). Other enzymes—Cell-free cellulases from organisms other than Trichoderma

Accessibility of Substrate and Rate of Saccharification. Let us reexamine the problem. Enzymes which catalyze the hydrolysis of simple soluble compounds may split a million (or more) bonds per minute per enzyme molecule. This is the ideal system towards which we must work.

It involves molecules which in solution pose no problem of accessibility to enzyme. What happens when the substrate molecule gets much larger —*i.e.*, as one moves from the monomer to polymer? The interaction between large enzyme and large substrate molecules may be somewhat impeded—*i.e.*, it is likely to be more difficult for the enzyme to accommodate itself to the site at which reaction is to occur. But even here the rate of catalysis is high, in the neighborhood of 18,000 bonds per minute per enzyme molecule for α-amylase and 240,000 bonds for β-amylase. Viscous polymer and branched polymers present additional impedence to enzyme attack, the first through restricting enzyme movement and the second through blocking hydrolysis sites. Substituents on the substrate may also bind enzyme through electrostatic charges, further lessening the rate of hydrolysis.

The above factors, however, limit hydrolysis far less than conversion of the soluble polymer to the solid state, and most resistant of all substrates are those in which molecules are in a highly oriented or crystalline state. Into these it is difficult even for small water molecules to penetrate. What chance, then, for the large enzymes?

This resistant condition is the state of most cellulose. Our problem, thus, is to lessen the resistance by converting it into one of the more susceptible forms. Some years ago, cellulose highly susceptible to enzymes was produced (22) by swelling it in acid, and subsequently precipitating the same in water. Other methods (Figure 2) which increase accessibility are physical size reduction (3, 4, 7, 8) and alkali swelling (1). Gamma rays in low doses decrease susceptibility, but in high doses exercise the opposite effect (14). Boiling in dilute acid produces a more resistant residue. All these methods yield an insoluble cellulose. Chemical modification on the other hand results in soluble derivatives of cellulose. When these have only sufficient substituents to lend solubility (CMC, HEC, C–SO$_4$), they are much more rapidly hydrolyzed than are any of the insoluble cellulosic forms. Greater substitution (DS1) may produce a more soluble product, but it is then resistant to enzyme attack.

Both the swollen and the soluble celluloses can be rapidly saccharified, and the former completely converted into sugar in a relatively short time. The problem with them is that the maximum concentration obtainable is quite low. A 5% suspension of swollen cellulose is a solid mass. Even complete conversion can produce only 5% plus glucose in the hydrolysate. We need to reach higher levels.

Our aim then is modified somewhat. We require not only a susceptible substrate, but one which has a high bulk density, one which can be used in suspensions of 30–50%. At first glance, this would seem to be an impossible requirement; for if swelling leads to increased susceptibility, it also results in decreased concentrations in a given volume of

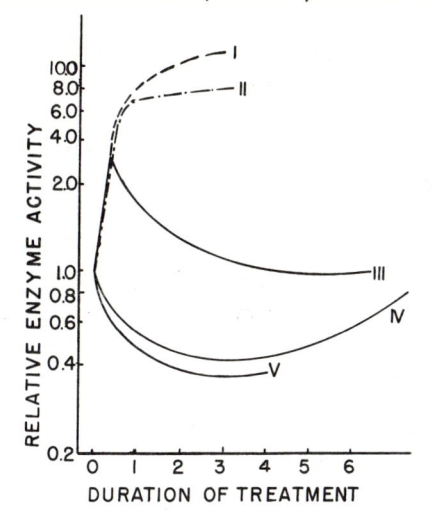

Figure 2. Effect of various treatments on the susceptibility of cotton cellulose to enzymatic hydrolysis (16)

I—Effect of 85% H$_3$PO$_4$, or 72% H$_2$SO$_4$, abscissa in hours;
II—Effect of ball milling (vibratory), abscissa in hours;
III—Effect of 30% NaOH, abscissa in days exposure to alkali swollen material to air;
IV—Effect of irradiation with cathode rays (van de Graaf), abscissa 0–60 megareps;
V—Effect of weak organic acids, abscissa in days of refluxing (loss of weight = 5 to 6% in 3 days)

enzyme. The recent Russian discovery (8) gets us out of this dilemma. It has been shown (Figure 3) that milling of wood (fir) pulp cellulose at 200°C. for 40 min. in a specially constructed grinding unit increases the capacity of the cellulose to be hydrolyzed by acid at a faster rate. The glucose yield reaches 105%—*i.e.*, almost equal to theoretical. The comparative yields of glucose from cellulose milled under heated conditions at various temperatures show that a temperature range between 200°–220°C. appears to be most favorable for the kind of heat transfer and size reduction systems which the cellulose is exposed to. At 230°C. the cellulose increases in susceptibility to hydrolysis (∼80%) up to 20 min. of milling and heating period and then falls off. Katz and Reese (4) have found that suspensions up to 50% w/v of ground and heat treated cellulose can be prepared; that this product is readily hydrolyzed by enzymes; and that concentrations of glucose of more than 30%

can be obtained. Our colleagues (Mary Rollins and Verne Tripp, U.S.D.A., SURDD, New Orleans, La.) have examined cellulose samples, heat treated and ground, which we sent them for examination. They report that the samples are probably slightly oxidized and their crystalline structure destroyed as evidenced by both infrared spectra and x-ray diffraction scans. The control samples (Solka Floc, a sulfite process spruce pulp) has an x-ray crystallinity index of 83 which is comparable with that of cotton (85–90). Milled, and milled and heated samples show severe loss of crystallinity (to a cryst. index of about 41 (and reduced degrees of polymerization (to DP 200 or less). In the enzyme digested sample of heated and milled cellulose there appears to be more fibrillar structure visible and an increase in crystallinity implying that non-fibrillar material may have been digested away. Staining tests indicated that the milled cellulose and subsequent samples were more accessible to methylene blue and to iodine than was the control sample.

More than 60% reduction in the degree of polymerization of cellulose (about 1150 to 440) in the first eleven hours of contact with *Penicillium variable* cellulase (Figure 4) has been reported (2). Further depolymerization was very slow, implying that the amorphous portion of the cellulose has reacted rapidly, leaving a more resistant crystalline structure represented by a DP of about 440. These conclusions strongly

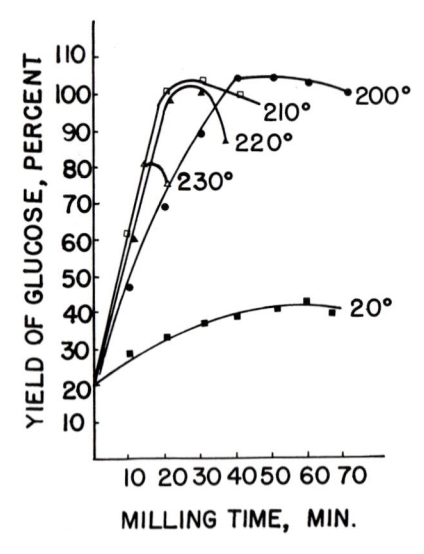

Gidroliznaya i Lesokhimicheskaya Promyshlennost

Figure 3. Influence of temperature and duration of dry milling of fir tree cellulose on subsequent hydrolysis with diluted sulfuric acid (8)

support our experimental data (Figures 15, 16)—*i.e.*, the activated cellulose showing increased digestibility. The compact nature of our product with a high bulk density (15–30 lbs./cu. ft.) is caused by both oxidative heat treatment and size reduction. The Russian report (8) claims that the readily hydrolyzable cellulose obtained by dry milling of wood pulp at 200°–240°C. is soluble in cold water; but such a high degree of solubility could not be accomplished by either Katz and Reese (4) or Ghose (3).

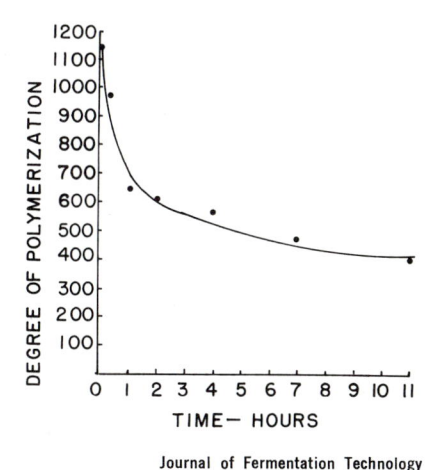

Journal of Fermentation Technology

Figure 4. Change of the degree of polymerization by enzyme action. Penicillium variable cellulase—wood pulp system (2)

In heated and milled cellulose, we have a material which is by far the best substrate yet found for cellulose saccharification, where the major goal is to obtain high glucose concentrations at a rapid rate.

Environmental Factors Affecting Rate. Temperature, pH, substrate concentration, and enzyme concentration are important factors in determining the rate of hydrolysis of any substrate. King (6) has summarized the advantage of several enzymes providing optimum activity at 50°C. (Figure 5). Li, Flora, and King (9) have demonstrated that the effect of temperature on reaction velocity in the cellulose *T. viride* enzyme system depends on the nature of cellulosic substrate used (Figure 6). For amorphous cellulose, the Arrhenius plot is linear over the temperature range from 10°–57°C. (activation energy—5.1 kcal./mole). Because of the phase change from a sol at temperatures above 37°C. to a gel at lower temperature the carboxymethylcellulose profile represents two activation energy values (6.4 kcal./mole from 37°–60°C. and 16.7

kcal./mole below 37°C.). King (5) showed a fragility effect by the cellulase (most probably C_1 component) acting on hydrocellulose. This is reflected in a marked increase in the total number of crystalline hydrocellulose substrate particles (assumed spherical) during the initial substrate enzyme contact (Figure 7), implying that the enzyme action is not exclusively a surface erosion effect, but fragmentation of the larger initial particles into smaller ones. The surface area provided by the particles is a linear function of the surface destroyed by the cellulase.

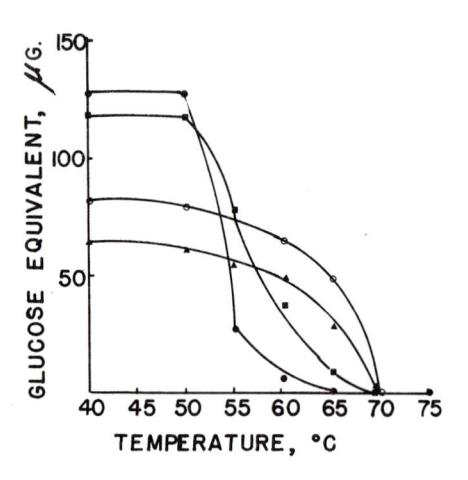

Biochemical Journal

Figure 5. Effect of heating on enzyme activity (6). Reaction mixtures contained substrate (2 mg.), 0.1M—acetate-buffer, pH—5.0 (2.0 ml.), and 0.1 ml. of an enzyme solution that had previously been held at the temperature shown for 10 min. Glucose equivalent was then estimated by the standard assay procedure described in the text

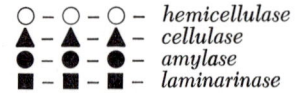

○ - ○ - ○ - *hemicellulase*
▲ - ▲ - ▲ - *cellulase*
● - ● - ● - *amylase*
■ - ■ - ■ - *laminarinase*

The culture filtrate used in the starch and laminarin experiments was diluted 1:10

For *T. viride* cellulase substrate system a temperature of 50°C. and a pH of 4.5–5.5 are reported (3) to be optimal. Over this pH range, loss of enzyme activity may be caused by adsorption on adsorbents such as Fuller's earth or on substrates. Enzyme alone does not appear to lose activity at the temperature and pH of the reaction (50°C., pH 4.8–5.0) mixture. Rates of hydrolysis increase with enzyme and substrate con-

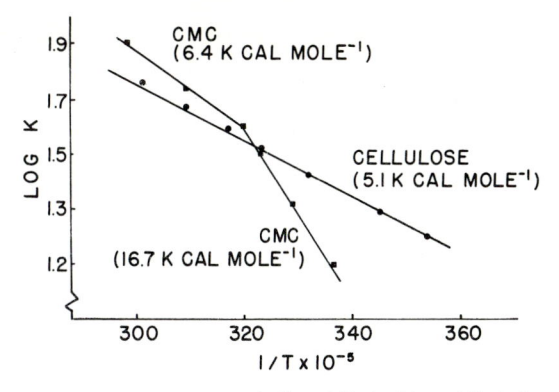

Figure 6. Arrhenius plot showing effect of temperature on the rate of hydrolysis of amorphous cellulose and carboxymethylcellulose by the endoglucanase (9)

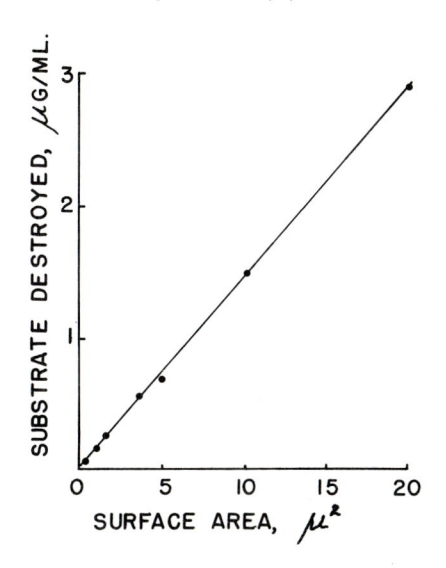

Figure 7. Relationship of the rate of degradation of crystalline hydrocellulose particles to their surface area (5)

centrations (3, 21). No one has yet reported a decrease in rate as the cellulose concentrations become very high. However, a decreased rate has been observed with carboxymethylcellulose at concentrations over 1% but not with cellulose sulfate with the same *T. viride* enzyme (10). Similarly, Nisizawa *et al.* (13) have shown that the rate of hydrolysis of cellopentaose and cellotetraose by *Irpex lacteus* cellulase decreased markedly when a particular substrate concentration had been exceeded (Figure 8). Cellopentaose inhibited strongly at a concentration as low as 1 mg./ml. of reaction mixture. It required much more of the tetramer (7 mg./ml.) and considerably more of the trimer (500 mg./ml.) to reach inhibitory levels. These data show that cellulase can be inhibited at high concentrations by some of its substrates, but as indicated above, no such inhibition has been reported in the case of solid unmodified cellulose.

Product Inhibition. In most microbiological and biochemical systems accumulation of end products exercises an inhibitory effect on the rate of the forward reaction. Stimulation by end product is thermodynamically improbable. One of the major products of hydrolysis of cellulose is cellobiose. There is a stimulation by cellobiose of C_x activity of *Streptomyces* spp. filtrates only when the substrate is solubilized by the introduction of various substituents—*e.g.*, CMC, hydroxycellulose, cellulose acetate, etc. Stimulation is absent when unsubstituted cellulose is used (*14*). On the other hand, product "inhibition" is common. Cellobiose inhibits the hydrolysis of cellulose by filtrates of most of the 36 organisms tested. This action of cellobiose is believed to be that of an end product inhibiting an enzymatic hydrolysis in much the same manner that maltose inhibits hydrolysis of starch by α-amylase. The inhibitory effect of products varies with the organism from which the cellulase is derived. Thus, lactose is a very good inhibitor of the enzyme from

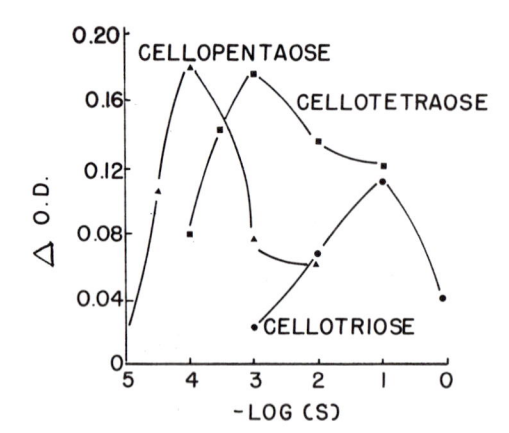

Figure 8. Activity—PS curves of cellulase II for cellotriose, cellotetraose and cellopentaose (13). Cellulase II: C_1 component obtained from a culture filtrate of Irpex lacteus *by starch–zone electrophoresis. Conditions: Plastic tray 2 × 5 × 45 cm., veronal buffer, pH 8.7, μ: 0.1, 24 hr., 5.6 V/cm., 2.0–3.0 mA./cm.^{2}*

Penicillium pusillum (Table I), but cellobiose is a good inhibitor of cellulases of many origins. Glucose inhibition is generally weak. For *T. viride* cellulases acting on heat treated cellulose, a concentration of 30% glucose gives only 40% inhibition (*3, 4*).

Product inhibition seems to be greatest for the larger soluble oligomers [cellopentaose (*13*)] and diminishes rapidly as the product molecular sizes become smaller. There would, therefore, seem to be a definite advantage in adding cellobiase to the cellulose hydrolysates for the conversion of cellobiose to glucose resulting in a decreased inhibition. In order further to cut down product inhibition and to keep the saccharification rate at its maximum, it would be of advantage to run the reaction in a steady state continuous system. A high bulk density of the substrate being only possible at a small particle size, a balance between the reaction time and glucose yield, may be drawn to reduce the inhibitory effects of glucose.

Table I. Inhibition of *Penicillium pusillum* Cellulase by Sugars (*10*)

Sugar 1%	CMC Visc.	Swollen Cellulose Weight Loss	Ball Mill Cotton Weight Loss	Cotton Swelling
Cellobiose	48	46	72	52
Lactose	88[a]	44	60	54
Maltose	5	4	1% stim.	NT
Glucose	5	NT	NT	37

[a] Lactose (0.1%) gave 52% inhibition; NT—Not tested.

Such a steady state continuous model has been favored for physiological studies in many microbiological systems primarily because environmental conditions can be kept constant, in sharp contrast with batch systems, where microorganisms are subject to continual change. The greatest advantage of a continuous biological system lies in the higher productivity than that for batch. Even a semi-continuous fermentation which eliminates a high incidence of microbial mutation appears advantageous. Engineering problems of aseptic control may also be avoided to some extent in a semi-continuous operation. Continuous culture is an excellent tool for studying mutation of microbial populations. Nevertheless, there are certain inherent practical problems associated with continuous fermentation. These are: (a) lack of homogeneity, especially in the case of thick and viscous media in large sizes, causing poor and fluctuating productivity at low dilution rates, (b) difficulty of maintaining a continuous fermentation under aseptic conditions over long period of time involving serious loads on the operations of sterilizing media and air, and (c) inadequate stability relating to microbial strain and mechanical operation. Long continuous microbial fermentations are subjected to invariable mutation problems; this difficulty is by-passed by having semi-continuous operation in which the first vessel in series is occasionally re-seeded.

All the disadvantages cited are, however, applicable to microbial fermentations and none to enzymatic systems. As long as mechanical possibilities permit agitating and pumping thick slurries with high substrate concentrations (we have done this with 15% suspensions without difficulty) and, allowing fine free-flowing cellulosic substrates into the hydrolyzer at constant rate and under aseptic conditions, we do not visualize any major obstacle to run the system continuously. Maintaining it under sterile state is rather a less bothering problem particularly because of the possibility of having high glucose concentrations (~30%) in single or multi-vessel system.

Among the candidates important as substrates for single cell protein (SCP) cellulose is conspicuous. Unlike petroleum, a substrate about which much is said but so little is available in most of the overpopulated countries, cellulose is abundant all over the world. Unfortunately, few laboratories and but small efforts are engaged today in the big mission of economically rendering cellulose into its building blocks, glucose—a perennial source of carbohydrate for man and microbes. Those engaged in cellulase research today have had no success in selling the idea; but the successful production of sugar from cellulose at a cost less than 8¢ a pound will call attention to many who are interested in producing SCP or other food components from unconventional sources. We seem to be closer today to the accomplishment of saccharification of cellulose in a manner comparable to that achieved by glucoamylase in the conversion of starch to glucose.

The present study is aimed at investigating the increase in the susceptibility of cellulose towards cellulase in order to increase both rate and yield of sugar. It demonstrates the possibility of regenerating enzyme digested-resistant cellulose into highly susceptible form. It also describes a stable model reaction system for semi- or fully-continuous system for saccharification of cellulose into glucose using untreated culture filtrates of *Trichoderma viride*.

Materials and Methods

The Substrate and Its Modification. Solka Floc—SW40A (Brown Co., Berlin, N. H., USA), a wood (spruce) pulp commonly used as a filter aid has been used as a basic cellulosic material for susceptibility tests and saccharification studies. The material was milled dry in a laboratory porcelain pot mill using glazed porcelain balls of 1 inch (2.65

kg.) and 1/2 inch (1.2 kg.) sizes over a period of time. Sweco 70 material used in several experiments was a dry milled product of Sweco Vibro Energy Mill received from Southwestern Engineering Company, Los Angeles, California. The dry milled products were screened down to 270 U.S. Standard Screens (opening 53μ). As heat treatment of the substrate was found to improve enzyme susceptibility further, the Solka Floc was heated to 200°C. for 25 min. either before or immediately after milling. To ascertain the extent of hydrolysis taking place during milling of cellulose in the presence of the cellulase, a few test runs were conducted with Solka Floc suspended in enzyme solution at room temperature (22°C.). Modification of Solka Floc was done by a combination of several kinds of treatments such as milling in pot mill, milling in Sweco Mill, heating in a laboratory forced draft air oven at 200°C. for 25 min., heating followed by milling, milling followed by heating, and milling in the presence of Tv cellulase. The following abbreviations are used in the text to designate various substrates:

SF = Solka Floc SW40A without any treatment
SF–M = Solka Floc pot milled
SF–H = Solka Floc heated at 200°C. for 25 min.
SF–MH = Solka Floc pot milled and heated as above
SF–HM = Solka Floc heated and pot milled as above
SF–MHD = Solka Floc pot milled, heated, and enzyme digested (~60%)
SF–EM = Solka Floc pot milled in presence of cellulase enzyme
SF–MHDWDr = Solka Floc pot milled, heated, enzyme digested, washed, and dried at 60°C.
SF–MHDWDrH = Solka Floc pot milled, heated, enzyme digested, dried, and heated as above
SF–MHDWDrM = Solka Floc pot milled, heated, enzyme digested, washed, dried, and milled as above
SF–MHDWDrMH = Solka Floc pot milled, heated, enzyme digested, washed, dried, milled, and heated

Screen Analysis of the Substrate. The milled cellulose was screen analyzed in a Ro-Tap Testing Sieve Shaker (The W. S. Tyler Company, Cleveland, Ohio) using a series of 40, 60, 100, 120, 170, 200, and 270 mesh U.S. Standard Screens sifting for 45 min. plus an additional 5 min. after the first weight to check the reproducibility of weights of the screened fractions. Both the weights checked very well. The same procedure was followed in the case of SF, SF–H, SF–HM, SF–MH, and Sweco 70 samples, all of which were subsequently tested. Summary of screen analyses reports is presented in Table II and Figure 9.

Enzyme and Its Assay. Culture filtrates from 10 liter batch or semi-continuous submerged fermentations of *Trichoderma viride* QM 6a were used in the saccharification studies after adjustment of pH ($\sim$3 → 5.0). Most of these enzyme solutions were prepared by Mary Mandels (*11*). Following harvests, the culture fluids were filtered off on glass wool and made free of the suspended mycelium, pH adjusted to 4.8–5.0, and stored at 2°C. The enzyme strengths expressed in terms of FP activity (*11*) of the batch and semi-continuous culture filtrates used in the studies varied generally between 0.75 to 1.26 and 0.9 to 1.86. For continuous saccharification experiments requiring large volumes of enzymes with constant activity, several harvests were mixed together to bring the enzyme strength to about a FP activity of 1.0. The filtrates were sparkling clear with a light yellow-green coloration and an agreeable smell. Assays of enzyme activity of all samples received from fermenters were performed immediately after each havest followed by filtration. For enzyme assay (FP activity) procedure, see Reference 11.

Table II. Screen Analysis of Solka Floc and Heated Solka Floc

Materials	Particle Size Distribution, %		
	$> 149\mu$	$< 149\mu$	$< 53\mu$
SF	19.4	80.5	11.6
SF-H (200°C. for 25 min.)	23.7	74.8	11.9
SF-HM (200°C. for 25 min. and pot milled for 24 hr.)	4.9	95.9	76.0
SF-MH (milled before heating)	8.6	91.9	54.9
Sweco 70	0.9	99.0	93.8
Sweco 70-H	1.1	99.3	93.8

Reducing Sugars. Estimation of total reducing sugars (RS) as glucose has been based on the DNS method (*18*) used in the cellulase assay. Actual procedure has been described in another communication (*3*).

Experimental

The factors affecting stability and activity of cellulase have been studied by several investigators. The present experiments include some of these factors particularly as they relate to a system being evolved for the production of high glucose concentrations.

Enzyme Studies. (1) Effect of Temperature and pH on the Inactivation of Cellulase. Details of experiments and results on the effects of temperature on the inactivation and adsorption of Tv cellulase, effect of pH on adsorption, activity, and inactivation of Tv cellulase, effect of enzyme concentration on saccharification of cellulose, and glucose inhibition of saccharification of cellulose have been reported in a separate communication (*3*). These data show that small enzyme losses (up to 10%) are difficult to detect. Tv cellulase alone does not appear to lose

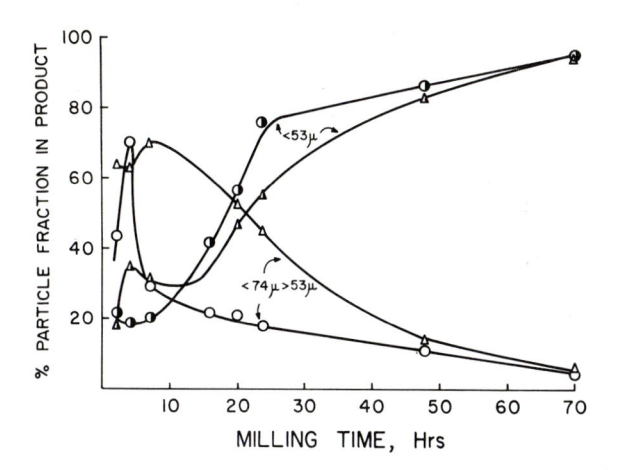

Figure 9. Effect of pre- and postheating of Solka Floc on particle size distribution owing to milling. Heating at 200°C. for 25 min. in forced draft air oven

Preheated
- –O–O–O– $<74\mu > 53\mu$
- –◐–◐–◐– $<53\mu$

Postheated
- –△–△–△– $<74\mu > 53\mu$
- –▲–▲–▲– $<53\mu$

any activity at 50°C. for one hour at pH 4.5. At 60°C. the inactivation amounts to 25% in one hour. In the presence of substrate (SF) at 50°C., the enzyme loss remains almost constant at about 13% up to one hour of contact. The cellulase alone remains most stable between pH 4.0 and 5.0 (50°C.; 4 hr.). Outside this range there appears a marked loss in the activity (about 94% at pH 3 and 37% at pH 6.5). In the presence of substrate, however, the range for maximum cellulase stability is slightly broader—e.g., 3.5 to 5.5. The best range for enzyme activity (pH— 4.8–5.0) falls within the range of maximum stability. It has further been demonstrated (3) that the highly heterogeneous nature of the suspension of the substrate enzyme system increases its initial rate of hydrolysis owing to agitation bringing the system close to homogeneous, even at a lower enzyme concentration. Secondly, at the same substrate level, increased concentrations of enzymes increases the rate of hydrolysis considerably. Inhibitory effect of product (glucose) accumulation in the system has also been shown.

(2) Loss of Cellulase (C_x) Activity in the Presence of Highly Susceptible Substrate and Absorbent. Temperature and pH affect the inactivation and adsorption losses of *T. viride* cellulase in contact with Solka Floc and milled Solka Floc (3). Since several saccharification studies were conducted with highly reactive cellulose (Sweco 70 heated), a knowledge of the difference, if any, between adsorption and other losses of *T. viride* cellulase in contact with such substrates (at 50°C., pH—4.8) was necessary. Accordingly, *T. viride* cellulase was incubated with Sweco

70 heated substrate and with Fuller's earth as adsorbent (Figure 10). The enzyme alone suffers practically no loss of activity. In contact with Fuller's earth, a progressive loss of activity took place. In the first four hours, because of adsorption alone on Fuller's earth, 44% of the activity was lost and more than 60% was lost in 22 hours (both at 50°C.). At 60° the loss increased from 13% on immediate contact to 42% in one hour (3). In contact with substrate (Sweco 70 heated) a loss of 40% activity takes place in the first hour at 50°C. No additional loss is observed in the next 21 hours.

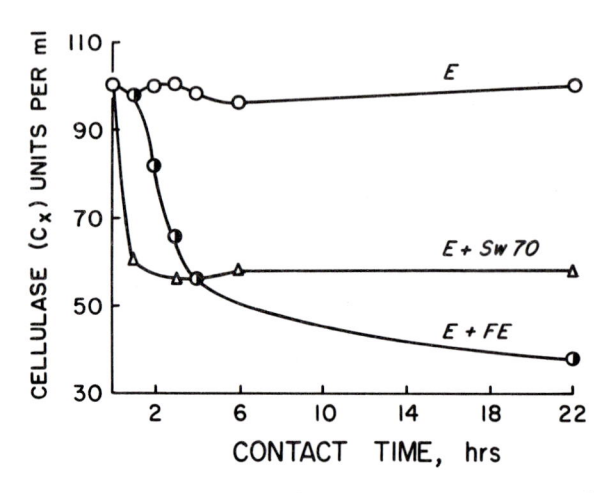

Figure 10. Loss of cellulase activity in contact with highly susceptible substrate (Sweco 70–93.8% < 53μ)

○-○-○-○- Cellulase alone (E)
△-△-△-△- Cellulase plus Sweco 70 (E + Sw 70)
◐-◐-◐-◐- Cellulase plus Fuller's earth (E + FE)

Cellulase activity—100 C_x units/ml.; substrate and adsorbent concentration—50 mg./ml. in tubes; temperature—50°C.; pH—4.8

(3) Effect of pH Levels on Rate of Saccharification. Studies on the rate of saccharification and cellulase inactivation have shown (3) that a pH range of 4.5–5.5 appears most effective for enzymatic hydrolysis of native cellulose. Since most of the experimental studies are conducted with modified cellulose (SF–HM, SF–MH) and under agitated systems, a verification of these facts was necessary. A 500 ml. glass stirred tank reactor was used for the study at 40°C. and 50°C. with SF–HM material and *T. viride* cellulase providing 1.26 FP activity. Each saccharification test was conducted for a period of 40 hr. at pH values of 3.0 to 6.5. The reducing sugar produced during this period is plotted against the pH values (Figure 11). As in all other experiments, each reaction mixture was provided with 1 mg./liter of Merthiolate to eliminate the possibility of contamination. It is clear from the two profiles representing two different temperatures of digestion that for both 40°C. and 50°C. a pH range between 4.5 and 5.5 appears most effective for high rates of con-

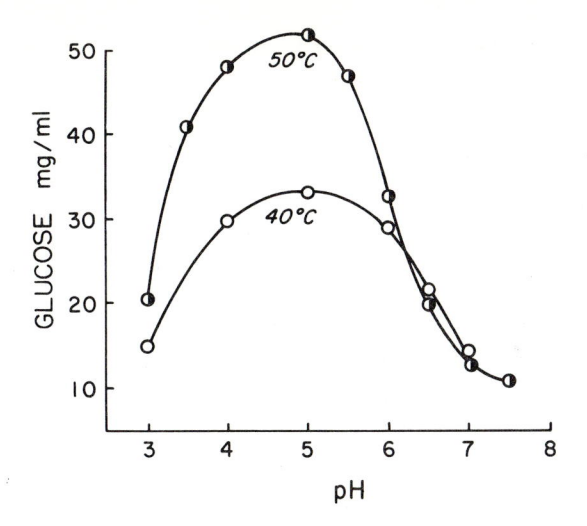

Figure 11. Effect of pH on the saccharification of modified cellulose (SF-HM, 76% < 53μ)

$-\bigcirc-\bigcirc-\bigcirc-\bigcirc-$ 40°C.
$-\talloblong-\talloblong-\talloblong-\talloblong-$ 50°C.

Cellulase activity—1.26 FP units; substrate concentration 10%; total reaction time—40 hr.; 0.5 liter glass stirred tank reactor

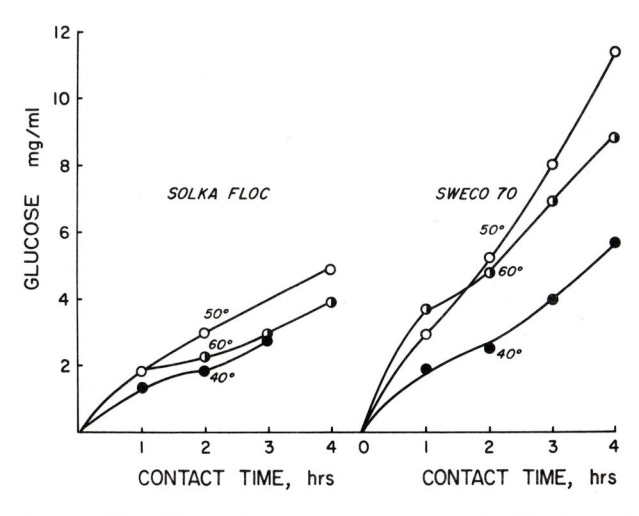

Figure 12. Effect of temperature on saccharification of native and modified cellulose

$\bigcirc-\bigcirc-\bigcirc-\bigcirc-$ 50°C.
$\talloblong-\talloblong-\talloblong-\talloblong-$ 60°C.
$\bullet-\bullet-\bullet-\bullet-$ 40°C.

Cellulase activity—36 C_x units/ml.; substrate concentration 50 mg./ml. in tubes; pH—4.8 version. The difference of more than 18 mg./ml. glucose production between 40°C. and 50°C. (at pH 5.0) indicates superiority of higher temperature.

(4) EFFECT OF TEMPERATURE ON SACCHARIFICATION OF NATIVE AND MODIFIED CELLULOSE. Tests were conducted to determine the optimum temperature for enzymatic saccharification of modified cellulose under the conditions employed in the present studies. Two series of tests, one with SF, and the other with Sweco 70 were made at 40°, 50°, and 60°C. in tubes containing 10 ml. of 10% substrate suspension in *T. viride* cellulase (activity = 36 C_x units/ml.) over a period of 4 hr. The enzyme acts at an initially higher rate of hydrolysis at 60°C. than at 50°C. or 40°C. in the first hour. The steady rate of reaction at 50°C., however, remains high after the first hour while the rate at 60°C. declines, keeping an increasing margin of reducing sugar yield between itself and those at 40° or 60°C. This trend is also applicable to Solka Floc. Under agitated conditions the rates at 50°C. give higher values than those illustrated here (Figure 12).

Substrate Studies. (1) EFFECT OF HEAT TREATMENT OF THE NATIVE CELLULOSE (NOT MILLED). In order to ascertain the effect of heat treatment of Solka Floc under an oxidizing atmosphere without any reduction in particle size of the substrate, a series of hydrolysis tests was conducted in test tubes using SF heated to various temperatures. Fifty grams of

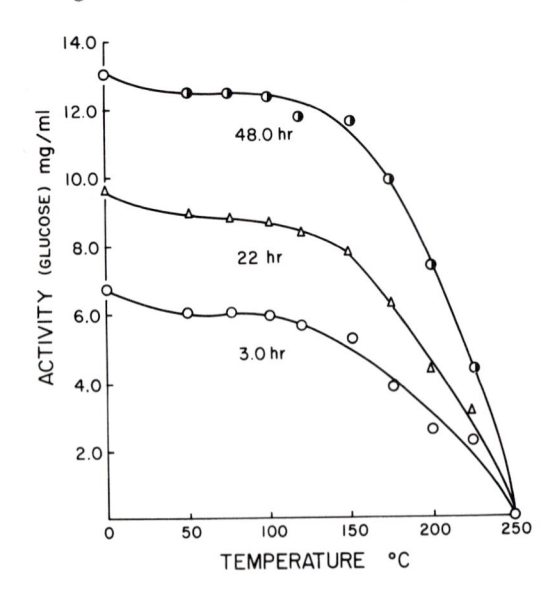

Figure 13. Susceptibility of Solka Floc (not milled) preheated at various temperatures. Hydrolysis conditions: Substrate—50 mg./ml.; Cellulase—0.75 FP activity; pH—4.8; Temperature—50°C.

○ - ○ - ○ - ○ -   3 hr. hydrolysis
△ - △ - △ - △ - 22 hr. hydrolysis
◐ - ◐ - ◐ - ◐ - 48 hr. hydrolysis

samples of SW40A Solka Floc were evenly spread in 1/4-inch layers on enamelled trays and exposed to various temperatures in a forced draft air oven for 25 min., and then quickly cooled and kept in bottles. These heat treated samples were tested at 50°C. for their relative susceptibilities to *Trichoderma* cellulase (Figure 13). The results indicate that heating alone exercises an inhibitory effect on the hydrolysis by cellulase. There is a progressively increasing loss of susceptibility as the temperature of heating is increased and the enzyme action on cellulose heated to temperatures above 225°C. ceases. This unusual inhibitory effect may result from drying out the fibers in such a way that they do not again readily take up water. It is an effect quite the reverse of that obtained by similarly heating finely ground Solka Floc (Table III) where a considerable increase in saccharification has been observed.

Table III. Effect of Heat Treatment of Finely Ground Substrate (Sweco 70)

| Substrate | Reducing Sugar Produced, mg./ml. | |
100 mg./ml.	In 2.75 hr.	In 7.0 hr.
Sweco 70	22.5	43.0
Sweco 70 heated to 200°C. for		
25 min.	38.0	50.0
40 min.	27.0	51.0
60 min.	29.0	50.0
120 min.	21.5	42.0

Cellulase activity = 111 C_x units/ml.; temperature—50°C.; pH—4.8.

In view of the differences in the digestibility between the heated cellulose, and the milled and heated celluloses, as observed in the previous experiments (Table III) towards *T. viride* cellulase, it was necessary to study the extent of the effect of heat treatment of cellulose under on oxidizing atmosphere with particle size reduced to a low level. This was accomplished by milling Solka Floc for 2 to 48 hr. in laboratory porcelain pot mill. The milled products were screened down to < 53μ size. This procedure was followed with both SF and SF–H materials to give SF–M and SF–MH. The results are presented in Figure 14. A comparison of the sugar values between the two substrates, SF–HM and SF–MH, suggests that the preheating provides a better substrate for cellulase than postheating. This may be owing to the presence of a larger proportion of smaller particles in the SF–HM material, particularly between 10 and 48 hr. of milling as evidenced from Figure 9. Variations in reducing sugar values in a few samples might have been owing either to difference in particle size of randomly weighed samples, or to non-uniform substrate enzyme contact, as the contents were not agitated, or to inadequate heat exposure of some of the particles, or a combination of all these effects. However, these samples represent less than 8% of all the samples tested.

(2) REACTIVATION OF RESIDUE (OF ENZYME SACCHARIFICATION) BY HEAT TREATMENT AND MILLING. Oxidative heat treatment of finely pulverized cellulose has been found to impart increased susceptibility to *T.*

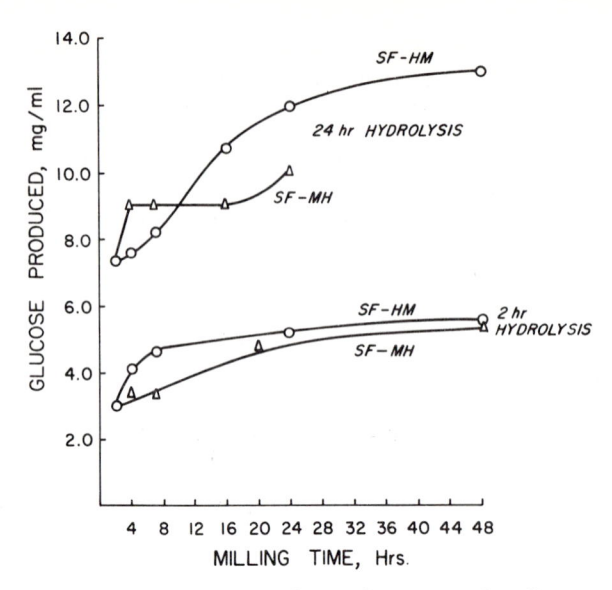

Figure 14. *Effect of order of heating and milling of Solka Floc on its enzymatic saccharification*

○ - ○ - ○ - ○ - SF-HM ($< 53\mu$)
△ - △ - △ - △ - SF-MH ($< 53\mu$)

Conditions of saccharification:
2 hr. at 50°C. at pH—4.8
24 hr. at 50°C. at pH—4.8
Cellulase—0.92 FP activity

Table IV. Effect of Various Treatments of Substrates on the Production of Reducing Sugars

Substrate (50 mg./ml.)	Nature of Treatment	Reducing Sugar Produced, mg./ml.		
		2.5 hr.	4.5 hr.	23.0 hr.
Solka Floc	None	6.6	8.5	17.0
Solka Floc	Milled for 48 hr. (82% $< 149\mu$)	8.7	12.7	22.2
Solka Floc	Milled and heated[a] in air (82% $< 149\mu$)	16.0	21.0	33
Solka Floc Sweco 70 (94% $< 53\mu$)	Milled and heated[a] in N_2 (O_2 free)	7.8	11.0	25.5
Sweco 70 (94% $< 53\mu$)	None	11.2	13.0	31.0
Sweco 70 (94% $< 53\mu$)	Heated[a] in air	17.1	20.0	34
	Heated[a] in N_2 (O_2 free)	10.2	13.2	28.5

[a] At 200°C. for 25 min.; cellulase—60 C_x units/ml.; temperature 50°C., pH—4.8.

viride cellulase (Table IV). It remains to be seen if similar treatments of the digested cellulose would produce any noticeable increase in its susceptibility. In other words, can the residues of the digested cellulose samples, which are not longer accessible to cellulase be modified by heat and milling in such a way that they become susceptible? Residues and several substrates were simultaneously tested with regard to their rates of saccharification at 50°C. at 10% substrate level using *T. viride* filtrates with 1:5 FP activity at pH 4.8. The substrates and enzyme digested residues tested were S_1 = SF; S_2 = SF–M ($< 53\mu$); S_3 = SF–HM (200°C. for 25 min., $< 53\mu$); S_4 = SF–MH ($< 53\mu$, 200°C. for 25 min.); S_5 = SF–MHDWDr (digested up to $\sim$60%, dried at 60°C.); S_6 = SF–MHDWDrHM (heating and milling as others).

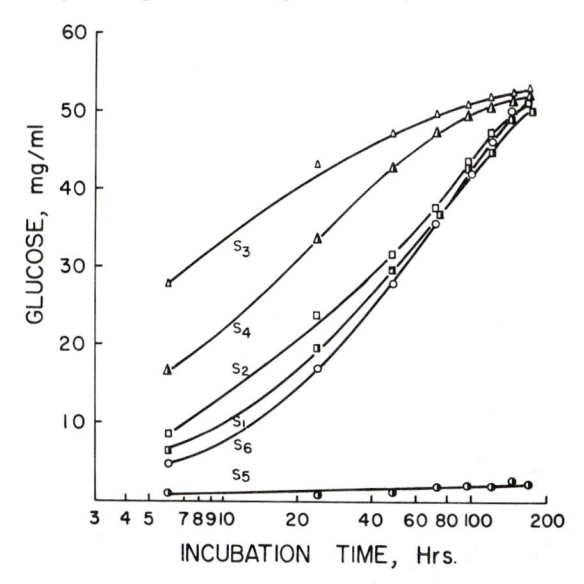

Figure 15. Effect of heat treatment and milling of cellulose on saccharification

S_1 ■ - ■ - ■ - ■ -   SF (11.6% $< 53\mu$)
S_2 □ - □ - □ - □ -   SF–M
S_3 △ - △ - △ - △ -   SF–HM (200°C. for 25 min., 76% $< 53\mu$)
S_4 ▲ - ▲ - ▲ - ▲ -   SF–MH (76% $< 53\mu$, 200°C. for 25 min.)
S_5 ◐ - ◐ - ◐ - ◐ -   SF–MHDWDr (Dried at 60°C.)
S_6 ○ - ○ - ○ - ○ -   SF–MHDWDrHM (200°C. for 25 min. $< 53\mu$)

Cellulase—1.80 FP activity; pH—4.8; substrate concentration 10%; temperature—50°C.

The results of tests carried out for 168 hr. (Figure 15) show conclusively that the residue of digested cellulose (S_5) which has been rendered extremely resistant to cellulase action can be reactivated into a highly reactive form (S_6) almost comparable to that of the initial Solka Floc (S_1). This has been possible by subjecting the fully digested residue to the same cycle of heating and milling as done to SF to obtain SF–HM (S_3). The results further show the beneficial effects of preheating cellulose prior to milling in order to have maximum susceptibility of the substrate (S_3) to *T. viride* cellulase, compared with SF–MH (S_4). However, the ultimate percentage conversion of cellulose in its various modified forms after 168 hr. of saccharification appears to become more or less equal in all cases.

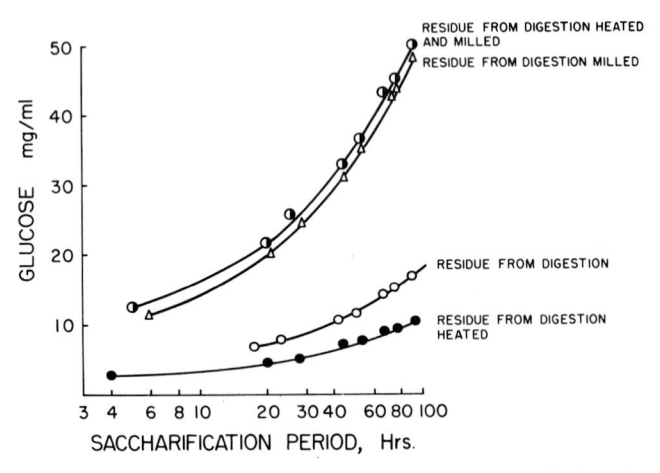

Figure 16. Reactivation of enzyme digested cellulose by heating and milling

○ – ○ – ○ – ○ – SF-MHDWDr ($< 53\mu$)—Residue from digestion
● – ● – ● – ● – SF-MHDWDrH ($< 53\mu$)—Residue heated for 25 min. at 200°C.
△ – △ – △ – △ – SF-MHDWDrM at 200°C. ($< 53\mu$)
◐ – ◐ – ◐ – ◐ – SF-MHDWDrHM at 200°C. milled ($< 53\mu$) heated as above

Substrate—10%; Cellulase—1.26 FP activity; pH—4.8–5.2; Temperature—50°C.; Reaction in 0.5 liter glass stirred tank reactor

Subsequently, four sample residues (SF–MHDWDr, SF–MHDWDrH, SF–MHDWDrM, and SF–MHDWDrHM) all comprising 100% $< 53\mu$ particles were separately hydrolyzed in 0.5 liter agitated glass reactors at 50°C. using 10% suspensions of the substrates in Tv cellulase (1.26 FP activity) at pH 4.8–5.2. The saccharification continued for nearly 90 hr. (Figure 16) indicating that a reactivation of the so-called resistant (digested) cellulosic substrate is possible by a combined treatment or milling and heating. The extent of increase in susceptibility brought about in the digested cellulose is comparable with that possible by similar treatments of SF. Much, however, depends on the fineness of the substrate particles. Susceptibility increases with increasing fineness of grind. It is thus possible to re-use the enzyme digested cellulose for saccharification in a closed cycle operation. The lack of difference between the residue heated and milled, and residue heated may be caused by the presence in the former material of a higher fraction of oxidized residue, which is seemingly inhibitory to enzyme action. As explained before, heating of SF helps particle size reduction in the subsequent milling; but apparently such an assistance is not needed by already finely ground particles of cellulose.

(3) SEMI-CONTINUOUS SACCHARIFICATION OF SF–HM AND SF–MH. Following the initial studies on the enzymatic hydrolysis under changed environments of pH, temperature, cellulase activity, and other factors reported earlier, a few semi-continuous saccharification experiments were conducted. Comparative information on the differences in the rates between SF–HM and SF–MH was also necessary. For these two systems, the reactions were carried out in 0.5 liter agitated glass reactors of the same dimensions, at 50°C. using 10% substrate levels in each case and the same Tv cellulase (FP activity = 1.8). After an initial 40 hr. of batch hydrolysis, 200 ml. (or 40%) of the reaction mixture was removed and replaced by 200 ml. of fresh enzyme plus an amount of substrate equal to the fraction of cellulose converted into glucose at that time, plus the amount present in the 200 ml. of effluent removed. Both the original reaction volumes (0.5 liter) and the subsequent feeds were provided with 1 mg. of Merthiolate per liter. The cycles of discharge and feeds were repeated and the reactions continued over several days (Figures 17 and 18).

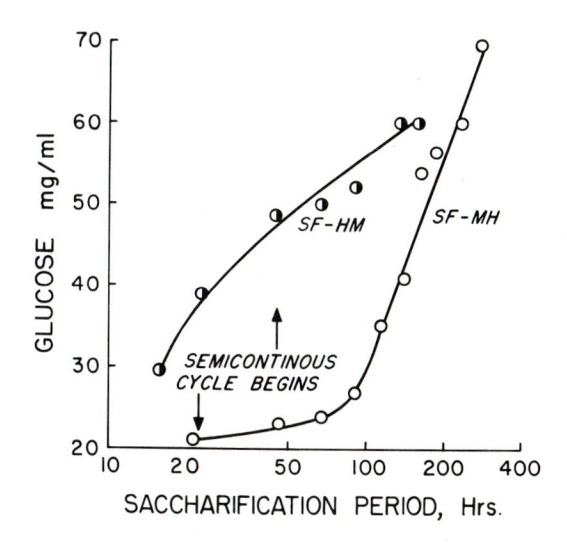

Figure 17. Semi-continuous saccharification of modified cellulose

$-\bigcirc-\bigcirc-\bigcirc-$ 10% suspension of a 54.9% < 53μ size SF-MH substrate
$-\bullet-\bullet-\bullet-$ 10% suspension of a 76% < 53μ size SF-HM substrate

Reactor conditions: 0.5 liter glass stirred tank; Cellulase—1.8 FP activity; pH—4.8; Temperature—50°C.; Substrate particle size SF-HM 76% < 53μ; SF-MH 54.9% < 53μ. Semi-continuous harvest and feeding started for SF-HM at 45th hr. and for SF-MH at 22.5th hr.

The difference of nearly 21% of fine ($< 53\mu$) particles (Table II, Figure 9) between the two substrates SF–HM and SF–MH makes a contribution to increased susceptibility of the SF–HM substrate over the other. Heating most likely renders the cellulose fragile and subsequent milling results in a disappearance of a large portion of its fibrillar structure, which is replaced by fine and more amorphous materials. This steady rate of hydrolysis after the first 40 hr., in the case of SF–HM,

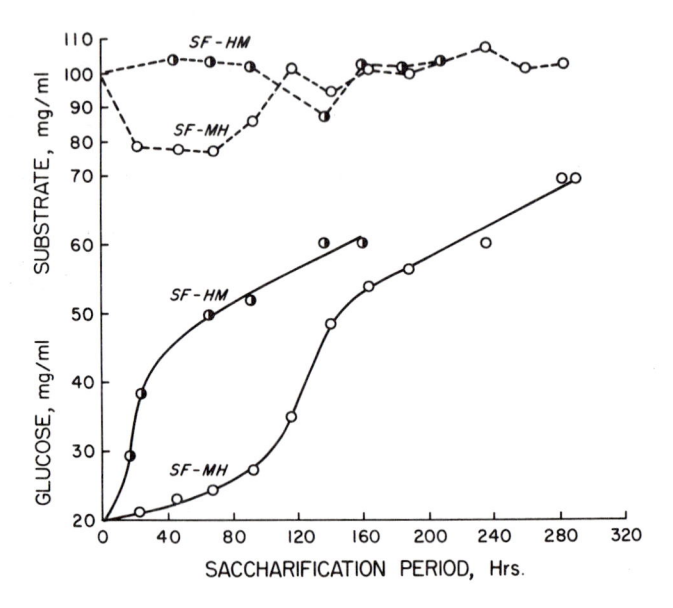

Figure 18. *Relationship between substrate concentration and saccharification rate of modified cellulose (SF-HM, SF-MH) in semi-continuous agitated reactors*

--◐--◐--◐--◐	SF-HM	Glucose level
--○--○--○--○	SF-MH	
-◐-◐-◐-◐	SF-HM	Substrate level
-○-○-○-○	SF-MH	

Reactor conditions: 0.5 liter glass stirred tank reactor; Substrate concentration 77–108 mg./ml. (av.—100 mg./ml.); Cellulase— 1.8 FP activity; pH—4.8; Temperature—50°C.; Substrate particle size—SF-HM—76% $< 53\mu$; SF-HM—54.9% $< 53\mu$; Semi-continuous feeding and harvest started at 45th hr. for SF-HM, and at 22.5th hr. for SF-MH

progresses steadily. The other substrate (SF–MH) containing a larger portion of large particles ($> 149\mu$) and smaller portion of fines keeps down the initial rate up to about 90 hr. and then progresses at a fast rate, almost equal to the initial rate exhibited by SF–HM. The two rates become equal after 160 hr. of hydrolysis. It was evident from these experiments that the hydrolysate in batch and semi-continuous saccharification of cellulose, a level of about 5% sugars is reached around 40 hr. in the SF–HM and Sweco 70, whose average particle size consisted of between 85–94% $< 53\mu$. Generally higher percent conversions were obtained in Sweco 70 which contained more than 93% of 53µ or smaller particles as against little over 40% for SF–MH and around 85% for SF–HM, both milled for 48 hr. under the same conditions.

(4) CONTINUOUS SACCHARIFICATION OF MODIFIED SUBSTRATE IN SINGLE STAGE. Because of limited supply of Sweco 70, most of the continuous studies were based on SF–MH and SF–HM. One of the major difficulties encountered in feeding the solid substrate into the reactor continuously was the small reactor volume and necessarily low feed rates. Constant feeding has therefore been made in the form of an enzyme substrate suspension kept under continuous agitation at 1°–2°C. in an ice-cooled jacketed bath (Figure 19). Details of the method are described in another communication (3). Rate data on reducing sugar yields in a 4 liter single stage stirred reactor using 10% substrate (SF–HM) and *T. viride* cellulase are illustrated in Figure 20. The unit was run

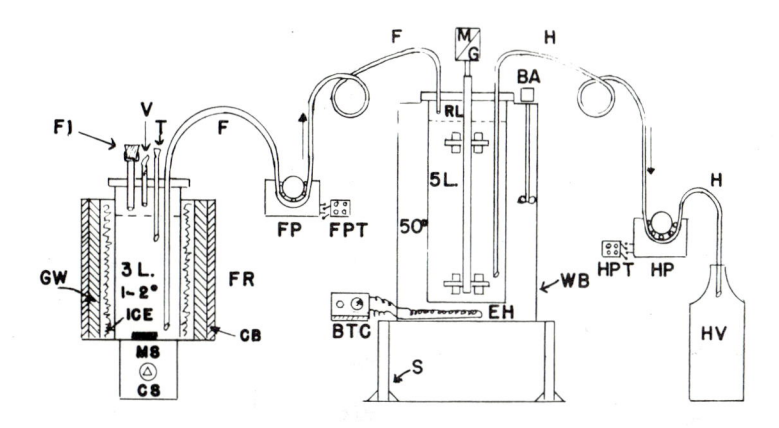

Figure 19. Continuous forward feed steady state single vessel saccharification system

Legend:
GW = Glass wool lagging
CB = Cardboard lagging
FR = Feed reservoir
FI = Feed (Enzyme + Solid Substrate) Inlet
V = Vent
T = Thermometer
MS = Magnetic stirrer
CS = Magnetic stirrer control switch
F = Feed line
H = Harvest line
M/G = Speed geared motor
BA = Bath agitator
RL = Reactor level
EH = Electrical heating element
FP = Feed pump
FPT = Feed pump timer
HP = Harvest pump
HPT = Harvest pump timer
WB = Water bath
S = Support
HV = Harvest vessel
BTC = Bath thermostat control

as a batch up to the first 40 hr. Continuous feeding of enzyme substrate slurry from the ice-cooled reservoir, and discharge of the effluent from the reactor, both at the rate of 100 ml. per hour, were started from this point onward and were maintained at this steady dilution rate of 0.025^{-1} hr. over a period of 170 hr. The effluent maintained a reducing sugar level between 5.1–5.6 mg./ml. during the continuous phase. Because certain mechanical troubles developed in the agitation system of the reactor, the experiment had to be discontinued beyond this period.

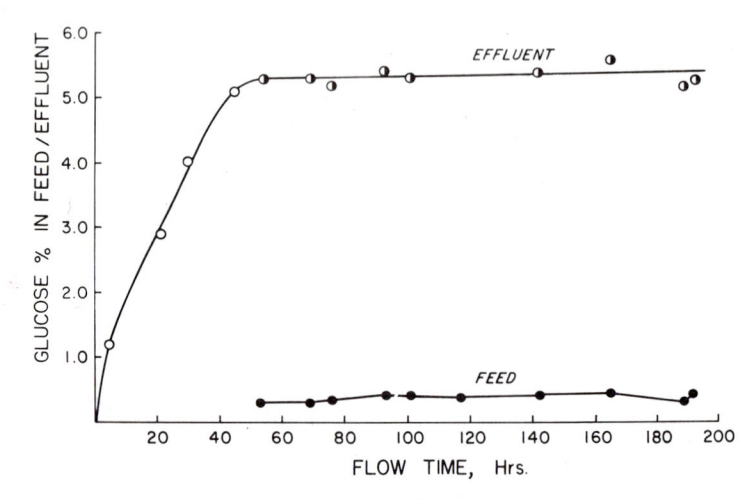

Figure 20. Continuous steady state saccharification of modified cellulose (SF-HM)

– ○ – ○ – ○ – ○ – Batch rate profile
– ◐ – ◐ – ◐ – ◐ – Continuous phase
– ● – ● – ● – ● – Reducing sugar in feed

Enzyme—1.0 FP activity; Substrate—SF-HM (76% < 53μ)—v0%; pH—4.05–5.2; Saccharification temperature—50°C.; Dilution rate in continuous phase—0.025^{-1} hr.; Reactor conditions: 4.0 liter glass stirred tank reactor; Feed temperature—1°–2°C.

(5) HYDROLYSIS DURING MILLING OF CELLULOSE IN THE PRESENCE OF ENZYMES. Since the size reduction operation takes the major portion of the total time needed in converting cellulose into reducing sugar, it seemed desirable to see if hydrolysis could be done simultaneously with the size reduction operation. No published information is available on such an approach. Cellulose was subjected to milling at room temperature (22°C.) in presence of cellulase in a porcelain pot mill containing porcelain balls, both previously steam sterilized. The substrate (SF–H) was kept at 10% consistency in 1.0 liter of *T. viride* cellulase. After 48 hr. of milling, 0.5 liter of the milled enzyme substrate slurry was transferred into a 0.5-liter stirred tank reactor, and further hydrolysis continued in both of the systems (mill and agitated reactor) at 22°C. and 50°C., respectively. The progress of saccharification is shown in Figure 21. The enzyme still remains active after 48 hr. of abrasive milling in contact with substrate. While this is in no way a conclusive study of the new

system, it illustrates the possibility of further experimentation on milling of substrate in contact with enzymes at 50°C. Although there is likely to be more loss or inactivation of enzyme in such a system, the gain in time and milling cost is worth looking into.

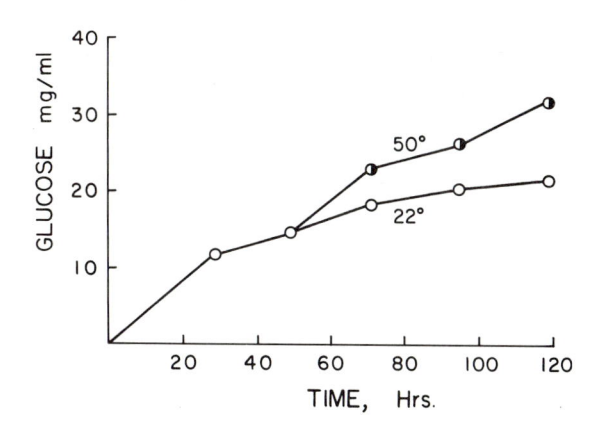

Figure 21. Effect of milling cellulose in contact with cellulase on saccharification

○ - ○ - ○ - Solka Floc milled in the presence of Tv cellulase at 22°C.—10% suspension
◑ - ◑ - ◑ - Solka Floc milled in contact with Tv cellulase subjected to further hydrolysis at 50°C. in agitated reactor—10% suspension

Enzyme—0.9 FP activity; pH—5.0; Mill size—1.0 liter in porcelain pot mill (2.5 gal.) previously steam sterilized

Discussion

Experimental results show that heating of cellulose exercises an "inhibitory" effect on the action of cellulase on the substrate and there is a progressive increase of inhibition of the rate of hydrolysis with increasing temperature (Figure 13). But a considerable increase in susceptibility of the substrate to the cellulase is possible by a two-step process of size reduction and oxidative heat treatment (Table III, Figures 14 and 15). Reduction in particle size and consequent exposure of hidden reactive points of the cellulose through their opening up plays a big role in the increase in susceptibility of the substrate. The combined effects of heat treatment and particle size reduction of cellulose make a distinct difference as a reactive substrate, either from size reduction alone, or from size reduction combined with heat exposure (Figure 15). Improvement in the quality of the substrate through preheating in air at 200°C. for 25 min. followed by milling is, in part, owing to the presence of high percentage of fines ($< 53\mu$ size) in the product. A highly resistant cellulose left as residue of enzyme digestion can be regenerated into a susceptible substrate by repeating the combined operation of heating and milling (Figure 16). It has been shown, however, that heating alone of the digested cellulose does not improve its quality. These results suggest that as far as cellulase action on cellulose is concerned, availability of active areas of the substrate is of primary importance. Heating assists the process of particle size reduction, but does not, in itself, make any contribution to increase susceptibility. The period of complete saccharification is only prolonged in the presence of increasing concentration of substrate molecules whose hidden reactive areas do not readily become available for further hydrolysis. Milling, or heating plus milling, make these areas available to the enzyme, and this so-called resistant substrate again becomes susceptible (3). As mentioned earlier, the infrared spectra and x-ray diffraction analyses of the treated cellulose (SF–M. SF–HM, SF–MH, SF–HMD) samples show that they are probably slightly oxidized and their crystalline structure destroyed. The control sample (SF) resembles that of cotton in terms of x-ray crystallinity index (83 against 85–90). Milled samples (SF–M, SF–MH, SF–HM) show severe loss of crystallinity and reduced degrees of polymerization (DP 200 or less). In the enzyme hydrolyzed samples of heated and milled cellulose, there appears to be more fibrillar structure visible implying that the amorphous nature so characteristic of the finely ground cellulose may have been hydrolyzed away. These conclusions strongly support our experimental data (Figures 15 and 16) which illustrate that a considerable portion of the fibrillar structure of the hydrolyzed cellulose residue can be reactivated by a heating and milling process. *T. viride* cellulase does not seem to suffer any detectable loss of activity (C_x) at pH 4.8 and 50°C. in 22 hr. (Figure 10). The loss of activity in contact with highly a reactive substrate is about 40% in the first half hour of contact with substrate and it remains almost at that level over the next several hours.

The results of stirred tank saccharification of modified cellulose (SF–HM) at various pH (Figure 11) confirms the optimum pH to be 4.5–5.5 as shown in the case of untreated Solka Floc substrate studied in unagitated tube experiments (3). It has been demonstrated (Figure 21) that saccharification of cellulose is possible while the latter is milled in contact with *T. viride* cellulase. Although no published information is available describing a solid substrate reacting with an enzyme solution under high impact and shear conditions, the preliminary data reported here illustrate the possibility that such an approach may be advantageous.

The saccharification of cellulose in crystalline (Solka Floc) or amorphous (finely pulverized Sweco 70) forms by *T. viride* cellulase has a maximal rate at 50°C. as compared with either 40°C. or 60°C. Up to the first hour the rate at 60°C. appears faster than at 50°C. and 40°C., but thereafter the 50°C. system progresses at a much higher rate than at 40° or 60°C. The inactivation of C_x at 60°C. appears to be the main reason for this.

The results of the semi-continuous saccharification (Figures 17 and 18) of the two substrates (SF–HM and SF–MH) can be summarized:

The SF–HM cellulase system in a 10% suspension reaches a maximum rate of saccharification at about 40 hr. The SF–MH cellulase system at the same consistency maintains a sluggish hydrolytic rate up to about 90 hr., then starts to rise and becomes almost equal to the rate exhibited by the SF–HM system beyond 100 hr. A comparison of the SF–HM and SF–MH (Figure 1) systems shows a distinct difference between the capacities of the two substrates to hydrolyze under the same conditions of enzyme concentration, temperature, and pH. Overfeeding or feeding prior to the time maximum rate is reached causes a falling of rate or temporary accumulation of substrate. Beyond 90 hr. of hydrolysis with agitation, rates of saccharification of both the substrates become equal. The difference in the susceptibility of the two substrates is attributed to the major difference in the particle size distribution in the two substrates (Figure 9, Table II)—*e.g.*, SF–HM material containing 76% of $< 53\mu$ particles and SF–MH material containing about 55% of $< 53\mu$ particles.

Acknowledgment

The senior author gratefully acknowledges the Visiting Scientists' award from the National Research Council, U. S. Academy of Sciences, U. S. Academy of Engineering, Washington, the facilities provided by the U. S. Army Natick Laboratories, Natick, Massachusetts and the leave granted by the Jadavpur University, Calcutta, India in connection with the present studies done at the Microbiology Division, Food Laboratory, U. S. Army Natick Laboratories. The authors would like to acknowledge with gratitude the helpful suggestions received from E. T. Reese in the preparation of the manuscript. Friendly assistance, continued support and encouragement received from H. M. El-Bisi and Mary Mandels were valuable.

Literature Cited

(1) Adams, E., *Textile Res. J.* **20**, 71 (1950).
(2) Amemura, Akinovi, Terui, Gyozo, *J. Ferm. Technol. (Japan)* **43**(5), 275 (1965).
(3) Ghose, T. K., *Biotech. Bioeng.* **11**, 239 (1969).
(4) Katz, M., Reese, E. T., *Appl. Microbiol.* **16**(2), 419 (1968).

(5) King, K. W., *Biochem. Biophys. Res. Comm.* **24**(3), 295 (1966).
(6) King, N. J., *Biochem. J.* **100**, 784 (1966).
(7) Krupnova, A. V., Sharkov, V. I., *Gidrolizn. Lesokhim. Prom.*, U.S.S.R. **16**(3), 8 (1963).
(8) *Ibid.*, **17**(3), 3 (1964).
(9) Li, L. H., Flora, R. M., King, K. W., *Arch. Biochem. Biophys.* **111**, 439 (1965).
(10) Mandels, M., Reese, E. T., "Advances in Enzymic Hydrolysis of Cellulose and Related Materials," E. T. Reese, Ed., p. 115, Pergamon Press, London, 1963.
(11) Mandels, M., Weber, Jim, ADVAN. CHEM. SER. **95**, 391 (1969).
(12) Miles Chem. Co., *Tech. Bull.* **No. 9-245**, Elkhart, Indiana (1963).
(13) Nisizawa, K., Hashimoto, Y., Shibata, Y., "Advances in Enzymic Hydrolysis of Cellulose and Related Materials," E. T. Reese, Ed., p. 171, Pergamon Press, London, 1963.
(14) Reese, E. T., Gilligan, W., Norkrans, B., *Phy. Plant* **5**, 379 (1952).
(15) Reese, E. T., *Appl. Microbiol.* **4**(1), 39 (1956).
(16) Reese, E. T., "Marine Boring and Fouling Organisms," Dixy Lee Ray, Ed., p. 265, University of Washington Press, Seattle, 1959. Interscience Press, New York, in press.
(17) Reese, E. T., Mandels, Mary, "Cellulose," revised ed., Ott and Spurlin, Interscience, New York, in press.
(18) Sinclair, P. M., *Chem. Eng.* **72**, 90 (1965).
(19) Sumner, J. B., Somers, G. F., "Laboratory Experiments in Biological Chemistry," p. 34, Academic Press, Inc., New York, 1944.
(20) Toyama, N., "Advances in Enzymic Hydrolysis of Cellulose and Related Materials," E. T. Reese, Ed., p. 235, Pergamon Press, London, 1963.
(21) Walseth, C. S., Ph.D. Thesis, Lawrence College, Appleton, Wisc., p. 109 (June 1948).
(22) Walseth, C. S., *Tappi* **35**(5), 233 (1952).
(23) Whitaker, D. R., Colvin, J. R., Cook, W. H., *Arch. Biochem. Biophys.* **49**(2), 259 (1954).

RECEIVED October 14, 1968.

Discussion

E. T. Reese: "The highest concentrations of glucose that we can get in a digest must compare favorably with the 30% or more that one gets in a glucamylase digest. One of our problems then is to get a lot of cellulose into the initial suspension. If we swell the cellulose, we can not get very much in, maybe a 5 to 10% suspension, and that is a mush. But the milling plus heating process that Dr. Ghose is using allows one to get suspensions of up to 50% cellulose. We need these high concentrations in order to reach 30% glucose concentrations in the digest."

J. Saeman: "If you take your digested material and remove all substrate and product, do you find the same cellulase activity that you had initially? In other words, is there a loss in cellulase activity as the digestion proceeds?"

T. K. Ghose: "It has been shown in one of the slides that in contact with Sweco 70 (cellulose) substrate we did find that 40% of the enzyme activity disappeared. We would designate this as adsorbed loss rather than inactivation. At 50°C. the loss in enzyme activity in the *absence* of substrate over four hours is practically nil."

J. Saeman: "Did you try treating cellulose with sodium hydroxide?"

T. K. Ghose: "We did try. In one case it was useful, in another case it was detrimental."

K. Selby: "Did you use any antiseptic in the system while you were doing the continuous digestion?"

T. K. Ghose: "Yes. We used 1 mg. of Merthiolate per liter of enzyme substrate suspension."

L. Baribo: "Dr. Reese, you made a comment that you could swell. How did you swell the cellulose?"

E. T. Reese: "Yes, we have used swollen cellulose in some of our earlier work. We were comparing various methods of swelling with the rate of subsequent hydrolysis. We used alkali swelling followed by washing out the alkali, but never drying the sample. Under these conditions there was a two-fold increase in susceptibility."

L. Baribo: "Well, then you might do what they are doing in the starch industry, that is to decrease the molecular weight of the cellulose so that you can get a higher concentration in your substrate media. They could not run a 30% gelatinized starch either, unless they degraded it to something like 10–12 DE. But if you do swell, then you can not get a high concentration in suspension."

T. K. Ghose: "With swelling you decrease the bulk density of the substrate in the reactor. In other words, if you have a solid fluffy material like Solka Floc you do not expect to charge the reactor in high concentrations. On the other hand when the particle size is reduced, not only does the total surface increase, the reactor capacity is also increased due to increased bulk density of the substrate."

E. T. Reese: "We have no way of hydrolyzing the cellulose readily to a soluble oligosaccharide fraction. If we use anything like the strong sulfuric acid treatments, then we may as well be doing acid hydrolysis again."

J. M. Leatherwood: "Did you try percolating the enzyme through a column of cellulosic substrate?"

T. K. Ghose: "Yes, in fact, our first experiment was done in a column containing Solka Floc and fresh enzyme and hydrolysate respectively, pumping from the top to the bottom. After a period of operation the material swelled so much that it was impossible to operate at a steady rate, and ultimately the whole thing plugged up."

J. M. Leatherwood: "Have you tried to measure adsorption loss in other susceptible substrates used by you?"

T. K. Ghose: "Do you mean in place of Sweco 70, whether we attempted to use Sweco 70-heated in the activity loss studies? No, we did not. We expect essentially the same trend of activity loss as in the case of Sweco 70 unheated."

J. M. Leatherwood: "What about the adsorption of enzymes on the digested cellulose?"

T. K. Ghose: "No, we did not do that."

M. Mandels: "In previous long term digestions of cotton, the cellulase was recovered approximately 100% after the cellulose had been completely hydrolyzed."

L. Underkofler: "May I say something further about the starch situation. It is correct that the amylases do not attack undamaged starch granules. However, if you ball-mill starch then the raw starch can be attacked by the enzymes. But the best way is to gelatinize the starch. Then, of course, it becomes susceptible. The thinning of the starch suspension is just a matter of reducing the viscosity for easier handling. In the laboratory one can make a 30% starch suspension, bring the enzyme in, and convert it to glucose. But in the plant you carry out a prehydrolysis step to reduce the load on the stirrers, and to get the enzyme mixed in. Glucoamylase does act very rapidly on the large starch molecules. If you can get the enzyme mixed in, then you get practically the same rate of hydrolysis whether the starch is thin or viscous."

25

Utilization of Bagasse

V. R. SRINIVASAN and Y. W. HAN
Department of Microbiology, Louisiana State University,
Baton Rouge, La. 70803

> *Bagasse has been variously used as fuel and agricultural mulch as well as a raw material for pulp in the manufacture of paper, acoustical boards, etc. Studies on cellulose fermentation by microorganisms indicate the feasibility of utilizing bagasse as a substrate for the production of single cell proteins. A microorganism isolated by elective culture technique from the sugarcane fields and characterized as* Cellulomonas *was shown to break down pre-treated bagasse at a rate suitable for development into large scale fermentations. The amino acid analysis of the cell protein showed a high lysine content and the quality of the protein compared favorably with that proposed by FAO. Mixed fermentation of* Cellulomonas *with* Alcaligenes *capable of growth on cellobiose increased the rate of increase in cell mass.*

Bagasse is the fibrous residue obtained after extraction of sugar from sugarcane. In the manufacture of sugar, the sugarcane stalks are chopped to small pieces by rotary knives, and then these pieces are extracted by crushing them through one or more roller mills. During this process more than 95% of the sucrose content of the cane is removed. The residual material from this operation is termed bagasse. Fresh bagasse from the sugar mills contains about 50% moisture and consists of 70–75% holocelluloses and 15% lignin. It varies in color but generally from pale yellow to dirty green. The particle size is nonuniform and depends upon the variety of sugarcane and the efficiency of the milling operation.

The annual production of bagasse in the world exceeds 100 million tons, more than half of which is produced in the Western Hemisphere (Table I). Since the operation of sugar mills is seasonal in different parts of the world, efficient storage of bagasse is essential if it is to be used in any manufacturing process. Because of the perishable nature of the material special techniques have been developed to preserve it during storage. Bagasse is usually baled and stored outdoors in piles. The piles are arranged in such a way to allow sufficient circulation of air for drying. They are covered simply with a metal sheet to keep out rainwater. The natural microbial activity during the early period of storage inside the pile is sufficient to reduce the sugar content and to sterilize the pile effectively and to render the residual fibers more suitable for further processing. Nevertheless this type of storage presents certain health hazards. Precautionary measures have to be taken to prevent an occupational hazard known as "bagassosis"—a disease of the lungs similar to "silicosis"—caused by the dust from the storage piles.

Table I. Availability of Sugarcane Bagasse
Million Tons Per Year

Western Hemisphere	60.0
Asia	26.4
Africa	11.1
Oceania	6.7
Total	104.2

Composition of the Bagasse. Raw bagasse consists of two major components of cellular constituents, namely pith and fiber. It can be classified under ligno-celluloses and is composed of interpenetrating systems of high polymers, and hence it is difficult to isolate or separate the pith from the fiber without significant modifications. Many analyses of the chemical composition of bagasse have been reported, but the methods of separation and analysis are rather empirical. This is reflected in the variation and nonuniformity of the composition. A typical analysis of bagasse is given in Table II (*12*). The entire bagasse is made up of nearly 79% holocelluloses and 19% lignin. The hemicelluloses which are a mixture of polysaccharides that are insoluble in water but readily soluble in alkali form 30% of the holocelluloses. The hemicelluloses are

Table II. Chemical Constituents in Bagasse

Constituent	Entire Bagasse % (dry wt.)	Bagasse Fiber % (dry wt.)	Bagasse Pith % (dry wt.)
Cellulose	46.00	56.60	55.40
Hemicelluloses (Araban, galactan, xylan, etc.)	24.50	26.11	29.30
Fats and Waxes	3.45	2.25	3.55
Ash	2.40	1.30	3.02
Lignin	19.95	19.15	22.30
Silica	2.00	0.46	2.42

more readily hydrolyzed by acids than α-cellulose. Bagasse pith which is derived from the thin walled cells of the ground tissue or parenchyma of the stalk contain more lignins and hemicelluloses than the bagasse fiber. The chemical composition of bagasse fibers determines the quality of the pulp in paper-making.

Present Uses of Bagasse. FUEL. Bagasse has been widely used not only as a fuel but also in paper manufacture as well as agriculture. It is used as a fuel mainly in the operation of cane sugar factories where it is produced. If burned efficiently, one metric ton of bagasse containing 50% moisture will produce heat equivalent to that from 0.333 tons of fuel oil. Its fuel value is approximately 10,000 kcal./kg. (2). It is not economical to utilize bagasse as fuel since its bulkiness requires the construction of special furnaces to operate efficiently. Due consideration has also to be given in the design of modern furnaces to burn bagasse to avoid problems of air pollution in the vicinity of the operating plants.

Paper and Pulp Manufacture. One of the largest uses of bagasse is the production of pulp and paper. One of the first steps in making paper from bagasse is separation of pith from the fiber—otherwise known as depithing. Successful utilization of bagasse in paper industry demands the development of economical methods of depithing. There are several chemical pulping processes which are in vogue in which lignin content of the material is either degraded or dissolved by hot aqueous alkali or sulfonated by bisulfite ion and made water soluble. Bagasse contains relatively large amounts of hemicelluloses which results in consumption of appreciable quantities of chemicals and affects the paper-making qualities of the fiber adversely. Recently a process of mechanical pulping without the use of chemicals has been developed and is known as Hawaiian Process (2). The paper produced by this process is sufficiently opaque and possesses the necessary strength without the addition of mineral fillers required in other processes for making bagasse newsprint. In another pulping operation, bagasse is softened initially and impregnated with chemicals in a saturated steam zone at elevated temperatures and pressures, and the pre-treated material is processed in a continuous digestor. This treatment was followed by mechanical defibering at the same temperature and pressure (17). Several advances have been made in pulping operations of bagasse; nevertheless, the paper produced from this material is of low grade quality with a low tear resistance and tensile strength possibly because of the reduction of individual fiber strength as a result of the decrease in the degree of polymerization of the cellulose molecules during pulping.

Other Uses in Structural Materials. Until recently, one-fifth of the total production of bagasse in Louisiana was used to make structural and acoustical wall-board and other building products (12). Panels of sugarcane bagasse and synthetic resins have been prepared by agglomerating bagasse fibers with urea-formaldehyde resins (3). These panels possess the necessary desirable properties, namely, a certain elasticity, mechanical resistance, and resistance to deterioration. Bagasse fibers have been mixed with hydraulic cement to the extent of approximately 10% (by wt.) of dried fibers, molded, and cured. These have been used as light weight concrete in structural materials in buildings (4).

Agricultural Uses. Bagasse is used to a limited extent in composting as soil conditioners. Raw bagasse containing 45–60% moisture is baled at 100–150 p.s.i.g. and stored in piles with efficient ventilation for 30 days, when the moisture content of the material drops to 20%. Then it is ground to sufficient fineness, and the water content is increased to 30–60%. The material is sprayed with nitrogen, phosphorus, and potassium compounds to provide N—0.6–1.5%, P_2O_5—0.3–1.0%, and K_2O—0.3–1.0% by dry weight. The product is compressed and stored in bags lined with polyethylene to retain the moisture (14). Raw bagasse is mixed with feed concentrate and used as roughage for cattle. Experiments have been conducted with successful results in which the bagasse is subjected to fermentation with known cultures of microorganisms prior to its supplementation in cattle-feeds. In such experiments it was found that the content of the protein was increased by approximately 10% (dry wt.) after fermentation process (7). Furthermore, investigations have been carried out to find out the economic utilization of bagasse hydrolyzates as substrates for producing riboflavin by *Candida utilis* (1).

Development of Single Cell Proteins from Bagasse. "Single cell protein" is a term applied to proteins obtained from microbial cells. The necessity for intensified research on developing feasible methods of producing and utilizing single cell proteins from unconventional substrates independent of agricultural land use has been sufficiently emphasized in a recent discussion in the UNESCO (16). Cellulose is one of the most widely naturally occurring organic compounds and forms roughly one-third of the plant products on this earth. Most of the cellulosic materials are non-digestible by animals as well as man. The digestibility of cellulosic feed supplements can be increased (and thereby the conversion index of plant to animal proteins) by pre-treatment of the feed by cellulolytic organisms. Large scale production of microbial proteins from cellulosic substrates such as bagasse requires a judicious selection of organisms and studies on their optimal conditions of cultivation. In our laboratories, we have initiated a project on the isolation of cellulolytic organisms and investigations on the effect of environments on their growth and cellulolytic activity. Furthermore, we have extended our studies on the feasibility of utilizing more than one organism to increase the rate of degradation of cellulose.

Experimental

Media. The growth media for cellulolytic organisms usually consists of a mixture of mineral salt solution with cellulosic substrates supplemented with 0.05–0.1% yeast extract. In later experiments yeast extract was replaced by thiamine (0.001%) w/v. The composition of the mineral salt solution is as follows: NaCl—6.0 grams; $(NH_4)_2SO_4$—1.0 gram; KH_2PO_4—0.5 gram, $MgSO_4$—0.1 gram, $CaCl_2$—0.1 gram, in one liter of distilled water.

Organisms that were capable of growth utilizing cellobiose as the sole carbon source were grown in media containing the same basal salts solution with 0.1% of cellobiose.

Isolation of Organisms. Since we were mainly interested in cellulolytic organisms as single cell proteins, our choice of microbes was restricted to the selection of bacteria or yeast. Our search for organisms capable of utilizing bagasse as substrates led us to screen for cellulolytic bacteria in rotting sugarcane stalks as well as the soil from sugarcane fields during the harvesting season. The usual method of "elective technique" was used to enrich for cellulolytic bacteria. Plating methods were used to isolate single colonies. Relatively pure cultures of cellulolytic bacteria were obtained by these procedures. However, it was noticed on careful examination of their cellulolytic activity, the isolates were always associated with some other species of bacteria on the solid media. Hence, a more stringent selection of a single clone of bacteria by the method of terminal dilution was carried out and the isolated clone was identified by the diagnostic tests according to the Manual of Microbiological Methods (*10*).

Several strains of yeast were tested and a few were selected for growth on cellobiose as the sole carbon source. Two gram negative rods were isolated from stored piles of bagasse, which exhibited the capability of growing in basal salt solutions with cellobiose.

Cellulosic Substrates. The following were used as sources of cellulose for our investigations:

(1) filter paper (Whatman No. 1 cellulose powder, Whatman Chromedia CFII, supplied by Reeve Angel Co., N. J.)

(2) (PAB-) Paraaminobenzoyl cellulose (Bio-Rad Labs, Calif.)

(3) CH-cellulose (Carl Schleicher and Schuell Co., Keene, N. J.)

(4) CM-cellulose Gum (Hercules Powder Co., Wilmington, N. J.)

(5) cotton fiber (Johnson & Johnson, New Brunswick, N. J.)

(6) paper towel (Garland, soft-knit, single folder towel, Fort Howard Paper Co.)

(7) Azo-cellulose was prepared by diazotizing PAB cellulose with sodium nitrite and coupling it with β-naphthol to form a red colored cellulose.

(8) Bagasse pith; bagasse fibers; Alkali treated bagasse; untreated bagasse. (All these were obtained from the Department of Chemical Engineering at Louisiana State University)

Determination of Cellulolytic Activity. The presence of cellulolytic activity of an organism was determined qualitatively by inoculating it into test-tubes containing the proper growth medium with a strip of filter paper and incubating the culture at 30°C. If the organism is positive for cellulolytic activity, degradation of filter paper at the air-liquid interface could be noticed within 48 hrs. of incubation. Two different methods were used for quantitative determination of the activity:

(i) The residual fibrous cellulose after the activity of the organisms was quantitatively analyzed by the gravimetric method described by Lembeck (13).

(ii) The amount of solubilized substrate after the degradation of cellulose by the organisms was determined colorimetrically by the method of Dubois et al. (8).

Results and Discussion

Organism. The isolated strain of cellulolytic bacterium is a gram negative rod, 0.8–1.2μ in length and 0.3–0.5μ in diameter. It is catalase

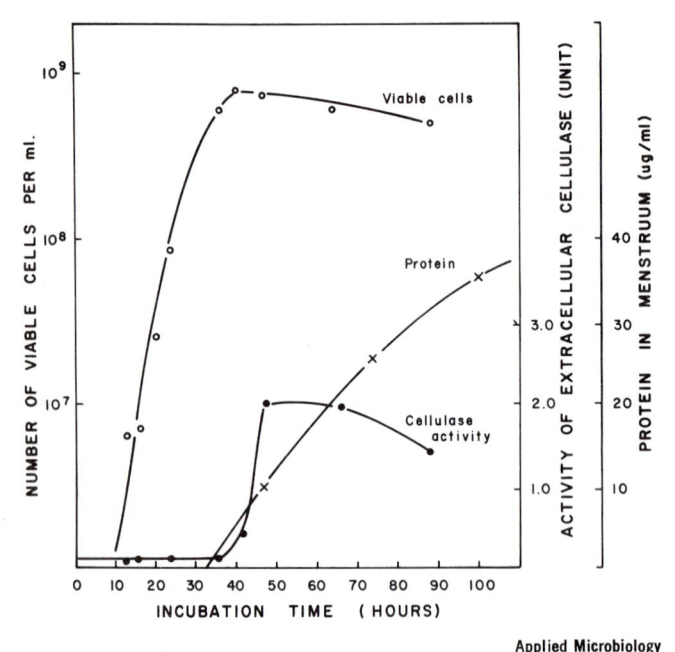

Applied Microbiology

Figure 1. The growth, excretion of protein, and the cellulase activity in the menstruum. The activity of the enzyme was expressed in terms of solubilized cellulose (10)

positive and hydrolyzes gelatin slowly. The morphological and cultural characteristics of the isolate were comparable with the description of the genus *Cellulomonas* in Bergey's Manual of Determinative Bacteriology.

Growth Conditions. Figure 1 represents the kinetics of growth and the excretion of cellulase in the medium. It is apparent from Figure 1 that the activity of the enzyme in the menstruum increased sharply at the early stationary phase of growth. Since the organism is capable of hydrolyzing cellulose even during the logarithmic phase of growth it is reasonable to assume that the enzyme is intramurally located and is released into the medium as a result of certain transformations in the structure of the cell wall of the organisms at the end of the logarithmic phase. Perhaps it is necessary for the stability of the enzyme that it has to be bound to a surface. It is known that the stability of many enzymes towards heat or extremes of pH can be increased by attachment of enzymes to solid matrices (6). This might explain the decline in cellulolytic activity despite the continued increase of protein in the menstruum on prolonged incubation of stationary cultures.

Investigations on the growth of this organism under different environmental conditions showed that the organism was able to grow well between the temperatures 25° and 35°C. in the medium described earlier. Below 20° and above 40°C. there was a marked decrease in the growth rate of the organism. The optimal growth of the organism was noticed in the pH region of 6–8.

Digestibility of Substrates. Table III presents the digestibility of several cellulosic materials by the organism. It is clear from the table that the isolate is incapable of breaking down native cellulose in the form of cotton fiber. Regenerated cellulose such as filter paper or cellulose powder is digested more easily. Entire bagasse, bagasse fiber, or pith is not broken down efficiently. However, pretreatment of bagasse with alkali results in its becoming a good substrate for the cellulolytic activity of the organism. Hence, efficient conversion of bagasse into single cell proteins entails a preliminary operation of partial delignification of bagasse by alkaline treatment. The optimal conditions of pre-treatment of bagasse are being studied in the laboratories of the department of chemical engineering for specific use of cellulose as substrate for the production of *Cellulomonas* in large scale fermentors.

Yield of Microorganisms. If microbial cells are to be produced in large quantity from bagasse or other cellulosic materials, it is necessary to obtain some information about the ratio of the amount of cellulose consumed to the production of cells. Such data obtained from experiments utilizing limiting amounts of CM-cellulose under optical conditions of temperature and pH are presented in Table IV. Fifty percent of the carbohydrate disappeared and can be accounted for by the yield of

Table III. Digestibility of Various Cellulosic Materials by the Organism

Substrate	Initial Wt. (mg.)	Residual Wt. (mg.)	Degree of Digestion (5)[a]
Filter paper	90.6	40.7	55.0
Cellulose powder	97.0	64.0	34.0
PAB cellulose	92.9	54.2	41.0
CM cellulose	91.8	66.0	28.0
Azo-cellulose	93.0	74.0	20.5
Cotton fiber	99.4	99.4	0
Paper towel	65.4	46.0	30.0
Bagasse pith	73.0	65.4	10.0
Bagasse fibers	90.0	87.0	3.4
Sorgo bagasse	109.2	93.0	15.0
Alkali treated sorgo bagasse	100.0	20.0	80.0

[a] Degree of digestion was measured gravimetrically after the organism was inoculated into 100 ml. of media containing each substrate as a sole source of carbon. Inoculum was incubated for 5 days at 30°C. on reciprocal shaker (10).

Table IV. Growth Yields of the Organism Grown on CM-cellulose

	Yields (mg./mg.) in 100 ml.	(%)
Cell mass[a]/CH_2O consumed[b]	13.0/26.0	50.0
Protein[c]/Cell mass	6.0/13.0	46.2
Non protein-N[d]/Cell mass	1.0/13.0	7.7
Protein/CH_2O consumed	6.0/26.0	23.0

[a] Cell crops are dried overnight at 110°C.
[b] Difference of CH_2O concentration in initial and final medium.
[c] By micro Kjeldahl method after extracting nucleic acid with 5% TCA at 90°C. for 30 min.
[d] Difference of N content in whole cells and hot TCA treated cell.

microbial cells. This is comparable with the cell yield of cultures of *Candida* grown on glucose (11). Although the cell yields of *Cellulomonas* seem to be encouraging, economic utilization of this process for producing single cell proteins requires a study of the rate of conversion of properly pre-treated bagasse into cell protein. In a few preliminary experiments, 10 liters of media containing 2% regenerated cellulose were inoculated with 1% inoculum of *Cellulomonas*. After 72 hrs. of incubation with forced aeration, the cells were harvested by centrifuging in a Sharples laboratory model continuous centrifuge. The undigested cellulose fibers were removed by filtration through cheesecloth prior to harvesting. The yield of cells varied between 5–8 grams per liter of the culture. Although the cell mass obtained in the laboratory was low, it is conceivable that further development of the process as a semi-continuous fermentation of cellulose may result in increasing the percentage of cell mass comparable with the yield of cells in a hydrocarbon fermentation.

One of the important steps in a large scale production of *Cellulomonas* is the recovery of the cell crop from the fermentation liquor. Undigested cellulosic material may be removed by passing the tank liquor through a filter press. Then the recovery of cells may be accomplished by concentration to a heavy slurry in a centrifugal separator. The cells can be effectively washed by diluting the slurry with water and recovered by processing through a centrifuge. The efficiency of the centrifugal method of harvesting the cells depends not only upon the concentration of the cell mass in the fermentor but also on the size of the organism. Since *Cellulomonas* is only $1-1.5\mu \times 0.5\mu$ in size, it may well be that modern flocculating agents such as polyelectrolytes of cationic or nonionic properties have to be examined to increase the rate of settling of the organisms, thereby reducing the cost of centrifugation. A realistic appraisal of the economics of the process can be arrived at only after studying the whole process on a pilot plant scale.

Table V. Essential Amino Acid Content of the Cell Protein[a]
(grams of amino acid per 100 gram protein)

Amino Acid	Cell Protein	FAO[b] Reference Protein	Wheat[c] Flour	Beef[c]	B. P[c] Protein
Arginine	9.21		4.2	7.7	5.1
Histidine	2.30		2.2	3.3	5.1
Isoleucine	4.74	4.2	4.2	6.0	4.6
Leucine	11.20	4.8	7.0	8.0	3.1
Lysine	6.84	4.2	1.9	10.0	6.0
Methionine	1.86	2.2	1.5	3.2	1.1
Phenylalanine	4.36	2.8	5.5	5.0	8.1
Tyrosine	2.67	2.8			
Threonine	5.37	2.8	2.7	5.0	11.0
Valine	10.71	4.2	4.1	5.5	7.0

[a] The sample was hydrolyzed with $6N$ HCl at 100°C. for 22 hours and determined with a Beckman model 116 amino acid analyzer.
[b] National Academy of Science–National Research Council.
[c] Reference 10.

Composition of the Protein. Under the experimental conditions of growth it has been found that *Cellulomonas* comprises 23% of its dry wt. as protein. As an essential step in evaluating the nutritional aspects of the product, the quality of the protein is determined by analyzing for the different amino acids present in the product. Table V presents the results obtained, and the amino acid content of the *Cellulomonas* cell protein is compared with those of other proteins of animal and plant origins. Also included in the table are the analyses of single cell proteins obtained by fermentation of petro-chemicals as well as the essential amino acid pattern of an ideal high quality protein proposed by the Food and Nutrition Board of the Food and Agricultural Organization of the United Nations. It can be seen that the cell protein contains lysine in more than sufficient quantity recommended by FAO. It does not have as much methionine as beef, but the level of the amino acid is comparable with that in wheat flour or single cell proteins produced from hydrocarbons.

Table VI. Estimated Cost of Production of 20 Tons Per Day of Single Cell Protein

Capital costs	
10 year depreciation at 10% interest on capital	$ 4.34/hr.
Operating cost	76.07/hr.
Capital plus operating costs	80.41/hr.
Capital plus operating costs of production per lb. of SCP	.048/lb.
Other costs (land, laboratory, taxes, etc.)	.01/lb.
Total production costs	.058/lb.
Cost per lb. of pure protein (50% of cell mass)	.116/lb.

Economic Aspects. The cost of production of single cell proteins from bagasse has been estimated to be 11.6 cents/lb. (5). A breakdown of the estimated cost of producing 20 tons per day is given in Table VI. Based upon the present cost of available proteins in the market the single cell proteins are highly competitive. For comparison, the cost of proteins from several sources is shown in Table VII (15), along with the estimated cost of single cell proteins from petro-chemicals as well as cellulosics.

Table VII. Comparison of the Cost of SCP with Commercially Available Proteins in U.S.

Product	Price per lb. U.S.$	Protein Content	Price per lb. Protein U.S.$
Peanut flour	0.07	59%	0.12
Soy flour	0.05	43%	0.12
Cotton seed flour	0.05	50%	0.10
Skim milk powder	0.15	36%	0.41
Torula yeast (food grade)	0.17	48%	0.36
Casein (food grade)	0.40	100%	0.40
Single cell protein			
(i) from Hydrocarbons	—	—	0.35 [a]
(ii) from Bagasse	—	—	0.12 [a]

[a] Estimated cost.

Mixed Fermentation of Bagasse. Thus far we have stressed the feasibility of producing *Cellulomonas* on a large scale as a source of single cell proteins. In the early experiments with *Cellulomonas* with cellulose substrates it was found that a small amount of yeast extract was absolutely necessary to ensure a good growth of the organism. Sub-

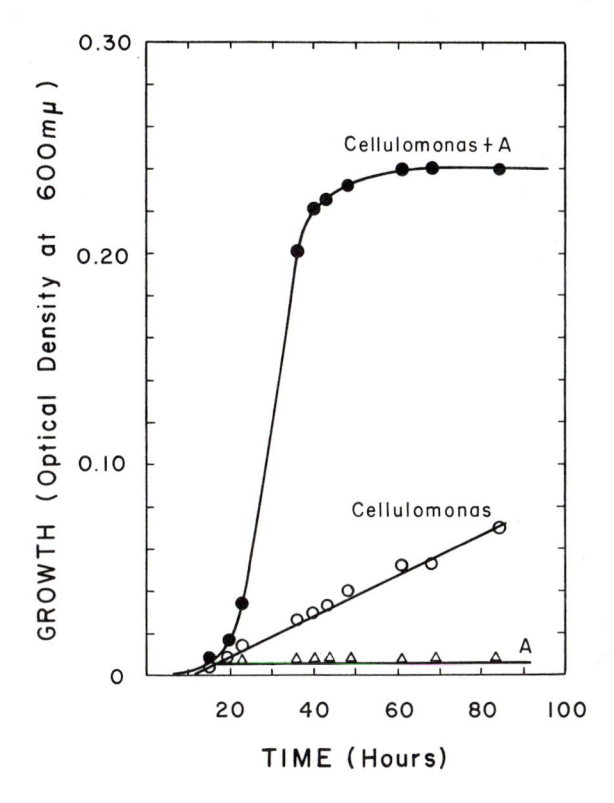

Figure 2. The kinetics of growth of organisms in CM-cellulose medium mixedly inoculated with Cellulomonas and Pseudomonas. A—Pseudomonas

sequent investigations carried out to find out the growth factor contributed by yeast extract showed that thiamine at a level of 10 μg./ml. enabled the organisms to grow in the presence of cellulosic substrates although the rate of growth was slower than that in the presence of yeast extract. If the method of production of microbial protein were to be competitive in cost with other proteins, it is imperative that a reasonable rate of degradation of cellulose has to be obtained. To this effect, studies were undertaken to ascertain the rate of increase of cell mass by mixed inoculation of *Cellulomonas* with another organism. It has been shown that the bacterial cellulase from *Cellulomonas* breaks down the cellulose to cellobiose (10). Hence, initially a number of organisms were surveyed for the ability to grow on cellobiose as the carbon source. One of such organisms, preliminarily characterized as belonging to the genus *Alcaligenes* was used to study its interaction with *Cellulomonas* in the breakdown of cellulose. Figure 2 depicts the kinetics of growth of *Cellulomonas* as well as the mixed culture. Growth was monitored by measuring turbidity in the culture medium containing CM-cellulose as the soluble substrate. The medium contained the basal salt solution supplemented with thiamine. The organism *Alcaligenes* unknown sp. did not show any visible growth in this medium; whereas *Cellulomonas* was found to grow rather slowly but at a steady rate. However, a mixed inoculation of both the organisms in the medium resulted in a marked increase in the growth rate of the organisms. A differential count of the two organisms was made during the growth (Figure 3) which showed that the majority of

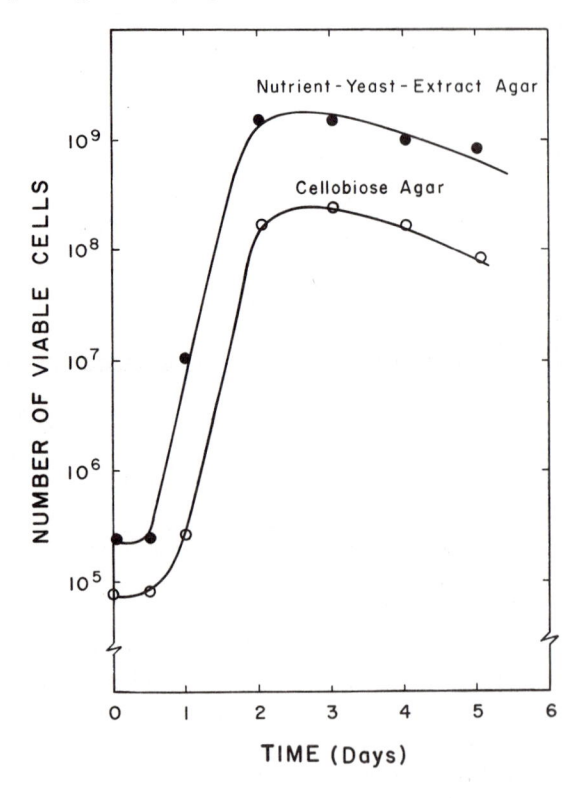

Figure 3. Differential count of the organisms during growth in CM-cellulose medium mixedly inoculated with Cellulomonas *and* Pseudomonas. *The counts* ·—· *give the total number of both the organisms.* o—o *represents the number of* Pseudomonas *in the mixture*

the population in the mixed culture consisted of *Cellulomonas*. Several explanations can be advanced to account for the behavior of the two organisms in the mixed culture. *Cellulomonas* perhaps is able to assimilate cellobiose only at a slow rate and the accumulation of cellobiose in the medium inhibits the activity of the cellulase and this results in retarding growth. Continuous partial removal of cellobiose by the second organism leads to a removal of the limiting factor of the growth of *Cellulomonas*, namely the cellulolytic activity of the organisms, and thereby enhances its growth. Or, it might be that the *Alcaligenes* supplies the necessary growth factor(s) required by the cellulolytic organism. It is also possible that the second organism which possesses an efficient β-glucosidase cross-feeds the *Cellulomonas* with glucose from the breakdown of cellobiose. Further experimentation is required before any conclusion can be drawn about the mechanics of interaction between the two organisms.

Conclusion

The development of microbial proteins from cellulosics offers distinct advantages over the manufacture of single cell proteins from petrochemicals. First, the availability of cellulosics surpasses the sources of hydrocarbons suitable for the production of SCP. Conversion of hydrocarbons into SCP requires a knowledge of an advanced technology of petroleum refining, which is at present lacking in developing countries, where the need for protein is mostly felt. In spite of the great strides the fermentation of hydrocarbons has taken in the last decades, the process faces a number of problems because of the immiscibility of the substrates in water. Since the hydrocarbons are in a lower state of oxidation than the usual carbohydrates, oxygen requirements in such systems are high and some progress has been reported in efficient mass transfer of oxygen in hydrocarbon fermentation. Microbial cells have a tendency to synthesize a large amount of fat in hydrocarbon substrates and unused hydrocarbons adhere tenaciously to the cell surface. In the case of cellulose substrates conventional techniques of fermentation can be used with proper selection of the organisms. Furthermore, since cellulosics form a large percentage of solid wastes in highly developed countries, the advancement in the technology of cellulose fermentation might lead to an important method of disposal of wastes and their conversion into useful products.

Acknowledgment

The authors thank S. P. Yang for the amino acid analysis of the cell protein and Joan Raymond for the valuable technical assistance.

Literature Cited

(1) Agarwal, P. N., Rawal, T. N., Gurtu, A. K., *Ind. J. Technol.* **2**, 246 (1964).
(2) Atchison, J. E., *Chem. Age India* **15**, 971 (1964).
(3) Borlando, Luis A., *Plast. Resinas.* **6**, 4 (1964); *C.A.* **67**, 34014f (1966).
(4) Bourlin, Gabriel P., *U. S. Patent* **3,264,125** (1966).
(5) Callihan, C. D. (personal communication).
(6) Crook, E. M., *Biochem. J.* **107**, 2p (1968).
(7) Cruz, R. A., Macaisa, R. E., Cruz, T. J., Prisag, C. C., *Sugar News* **43**, 15 (1967).
(8) Dubois, M., Gilles, K. A., Hamilton, J. K., Rebers, P. A., Smith, F., *Anal. Chem.* **28**, 350 (1956).
(9) "Gilmore Louisiana-Florida Sugar Manual," p. 264, Hauser-American, New Orleans, 1965.
(10) Han, Y. W., Srinivasan, V. R., *Appl. Microbiol.* **16** (1968).
(11) Johnson, Marvin J., *Science* **155**, 1515 (1967).
(12) Keller, A. G., *Kirk-Othmer Encycl. Chem. Technol.* **3**, 36 (1964).
(13) Lembeck, W. J., Colmer, A. R., *Appl. Microbiol.* **15**, 300 (1967).
(14) May, Harry M., Nadler, Jr., H. A., *U. S. Patent* **3,337,326** (1966).
(15) McNab, J. G., Rey, L. R., *Chem. Eng. News* **45**, 46 (Jan. 9, 1967).
(16) UNESCO Report **E/4343** on "Increasing the Production and use of Edible Protein" (1967).
(17) Villavicencio, E. J., Rojas, Mario Sierra, Escobar, Salvador, *U. S. Patent* **3,238,088** (1966).

RECEIVED October 14, 1968. Supported by funds from the Graduate Research Council of the Louisiana State University, Baton Rouge, Louisiana 70803.

INDEX

A

Accessible surface area 229
Accessibility 149, 221, 234
  of bacterial cellulose membranes 146
  of carbohydrates 147
  of native cellulose 145
  of substrate vs. rate of saccharification 417
Acetobacter xylinum 14
Acid treatment 282
Acorns 299
Actinomycetes 394
Activation energy in degradation of celluloses by enzymes 8
Activities of cellulase preparations 396–7
Activity of cellulase, chemical modification of cellulose vs. 10
Additives 403
Adenostoma fasiculatum 281
ADF 288
Adsorption on different cellulose columns 95
Aeration 414
Affinity factor 56
Agar–agar 385
Agitated systems, batch and semicontinuously 415
Agrobacterium spp. 108
Alcaligenes faecalis var. *myxogenes* 108
Alder sawdust 286–7, 289, 295
Alfalfa ...246, 248, 256–7, 265, 268, 305
  hay 200
Alkali, dilute 197
Alkali treatment 282
Amino acid composition of the endo and exoglucanases of *Trichoderma viride* 17
Ammonia, liquid 197
Ammonium sulfate treatment 281
Amorphous cellulose by endoglucanase, hydrolysis of 423
Amylase 345
α-Amylase 125
β-Amylase 111
Amyloglucosidase 26
Amyloids 108
Anhydroglucose units 179
Apple 362
Arabinose 109
Aryl-β-glucosidase 93
Ascomycetes 91
Aspen 207, 248–50
  silage 252

Aspergillus 91, 93
  awamori 348
  foetidus 30, 348, 355
  fonsecaeus 191
  fumigatus 191
  niger 11, 114, 118–9, 126–7, 172,
      346, 348, 359–60, 379, 382,
      388, 391, 396
  oryzae 119, 346, 382
  phoenicis 348
  terreus 394
  versicolor 192
  wentii 106
Assay of cellulase activity 86
Astasia ocellata 108
Aureobasidium cell wall glucan .. 108
Aureobasidium pullulans 108
Avena sativa 330
  endosperm 108
Avicel 39, 53, 73

B

Bacillus
  cereus 69
  circulans 119
  licheniformis 69
  stearothermophilus 64
  subtilis 29, 120
Bacterial cellulose 141
  membranes, accessibility of 146
Bacteroides succinogenes 285
Bagasse, mixed fermentation of .. 457
Bagasse, utilization of 447
Balled filter paper 53
Balsam fir 249, 257
Bamboo shoots 281
Barley 265
  glucan 108
Basidiomycete 31, 394, 396
  enzyme 116
Batch and semi-continuously agitated systems 415
Beech 207
  leaves 299
Beef cattle rations, wood as a source of energy in 279
Bermuda grass 280
  hay, sawdust vs. coastal 321
Birch 241
  fibers 159
  leaves 299
  pulps 316
  wood 246

Black spruce 249
Boron vs. cellulose digestibility .. 255
Bovine mycoplasma 108
Breakdown of cellulose, enzymes
    involved in 16
Brush wood 299

C

C_1 .... 34, 42, 56, 84
    cotton 393
    glucanases 23
    reaction 4, 8–9
C_x enzyme 56, 84, 192, 194
$C_x\beta(1 \rightarrow 4)$ glucanase 393
Cabbage 362
Calcium vs. cellulose digestibility 255
Callose 108
Cambium 146–7
Candida utilis 450
Cane molasses 299
Capillary structure of cellulose ... 168
Carbohydrates, accessibility of ... 147
Carbohydrates, readily digestible .. 255
Carboxymethylcellulase 37, 42, 285
Carboxymethylcellulose 106, 189
    change in chain length vs. change
      in end groups during hydrolysis of 14
    by endogluconase, hydrolysis of 423
    enzymatic hydrolysis of 12
    enzyme concentration and temperature vs. hydrolysis of .. 398
    intrinsic viscosity of 11
    random hydrolysis of 13
    time and temperature vs. hydrolysis of 399
Carboxymethylpachyman 106
Carrot 362, 365, 369
Cathode rays on sprucewood and
    cotton linters, effects of 283
Cattle, fattening of mature 315
Caulerpa brownii 118
Cell-bound cellulase components of
    bacteria 60
Cell-bound cellulases 63
Cellobiase 37, 42
Cellobiose ...27, 61, 73, 78, 97, 365, 400
Cellodextrin 63
Cellodextrinases 285
Cellohexaose
    -^{14}C, end labelled 15
    hydrolysis of 16
    reduced 16
Cello-oligosaccharides 63, 73–75
Cellopentaose 424
Cellotetraose 21, 27, 424
    optical rotation vs. hydrolysis of
      reduced 19
Cellotriose 21, 78, 424
Cell protein 455
    cost of production of 456
Cell separating enzyme 359
Cellulase 37, 70, 93
    activity 90

Cellulase (Continued)
    assay of 86
    chemical modification of cellulose
      vs. activity of 10
    complex 23, 53
      C_1-component of the 34
      enzymes of the 7
      of Ruminococcus 53
    components of bacteria, extracellular and cell-bound 60
    components, discrimination tests
      of 76
    formation by Ps. fluorescens ... 66
    inactivation of 401
    induction of 399–400
    inhibition of Penicillium pusillum 425
    molecules, size and shape of ... 170
    onozuka
      F1500 386
      F400 376
      P1500 379
      P500 ....360, 374–5, 381, 387
      PE 360–1, 366, 371
    patterns of synthesis of 67
    perturbation spectra of 97
    P. funiculosum 150
    P. notatum 100–1
    preparations, activities of 396–7
    production, concentration of Solka
      Floc, proteose peptone, and
      Proflo vs. 404–5
    production of 403
      T. viride 409
    components, properties of purified 75
    research, perspective on 1
    solutions exo-β-1 $\rightarrow$ 4-glucanase
      in crude 26
Cellulases
    characterization of 95
    in Japan, application of 359
    new methods for 83
    production of 391
    pseudomonas 72, 73
    specificities of pseudomonas ... 73
Cellulase, T. viride 149
Cellulolytic
    components, mode of action of .. 14
    enzymes 83, 152
    enzymes, superficial 191
    enzymes of Diplodia zeae, constitutive 188
    filtrates 31
    systems, supramolecular structure
      vs. activity of 7
Cellulomonas 453–4
Cellulose 61, 108, 245, 315
    accessibility of native 145
    vs. activity of cellulase, chemical
      modification of 10
    bacterial 141
    capillary structure of 168
    –cellulase system 233

INDEX

Cellulose *(Continued)*
  with cellulase *vs.* saccharification, milling 441
  chains, conformation of 140
  columns, adsorption on different 95
  continuous saccharification of modified 440
  degradation, new mechanism for 53
  degree of polymerization of ... 179
  by endo-gluconase, hydrolysis of amorphous 423
  enzymatic hydrolysis of 392
  enzymatic saccharification of ... 415
  enzymes in breakdown of 16
  fibers, chemical constituents of 159
  fibers, structure of 155
  growth of yeasts on enzymatic digests of 411
  hydrolyzates, use of 410
  membranes, accessibility of bacterial 146
  microfibrils, native 141
  model of 10
  oligosaccharides
    by *Cellvibrio gilvus*, metabolism of 19
    as donors to glucose 21
    phosphorolysis of 22
  series 15
  pH *vs.* saccharification of modified 431
  polymer series 18
  reactivation of enzyme digested 436
  by ruminants, utilization of 242
  saccharification rate of modified 438
  semi-continuous saccharification of modified 437
  structure and morphology of .. 139
  temperature *vs.* saccharification of 431
  unit cell structures of 178
Cellulosic
  materials, digestibility of 454
  materials *vs.* enzymatic hydrolysis, structural features of .. 152
  substrates, pH *vs.* enzymatic hydrolysis of 12
Cellulosin AP 360–1, 366, 371, 374, 376, 381
Cellvibrio gilvus ... 7, 14, 22, 71, 80, 114
  cellulases 75
  metabolism of cellulose oligosaccharides by 19
Cell wall constituents
  digestibility of 256
  in lignocellulose 246
Cereal straw 282
Cetraria icelandica 108
CFE 292
Chaetomium globosum 394
Chamise 281
Change in chain length *vs.* change in end groups during hydrolysis of carboxymethylcellulose 14

Change in number of end groups
  vs. change in chain length .. 13
Characterization of cellulases 95
Chemical
  constituents of cellulose fibers .. 159
  modification of cellulose *vs.* activity of cellulase 10
  processing *vs.* digestion ceiling .. 257
  treatment 281
Chitin 124
Chitinase 367
Chrysolaminarin 108
Chrysosporium lignorum 58, 85, 91, 93, 172
Chrysosporium pruinosum 30, 59, 394–6
Citrus natsudaidai 380
Citrus unshiu 380
Claviceps glucan 118
Clostridium thermocellum 15, 20, 22, 80
CMC 88, 395
CMC-saccharifying activities of *pseudomonas* cellulase components 77
CMCase 78, 365, 368, 381
Coastal Bermuda 268
Coastal hay 321–3
Cobalt *vs.* cellulose digestibility .. 255
Coconut cake 383
Commercial enzyme process: glucoamylase 343
Comminution *vs.* particle size 249
Commission on Enzymes 86
Complex of *Ruminococcus*, cellulase 53
Composition
  vs. digestibility 268
  of forages 262
  of wood products 299
Conduritol B-epoxide 106
Conformation of cellulose chains .. 140
Coniferous woods 241
Constitutive cellulolytic enzymes of *Diplodia zeae* 188
Copper *vs.* cellulose digestibility .. 255
Corn 265, 304, 323–4
  steeped 384
  steep liquor 349
Cost of commercial proteins 456
Cost of production of cell protein 456
Cotton 223
  cellulose to enzymatic hydrolysis, susceptibility of 419
  duck 400
  fibers 159
  linters, effects of cathode rays on 283
  in phosphoric acid, swelling of .. 224
Cottonseed hulls 304–5, 308–9
Cottonseed flour *vs.* cellulase production, concentration of ... 404–5
Cotton sliver 400
Crown gall polysaccharide 108
Crystallinity, degree of 176
CSE 359, 362–3, 367, 378–80, 388
Cultural conditions 402

Curdlan 108

D

DE 247, 333
Degradability of wood waste 326
Degradation
  of celluloses by enzymes, activation energy in 8
  of crystalline hydrocellulose vs. surface area 423
  new mechanism for cellulose ... 53
Degree of crystallinity 176
Degree of polymerization of cellulose 179
Delignification 329
Deuterium oxide 145
Dextran 239–40, 283
Diatoms 108
Dicotyledon seed 108
Difference spectrophotometric studies 96
Digestible
  carbohydrates 255
  dry matter, lignin content vs. ... 270
  energy 247
  of laurel hemicellulose 307
Digestibility 219, 241, 251, 320, 329
  of cellulosic materials 454
  of cell wall constituents 256
  composition vs. 268
  of hemicellulose constituents ... 257
  hemicellulose vs. 312
  of lignocellulosic materials 197
  of spruce sulfite pulps 230
Digestion ceiling concept 247
Digestion, change in wood fiber during 325
Digestive system of the cow 243
Dilute alkali 197
Diplodia zeae, constitutive cellulolytic enzymes of 188
Discrimination tests of cellulase components 76
Distribution of protein and enzymic activities 92
DM 246
Dolabella 124
Douglas fir sawdust 283
Dry matter 245

E

Eastern hemlock 249
Eisenia bicyclis 108, 118
Electrolytic reduction 98
Electrophoresis, free-zone 100
Electrophoresis, zone 63, 79
Elm leaves 299
EMC 205
Enzymatic
  digestion 224

Enzymatic *(Continued)*
  digests of cellulose, growth of yeasts on 411
  hydrolysis 154
  of carboxymethylcellulose ... 12
  of cellulose 392
  structural features of cellulosic materials vs. 152
  susceptibility of cotton cellulose to 419
  saccharification of cellulose ... 415
Enzyme
  activity, heat vs. 422
  activity, measurement of 395
  concentration vs. hydrolysis of carboxymethylcellulose and filter paper 398
  concentration vs. hydrolysis of cellulose 407
  process: glucoamylase, commercial 343
  –substrate relationships 105
Enzymes
  activation energy in degradation of celluloses by 8
  in the breakdown of cellulose .. 16
  of the cellulase complex 7
  Commission on 86
  hydrolyzing β-glucosidic linkages 107
  superficial cellulolytic 191
  synthetic action of 24
Enzymic activities, distribution of 92
Enzymic activities from isoelectric separation, distribution of ... 94
End-labeled cellohexaose-^{14}C 15
Endoglucanase, hydrolysis of amorphous cellulose and carboxymethylcellulose by 423
Endoglucanases, amino acid composition of 17
Endogluconoses of *Trichoderma viride* 18
Endomyces 347
Endotype synthesis 65
Energy in beef cattle rations, wood as a source of 279
Energy of the glucosyl bonds 20
Eremoplastron bovis 285
Erwinia aroideae 378
Escherichia coli 68
Euglena gracilis 108, 118, 126
Excreta 387
Exo-β-1 $\rightarrow$ 4-glucanase in crude cellulase solutions 26
Exoglucanases 14, 24
  amino acid composition of the ... 17
  of *Trichoderma viride* 18
Exo-hydrolases, β-glucan 111
Exo-polysaccharases 26
Exo-type synthesis 65
Extracellular cellulase components of bacteria 60
Extracellular cellulases 63

INDEX

F

Fabospora fragilis	119
Fat *vs.* cellulolytic activity	255
Fattening of mature cattle	315
Feed efficiency, growth response *vs.*	341
Feeding level	251
Feed for ruminants, hardwood sawdust as	315
Fermentation of bagasse, mixed	457
Fermenter, semi-continuous laboratory	409
Fiber saturation	231
Fibers, chemical constituents of cellulose	159
Fibers, structure of cellulose	155
Filter paperase	365, 368, 376
Filter paper	393, 400
degrading enzyme activity	366
enzyme concentration and temperature *vs.* hydrolysis of	398
Fir tree cellulose, hydrolysis of	420
Fistulated steer	223, 291–2, 320, 330
Fluidity *vs.* reducing power during the hydrolysis of CMC	74
Food, source of	3
Fomus annosus	91, 93, 172, 191
Fractionation of phosphorylase and hydralase activities	20
Free-zone electrophoresis	100
Fruits, unicellular	362
FSP	206
Fungal koji	345
Fungi	394
Fusarium moniliforme	394
Fusarium roseum	394

G

β-Galactans	124
Galactose	109
Galactoside, lactose $a\beta(1 \to 4)$	401
Gelatinization of seaweed	375
Gel filtration	90, 100
Gelidium amansii	384–5
Gentiobiose	64
Geotrichum candidum	192
Glucoamylase	26, 111
commercial enzyme process:	343
β-Glucomannans	124
Glucanase β-1,3-	367
$C_x\beta(1 \to 4)$	393
in crude cellulase solutions, exo-$\beta(1 \to 4)$	26
β-1,4-Glucan cellobiosyl hydrolases	114
β-Glucan endo-hydrolases	128
β-Glucan exo-hydrolases	111
β-1,3-Glucan glucosyl hydrolases	115
β-1,4-Glucan glucosyl hydrolases	113
α-1,4-Glucan hydrolases	111
α-1,6-Glucan hydrolases	111
β-Glucan hydrolases	105
β-Glucans, structure and occurrence of	109

Glucans, substituted	121
Glucose	61, 400
cellulose oligosaccharides as donors to	21
model of	10
β-Glucosidase	67, 70, 93
Glucosidases	23
Glucoside, sophorose $a\beta(1 \to 2)$	401
β-Glucosidic linkages, enzymes hydrolyzing	107
Glucostat reagent	192
Glucosyl bonds, energy of	20
Glucosyl bonds *vs.* glucotransferase	21
Glucotransferase	20
glucosyl bonds *vs.*	21
Glycerol	400
Gracilaria convervoides	385
Gracilaria verrucosa	386
Grass, orchard	246
Green tea component	367–8
Ground corn	305, 349
Growth response *vs.* feed efficiency	341

H

Hansenula anomala	119
Hardwoods	199, 241
Hardwood sawdust as feed for ruminants	315
HC	302
Heat *vs.* enzyme activity	422
Helix pomatia	122, 396
Hemicellulose	245, 302, 304, 306, 308–9
constituents, digestibility of	257
digestible energy of laurel	307
vs. palatability and digestibility	312
Hemlock	257
sawdust	291–2, 294
Histidyl groups	102
Holocellulose	147
Hordeum vulgare endosperm	108
Hungate technique	53
Hyaluronic acid	124
Hydrocellulase	39
vs. surface area, degradation of crystalline	423
Hydrogenbondase	34
Hydrolase activities, fractionation of phosphorylase and	20
Hydrolases, β-glucan	105
Hydrolysis	
of amorphous cellulose and carboxymethylcellulose by endogluconase	423
of carboxymethylcellulose and filter paper, enzyme concentration and temperature *vs.*	398
of cellohexaose	16
of cellulose, enzyme concentration *vs.*	407
long term	404
structural features of cellulosic materials *vs.* enzymatic	152

Hydrolysis *(Continued)*
  susceptibility of cotton cellulose
    to enzymatic 419
  by *T. viride* cellulase, substrate
    vs. extent of 405–6
  by *T. viride*, pH *vs.* long term .. 408
Hydrolytic factor 56
Hydrolytic treatments 282
Hydrolyzates, use of cellulose ... 410
Hydroxyethylcellulose 106
Hyphae 163–4, 166–7, 175, 182, 190, 192, 194

I

Inactivation of cellulase 401
Induction of cellulase 399
Inhibition of *Penicillium pusillum*
  cellulase 425
Inoculum size 413
Inoculum, source of 252
Intracellular enzymes 69
Intrinsic viscosity of carboxymethylcellulose 11
Iron *vs.* cellulose digestibility ... 255
Irpex lacteus 121, 124, 127–8
  cellulase 424
Irradiation *vs.* digestion ceiling .. 257
Islandic acid 108
Isoelectric focusing 92
Isoelectric separation, distribution
  of enzymic activities from ... 94
Isomaltose 358

J

Japan, application of cellulases in 359
Jute 200

K

Koji 365
  fungal 345
Konbu 374
  process 407

L

Laboratory fermenter, semi-continuous 409
Lactose 400
Lactose a,β(1 → 4) galactoside .. 401
Lambs 322, 332
Laminaria 375
  cloustoni 108
  digitata 108
  hyperborea 108
  saccharina 108
Laminaribiose 20
Laminaridextrins 24
Laminarinase, *Rhizopus arrhizus* .. 125
Laminarins 108
Laurel hemicellulose 308–9
Leucosin 108

Lentinus edodes 365
Lenzites trabea 200
Lichenin 108, 125, 400
Light intensity *vs.* composition of
  rye grasses 274
Lignification 328
Lignin 245, 283
  content *vs.* digestible dry matter 270
Lignocellulose
  cell wall constituents in 246
  in ruminant nutrition 245
  utilization, minerals *vs.* 254
Lignocellulosic materials, digestibility of 197
Live oak 281
Liquid ammonia 197
Locust beans 299
Lodgepole pine 249, 257
Long term hydrolysis 404
  by *T. viride*, pH *vs.* 408
Luteic acid 108
Lysozyme 105, 128
  -catalyzed reactions 129

M

Macerozyme 360–1, 366–7, 371, 374–5, 381
Macromolecules, properties of ... 227
Magnesium *vs.* cellulose digestibility 255
Malonic acid 109
Malt 345
β-Mannans 124
Mannanase 93
Mannitol 109
Mannosidase 93
Maturity of forages 262
Maturity of the plant 256
Mechanism for cellulose degradation, new 53
Median pore sizes 232
Meicelase 383
  P 360–1
Merthiolate 190, 193, 445
Metabolism of cellulose oligosaccharides by *Cellvibrio gilvus*. 19
Michaelis constants 18
Microbial fermentation of wood
  products 310
Microfibrils 156
  native cellulose 141
Milling cellulose with cellulase *vs.*
  saccharification 441
Minerals *vs.* lignocellulose utilization 254
Mixed fermentation of bagasse 457
Molasses 299
Molecular sieves 237
Molybdenum *vs.* cellulose digestibility 254
Monodus glucan 108
Monodus subterraneus 108

INDEX

Morphology of cellulose, structure
 and 139
Multiple attack 125
Mushroom 365, 369
Mycodextranase 121
Mycoplasma glucan 108
Myrothecium 29
 verrucaria 13, 35, 84, 90, 110, 122,
 172, 236, 394, 396

N

Native cellulose, accessibility of .. 145
Native cellulose microfibrils 141
NaOH-treated oat straw, urea supplementation *vs.* nutritive
 value of 328
NaOH-treatment 329
NE 265
Neurospora crassa 78, 110
Never-dried pulps ..223, 229–32, 238–41
NFE 263
Nicotiana glutinosa 120
Nitrogen free extract 263
Nitrogen as a supplement 253
p-Nitrophenyl-β-glucoside 73
Nutritive value
 of forages 262
 index 336
 of NaOH-treated oat straw, urea
  supplementation *vs.* 328
 of wood products 299
NVI 333, 336

O

Oak leaves 299
Oak sawdust 280, 318–9, 322–4
 vs. oyster shell 317
Oat glucan 108
Oats 265, 268
Oat straw 265, 330
 urea supplementation *vs.* nutritive value of NaOH-treated 328
Ochromonas malhamensis 108
OH-stretching frequency bands .. 141
Oligosaccharide end products 126
Oligosaccharides 14
 by *Cellvibrio gilvus*, metabolism
  of cellulose 19
 as donors to glucose, cellulose .. 21
 phosphorolysis of cellulose 22
 of 7 and 14 monomers, model of 10
Onion 362
Optical rotation *vs.* hydrolysis of reduced cellotetraose 19
Orange peel 362
Orchard grass 246, 257, 281
ORD studies 101
Organisms 394
Organoleptic tests 376
Oyster shell, oak sawdust *vs.* 317

P

Pachyman 108
Palatability 302, 304
 hemicellulose *vs.* 312
 of rations 306
Pancellase 364
Panicum maximum 250
Papaya 362
Paramylon 24, 108
Parenchyma 360
 cells 377
Particle size, comminution *vs.* ... 249
Patterns of synthesis of cellulase .. 67
PDE 247
Peanut hull 280
Pectic acid 365
PEG 205
Penicillium 108
 funiculosum 41–4
  cellulase 150
 fluorescens 82
 notatum 92, 97–8, 172
  cellulase 100–2
 pusillum 30, 394–5
  cellulase, inhibition of 425
 variable cellulase 420–1
Peranema trichophorum 108
Perturbation spectra of cellulase .. 97
Pestalotiopsis westerdijikii 390, 394
Phaeodactylum tricornutum 108
pH *vs.* enzymatic hydrolysis of
 cellulosic substrates 12
pH *vs.* saccharification of modified
 cellulose 431
pH *vs.* long-term hydrolysis by
 T. viride 408
Phosphoric acid, swelling of cotton
 in 224
Phosphorolysis of cellulose oligosaccharides 22
Phosphorus *vs.* cellulose digestibility 255
Phosphorylase and hydrolase activities, fractionation of 20
Pinus mugo 108
PMV 286
Pneumococcal type III 124
Polyplastron multivesiculatum ... 285
Polyporus
 cinnabarinus 30
 annosus 65
 versicolor 85, 90, 172, 200
Polysaccharides 124
 in wood fibers 162
Poplar leaves 299
Populus tremuloides 252
Pore size 222
Poria 395–6
 cocus sclerotia 108
Porosity 226, 238
Potato 365, 369
 degradation 366
Potentially digestible energy 247
Potentiometric titration 101

Production of cellulases 391, 403
Production of *T. viride* cellulase .. 409
Proflo *vs.* cellulase production, concentration of 404–5
Protein activities, distribution of .. 92
Protein–protein interaction 53
Proteins, cost of commercial 456
Proteolytic enzymes 99
Proteose peptone *vs.* cellulase production, concentration of 404–5
Pseudomonad 61, 63
Pseudomonas
  cellulase 72–3, 75–6, 78
    specificities of 73
    components, CMC-saccharifying activities of 77
  fluorescens 61, 69–71, 75, 78, 124, 128
    cellulase formation by 66
  saccharophila 64
Pullulanase 111
Pulps, never-dried ..223, 229–32, 238–41
Pustulan 108
Pyrenochaeta terrestris 65, 188

Q

Quercus wislizenii 281

R

Random hydrolysis of carboxymethlycellulose 13
Raw wood in ruminant rations ... 312
Reactivity 221
Red clover hay 299
Reduced cellohexaose 16
Reducing sugars 428
Reed canary grass 273
Relative intake 235
RH 205
Rhizoctonia solani 192
Rhizopus
  albus 285
  arrhizus 119
    laminarinase 125
  CSE 360, 381, 388
    Koji 370–1
  delamar 118, 127, 346, 348
    Koji 360–1
  oryzae 346
  sp. 359–60, 379–80
  stolonifer 378
Rhodophyceae 384
Rice 365, 369
Ruminant 329
  nutrition, wood derived products in 298
  nutrition, lignocellulose in 245
  rationing systems 299
  rations, raw wood in 312
Ruminants, hardwood sawdust as feed for 315
Ruminants, utilization of cellulose by 242

Rumen
  inoculum 229–30, 241
  microflora 252
  microorganisms 315
  parakeratosis 319, 321
Ruminococcus
  albus 53, 55
    cellulase complex of 53
  flavefaciens 80, 285
Rye 256–7
Ryegrass 246, 268

S

Saccharification
  accessibility of substrate *vs.* rate of 417
  of cellulose, enzymatic 415
  of cellulose, temperature *vs.* ... 431
  milling cellulose with cellulase *vs.* 441
  of modified cellulose, continuous 440
  of modified cellulose, pH *vs.* .. 431
  of modified cellulose, semi-continuous 437
  rate of modified cellulose 438
  system 439
Saccharomyces cerevisiae 116, 119
Sawdust 283, 300, 317
  vs. coastal Bermudagrass hay .. 321
  as feed for ruminants, hardwood 315
  from pine 299
Schizophyllum commune 119
Sclerotinia libertians 116
Seaweed
  components 377
  gelatinization of 375
  jelly 374
Seaweeds 385
Selenium *vs.* cellulose digestibility 255
Semi-continuous laboratory fermenter 409
Semi-continuously agitated systems, batch and 415
Semi-continuous saccharification of modified cellulose 437
Sephadex 36, 85, 91
Septic tank 387
Sewage sludge 280
Shaddock peel 362
Shape of cellulase molecules 170
Sheep 316, 318
Shelled corn 303
Shiitake mushroom 365, 367
Silage, aspen 252
Silica in wood 278
Size of cellulase molecules 170
Smith degradation 121
Sodium citrate 376
Sodium phosphate treatment 281
Softwoods 241
Solka Floc 400, 420, 426–7, 428–9, 431–3, 434–5, 445
  vs. cellulase production, concentration of 404–5

Soluble starch	365
Solute exclusion	222, 225
Sonification	190, 194
Sorghum	268
Sophorose	61, 64, 400
$\alpha,\beta(1 \rightarrow 2)$ glucoside	401
Soybean	369, 371
cake	382
meal	303, 305
protein	367
Spectrophotometric titration	101
Spinach	362
Sporotrichum pruinoides	59
Sporotrichum pruinosum	117
Spruce fibers	159
Sprucewood	220, 246
effects of cathode rays on	283
Stachybotrys atra	110, 114, 394
Staphylococcus aureus	192
Starch	349, 372–4, 383, 400
Steeped corn	384
Steer, fistulated	223, 291–2, 320, 330
Steers	319, 322–3
Stereum sanguinolentum	58, 65, 85, 91–3, 127–8, 172
Straw	281
Streptomyces	30, 424
sp.	394, 396
Structural features of cellulosic materials *vs.* enzymatic hydrolysis	152
Structure of cellulose fibers	155
Structure and morphology of cellulose	139
Substituted glucans	121
Substrate *vs.* extent of hydrolysis by *T. viride* cellulase	405–6
Substrate *vs.* rate of saccharification, accessibility of	417
Substrates	427
Sucrose	337
Sudan grass	274
Sugar maple	200, 207
Sugarcane bagasse	448
Sulfite pulp	387
Sulfur *vs.* cellulose digestibility	255
Superficial cellulolytic enzymes	191
Supramolecular structure *vs.* activity of cellulolytic systems	7
Surface area	222
degradation of crystalline hydrocellulose *vs.*	423
Susceptibility of cotton cellulose to enzymatic hydrolysis	419
Sweco	430–1, 432–3, 434
Sweet potato	362–3, 365, 369, 372–4
starch	370
Swelling	223
of cotton in phosphoric acid	224
Synthesis of cellulase, patterns of	67
Synthetic action of enzymes	24

T

Tamarindus amyloid	122
TDN	264, 265

Temperature	
vs. composition of rye grasses	274
vs. hydrolysis of carboxymethylcellulose and filter paper	398–9
vs. saccharification of cellulose	431
Time *vs.* hydrolysis of carboxymethylcellulose and filter paper	399
Timothy	220, 248, 303
Tomato	362
Total digestible nutrients	264
Trametes sanguina	121, 396
Transglycosylation	127
of cellotriose	22
Trichoderma	
enzyme	417
koningi	84, 124, 172
viride	8, 16, 17, 19, 30, 35, 38, 44, 61, 65, 67, 78, 82, 84, 113, 119, 121, 124, 172, 178–9, 222, 224, 359–61, 364, 376, 379–83, 385–6, 388, 390–2, 394–6, 400, 402, 404–5, 410, 416
amino acid composition of the endo and exoglucanases of	17
cellulase	149, 398, 407, 409
cellulase, substrate *vs.* extent of hydrolysis by	405–6
Tritiated water	146
Tryptophyl groups	101
Tyloses	215
Tyrosyl groups	102

U

Umbilicaria pustulata	108
Unicellular fruits and vegetables	362
Unicellular vegetables	380
Unit cell structures of cellulose	178
Urea	253, 303–4, 334
supplementation *vs.* nutritive value of NaOH-treated oat straw	328
Uronic acids	109, 256
Utilization of cellulose by ruminants	242

V

Vegetables, unicellular	362, 380
Vitis vinifera	108
Voluntary intake	329

W

Waste materials, disposal of	153
Wheat straw	248, 257–8, 265, 282
White clover	256–7, 268
Wood	280
-derived products in ruminant nutrition	298
fiber	316
during digestion, change in	325
polysaccharides in	162

Wood *(Continued)*
  molasses 299
  products, composition and nutritive value of 299
  products, microbial fermentation of 310
  pulps 222, 233, 421
  shavings 280
  as a source of energy in beef cattle rations 279
  waste, degradability of 326
Wood by-products 279
World population 279

X

Xylan 256, 365
Xylanase 93, 368–9
β-Xylans 125
Xylose 109

Y

Yeasts on enzymatic digests of cellulose, growth of 411

Z

Zinc *vs.* cellulose digestibility 255
Zone electrophoresis 62, 79